Limitations to Efficient Water Use in Crop Production

Limitations to Efficient Water Use in Crop Production

Edited by

HOWARD M. TAYLOR

WAYNE R. JORDAN

THOMAS R. SINCLAIR

Managing Editor: RICHARD C. DINAUER

Assistant Editor: CATHERINE HARP

Editor-in-Chief ASA Publications: DWAYNE R. BUXTON

Published by

American Society of Agronomy, Inc.

Crop Science Society of America, Inc.

Soil Science Society of America, Inc.

1983

Cover design by Patricia Jeffson

American Society of Agronomy, Inc.
Crop Science Society of America, Inc.
Soil Science Society of America, Inc.
677 South Segoe Road, Madison, Wisconsin 53711 USA

Library of Congress Cataloging in Publication Data

Limitations to efficient water use in crop production.

Includes bibliographical references and index.
1. Crops and water. I. Taylor, Howard M., 1924– II. Jordan, Wayne R., 1940– . III. Sinclair, Thomas R., 1944– . IV. American Society of Agronomy.
S494.5.W3L55 1983 631.5′87 83-3902
ISBN 0-89118-074-5

Printed in the United States of America

CONTENTS

2B Soil Properties and Efficient Water Use: Evaporation of Water from Bare Soil

H. R. GARDNER

2C Soil Properties and Efficient Water Use: Water Management for Salinity Control

GLENN J. HOFFMAN AND M. TH. VAN GENUCHTEN

3A Managing Root Systems for Efficient Water Use: An Overview

HOWARD M. TAYLOR

3B Managing Root Systems for Efficient Water Use: Axial Resistances to Flow in Root Systems—Anatomical Considerations

BETTY KLEPPER

7B Whole-Plant Response to Water Deficits: Carbon Assimilation and Utilization

DANIEL R. KRIEG

7C Whole-Plant Response to Water Deficits: Water Deficits and the Nitrogen Economy

ANDREW D. HANSON AND WILLIAM D. HITZ

8A Crop Manipulations for Efficient Use of Water: An Overview

ROBERT S. LOOMIS

8B Crop Manipulation for Efficient Use of Water: Inadvertent and Intended Manipulations of Crop Water Use

GARY H. HEICHEL

8C Crop Manipulation for Efficient Use of Water: Constraints and Potential Techniques in Breeding for Efficient Water Use

R. W. ZOBEL

9A Yield and Water-Use Relationships: An Overview

R. J. HANKS

9B Yield and Water-Use Relationships: Some Problems of Relating Grain Yield to Transpiration

E. T. KANEMASU

10A Soil Management for Efficient Water Use: An Overview

PAUL W. UNGER AND B. A. STEWART

11B Irrigation Options to Avoid Critical Stress: Irrigation Scheduling—Some Applied Concepts

EARL C. STEGMAN

11C Irrigation Options to Avoid Critical Stress: Optimization of On-Farm Water Allocation to Crops

JAMES W. JONES

12 Water Resources Management in a Growing Society

J. VAN SCHILFGAARDE AND S. L. RAWLINS

FOREWORD

This book deals comprehensively with practically every aspect of water from the time it falls as rain, passes through the soil and plant, and emerges again into the atmosphere. It also includes detailed consideration of the many interactions of water with plant tops, roots, soil and soil organisms, especially as they affect crop yield. It particularly focuses upon factors likely to be involved in research that affect water as a limiting resource.

Historically, water has received attention from plant and soil scientists since early times. Its quantitative consideration along with energy processes, however, and detailed description of its physiological role, is a product largely of the present century. This is particularly true of the past few decades, as it has become possible to measure the energy state of water in detail over the entire soil-plant-atmosphere continuum. In recent years greater attention also has been given to the dissolved and suspended substances carried by water and to their influence, in turn, upon crop production and environmental problems.

Water, apart from its undenied importance in practically every process of nature and of civilization, serves one additional function, that of a catalyst in strengthening the bonds among three major scientific societies: The American Society of Agronomy, the Crop Science Society of America, and the Soil Science Society of America. Nowhere is this more evident than in this book, *Limitations to Efficient Water Use in Crop Production,* which brings together many of the best scientists from the three societies to deal comprehensively with water use in crop production.

The three cooperating societies commend this book to every scientist and student interested in water resources and crop production.

March 1983

CHARLES F. ENO
president, American Society of Agronomy
WILLIAM L. COLVILLE
president, Crop Science Society of America
WALTER H. GARDNER
president, Soil Science Society of America

PREFACE

Life as we know it cannot exist without water—the solvent in most physiological processes. Vegetation in particular has evolved along a multitude of pathways to specifically deal with the vagaries of water availability. It is no wonder that mankind, since early recorded times, has attempted to understand and describe the functional relationships between water use and crop production. This book attempts to present our current state of knowledge in this quest and to pinpoint some lines of research likely to contribute to efficient use of the limiting water resource during the next one or two decades.

Today's agricultural scientists have a profound obligation to devise means for increasing production of plant material in order to feed and clothe an ever-increasing population of the world. This increased production must be achieved at a time when it is being recognized that the water resource is not unlimited. Depleted irrigation water sources and competition with industry for water signal increasing significance for efficient water use in crop production. At the same time, there is an increasing need to conserve and improve the environment to ensure our continued existence on this planet. These challenges require that mankind recognize the practical problems associated with water management and its impact on the physiology and growth of biological organisms, including humans. This book addresses the more limited topic of the efficient use of water in crop production.

A symposium on "Plant Environment and Efficient Water Use" was held at Ames, Iowa in late 1965. The sales of the book that arose from that meeting, jointly published by the American Society of Agronomy and the Soil Science Society of America, continued briskly until 1977. At that time, officers of the associated societies had to decide whether to reprint that 12-year-old book or to authorize a complete revision. In view of the major increases in knowledge about water relations and plant growth, they decided to publish a new book "Limitations to Efficient Water Use in Crop Production." By late 1978, editors were selected, the subject matter was divided into chapters, and authors of chapters had accepted their assignments. The authors were asked to prepare papers describing the state of current knowledge and pointing out major gaps presently amenable to research attack. In November 1979, the authors described their chapters in a working session at Cornell University, Ithaca, New York. At the same time, additional scientists were asked to prepare short critiques or supplements to the overview chapters. This book contains, with two exceptions, the presentations at that meeting.

We believe that this book will be of interest not only to the scientists working directly in plant water use but also to students, teachers and others interested in increasing the productive potential of our planet.

March 1983

HOWARD M. TAYLOR, *co-editor and chairman Editorial Committee*
Texas Tech University, Lubbock, Texas

WAYNE R. JORDAN, *co-editor*
Texas A&M University, Temple, Texas

THOMAS R. SINCLAIR, *co-editor*
USDA-ARS, Gainesville, Florida

CONTRIBUTORS

Donald K. Barnes	Research Geneticist, Agricultural Research Service, U.S. Department of Agriculture, and Department of Agronomy and Plant Genetics, University of Minnesota, St. Paul, Minnesota
Kent J. Bradford	Assistant Professor, Department of Vegetable Crops, University of California, Davis, California
Ronald M. Cionco	Research Meteorologist, U.S. Army Atmospheric Sciences Laboratory, White Sands Missile Range, New Mexico
J. Robert Cooke	Professor of Agricultural Engineering, Department of Agricultural Engineering, Cornell University, Ithaca, New York
Edwin L. Fiscus	Plant Physiologist, Agricultural Research Service, U.S. Department of Agriculture, Colorado State University, Fort Collins, Colorado
H. R. Gardner	Soil Scientist, Agricultural Research Service, U.S. Department of Agriculture, Fort Collins, Colorado
W. R. Gardner	Head, Department of Soils, Water, and Engineering, and Professor of Soil Physics, University of Arizona, Tucson, Arizona
R. J. Hanks	Professor, Soils and Biometeorology Department, Utah State University, Logan, Utah
Andrew D. Hanson	Associate Professor, MSU-DOE Plant Research Laboratory, Michigan State University, East Lansing, Michigan
Gary H. Heichel	Plant Physiologist, Agricultural Research Service, U.S. Department of Agriculture and University of Minnesota, St. Paul, Minnesota
Edward A. Hiler	Professor and Head, Department of Agricultural Engineering, Texas A&M University, College Station, Texas
William D. Hitz	Formerly MSU-DOE Plant Research Laboratory, Michigan State University, now with E. I. duPont de Nemours & Company, Wilmington, Delaware
Glenn J. Hoffman	Research Leader, U.S. Salinity Laboratory, Agricultural Research Service, U.S. Department of Agriculture, Riverside, California
Terry A. Howell	Agricultural Engineer, Water Management Research Laboratory, Agricultural Research Service, U.S. Department of Agriculture, Fresno, California
Theodore C. Hsiao	Professor, Department of Land, Air, and Water Resources, University of California, Davis, California
Wayne R. Jordan	Resident Director and Professor, Blackland Research Center, Texas Agricultural Experiment Station, Temple, Texas
James W. Jones	Professor, Agricultural Engineering Department, University of Florida, Gainesville, Florida
Edward T. Kanemasu	Professor of Agronomy, Department of Agronomy, Kansas State University, Manhattan, Kansas
Bruce A. Kimball	Soil Scientist, U.S. Water Conservation Laboratory, Agricultural Research Service, U.S. Department of Agriculture, Phoenix, Arizona

Betty Klepper Plant Physiologist, Columbia Plateau Conservation Research Center, Agricultural Research Service, U.S. Department of Agriculture, Pendleton, Oregon

Daniel R. Krieg Professor of Crop Physiology, Plant and Soil Science Department, Texas Tech University, Lubbock, Texas

Edgar Lemon Formerly Soil Scientist, Agricultural Research Service, U.S. Department of Agriculture and Professor of Soil Science, Cornell University, Ithaca, New York. Now retired, Tortola, British Virgin Islands, West Indies

Robert S. Loomis Professor, Department of Agronomy and Range Sciences, University of California, Davis, California

Robert W. Pearcy Associate Professor, Department of Botany, University of California, Davis, California

James F. Power Research Soil Scientist, Agricultural Research Service, U.S. Department of Agriculture, University of Nebraska, Lincoln, Nebraska

John W. Radin Plant Physiologist, Western Cotton Research Laboratory, Agricultural Research Service, U.S. Department of Agriculture, Phoenix, Arizona

Stephen L. Rawlins National Research Program Leader for Soil Tillage and Management, Agricultural Research Service, U.S. Department of Agriculture, Beltsville, Maryland. Formerly, Research Leader, U.S. Salinity Laboratory, Riverside, California

Donald C. Reicosky Soil Scientist, North Central Conservation Research Laboratory, Agricultural Research Service, U.S. Department of Agriculture, Morris, Minnesota

Joe T. Ritchie Soil Scientist, Grassland Soil and Water Conservation Research Center, Agricultural Research Service, U.S. Department of Agriculture, Temple, Texas

Thomas R. Sinclair Plant Physiologist and Professor, Agricultural Research Service, U.S. Department of Agriculture, and Agronomy Department, University of Florida, Gainesville, Florida

Earl C. Stegman Professor, Agricultural Engineering Department, North Dakota State University, Fargo, North Dakota

B. A. Stewart Soil Scientist and Laboratory Director, Conservation and Production Research Laboratory, Agricultural Research Service, U.S. Department of Agriculture, Bushland, Texas

C. B. Tanner Professor of Soil Science, Department of Soil Science, University of Wisconsin, Madison, Wisconsin

Howard M. Taylor Professor, Department of Plant and Soil Science, Texas Tech University, Lubbock, Texas

Paul W. Unger Soil Scientist, Conservation and Production Research Laboratory, Agricultural Research Service, U.S. Department of Agriculture, Bushland, Texas

M. Th. van Genuchten Research Soil Scientist, U.S. Salinity Laboratory, Agricultural Research Service, U.S. Department of Agriculture, Riverside, California

Jan van Schilfgaarde Agricultural Engineer, U.S. Salinity Laboratory, Agricultural Research Service, U.S. Department of Agriculture, Riverside, California

William Wenkert Formerly Plant Physiologist, Agricultural Research Service, U.S. Department of Agriculture, Columbus, Ohio. Presently, self employed.

R. W. Zobel Research Geneticist, Agricultural Research Service, U.S. Department of Agriculture and Cornell University, Ithaca, New York

CHAPTER 1A

Efficient Water Use in Crop Production: Research or Re-Search?[1]

C. B. TANNER

University of Wisconsin
Madison, Wisconsin

T. R. SINCLAIR

Agricultural Research Service, USDA
Gainesville, Florida

I. INTRODUCTION

For centuries man has been concerned to some degree with the efficient use of water in the production of his crops. Indeed, the ability to grow crops and manage their needs for water probably was a requisite for civilization. Yet only in the past century have scientific studies been undertaken on the role of water in crop growth. This book is intended to review, to give perspective, and to explore the future of research into the efficient use of water in crop production.

In this chapter, we hope to give general historical and current perspective to the detailed and specific discussions contained in the chapters to follow that are essential to understanding efficient water use.

We will begin with the title of this book. The phrases "efficient water use" or "water-use efficiency" are intrinsically ambiguous in relation to crop production. We may mean saving water from a given supply for crop use; we also may mean increasing production per hectare (yield) per unit of water evaporated from the soil and/or transpired from the plants in the field.

Saving water by decreasing the 59% loss from storage and conveyance of irrigation water (Ackermann et al., 1978), decreasing surface runoff so that more water is stored in the soil and water table for future use, decreasing evaporative losses by fallowing and mulching, or avoiding water losses from excessive irrigation, allows more hectares of crops to be grown. These savings of water can increase total production per unit of water available to agricultural use, and are "efficient uses of water." However, they do not

[1] Contribution from the University of Wisconsin–Madison, College of Agricultural and Life Sciences, and the USDA Microclimate Project, Ithaca, New York.

Limitations to Efficient Water Use in Crop Production.

improve the biomass production per unit cropped area for a unit of water evaporated and transpired (ET) from the same area—the "water-use efficiency" as used by Viets (1962). We wish here to emphasize that water-use efficiency can be based either on the ET (we will call this the ET efficiency) or on the crop transpiration (we will call this the *T* efficiency). The difference is important since suppression of soil evaporation and prevention of weed transpiration can improve the ET efficiency; however, it need not improve the *T* efficiency, which is a measure of crop performance. These two water-use efficiencies also may be based on either the total dry matter production or the marketable yield, and the yield base should be given.

Importantly, these esoteric distinctions between efficient water use and water-use efficiencies are significant in the context in which farmers operate. The water-use efficiency of a crop or cropping system is a basic constraint in the agronomic decisions on the farm; i.e., it defines the achievable biomass production for the water to be transpired. Yet, to the U.S. farmer the ultimate decisions are generally made in response to the criterion of increased profitability. Accordingly, a farmer regards efficient water use as the improvement of the economic return for the investment in water. The farmer is only indirectly interested in total biomass productivity. His direct concern is to increase the harvest of that component of the crop which improves his economic position. Similarly, he has a concern for minimizing the cost of the water use in crop production.

The economic view of efficient water use is most easily recognized in arid-region irrigation agriculture where water is scarce. Costs of the water are minimized by improving irrigation techniques, reducing soil evaporation, increasing water recovery by the crop, etc. The return on the investment in water is maximized by the selection of the crop.

While less obvious, these economic considerations of efficient water use are equally as important as rainfed lands in humid regions. The farmer does not pay directly for his water, but the capital investment in land, in large part, reflects the rain catchment and storage potential of that land. Certainly, the high prices of farmland in the midwestern United States could not be sustained without the historical assurances of adequate and timely precipitation. Since the farmer holds a deed to land rather than to the precipitation it represents, the return on the capital investment per acre receives the economic attention. Nevertheless, in a real sense the fundamental consideration is the return on the capital investment in precipitation, i.e., efficient water use.

Viets (1962) recognized economic aspects in pointing out that attainment of maximum ET efficiency is not desirable on two bases. First, maximum ET efficiency is contingent on maximum yields and the cost of marginal returns dictates stopping short of maximum production. His second argument relates to dryland production where water is scarce and cultural operations aim at stretching water as far as possible for production. Evapotranspiration efficiency, which may be higher under this regime than under those with plentiful water, is attained at much lower, "sometimes disastrous," yields.

The succeeding chapters of this book review in detail the various approaches to improving efficiency of water use. To put these reviews in perspective, we have chosen to re-evaluate the fundamental concept of water-use efficiency. The central question we have chosen to examine is the nature of the ratio of biomass productivity to evapotranspiration, and its inherent variability.

II. THE PIONEERING EXPERIMENTS

At the opening of this century considerable work was in progress measuring "water requirement" (WR), the inverse of *T* efficiency. King at the University of Wisconsin was the first in the United States to research the water required to produce field crops; this work, using small lysimeters in glasshouse and fields, extended from 1890 to 1902.[2] Widtsoe at the University of Utah and Kiesselbach at the University of Nebraska initiated research in 1902.[2] Briggs and Shantz (1913a), based at Akron, Colo., compiled much of the early water-requirement data from container experiments beginning in 1910. The WR work of Kiesselbach and of Briggs and Shantz was concerned with total dry matter and transpiration on a plant rather than land area basis. Briggs and Shantz (1913b) and Kiesselbach (1916) reviewed the work before them—omitting water-culture experiments and those on seedlings, 14 researchers had worked on water-use efficiency prior to 1900. Since that time, there has been an increasing amount of work on water use and water-use efficiency, all done with more elaborate equipment and experiments. However, these early experiments still remain the foundations of our "modern" understanding of crop production. In fact, it has occurred to us that the more recent efforts with increasing sophistication in equipment and manpower have not appreciably advanced our understanding of water-use efficiency. We may have been ahead to have followed a proposal analogous to Stanhill's (1970), using most of the scientific papers as a mulch!

A. Soil Water Content

Briggs and Shantz (1913b) reviewed 18 container experiments conducted prior to 1913, to determine effects of soil water content. Their summary on container experiments that attempted to control levels of soil water is quoted:

> In general, the results of the experiments herein recorded show an *increase* in the water requirement when the water content of the soil approaches either extreme. . . . With a high water content, the areation would be reduced and the soil solution would be diluted. In the case of soils kept at a low water content, the small amount of water required from time to time to bring the pot to normal weight is sufficient to moisten only a comparatively small portion of the soil

[2] Refer to the experiment station annual reports for these states.

mass. In effect, then, the low soil-moisture plants are growing in a very restricted soil mass as compared with plants grown with an optimum soil-moisture content. This condition might cause an increase in the water requirement of plants grown in a comparatively dry soil due to a deficiency of plant food. In view of these considerations the writers believe that *the direct influence of soil-moisture content in the water requirement cannot be established without further investigation* [emphasis added with italic].

However, in many of the experiments Briggs and Shantz reviewed, containers were open to evaporation. In those experiments measuring only transpiration, the WR usually *decreased* a little when growth was limited by water deficits. This is illustrated best by Kiesselbach's (1916) experiments on corn. With a deficient water regime that decreased the dry matter of corn stalks 37%, ears 28%, and leaves 10%, the WR based on ear weight decreased only 4.3%, and that based on the total dry matter only 10%. Later Briggs and Shantz (1917) showed that the WR of wheat was essentially the same under deficient soil water (dry matter decreased 40%) as when water was adequate. It is worth noting here that King had reached the same conclusions from experiments with corn, oats, and poatto (see King, 1914, for summary). de Wit (1958) analyzed several later experiments showing similar T efficiencies regardless of water shortages.

B. Climate

The seasonal changes in both T and ET efficiencies found with exposures to different radiation, temperature, and humidity regimes indicated to many early European and United States researchers that climate exerted a major influence on water-use efficiencies. However, no systematic relations between T and ET efficiencies and the climate were developed until 1910. Kiesselbach (1910), Kiesselbach and Montgomery (1911), and Montgomery and Kiesselbach (1912) showed that daily and weekly transpiration from corn was correlated with evaporation from 1-gal (3.8-liter) glazed, stone jars filled with water. Although most of their experiments were out-of-doors, one attempt to control environmental conditions is of interest. Montgomery and Kiesselbach (1912) grew corn in two glasshouses with one kept dry and ventilated and the other kept humid by wetting floors and atomizing water into the air; these treatments extended through June and July. The relation of WR to saturation deficit deduced from Montgomery and Kiesselbach's data by Briggs and Shantz (1913b), as shown in Table 1, was not recognized by Montgomery and Kiesselbach, who sought a relation with relative humidity. It was thus clear very early, that with well-watered plants, transpiration variation with climate caused changes in water-use efficiency as the season progressed, between years, and between locations, and that "free-water" evaporation and saturation deficit could be used to normalize the transpiration component. This was illustrated by Briggs and Shantz (1917) for alfalfa, as shown in Table 2.

Following the demonstration by Kiesselbach and Montgomery that climate affected WR via the influence on transpiration, Briggs and Shantz

Table 1—Data recalculated from Montgomery and Kiesselbach (1912).

Parameter	Humid	Dry	Humid/dry
Water requirement	214	340	0.63
"Free-water" evaporation, cm	9.4	16.8	0.56
Mean saturation deficit, mbar	19.0	31.3	0.61

Table 2—Water requirement of Grimm alfalfa at different stations in the Great Plains (Briggs & Shantz, 1917).

Location	Growth period (1912)	WR	Pan average (mm day^{-1})	WR/Pan
Williston, N. Dak.	29 July–24 Sept.	518 ± 12	4.04	128
Newell, S. Dak.	9 Aug.–6 Sept.	630 ± 8	4.75	133
Akron, Colo.	26 July–6 Sept.	853 ± 13	5.74	149
Dalhart, Tex.	26 July–31 Aug.	1,005 ± 8	7.77	129

did considerable work correlating climate parameters and transpiration from their container-grown plants. Their results obtained under advective conditions, indicated that for daily measurements, shallow tank evaporation and saturation deficit (or wet-bulb depression) were preferred for normalizing the transpiration component of WR.

Before leaving this account of early *T* efficiency work, two quotes are of current interest. In Briggs and Shantz (1916) we find:

> The idea is often advanced that wheat and similar crops increase their water demand suddenly at or just before the time of heading. The measurements of the writers, however, lend no support to this conclusion. *Aside from fluctuations due to weather the transpiration increases uniformly during this period* [emphasis added with italic].

In their report (1913b), Briggs and Shantz voiced aspirations often heard today:

> One of the most striking features of water requirement measurements is the marked difference in efficiency exhibited by different plants in the use of water. The millet, sorghum, and corn groups have been found the most efficient, while alfalfa and sweet clover are least efficient in producing dry matter with a given amount of water. The small-grain crops have a water requirement intermediate between the legumes and corn. Measurable differences in the water requirement also exist between different varieties of the same crop, and this *suggests the possibility of developing through selection strains which are still more efficient in the use of water* [emphasis added with italic].

C. Plant Nutrition

Briggs and Shantz (1913b) reviewed data from about 20 experiments showing the effect of varying fertility on WR, and concluded that with poor soils, the WR may be reduced one-half to two-thirds by increasing fertility. In all but five experiments, evaporation from the soil was included in the water use and could bias seriously the WR for smaller plants with less transpiration. In experiments where evaporation was prevented, the WR did not increase significantly as the fertility and yield decreased until the dry weight of the plants had decreased approximately 50% because of malnutrition. It appears from these data that unless malnutrition is fairly severe, *T* efficiency based on total dry matter is not greatly affected; however, with severe nutrient deficiencies, *T* efficiency decreases markedly. Kiesselbach (1916) provided extensive data from corn grown in large, sealed containers filled with well-fertilized to severely deficient soils showing the same trend; the *T* efficiency changed little, if at all, until nutrient deficiency reduced yield to about half that on the well-fertilized soil. de Wit (1958) arrived at similar conclusions from other data.

D. Pests

There are few data on the effects of pests on water-use efficiency. Briggs and Shantz (1913b) cited Wimmers' data showing ET efficiency and yield of sugar beet and celery decreased 25–50% with nematode injury; however, evaporation from the soil could have affected results significantly. Recent data (Elkins et al., 1979) on nematode injury to two tall fescue genotypes, one small-rooted and the other large-rooted, indicated that while yield was suppressed 9 and 50%, respectively, there was no change in ET efficiency; evaporation would not be as significant in this study.

The rusts decrease transpiration in the early stage of infection, but on sporulation cause epidermal leakage and increased water loss in light and dark (Yarwood, 1947). The powdery mildews cause a similar decrease and then increase of " transpiration" (Yarwood, 1936; Graf-Marin, 1934). These "transpiration" trends, which have been confirmed by many others, have been found to decrease the *T* efficiency (Weiss, 1924; Graf-Marin, 1934). As an example, Johnston and Miller (1934) found the *T*-efficiency (total dry matter base) of wheat was halved with early rust infection and when based on grain yield was 25-fold smaller; infection at later stages of development produced correspondingly smaller decreases.

The vascular wilt diseases appear to cause increased xylem resistance to water transport; a decrease in stomatal conductance and concomitant loss in transpiration and photosynthesis ensues. Duniway and Slatyer (1971) in a short-term experiment found that net photosynthesis is decreased slightly more than transpiration. We found no long-term water-use efficiency data for vascular wilts, and it is probable that *T* efficiency is unaffected despite large decreases in production.

Burkholder (1920) found the *T* efficiency of beans to be little affected by *Fusarium* root rot; seed yield and transpiration decreased about equally. It would appear that some pathogens not only decrease total yield, but decrease yield more than they decrease transpiration; others may affect yield with little change in *T* efficiency.

We found no reports on how water-use efficiency was affected by insect pests.

III. ANALYSIS OF FIELD CROP WATER-USE EFFICIENCY

We would like some reasonably simple assessment of the effect of water deficits on field yields. The transpiration per plant obtained with container experiments is different than in the field because of different exposures and energy balances. The historic data showing that transpiration and T efficiency varied with climate and that the T efficiency was essentially the same for droughted and undroughted crops were not applied usefully to water management of field crops, and in fact often were applied detrimentally by using efficiency data gathered in arid regions to predict the water needed to produce a given yield in humid regions. Nevertheless, an impressive amount of data showed that yield and ET were well-correlated (see de Wit, 1958; Arkley, 1963). A large step forward was made by de Wit (1958), who, after a monumental review of remotely applicable data, found "that the relation between transpiration and total dry matter production of plants under field conditions must often be the same as for plants in containers provided the dry matter production of the plants in the field is limited by the availability of water."

A. de Wit's (1958) Analysis

de Wit showed that for dry, high-radiation climates, yield and transpiration were related as

$$Y/T = m/T_{max}, \qquad [1]$$

where Y = total dry matter mass per area, T = total transpiration per area during growth to harvest, and T_{max} = mean daily "free water evaporation" for the same period. de Wit used either pan data or the Penman (1948) combination formula to estimate T_{max}. The "constant" m is related to the WR/pan used by Kiesselbach and by Briggs and Shantz ($1/m \sim$ WR/pan), but as de Wit showed, is preferable for yield and water-use analysis. de Wit showed that m was governed mainly by species and "at first approximation" was independent of soil nutrition and water availability unless seriously nutrition-limited or unless soil water was "too high" (due to aeration, leaching, etc.).

de Wit proposed this relation should hold until T approaches a maximum production governed by growing conditions. The relation in Eq. [1] could be simplified for humid regions because when water was not limiting, fluctuations in intercepted radiation, although reflected in transpiration and growth, would not affect appreciably the ratio T/T_{max}. de Wit found under these conditions that

$$Y/T = n, \qquad [2]$$

where n is a constant, gave a better description.

Table 3—Experimental estimates of m (kg ha^{-1} d^{-1}) from data of Briggs and Shantz (1914) and more-recent field experiments.

Crop	Briggs & Shantz†	Recent field data	Source
Corn	213 ± 14	215 ± 20‡	Logan, Utah; Stewart et al. (1977)
		258 ± 1‡	Ft. Collins, Colo.; Stewart et al. (1977)
		262 ± 46‡	Yuma, Ariz.; Stewart et al. (1977)
		314 ± 12‡	Davis, Calif.; Stewart et al. (1977)
Grain sorghum	240 ± 10	141 ± 6§	Hanks et al. (1969)
Millet	260 ± 35	150 ± 18§	Hanks et al. (1969)
Wheat	158 ± 10	125 ± 15§	Hanks et al. (1969)
Potato	160 ± 8	217 ± 24¶	Wisconsin; Tanner (1976, unpublished data)
Alfalfa	90 ± 11	214 ± 26¶	Wisconsin; Tanner (1977, unpublished data)
Soybean	102 ± 7	128 ± 34¶	Teare et al. (1973)

† Variation expresses differences among best-yielding cultivars.
‡ Variation expresses 2 years of data (mean given).
§ Variation expresses uncertainty in evaporation estimates.
¶ Variation expresses standard deviation among trials.

The value of m in Eq. [1] can be approximated with Eq. [3] from water-use efficiency and mean daily pan evaporation (E_{pan}):

$$m = (Y/T)\,E_{pan}. \qquad [3]$$

In Table 3 we compare the values of m derived from the experiments of Briggs and Shantz and from more recent field observations. Since the experiments of Briggs and Shantz were conducted in screened enclosures and the evaporation was measured with a BPI (Bureau of Plant Industry) pan, there is some ambiguity. Nevertheless, the approximately 20% decrease in water requirement in the screened enclosures is offset within 5% by the lower evaporation of the BPI pan as compared with the Class-A pan. In Table 3 the m's developed from the data of Briggs and Shantz and Hank et al. do not include root dry matter, whereas an estimate of root yield was made for other data. Also, pan evaporation was used directly with no correction to free water evaporation as made by de Wit. The corrections for roots and free water evaporation are discussed by Hanks (Chapter 9A); they are not large enough to void the comparisons we wish to make.

Except for the Wisconsin data, the relative ranking of corn, sorghum, and millets giving the highest values of m, followed by the grain cereals, potato, and then the legumes, has remained consistent. Yet two interesting features appear in these comparisons. First, the independence of m for a species among locations, originally hypothesized by de Wit, seems to be inconsistent with the recent experience with corn. A high level of variability in m is evident among the four locations where the crops were subject to nearly the same experimental treatments. Note also that de Wit, using data of Briggs and Shantz and a number of other sources, found m was equal to 55, 115, and 207 kg ha^{-1} day^{-1} for Grimm alfalfa, Kubanka wheat, and Red

Amber sorghum, respectively. The variability among measurements indicates some constraint on the crop yield other than by water, or an inadequacy in de Wit's approach. The high m for potato and alfalfa in Wisconsin may indicate that Eq. [2] rather than Eq. [1] is applicable to these more humid conditions. If so, the choice of using Eq. [1] or [2] is uncertain for the gradation of humidity that occurs both between locations and seasonally at a given location. Note that de Wit found $m = 55$ for Grimm alfalfa, and the data in Table 2 yields $m = 74 \pm 5$ kg ha^{-1} day^{-1}. R. J. Hanks, in Chapter 9A, provides further discussion of the variability of m. Second, to the extent that m is a measure ot T efficiency for total biomass production, there is no consistent improvement in m between the crops grown in 1912–1913 and recently (University of Wisconsin data excluded for reasons discussed above).

B. Other Simple Water-Use Efficiency Models

1. STEWART MODEL

Stewart (1972) modified Eq. [1] to include the maximum production visualized by de Wit and to provide a single yield and water-use relation using ET rather than T. His model, discussed in greater detail in Chapter 9A, decrements yield below a Y_{max} in proportion to the decrement of ET below a corresponding ET_{max}. Stewart developed weightings to account for marketable yield rather than total dry matter and to account for differences in sensitivity to water deficits at different stages of crop development.

When ET is used rather than transpiration, reliable correlations will depend on evaporation being either small or a relatively constant fraction of ET. With annual row crops in humid regions, soil evaporation (E), which depends strongly on precipitation patterns, is highly variable until leaf area index is about 2.5–3. We thus would expect correlations between yield and ET for annual crops in humid regions to be poorer than with transpiration.

2. HANKS' MODEL

Hanks (1974) also included a maximum yield corresponding to maximum transpiration so that

$$Y/Y_{max} = T/T_{max}, \qquad [4]$$

where Y_{max} is associated with T_{max}. This method, including means of accounting for differing sensitivities of yield and marketable yield to water deficits at different growth stages, is discussed in greater detail in Chapter 9A.

3. ARKLEY MODEL

Arkley (1963) believed that de Wit required two formulas because he used pan evaporation to normalize climate differences. He argued that in humid regions such as the Netherlands, pan evaporation is mainly influenced by solar radiation, and that the effect of relative humidity (RH) on pan evaporation is much smaller and less variable from year-to-year than in the subhumid regions where Eq. [1] applied. Arkley proposed the relation

$$Y/T = \alpha/(1 - \text{RH}) = \alpha e^*/(e^* - e), \qquad [5]$$

where α is a constant, RH is the mean fractional relative humidity, e is the vapor pressure, and e^* is the saturation vapor pressure at mean temperature. He found that the WR data from humid Europe and the western United States coalesced for oats and were within 30% for peas. Arkley tried saturation deficit as a weighting parameter, but he found for corn that the Y vs. $T/(e^* - e)$ data fit families of lines, with each line corresponding to a mean temperature. Since the data coalesced about a single line with Eq. [5], Arkley concluded the (1 − RH) compensated for increasing growth with increasing temperature. The low-temperature corn data Arkley used for his temperature argument were gathered at below-optimum growing temperature for corn. There are cool-season plants that exhibit decreasing yield as temperature rises above the optimum, and Arkley's argument that growth increases exponentially with temperature could not apply to these. We might expect Arkley's relation to degrade if the optimum temperature for the crop were below that experienced during growth. Thus, it would seem more reasonable to keep separate, if possible, the climatic terms that may influence growth and those that express differences in transpirational demand.

4. BIERHUIZEN AND SLATYER MODEL

Bierhuizen and Slatyer (1965) used the descriptions for net photosynthesis and transpiration of a leaf to obtain its leaf water-use efficiency. They expressed photosynthesis per unit leaf area (N_L) as

$$N_L = \Delta C/(r_b' + r_s' + r_m'), \qquad [6a]$$

where ΔC = CO_2 concentration difference between the atmosphere (C_a) and the CO_2 compensation point (Γ); r_b' and r_s' are the boundary layer and stomatal resistances, respectively, to CO_2 diffusion into the leaf; and r_m' describes the diffusion of CO_2 into the cells to the chloroplasts. The transpiration per unit leaf area (T_L) is

$$T_L = (\varrho\epsilon/P)(e_L^* - e)/)r_b + r_s), \qquad [6b]$$

where ϱ and P are the air density and atmospheric pressure, respectively; ϵ is the ratio of the mole weight of water vapor to air; $(e_L^* - e)$ is the vapor pressure gradient from the leaf e_L^* = saturation vapor pressure of the leaf); and r_b and r_s are vapor diffusion resistances. From these:

$$N_L/T_L = (P/\varrho\epsilon)[\Delta C/(e_L^* - e)][(r_b + r_s)/(r_b' + r_s' + r_m'). \qquad [6c]$$

From transport theory, $r_s' = 1.56\ r_s$ and $r_b' = 1.34\ r_b$. Assuming that in the external environment leaf temperatures often do not depart seriously from air temperature, Bierhuizen and Slatyer approximated $(e_L^* - e)$ by the saturation deficit $(e^* - e)$, where e^* is the saturation vapor pressure of the air. Since the atmospheric CO_2 concentration is essentially constant, Bierhuizen and Slatyer used the above approximations and argued additionally that, over the day, effects of light on r_s', r_m', and r_s would not seriously affect the ratio $(r_b + r_s)/)r_b' + r_s' + r_m')$. Then Eq. [6c] becomes, for $\Gamma = 0$ (corn), $C_a = 0.59$ g m^{-3} (333 ppm) and 330°K at 1 bar pressure:

$$N_L/T_L \approx k'/(e^* - e)\,(\Sigma r/\Sigma r'), \qquad [6d]$$

where $k' = 0.80$ mbar. Assuming $\Sigma r/\Sigma r'$ remained relatively constant except under drought, and that on a field basis Y/T is proportional to N_L/T_L, they concluded

$$Y/T \approx k/(e^* - e). \qquad [7]$$

Note that Eq. [7] is essentially the saturation deficit weighting of WR used by Briggs and Shantz, as reviewed in Table 1. Bierhuizen and Slatyer's tests of Eq. [7] against early WR data using crude estimates of monthly mean saturation deficit showed promise. Although they questioned the stability of $\Sigma r/\Sigma r'$ under drought, the early WR work and the analysis of field data by de Wit indicated insignificant overall effect. Reminiscent of Arkley, Bierhuizen and Slatyer believed that the two formulas de Wit found necessary for the water-limiting and water-adequate regimes probably were due to different saturation-deficit regimes.

Rijtema and Endrodi (1970), using an estimated transpiration ($<T>$), found an excellent correlation between Y (total dry matter) and the seasonal integral of $<T>/(e^* - e)$ for four potato cultivars over 7 years, with yield varying threefold because of water deficits ($k \approx 0.015$); there was a poor correlation between Y and $<T>$ for these data. Slabbers et al. (1979) reviewed three yield vs. water-use models that make use of Eq. [7]. However, the models contain several empirical coefficients, accounting for photosynthetic efficiency, development of foliage, and partitioning of dry matter; these coefficients are used as independent products, and are evaluated by regression. Since Slabbers et al. (1979) employed ET rather than transpiration, the empirical coefficients must share the adjustment for evaporation as well as other uncertainties arising from interaction among coefficients.

C. Re-Search of Crop Water-Use Efficiency

There is an innate attractiveness in Eq. [7] because of its extreme simplicity and its compatibility with de Wit's empirical relationship. To the extent that T_{max} (usually estimated from pan evaporation) varies as $(e^* - e)$, m of Eq. [1] and k should be in the same order for different crops. In addition, the derivation of Bierhuizen and Slatyer suggests that the value of

k should be explicitly definable from various crop characteristics, eliminating some of the empiricisms employed by Slabbers et al. (1979).

In this section we use mechanistic arguments to re-search for an expression of k. The goal is not to derive an exact value for k for a crop but to see if simple arguments concerning crop transpiration and photosynthesis lead to an expression of k that is consistent with the observed differences in transpiration efficiency among crop species and can provide some insight into the causes of minor variations within a species. The approach is to use simplifying assumptions whenever possible that give first-order approximations to crop behavior.

1. BIOMASS YIELD

The obvious starting point for estimating biomass yield is the photosynthetic rates of individual leaves. We will use an equation equivalent to Eq. [6a], eliminating Γ and r_m by using the intercellular CO_2 concentration in the leaf, C_i. Wong et al. (1979) have shown for usual conditions of crop growth, including slowly imposed water stress, $c = (C_a = C_i)/C_a = 1 - C_i/C_a$ is essentially constant for a crop. We can then write for leaves of a crop:

$$N_L = (C_a - C_i)/(r_b' + r_s') = cC_a/(r_b' + r_s'), \qquad [8a]$$

where the constant c = about 0.3 for C_3 crops and 0.7 for C_4 crops (Wong et al., 1979).

To express the CO_2 assimilation for all the leaves in a canopy, with leaf area index LAI, we will segregate the leaves into those exposed directly to incident radiation (subscript D) and those in shade (subscript o). Since $r_b' = 1.34\ r_b$ and $r_s' = 1.56\ r_s$, we can approximate $r_b' + r_s' = 1.5\ (r_b + r_s)$ with little error. Accordingly, Eq. [8a] becomes for a canopy:

$$N_C = (cC_a/1.5)[\mathrm{LAI}_D/(r_b + r_s)_D + (\mathrm{LAI} - \mathrm{LAI}_D)/(r_b + r_s)_o], \qquad [8b]$$

where N_C is photosynthesis of total crop canopy.

For a leaf canopy with randomly distributed leaves,

$$\mathrm{LAI}_D = 1.4\,[1 - \exp(-0.7 \bullet \mathrm{LAI})]. \qquad [9]$$

It is possible to simplify the expression for LAI_D by assuming a fairly well-developed crop canopy. If LAI is at least 3, no more than a 10% error is introduced by assuming that LAI_D is 1.4. While not critical, we will assume LAI $\gtrsim$ 3 and a constant value for $\mathrm{LAI}_D = 1.4$ throughout this analysis.

The expression in Eq. [8b] for CO_2 uptake needs to be converted to hexose production. This is done by multiplying by the constant ratio, a, of molecular weights ($a = [CH_2O]/[CO_2] = 0.68$).

The availability of hexose for supporting new biomass growth is obtained from Eq. [8b] only after subtracting the consumption of photosynthate by respiration required to maintain the crop. Estimating the maintenance respiration rate is still difficult and ambiguous. Penning de Vries (1975a) concluded that one of the main processes in maintenance respiration is the turnover of proteins that is continuously occurring in the plants. Although this suggests that maintenance respiration should be a function of N content of the crop, Penning de Vries chose to express maintenance respiration rate as a function of dry weight. In an analysis of soybean growth, Sinclair and de Wit (1976) assumed a maintenance respiration of 20 mg glucose d^{-1} g^{-1} dry wt, and then multiplied it by the specific leaf weight of the crop and by the ratio of total crop weight to leaf weight. As a result they expressed maintenance respiration rate as a function of crop leaf area index such that it equalled LAI multiplied by 2.8×10^{-8} g glucose s^{-1} cm^{-2}. Subtracting maintenance respiration rate from Eq. [8b] gives the daily rate that photosynthate is made available for biomass growth. Using the expression of maintenance respiration as a function of leaf area assumed by Sinclair and de Wit and comparing it with the shade leaf photosynthate production, it is found that these two terms are approximately equal in the range of the usual full-canopy crop (i.e., LAI = 3–6). At very large leaf area indices the two terms diverge because there will be no further increase in photosynthate production yet maintenance respiration will continue to increase. Rather than actually subtracting an estimate of maintenance respiration, we have assumed that on a daily basis the photosynthate produced by the shade leaves and the photosynthate consumed in maintenance respiration are roughly equal. In the normal leaf area range, the error resulting from this assumption is relatively small compared with the photosynthate produced by those leaves exposed directly to incident radiation (LAI_D). With this assumption, the rate that hexose is available for supporting biomass growth is approximately

$$N_C(\text{hexose}) \approx ac\text{LAI}_D C_a/1.5(r_b + r_s)_D. \qquad [10]$$

Finally, the hexose needs to be converted to final product biomass in the plant. Through a careful analysis of the assimilatory pathways, Penning de Vries (1975b) estimated that from 1 g of hexose, either 0.83 g of carbohydrates, 0.40 g of protein (assuming a nitrate source of N), or 0.33 g of lipids could be produced. Therefore, in principle the conversion coefficient, b, for taking hexose to biomass could range from 0.33 to 0.83. Sinclair and de Wit (1975) examined seed production in 24 crop species and found a range for b from 0.42 in sesame to 0.75 in barley and rice. Introducing this conversion coefficient, the biomass yield is

$$Y = bN_C(\text{hexose}) = abc\text{LAI}_D C_a/1.5(r_b + r_s)_D. \qquad [11]$$

2. TRANSPIRATION

Individual leaf transpiration rate (T_L) is unambiquously defined in Eq. [6b] as the vapor pressure deficit between the leaf interior and the atmosphere, divided by the resistance to water vapor diffusion. If the temperatures and resistances of all the individual leaves in the crop canopy were known, they could be summed to find crop transpiration. This is essentially an experimental impossibility, and only very complicated models (such as the SPAM model developed by Lemon et al., 1971) can approximate this situation. Fortunately, Sinclair et al. (1976) found in computer-simulation experiments of a maize crop, that crop transpiration rates were predicted reasonably well by lumping together the leaves exposed directly to incident radiation and lumping the remaining shaded leaves. For a canopy (T_C),

$$T_C = (\varrho\epsilon/P(\left[\mathrm{LAI}_D \frac{(e_D^* - e)}{(r_b + r_s)_D} + (\mathrm{LAI} - \mathrm{LAI}_D) \frac{(e_o^* - e)}{(r_b + r_s)_o} \right] . \qquad [12a]$$

In order to obtain a relation in the form of Eq. [7] using Eq. [11], we would like to define an "effective transpiration leaf area," L_T, such that $T_C = (\varrho\epsilon/P)L_T(e^* - e)/(r_b + r_s)_D$. Equation [12a] can be modified to this form. We will assume that the shade leaves are at air temperature, i.e., $(e_o^* - e) \approx (e^* - e)$, and that the variation of e with height in the canopy is small relative to $(e_D^* - e)$ or $(e_o^* - e)$. While there is some deviation of shade leaves from air temperature, this is a much more reasonable assumption than the frequent one that all canopy leaves are at air temperature. With these assumptions, Eq. [12a] can be written.

$$T_C = (\varrho\epsilon/P) \frac{(e^* - e)}{(r_b + r_s)_D} \left[\frac{\mathrm{LAI}_D(e_D^* - e)}{(e^* - e)} + \frac{(\mathrm{LAI} - \mathrm{LAI}_D)(r_b + r_s)_D}{(r_b + r_s)_o} \right] . \qquad [12b]$$

The bracketed term in Eq. [12b] is the effective transpiration area, L_T. We can evaluate L_T for a well-developed canopy for the condition of $(e_D^* - e)/(e^* - e) = 1$. For LAI = 4, a reasonable value of $(r_b + r_s)_D/(r_b + r_s)_o$ is about 0.3, giving $L_T = 2.2$. If LAI increases $(r_b + r_s)_o$ increases because of poorer light in the lower canopy, thus tending to stabilize the second term in brackets. A more convenient expression for a well-developed canopy is to normalize Eq. [12b] with an L_T to give

$$T_C \approx L_T(\varrho\epsilon/P) \, [(e^* - e)/(r_b + r_s)_D] \, B; \qquad [13a]$$

where the correction term on mean $L_T \approx 2.2$ is

$$B = \left[\frac{\mathrm{LAI}_D(e_D^* - e)}{L_T(e^* - e)} + \frac{(\mathrm{LAI} - \mathrm{LAI}_D)(r_b + r_s)_D}{L_T(r_b + r_s)_o} \right] , \qquad [13b]$$

with $B = 1 \pm 0.2$ for a range of LAI > 3 unless $(e_D^* - e)/(e^* - e)$ departs appreciably from unity.

Equation [13] is a simple and fairly complete expression of daily crop transpiration when $(e^* - e)$ and $(e_D^* - e)$ are integrated over the period of stomatal opening. The first term in brackets states explicitly the effect of the directly radiated leaf temperature differing from air temperature. The second term accounts for the fraction of transpiration from the shaded leaves. Changes in $(e_D^* - e)/(e^* - e)$ are damped somewhat by the second term, and we expect important effects mainly at very low $(e^* - e)$ (high relative humidity) when the temperature rise of radiated leaves is most likely and produces a largest relative change in $(e_D^* - e)/(e^* - e)$.

3. TRANSPIRATION EFFICIENCY

The *daily* transpiration efficiency is obtained from Eq. [11] and [13]:

$$Y/T = (abc/1.5)(P/\varrho\epsilon)(\mathrm{LAI}_D/L_T)\,[C_a/B(e^* - e)]. \qquad [14a]$$

Comparing with Eq. [7], we have a *daily crop k, k_d*:

$$k_d = (abc/1.5)(PC_a/\varrho\epsilon)(\mathrm{LAI}_D/L_T)/B. \qquad [14b]$$

Since T, B, and $(e^* - e)$ certainly will change with time as sunlight, $(r_b + r_s)_D$, and advection change from day to day, Eq. [14] must be integrated to obtain a mean k_d for a growth period. Using $T_T = \int T\mathrm{d}t$ and $Y_T = \int Y\mathrm{d}t$ over a period of growth,

$$Y_T/T_T = (1/T_T)\int [k_d T/B(e^* - e)]\mathrm{d}t. \qquad [15]$$

The integral $\int[T/B(e^* - e)]\mathrm{d}t$ will not equal $T_T/\overline{B}\overline{(e^* - e)}$ unless T, B, and $(e^* - e)$ are uncorrelated—an absurdity. Using $\delta = (e^* - e)$ for notation convenience, it can be shown that to a first approximation $\int T/B\delta = T_T/\overline{B}\,\overline{\delta}(1 - r_{B\delta}s_B s_\delta/\overline{T}\overline{B}\,\overline{\delta} - r_{TB}s_T s_B/\overline{T}\overline{B} - r_{T\delta}s_T s_\delta/\overline{T}\,\overline{\delta})$, where r is the correlation coefficient between subscripted variables and s is the standard deviation. Because T is positively correlated with $\delta = (e^* - e)$ and B, k estimated from T_T, $\overline{(e^* - e)}$, and $\overline{B}$ will be smaller than when estimated from the integral. Worse errors may be expected where $\overline{(e^* - e)}$ and $\overline{T}$ are small and where daily T, $(e^* - e)$, and B, influenced by $(e_D^* - e)/(e^* - e)$, fluctuate greatly as in very humid climates.

Although Eq. [15] is very much simplified over reality, it illustrates several considerations. First, it indicates the effect of sunlit leaf-temperature rise above air temperature at low temperatures, or leaf-temperature depression at high temperature (Linacre, 1964; Sumayao & Kanemasu, 1979). The increase in T is not in proportion to $(e_D^* - e)/(e^* - e)$, but is modified by transpiration from shaded leaves. We have no simple model for estimating this temperature effect; more detailed canopy energy balance models are necessary. We speculate that the empirical success of Arkley's formula, Eq. [5], could be due to leaf-temperature rise. The hyperbolic $1/(1 - \mathrm{RH})$ becomes increasingly sensitive at high RH, where leaf-temperature rise would require increasingly important adjustments.

Although in our use of Eq. [15] we have approximated $\mathrm{LAI}_D = 1.4$, $\mathrm{LAI} = 4$ to represent well-developed canopies, it is clear that as LAI decreases LAI_D/BL_T approaches unity; certainly this would be the case with a seedling dicot with leaves fully exposed. There will be other adjustments that require more detailed modeling at low LAI. For example, the $(e^* - e)$ at seedling level, with a dry, hot soil will be much greater than $(e^* - e)$ at screen-level; respiration, LAI_D and LAI needs to be accounted for explicitly with more detailed modeling. We believe, however, that T efficiency for a growing season is determined in most field crops mainly during the period when /LAI > 3/, and that Eq. [15] should apply. While T efficiency is difficult to estimate for smaller LAI, it should not be too different than later in the season unless b is profoundly different (e.g., soybeans).

4. ET-EFFICIENCY

Equation [15] can be written for ET rather than T. Using time integrals for a crop during growth,

$$\int Y/\int ET = [\int k_d ET/(e^* - e) - \int k_d E/(e^* - e)]/\int ET. \qquad [16a]$$

If we consider only those growth periods where k is relatively constant, and assume conditions where $B \approx 1$ and both $(e^* - e)$ and E are reasonably constant, we can approximate Eq. [16a] as

$$\int Y/\int ET \approx [k_d/(e^* - e)][1 - E/ET]. \qquad [16b]$$

If cumulative Y of a crop is plotted against cumulative ET for a season, E may be approximated empirically as the intercept on the ET axis. This does not apply to a graph of seasonal Y vs. seasonal ET using different field plots that have been differentially irrigated to provide different Y and ET, since E will differ between treatments, as will other terms in Eq. [16].

5. TESTS

We have calculated the field k for several crops for which experimental data were available, as shown in Table 4. This is compared with estimates of k_d from Eq. [14b] using LAI = 4, $L_T = 2.2$, $B = 1$, and $PC_a/\varrho\epsilon = 0.80$ mbar. In most instances the experimental k could only be approximated as $Y_T\overline{(e^* - e)}/T_T$.

The 1974–1975 Y/T data for corn were obtained using Y/ET data from Table 1 of Stewart et al. (1977), corrected using evaporation data from Hanks' analysis in the same report. Yield data are further corrected using (total dry matter)/(aboveground dry matter) = 1.2; data in the literature range from 1.08 to 1.5, but our own data on corn indicate that 1.2 is reasonable. Stewart et al. give data from which daily $e^*(T_{max}) = e^*_{max}$, $e^*(T_{min}) = e^*_{min}$, and e can be calculated. We estimated e^* as three-quarters the distance between e^*_{min} to e^*_{max}; this is reasonable for locations with a diurnal temperature range of 20°C and low T_{min}, assuming the daily e^* should be integrated from about 0900 hours to evening, when net radiation (R_N) becomes negative. Because T_{min} was high at Yuma, Ariz., we used an e^* about two-thirds

Table 4—Experimental values for k for total dry matter and calculated values for five crops. See text for sources of data.

Crop	Location	Year	Experimental			Eq. [14b]	
			$\overline{Y/T}$	$\overline{(e^* - e)}$	k	b	k
				mbar			mbar
Corn	Calif.	1974	0.0046	22.0	0.100	0.74	0.118
Corn	Calif.	1975	0.0054	18.2	0.098		
Corn	Colo.	1974	0.0055	21.7	0.120		
Corn	Colo.	1975	0.0042	20.1	0.084		
Corn	Utah	1974	0.0037	22.1	0.082		
Corn	Utah	1975	0.0046	18.7	0.086		
Corn	Ariz.	1974	0.0020	47.7	0.097		
Corn	Ariz.	1975	0.0026	39.3	0.104		
Corn	Nebr.	1912	0.0047	19.0	0.089		
Corn	Nebr.	1912	0.0029	31.3	0.091		
Sorghum	Kans.	1970	0.0058	23.8	0.138	0.74	0.118
Potato	Wis.	1972, 1973, 1976	--	--	0.065	0.77	0.055
Potato	Netherlands		--	--	0.015		
Alfalfa	Wis.	1977	0.0041	10.4	0.043	0.70	0.050
Soybean	Kans.	1970	0.0017	23.8	0.040	0.58	0.041

of the distance between e^*_{min} and e^*_{max}, representative of the positive R_N period. The 1912 corn data are from Table 1 of this chapter, and are uncorrected for root biomass.

The sorghum and soybean dry matter data are from Teare et al. (1973) for periods when $L \geq 3$; environmental data were provided by E. T. Kanemasu (1979, personal communication). Transpiration was found from T/ET ratios given in Kanemasu et al. (1976).

The Wisconsin potato data were obtained by summing daily $T/(e^* - e)$ over 7- to 10-day periods for which periodic yields of haulm, root, and tuber dry matter were sampled; T was found by correcting daily lysimeter data using T/ET ratios calculated as described by Tanner and Jury (1976) and using daily integrated $(e^* - e)$ from 0830 to 1800 CST, when over 85% of the transpiration occurs. The Netherlands potato data (Rijtema & Endrodi, 1970) were obtained from seasonal dry matter yields and integrals of $<T>/(e^* - e)$ over the season, where $<T>$ is transpiration estimated as outlined by Endrodi and Rijtema (1969).

The alfalfa data were from the last two of four main harvests for which interim dry matter samples were taken 2 weeks after the preceding harvest. The dry matter increased from the second to fifth week after cutting was used. The first 2-week period of regrowth was excluded since interception of radiation by alfalfa is very low (Stanhill, 1962; Hunt et al., 1970) and evaporation would have comprised a large part of the ET during this period; also, Eq. [14b] is not applicable. Evapotranspiration was measured (neutron probe), and evaporation was assumed to be 0.2 ET; $(e^* - e)$ was measured for the yield period. The dry matter accumulation by roots was taken as 25% of the total dry matter accumulation (Bray, 1963); this is compatible with measurements by Hunt et al. (1970) for these last two harvests of the growing season.

Except for the low experimental k of potato from Rijtema and Endrodi (1970), the agreement between the experimental and calculated k_d is within the uncertainty of the experimental determination and/or the assumptions in the calculation. Importantly, the experimental and calculated k for all species are ordered consistently. We believe the experimental k for potato derived by Rijtema and Endrodi is unduly low for two reasons. The main reason is that Rijtema and Endrodi used "24 hours mean values for vapour pressure deficit" (1976, personal communication); inclusion of nighttime $(e^* - e)$ when stomata are closed could seriously underestimate the required daytime integral. Second, if the daytime $(e^* - e)$ were small and $T_a \lesssim 30°C$, $(e_D^* - e)/(e^* - e)$ could be markedly greater than unity during sunny hours because of leaf temperature rise; this would make $B > 1$. It is of interest that their experimental k for one very dry year was 0.026—1.7-fold higher than other years. They attribute this to an underestimate of transpiration, but it also is likely that the 24-hour mean saturation deficit was closer to the daytime average.

The experimental k values for corn in Table 4 vary less between locations than the m values in Table 3, and the k for alfalfa and potato are ordered consistently whereas the m values are not. This reflects the fact that pan evaporation rates do not weight the transpiration of the sunlit leaves as well as the daytime saturation deficits, and that Arkley (1963) and Bierhuizen and Slatyer (1965) are correct in believing the two equations [1] and [2] were necessary for this reason. For example, the ratio of pan evaporation to daytime saturation deficit varies from 0.54 mm d^{-1} $mbar^{-1}$ at Hancock, Wis., to 0.40 at Davis, Calif., and 0.24 at Yuma, Ariz., as the relative radiation and convective heat supply to the pan changes. Pan evaporation changes with both siting and the style of pan. We would expect $k \approx Y(e^* - e)/T$ to be more stable between crops grown in containers and in the field than $m = Y E_{pan}/T$. The experimental k for corn grown in 1912 in containers is close to that of the field corn.

D. Discussion

Equations [15] and [16] are consistent with dry matter production and water-use data, which show to first-order that while yields of crops may vary with growing conditions, dry matter is decreased by water deficits in proportion to the decrease in transpiration caused by the deficits. They are consistent also with the models of Hanks (1974) and Stewart (1972), but the estimates of water-use efficiency from Eq. [15] and [16] require no T_{max} and Y_{max}. These results also indicate that some empiricism can be removed from the coefficients used by Slabbers et al. (1979).

In many instances the reductions of yields of grain and other marketable parts of crops are in rough proportion to the decreases in transpiration induced by water deficits; however, often there are stages of development, such as pollination, at which marketable yield may be extraordinarily affected. Also water deficits may be necessary to increase some yield qualities such as sugar content or storage quality. It appears to us that the main concern of water-use models should be to improve partitioning of ET into E and T, to seek reasonable but tractable estimates of $(e_D^* - e)/(e^* - e)$, and to improve our understanding of dry matter partitioning into roots, tops, and marketable yield. We doubt that stochastically derived multiple coefficients will provide real progress.

We believe that Eq. [15] provides considerable insight into water-use efficiency experiments and means of modification that is not very dependent on the assumptions made. The difference in T efficiency of different crop species can be attributed to differences in carboxylation pathways (different c for C_3, C_4 plants) and energy requirements to produce different biomass compositions (different b). If plant breeding is to effect appreciable changes in T efficiency of total dry matter production, then $c = 1 - C_i/C_a = (r_b' + r_s')/\Sigma r'$ must be modified substantially. This appears to be a stable parameter, and modifying it greatly appears unlikely, contrary to the hopes of Briggs and Shantz, quoted earlier. In most crop plants the value of r_m' is not too different from $(r_a' + r_s')$, and only large decreases in r_m' can drastically alter c and hence k. It is noteworthy that the k in Table 4 for corn grown in 1912 and 1975 are essentially the same. Breeding can exert greater influence on the partitioning of dry matter to marketable yield; this partitioning (harvest index) may have improved somewhat since the early water-requirement experiments, but a more likely source of improved field yields is that populations have been increased. If higher protein forage crops are developed, it will be at the expense of total dry matter and decreased T efficiency. Changing plant architecture is not likely to change k and T efficiency significantly in canopies achieving LAI = 3; however, canopy structure and population can modify the evaporation loss relative to the transpiration, and therefore can affect ET efficiency much more strongly. Certainly crop breeding can change maturation rates to take advantage of seasonal water availability and perhaps change rooting habits to increase soil water supply or change the timing of withdrawal. Such advances may aid efficient use of water and ET efficiency without changing the T efficiency.

Early water-requirement studies and work reviewed by de Wit indicate the T efficiency is not seriously affected by plant nutrition unless nutrition deficiency is so great as to reduce yields to about half those of adequately fertilized crops. Container experiments of Anderson and Read (1966) provide similar data. More recently, Yoshida and Coronel (1976) found that $c = 1 - C_i/C_a$ of Eq. [12a] changed little in rice with varying levels of N in culture solution over a wide range of N_L, and Wong et al. (1979) found similar results with maize when N and P fertilization was varied. On the other hand, Viets (1962) suggested that the ET efficiency of crops is increased by fertilization. This conclusion also is quite consistent with Eq. [15] and [16]. Fertilization that promotes greater leaf area and reduces evaporation relative to transpiration results is an increased ET efficiency.

Continued increases in fertilization and biomass yield will result in decreased E/ET and greater ET efficiency. As pointed out by Viets in his model B, the Y_T/ET_T will continue to increase asymptotically. In contrast to Viets, we conclude that the asymptote has a physical significance, in that it is $k\int T/B(e^* - e)/\int T_T \approx k/(e^* - e)$ at adequate fertilization. However, Viets' model D is inconsistent with our analysis because it shows no increase in evapotranspiration with increasing dry matter yields after reasonably complete cover is attained, implying $dY/dT \rightarrow \infty$. Viets presented data from four studies as examples where ET was not increased due to management practices that increased yield. The ET data in the study by Penman (1956) were not measured, but calculated based on meteorological data, so no treatment differences in ET were considered. The data of Carlson et al. (1959) allow comparisons of only two points for any treatment. While water-use efficiency increased with yield and fertilization for the irrigated corn (not for unirrigated), it is impossible to determine if Y/T was altered. Moreover, with the irrigated corn, Carlson et al. found substantially *less* ET at high fertility and corn population, indicating unreliable ET measurements, possibly because of undetermined drainage. The study of Weaver and Pearson (1956) used small plots of shallow clay and considered changes as sudangrass (var. 'sudanense') grew from 5 cm high over a 4-week period. The experimental conditions were those Viets would class as his model B. Surprisingly, there was little difference in ET with plant population during any of the 4 weeks, suggesting evaporation from the clay may have played a dominant role, and that model C may be most appropriate. In the fourth study (Holmen et al., 1961), equal ET at high and very low yields (ground cover) and much higher ET from the irrigated plots than from unirrigated at comparable yields indicated considerable evaporation from the soil. Model C probably is more appropriate; at least the Y/T relationship is obscured by high evaporation.

We agree with Fischer and Turner's (1978) conclusions that when increased water-use efficiency is found as a result of improved management, the increases result from increased transpiration as a fraction of the ET; ET efficiency is increased although T efficiency is changed little, if any. Conditions such as low fertility, water stress, plant disease, or insects that lower the leaf area so that the canopy is not longer closed will increase soil evaporation and thereby lower Y/ET. Also, factors such as poor growing temperatures and extreme infertility could cause $c = 1 - C_i/C_a$ of Eq. [15] to decrease markedly and lower both Y/T and Y/ET. Nevertheless, both a decrease in LAI or an increase in r_m has to be rather severe before substantial changes in Y/T will be observed.

To summarize, it appears to us that two avenues exist to significantly modify the T efficiency (total dry matter) of crops. First, crops can be grown in humid climates where $(e^* - e)$ is small and advection minimal; however, sunlight usually is less and total yields may be smaller. Also, $(e_D^* - e)$ may be decreased by local climate modification such as reflective coatings on plants or mist irrigation, provided there is economic justification. Second, the chemistry of the plant can be changed, but the modification must be large and is improbable. Changing the partitioning of total dry matter to more marketable product is a viable option to increase T efficiency of marketable yield. If the composition is changed to more proteins and lipids, however, T efficiency will be sacrificed.

Modifying ET efficiency has more options, as discussed previously. Shifting growing season or maturity periods can take advantage of the annual climate cycle to decrease T_T. The crop can be managed (e.g., population and fertility) to increase or decrease LAI, thus changing the partitioning of E and T and ET efficiency. Preventing evaporation from the soil and needless transpiration from weeds also modifies the partitioning of E and T. however, there are limits to the improvements such manipulations can provide; the ET efficiency can only approach T efficiency as the upper limit.

Although this discussion has been devoted to water-use efficiency, we need to emphasize the cautions of Viets, mentioned early in this chapter. Increased water-use efficiency is only one measure of management, and often it is a poor one. Optimizing productivity with respect to dollars or resource inputs other than water (e.g., energy) may be much more important. The other chapters in this book, in fact, address these important questions in the efficient use of water.

IV. IRRIGATION

In the previous section, the analysis of crop water-use efficiency results in a rather pessimistic view of substantially improving it by either plant breeding or local agronomic practices. Yet in a broader geographical sense, Eq. [15] and [16] clearly suggest that substantial improvements in water-use efficiency can be achieved by matching crop production with areas of low vapor pressure deficits. In general, solar radiation, saturation deficits, and evapotranspirational demand in a region increase as precipitation and cloudiness decrease. Figure 1 illustrates climatic divisions of the contiguous United States in which saturation deficits and evapotranspiration increase with regional aridity. Although potential productivity under irrigation can also increase with aridity because of increased sunshine and growing period, advection adds to the evapotranspiration of irrigated fields without a concomitant increase in productivity. The data in Table 5 illustrate the water-use efficiency advantage of the more humid areas.

Since irrigation often produces yield increases in the humid regions, we conclude that the greatest gains in improving crop water-use efficiency for the entire nation would be to put greater emphasis on water management and irrigation technology in these humid areas. In addition, these gains in yield could be achieved in the humid regions with much smaller water input than in the more arid areas. Rainfall would still provide much of the required water, and the transpiration losses are smaller in humid regions. Finally, and of increasing importance, in the humid regions where there is a much greater availability of water—and it is usually of high quality—salinization is not a problem.

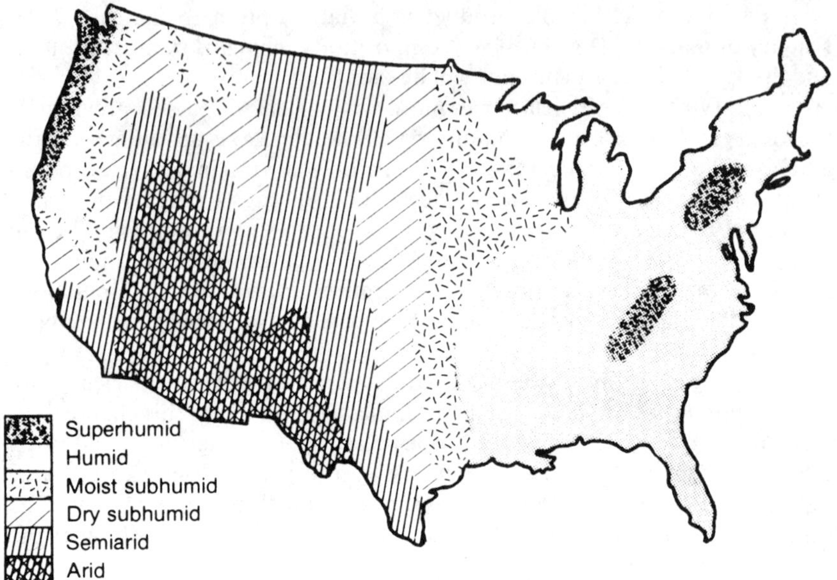

Fig. 1—Climatic regions of the contiguous United States based on the precipitation effectiveness index (Thornwaite, 1941).

However, both institutional and technological problems inhibit the use of irrigation in the humid regions. Many of the institutional problems arise because there is a long history of nonagricultural water use and a short history of irrigation. Agriculture's priority for water allocations often is low. Rapid increases in irrigation usually have followed water legislation that provided increased access to water by irrigators. Development of water supplies in humid regions, unlike the western United States, is left to be the individual. The main tax dollars from humid regions expended on water development for irrigation have been those transferred to the West; even tax outlays to use municipal wastes primarily are devoted to "clean-up" rather than irrigation. The role of irrigation in humid regions needs better institutional and economic evaluation, which in turn, requires better technical understanding.

Irrigation research also has suffered neglect in the humid regions, where unpredictable rainfall imposes considerably different technical problems than those experienced in the dry regions where most irrigation technology has been developed. The soil fertilizing practices in humid regions have been developed with little regard to the interaction between N-use efficiency and irrigation efficiency. Considerable research must be done to develop sound fertilizer/irrigation practice that recognizes this interaction, and minimizes leaching while maintaining an adequate soil water supply for the crop. This becomes increasingly important with shallow-rooted crops and sandy soils. This research, which involves irrigations that do not bring the root zone to full water supply, is essential to reduce the energy use in irrigation and to avoid nitrate contamination of ground water.

Table 5—ET efficiency for corn grain from locations with high and moderate saturation deficits.

Location	ET	Grain yield	Y/ET	Source
	cm	ton ha^{-1}		
Davis, Calif.	60	12.0	0.20†	Stewart et al. (1977)
Ft. Collins, Colo.	49	8.8	0.18†	Stewart et al. (1977)
Logan, Utah	53	8.2	0.16†	Stewart et al. (1977)
Hancock, Wis.	43	10.3	0.24	Unpublished data
Lafayette, Ind.	44	12.0	0.27	Stuff & Dale (1978)‡
Durant, Iowa	44	10.0	0.25	R. H. Shaw (personal communication)

† Best in 2-year trial.
‡ Mean of 3-years; yield data supplied by R. F. Dale.

Irrigation practices on sloping land must be researched to integrate them with sound conservation measures. Erosion hazards are greater in humid regions where unpredicted precipitation can follow an irrigation. Plant disease, insects, and weeds pose problems under irrigation in humid regions not found in dry regions. Trafficability for field operations is often a problem on fine-textured soils under natural rainfall, and the problem worsens when crops are irrigated. Increasing energy costs will change the feasibility of irrigation in both arid and humid regions. The impact in humid regions will vary greatly depending on availability of water supply and probability of yield increase; this needs assessment. Electric power use for pumping is seasonal, and with the weather-induced uncertainties on timing irrigation in humid regions, management of power generation and distribution for load peaks is an increasing problem. In short, scheduling irrigation must help optimize fertilizer use, soil conservation, pest control, power requirement, electric load management, and other crop management

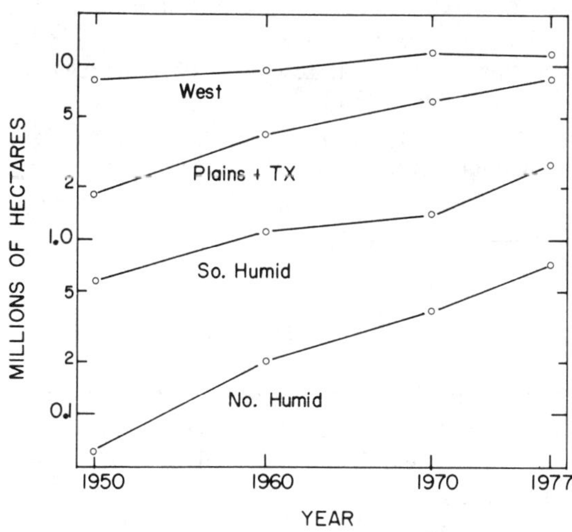

Fig. 2—Area of irrigated land in the contiguous United States from 1950 to 1977 as derived from the *Irrigation Journal.*

practices for high productivity and minimal environmental damage (ground-water pollution, soil loss) in the presence of unpredictable rainfall. Of course, better mesoscale forecasting of rainfall would help, but in view of forecasting advances since 1900, offers small hope.

In spite of these institutional and technological problems there has been a considerable expansion of irrigation in the humid regions. In fact, in some cases the greatest expansion is now occurring in the humid areas. Data gathered from issues of the *Irrigation Journal* and plotted in Fig. 2 indicate there has been a 1.6-fold increase in irrigated land in the western states between 1950 and 1977; this increase, which started from a high base area, has now turned downward because of limited water resources. In the Plains states (excluding the high plains of Texas), there has been a little over an eightfold increase in irrigated area for this period; this expansion, largely in Kansas and Nebraska, is mainly from wells, and continued increase at the same rate is unlikely. The irrigated land in the south-humid states has increased about 4.5-fold and in the north-humid about 9.5-fold since 1950. The South started off with considerable irrigation in Louisiana, Florida, and Arkansas, and irrigation has expanded substantially except in Louisiana. The north-humid region started very low and irrigation has increased greatly; two-thirds of the increase is in Iowa, Michigan, Minnesota, Missouri, and Wisconsin, where a combined 20-fold increase has occurred. This increase in humid-region irrigation probably will continue. Yields of all crops on sandy soils are decreased almost every year. On fine-textured soils, short-rooted horticultural crops suffer frequently. Even field crops on good soils that only occasionally suffer severe yield decreases are irrigated to provide production stability, which is important with high land values and other high operational costs.

Water resources in the subhumid and arid regions are limited. In humid regions, water resources are available and irrigation often produces yield increases. If we are to increase national food production, the greatest increase per investment is likely to derive where productivity is already high but limited by management rather than resource. It seems reasonable to suggest that should national policy really be concerned with increasing food production, instead of devoting major federal monies to support agricultural water management in arid regions where water resources are limited we might best make expenditures to learn how to manage water well in the humid regions where there is water.

V. SUMMARY

Our re-search of the factors influencing water-use efficiency indicates that considerable progress has been made in resolving the apparent ambiguities in this parameter. This is especially true when crops have closed leaf canopies and soil evaporation is small. Both the limited data (Table 4) and the theoretical arguments we presented suggest that k is a very stable term in the cropping system. Since the vapor pressure deficit is essentially defined by the local climate, it appears that transpiration efficiency for total biomass is a relatively unmanipulative variable. Clearly, from examination of Eq. [16], *Y*/ET can be improved; but as Viets (1962) suggested, it asymptotically approaches a constant. We have suggested that this asymptote is $k/(e^* - e)$. Consequently for intensive cropping situations, we conclude that *Y*/ET also is a fairly stable, bounded term and that the likelihood of significant improvements is small.

Nevertheless, stability in water-use efficiency does not mean that we cannot improve on efficient water use as we previously defined it. In fact, this predictability of water-use efficiency makes the research objectives for improving on efficient water use more clear-cut. The objective of this book, in fact, is to review the various approaches to improve efficient water use. First, methods for making more water available to the transpiration process are explored. In Chapters 2, 3, and 4 the processes of water movement in the soils, its absorption by roots, and its movements through the plant are considered in detail as possible approaches to maximizing the availability of water to crops. In Chapter 5 the physical processes controlling crop evapotranspiration are considered as keys to managing and minimizing water loss. The problem of water deficits developing in crops, causing severe limitation in the ability of crops to produce an economic yield, is analyzed in Chapters 6 and 7. Management options to increase the amount of water available for crop production and to improve crop-growth characteristics for increasing the economic yield are discussed in Chapters 8 and 10. Chapters 9 and 11 deal with two obvious economic considerations of improving efficient water use. In Chapter 11 improvement of irrigation management for economic yield of crops is considered, and in Chapter 9 techniques for predicting the yield of the economic component of the crop for the water invested is analyzed. Finally, in Chapter 12 the priority for increased efficient water use in crop production within the context of modern society is examined.

LITERATURE CITED

Ackermann, W. C., D. J. Allee, J. Amorocho, Y. Y. Haimes, W. A. Hall, R. A. Meserve, R. Patrick, and P. M. Smith. 1978. Scientific and technological considerations in water resources policy. EOS Trans. Am. Geophys. Union 59:516–527.

Anderson, C. D., and D. W. L. Read. 1966. Water-use efficiency of some varieties of wheat, oats, barley, and flax grown in the greenhouse. Can. J. Soil Sci. 46:375–378.

Arkley, R. J. 1963. Relationships between plant growth and transpiration. Hilgardia 34:559–584.

Bierhuizen, J. F., and R. O. Slatyer. 1965. Effect of atmospheric concentration of water vapor and CO_2 in determining transpiration-photosynthesis relationships of cotton leaves. Agric. Meteorol. 2:259–270.

Bray, J. R. 1963. Root production and estimation of net productivity. Can. J. Bot. 41:65–72.

Briggs, L. J., and H. L. Shantz. 1913a. The water requirement of plants: I. Investigations in the Great Plains in 1910 and 1911. USDA Bureau Plant Industry Bull. 284.

Briggs, L. J., and H. L. Shantz. 1913b. The water requirement of plants: II. A review of the literature. USDA Bureau Plant Industry Bull. 285.

Briggs, L. J., and H. L. Shantz. 1914. Relative water requirement of plants. J. Agric. Res. (Washington, D.C.) 3:1–63.

Briggs, L. J., and H. L. Shantz. 1916. Daily transpiration during the normal growth period and its correlation with the weather. J. Agric. Res. (Washington, D.C.) 7:155–212.

Briggs, L. J., and H. L. Shantz. 1917. The water requirement of plants as influenced by environment. Proc. 2nd Pan-Am Sci. Congr. 3:95–107. Washington, D.C.

Burkholder, W. H. 1920. The effect of two soil temperatures on the yield and water relations of healthy and diseased bean plants. Ecology 1:113–123.

Carlson, C. W., J. Alessi, and R. H. Mickelson. 1959. Evapotranspiration and yield of corn as influenced by moisture level, nitrogen fertilization, and plant density. Soil Sci. Soc. Am. Proc. 23:242–245.

de Wit, C. T. 1958. Transpiration and crop yields. Versl. Landbouwk. Onderz. 64.6 Inst. of Biol. and Chem. Res. on Field Crops and Herbage, Wageningen, The Netherlands.

Duniway, J. M., and R. O. Slatyer. 1971. Gas exchange studies on the transpiration and photosynthesis of tomato leaves affected by *Fusarium oxysparum f. sp. lycopersici.* Phytopathology 61:1377–1381.

Elkins, C. B., R. L. Haaland, R. Rodriguez-Kabana, and C. S. Hoveland. 1979. Plant-parasitic nematode effects on water use and nutrient uptake of a small- and a large-rooted tall fescue genotype. Agron. J. 71:497–500.

Endrodi, G., and P. E. Rijtema. 1969. Calculation of evapotranspiration from potatoes. Neth. J. Agric. Sci. 17:283–299.

Fischer, R. A., and N. C. Turner. 1978. Plant productivity in the arid and semiarid zones. Annu. Rev. Plant Physiol. 29:277–317.

Graf-Marin, A. 1934. Studies on powdery mildew of cereals. Cornell Agric. Exp. Stn. Memoir no. 157. Ithaca, N.Y.

Hanks, R. J. 1974. Model for predicting plant yield as influenced by water use. Agron. J. 66: 660–665.

Hanks, R. J., H. R. Gardner, and R. L. Florian. 1969. Plant growth–evapotranspiration relations for several crops in the central Great Plains. Agron. J. 61:30–34.

Holmen, H., C. W. Carlson, R. J. Lorenz, and M. E. Jensen. 1961. Evapotranspiration as affected by moisture level, nitrogen fertilization, and harvest method. Trans. ASAE 4: 41–44.

Hunt, L. A., C. E. Moore, and J. E. Winch. 1970. Light attenuation coefficient and productivity in "vernal" alfalfa. Can. J. Plant Sci. 50:469–474.

Johnston, C. O., and E. C. Miller. 1934. Relation of wheat rust to yield, growth, and water economy of two varieties of wheat. J. Agric. Res. (Washington, D.C.) 49:955–981.

Kanemasu, E. T., L. R. Stone, and W. L. Powers. 1976. Evapotranspiration model tested for soybean and sorghum. Agron. J. 68:569–572.

Kiesselbach, T. A. 1910. Transpiration experiments with the corn plant. Nebraska Agric. Exp. Stn. 23rd Annu. Rep. p. 125–139. Lincoln.

Kiesselbach, T. A. 1916. Transpiration as a factor in crop production. Nebraska Agric. Exp. Stn. Res. Bull. no. 6. Lincoln.

Kiesselbach, T. A., and E. G. Montgomery. 1911. The relation of climatic factors to the water used by the corn plant. Nebraska Agric. Exp. Stn. 24th Annu. Rep. p. 91–107. Lincoln.

King, F. H. 1914. Soil management. Orange Judd Co., London.

Lemon, E. R., D. W. Stewart, and R. W. Shawcroft. 1971. The sun's work in a corn field. Science 174:371–378.

Linacre, E. T. 1964. A note on a feature of air and leaf temperature. Agric. Meteorol. 1:66–72.

Montgomery, E. G., and T. A. Kiesselbach. 1912. Studies in water requirements of corn. Nebraska Agric. Exp. Stn. Bull. no. 128. Lincoln.

Penman, H. L. 1948. Natural evaporation from open water, bare soil, and grass. Proc. R. Soc. London Ser. A. 193:120–146.

Penman, H. L. 1956. Weather and water in the growth of grass p. 170–177. *In* F. L. Milthorpe (ed.) The growth of leaves. Butterworths, London.

Penning de Vries, F. W. T. 1975a. The cost of maintenance processes in plant cells. Ann. Bot. (London) 39:77–92.

Penning de Vries, F. W. T. 1975b. Use of assimilates in higher plants. p. 459–480. *In* J. P. Cooper (ed.) Photosynthesis and productivity in different environments. Cambridge University Press, Cambridge.

Rijtema, P. E., and G. Endrodi. 1970. Calculation of production of potatoes. Neth. J. Agric. Sci. 18:26–36.

Sinclair, T. R., and C. T. de Wit. 1975. Photosynthate and nitrogen requirements for seed production. Science 189:565–567.

Sinclair, T. R., and C. T. de Wit. 1976. Analysis of the carbon and nitrogen limitations to soybean yield. Agron. J. 68:319–324.

Sinclair, T. R., C. E. Murphy, Jr., and K. R. Knoerr. 1976. Development and evaluation of simplified models for simulating canopy photosynthesis and transpiration. J. Appl. Ecol. 13:813–839.

Slabbers, P. J., V. S. Herrendorf, and M. Stapper. 1979. Evaluation of simplified water-crop yield models. Agric. Water Manage. 2:95–129.

Stanhill, G. 1962. The effect of environmental factors on the growth of alfalfa in the field. Neth. J. Agric. Sci. 10:247–253.

Stanhill, G. 1970. A new method of reducing evaporation. J. Irreproducible Results 18:55–57.

Stewart, J. I. 1972. Prediction of water production functions and associated irrigation programs to minimize crop yield and profit losses due to limited water. Ph.D. Thesis. University of California-Davis. Univ. Microfilms #73-16, 934.

Stewart, J. I., R. M. Hagan, W. O. Pruitt, R. J. Hanks, J. P. Riley, R. E. Danielson, W. T. Franklin, and E. B. Jackson. 1977. Optimizing crop production through control of water and salinity levels in the soil. Utah Water Res. Lab., Utah State University, Logan. Pub. no. PRWG151-1.

Stuff, R. G., and R. F. Dale. 1978. A soil moisture budget model accounting for shallow water table influences. Soil Sci. Soc. Am. J. 42:637–643.

Sumayao, C. R., and E. T. Kanemasu. 1979. Temperature and stomatal resistance of soybean leaves. Can. J. Plant Sci. 59:153–162.

Tanner, C. B., and W. A. Jury. 1976. Estimating evaporation and transpiration from a row crop during incomplete cover. Agron. J. 68:239–243.

Teare, I. D., E. T. Kanemasu, W. L. Powers, and H. S. Jacobs. 1973. Water-use efficiency and its relation to crop canopy area, stomatal regulation, and root distribution. Agron. J. 65:207–211.

Thornthwaite, C. W. 1941. Atlas of the climatic types in the United States, 1900–1939. USDA Soil Conserv. Serv. Pub. no. 421.

Viets, F. G., Jr. 1962. Fertilizers and the efficient use of water. Adv. Agron. 14:223–264.

Weaver, H. A., and R. W. Pearson. 1956. Influence of nitrogen fertilization and plant population density on evapotranspiration of sudangrass. Soil Sci. 81:443–451.

Weiss, F. 1924. The effect of rust infection upon the water requirement of wheat. J. Agric. Res. (Washington, D.C.) 27:107–118.

Wong, S. C., I. R. Cowan, and G. D. Farquhar. 1979. Stomatal conductance correlates with photosynthetic capacity. Nature 282:424–426.

Yarwood, C. E. 1936. The diurnal cycle of powdery mildew, *Erysiphe polygani.* J. Agric. Res. (Washington, D.C.) 52:645–657.

Yarwood, C. E. 1947. Water loss from fungus cultures. Am. J. Bot. 34:514–520.

Yoshida, S., and V. Coronel. 1976. Nitrogen nutrition, leaf resistance, and leaf photosynthetic rate of the rice plant. Soil Sci. Plant Nutr. 22:207–211.

CHAPTER 1B

Efficient Water Use in Crop Production: Discussion on the Generality of Relations Between Biomass Production and Evapotranspiration

JOE T. RITCHIE

Agricultural Research Service, USDA
Temple, Texas

I. INTRODUCTION

There is little question that water supply for transpiration causes major variation in crop yields. Therefore, it is understandable why scientists have wanted to relate yield to water use through studying the water requirements of crops. Modern technology attempts to minimize a crop's exposure to soil water deficits through various management strategies, such as supplemental irrigation, proper fertilizer use, plant population and row-spacing, and use of maturity genotypes that match available soil water supply. When a producer is successful in minimizing water deficiency in a crop, factors other than water supply can become limiting. In such cases, evaluation of production as related to evapotranspiration may not be a useful quantitative concept.

This paper is a discussion of difficulties in making generalizations about crop production as related to evapotranspiration (ET) or transpiration (T), and offers an alternative approach to calculating biomass production independent of ET and T that is believed to be general.

II. CROP PRODUCTION AND EVAPOTRANSPIRATION

A. Evapotranspiration Efficiency and Leaf Area Index

"Re-search" of long-established data sets on the question of water-use efficiency discussed by Tanner and Sinclair in Chapter 1A, and recent new evaluation on the same question reported in this volume by Hanks in Chapter 9A, have established more general relationships between ET and biomass production. Authors of both chapters have pointed out that a key to generalization of the biomass-ET relationships is through separation of

Limitations to Efficient Water Use in Crop Production.

soil evaporation (E) and T. In Chapter 1A it was pointed out that when T is known or can be reasonably estimated, "T efficiency is a reasonably stable quantity for a green crop, provided a local climate normalization is used to reflect the differences in transpiration rates that occur between locations and time of the year." de Wit (1958) used average pan evaporation for normalization, Arkley (1963) claimed improvement by normalizing with relative humidity, and Tanner and Sinclair (Chapter 1A) use theoretical arguments and field crop calculations (their Table 4) to demonstrate that vapor pressure deficit ($e^* - e$) of the daylight period is an even more general normalization. They demonstrated that a rather stable constant of proportionality (k) when divided by ($e^* - e$) described the relationship between biomass production and T for crops that have closed leaf canopies. They further demonstrate that the k values vary among crop species consistently with differences in carboxylation pathways in C_3 and C_4 plants.

In their evaluation, Tanner and Sinclair used vapor pressure data from standard screen level, and recognized that adjustments requiring more detailed modeling would be necessary for low leaf area index (LAI) values. They recognized that dry, hot soil can cause a significantly greater vapor pressure deficit at seedling level when a low LAI exists than would be inferred by screen-level measurements. Their emphasis is on T-efficiency for growing seasons where LAI > 2.5. In the absence of a more detailed model or availability of actual data on vapor pressure profiles near the ground surface, it seems important to consider less-than-full-cover cases when considering efficient water use on a seasonal basis.

B. Transpiration, Soil Evaporation, and Soil Surface Wetness

For many years, it has been recognized that T can be influenced by local sources of sensible heat from nonevaporating surfaces such as dry soil surrounding transpiring plants. Marlatt (1961) found that ET from orchard grass with an adequate root-zone water supply was as great with 50% cover as with full cover, screen-level meteorological conditions being similar. In the study, the soil between rows was dry at the surface, thus contributing little to the ET. As a consequence, T per unit covered area was about double that of a full-cover crop. This phenomenon, called the *clothesline effect* (Tanner, 1957), occurs when hot, dry air from between rows passes through plants in the rows, increasing T. Hanks et al. (1971) found that about 64% of the sensible heat flux from a dry soil surface below a sorghum crop with an LAI of 1.2 was used for T.

Ritchie and Burnett (1971) quantified the influence of partial cover on ET and found that LAI of sorghum and cotton were more generally related to T as fractions of maximum ET (ET_{max}) than ground cover or plant dry weight. The relationship was obtained under conditions when the soil surface between rows was dry most of the time. Thus, the clothesline effect should have been significant at least for low LAI. A relatively large difference between T when the soil surface was dry as compared with when

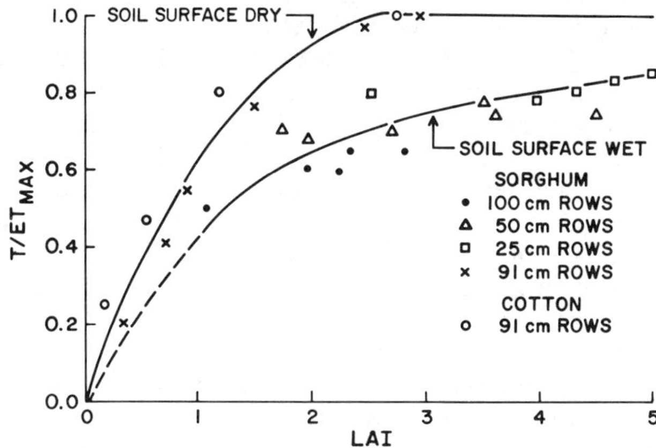

Fig. 1—Transpiration (T) relative to estimated ET_{max} as influenced by LAI when soil water in the root zone is nonlimiting for dry and wet soil surfaces. Data from Ritchie (1971), Ritchie and Burnett (1971), and Adams et al. (1976).

it was wet became apparent when separately calculating E and T for a partial-cover ET model (Ritchie, 1972), with the sum of E and T always exceeding calculated ET_{max} for wet soil surface conditions. The T value had to be reduced in such cases for the model to agree with lysimeter-measured ET.

Later interest in this problem led to independent measurements of E from wet soil using special evaporation plates designed to measure E under a canopy (Arkin et al., 1974). Using the plates, Adams et al. (1976) found that for LAI values as high as 5, 15% of the ET_{max} was E. The lower curve in Fig. 1 is a plot of T/ET_{max} calculated from the Adams et al. data (their Table 2 and Fig. 5), assuming $T = ET_{max} - E_{measured}$. Values of ET_{max} were taken with evaporation plates placed in the center of a large, bare area that was wet at the surface.

The upper curve of Fig. 1 shows T/ET_{max} values obtained from lysimeter ET data of Ritchie (1971) and LAI data of Ritchie and Burnett (1971). Measured ET/E_{max} values were selected for the upper curve using only days when the soil surface was dry following a relatively long period without rainfall. Values of ET_{max} for these measurements were estimated from 24-hour net radiation, which proved to be reliable for Temple, Tex., at that time of the year. Although ET_{max} for the wet and dry soil surface comparison were measured using different estimates and in different years, both ET_{max} estimates were close to 6 mm d^{-1} during the July and August period when measurements were made. Cotton showed evidence of being influenced by the clothesline effect somewhat more than sorghum at a given LAI when the soil surface was dry. In the wet-surface tests with sorghum, there was a demonstrated difference in T for a given LAI as influenced by row-spacing because of different amounts of radiation interception per unit leaf area for the different canopy configurations (Adams et al., 1976).

The dependence of T/ET_{max} on soil surface wetness, as implied in Fig. 1, suggests that if biomass production, for a given radiation and temperature condition above the crop canopy, is similar for a wet or dry soil surface, then T efficiency could vary by somewhat less than a factor of 2 for LAI $<$ 1, depending on the soil surface wetness. If the increase in T as E becomes smaller due to surface dryness is similar for most crops, then it is expected that ET will be equal to ET_{max} for dry *or* wet soil surfaces when cover is sufficiently established to obtain LAI values of 2.0–2.5, so long as T is not limited by a soil water deficit. Thus, a relatively constant T efficiency, when weighted for saturation deficit measured at screen level, should more precisely be a constant ET efficiency when LAI is about 2 or greater. For LAI less than about 1, T efficiency or ET efficiency is highly dependent on the soil surface wetness.

Leaf area expansion of many crops takes place very rapidly once a "grand growth" period is reached in plant ontogeny. Therefore, the period when LAI values are between 0.5 and 2.0 is quite short relative to an entire growing season, unless unusually small plant populations exist. For example, the 1968 cotton crop reported in Ritchie and Burnett (1971) had approximately 45 days with LAI $<$ 0.5, 25 days with 0.5 $<$ LAI $<$ 2.0, and 70 days with LAI $>$ 2.0. Thus, as a practical generality, crops go through two primary stages: (i) E and duration of soil surface wetness almost totally dominate ET (LAI $\lesssim$ 1.0), and (ii) ET is practically independent of soil surface wetness (LAI $\gtrsim$ 1.0). This is why plots of yield vs. ET for fixed climate and management, where variable water supply regulates cumulative ET, have a positive intercept on the ET axis. The intercept is often interpreted as being equal to the soil evaporation fraction of total crop ET. In my opinion, it would be more appropriately interpreted as the cumulative E from planting to the time when LAI is about 1.0. Likewise, the seasonal ET after LAI $>$ 1 is *both* T and E, with the surface wetness determining to some extent the value of T.

A daily incrementing model that uses a relationship similar to the upper curve of Fig. 1 for estimating T (Ritchie, 1972), or a slight deviation from it, has provided reasonable estimates of ET for several tests when LAI is known (Tanner & Jury, 1974; Kanemasu et al., 1974; Al-Khafaf et al., 1976). The model was used to simulate E and T for a cotton crop grown under average temperature and radiation conditions of Birmingham, Ala., where the precipitation was assumed to occur once every 7 or 18 days, and the amount was adequate to refill the soil storage (Ritchie, 1974). The soil storage was assumed to be great enough to prevent any stress in the plants between precipitation events. Maximum LAI was about 3. Total ET for the season was 641 and 572 mm for the 7- and 18-day rainfall frequency, respectively. For the 7-day frequency, calculated E was 46% of the ET and for the 18-day frequency, E was 34% of the ET. However, for the part of the season when LAI $>$ 2, ET totaled almost the same for the 7- and 18-day frequency, being 401 and 390 mm, respectively. Transpiration was 49 and 73%, respectively, of the ET during that period, demonstrating the dependence of T on E in moderately dense canopies. Assuming the crops would have produced equal biomass for each frequency, the ET-efficiency would be about the same, but the T-efficiency would be considerably less for the 18-day frequency than for the 7-day frequency.

C. Evapotranspiration Efficiency and Plant Nutrition

Because of the dependence of T on E for crops with a wide range of LAI values, the conclusion of de Wit (1958) and Tanner and Sinclair (Chapter 1A) that the major mechanism for decreased ET efficiency of nutrient-deficient plants is through reduced T from lower LAI is, in my opinion, not justified as a generality. Consider, for example, two crop canopies with maximum LAI values of 2.5 and 4.5, representing equal populations of plants with inadequate and adequate fertilization. Information in Fig. 1 suggests that the two canopies would have similar ET values, regardless of the soil surface wetness. However, biomass production per unit ground area should be considerably greater in the high-LAI crop. For this reason, the conclusion of Viets (1962) seems reasonable that, in most cases when water supply is fixed, any management factor that increases yield will increase ET efficiency because ET will be little affected by the management. The evidence Viets used for this Model D and almost all other relevant field studies done since his paper was published supports his hypothesis. The data of Carlson et al. (1959), for example, showed that irrigated maize biomass yields from a population of 34,600 plants ha^{-1} in 1956 were increased from 5,241 to 10,540 kg ha^{-1} primarily by N fertilizer, whereas ET varied from only 35.3 to 37.6 cm, respectively.

The ET estimates for irrigated maize data of Carlson et al. (1959) had considerable variation; the standard deviation for 2 years averaging about $\pm$ 3 cm. This deviation of about 10% of the total ET estimate is not unusual for water balance estimates of ET. Sources of possible deviation include inaccuracies in irrigation application and soil water content measurement, and spatial variation in soil water storage capacity. Jensen and Wright (1978) reported that the error in measuring soil water is generally smaller than irrigation application errors. My conclusion from evaluating several sets of field measured ET data is that, in practically all cases where water supply for ET is fixed and management is varied with fertilizers or plant population, ET variations among management treatments are no greater than the variations one would expect where management was not varied at all. The main exception to this conclusion is in cases where yield and LAI were very low.

Evidence Viets used to support the conclusion that ET efficiency increased when management increased yields was based on total seasonal ET. Evapotranspiration efficiency using seasonal ET are subject to a great deal of variation when comparing different locations and years, mainly because of the great variation of E during the part of the season when LAI is low. During that time, ET is strongly influenced by frequency and duration of surface wetness. Thus, for a more quantitative evaluation of the validity of Viets' conclusions, it is appropriate to examine data sets where the early-season E can be evaluated and removed from ET efficiency calculations.

Biomass yield and ET data of Power et al. (1961) for wheat grown in North Dakota are plotted in Fig. 2. The data were taken four times during

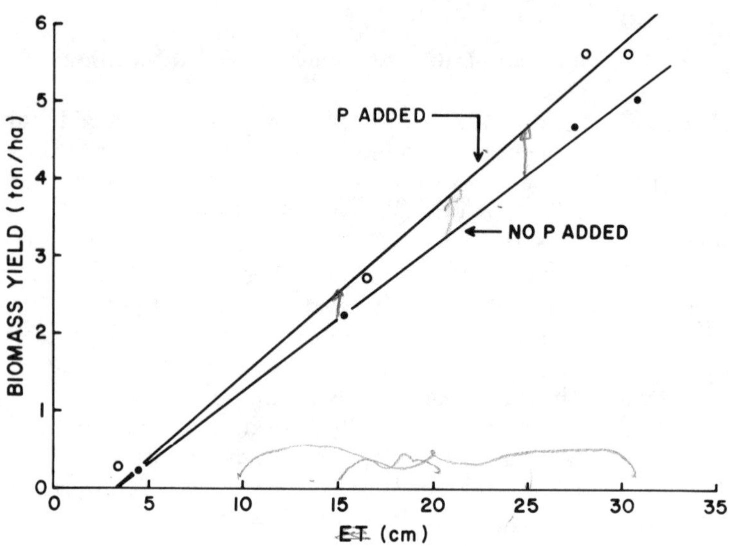

Fig. 2—Biomass yields of wheat as related to ET and P fertilization. Data from Power et al. (1961).

the season at tillering, heading, dough, and harvest, and represent a comparison between P fertilization rates for irrigated wheat (their treatment D4S2). There were significant and consistent yield differences between the treatments, especially at the last two harvest dates. The difference between ET for the P treatments never exceeded 1.2 cm on any sampling date. Using relations from Aase (1978) to convert biomass to leaf area, values of maximum LAI in the Power et al. study should have been about 2.8 and 3.3, respectively, for the unfertilized and fertilized treatments. Assuming validity of the relationship of Fig. 1 to this case, ET for both treatments should be about the same, with only a small early-season difference expected when both treatments had LAI values < 2. The linear curves fit through the data in Fig. 2 for the two treatments indicate that early-season *E* was about 3.2 cm (the ET intercept) and the later-season ET efficiency (the slope of the lines) was about 1.89 and 2.13 kg m^{-3} (1 kg m^{-3} = 10 metric tons ha^{-1} cm^{-1}), respectively, as a result of the fertilization. Assuming the validity of the assumption that the ET efficiency differences were caused by differences in *E* (Tanner & Sinclair, Chapter 1A) the seasonal *E* for the unfertilized treatment would have been about 4 cm more than the fertilized treatment at a yield level of 5.2 metric tons ha^{-1}. Considering that *E* is limited by soil water flow when the surface is not wet regardless of LAI, and the small difference in LAI and biomass between the two treatments, then a seasonal *E* of an extra 4 cm for the unfertilized treatment seems unrealistically high.

A comparison of biomass yield and ET as influenced by N fertilization of wheat (Jensen & Sletten, 1965a) is shown on the solid lines of Fig. 3. The comparison is for N fertilization rates of 90 and 134 kg ha^{-1} (their treatments F4 and F5). Variable ET data were from plants receiving from none to three post-emergence irrigations at Bushland, Tex. (their treatments M2,

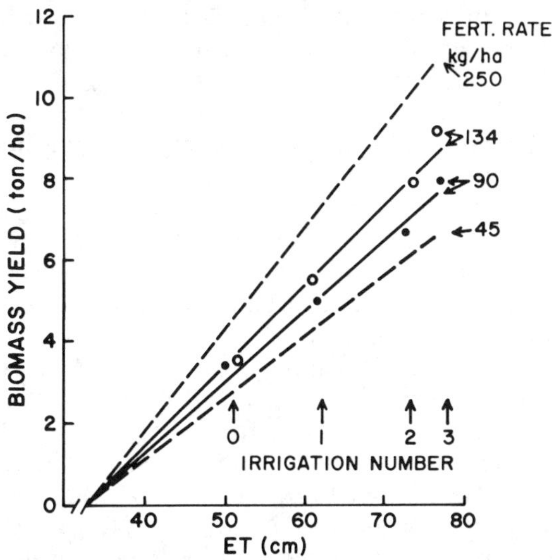

Fig. 3—Biomass yield of wheat as related to ET, N fertilization rate, and number of post-emergence irrigations. Dashed lines are assumed, using logic discussed in the text. Data from Jensen and Sletten (1965a).

M3, M6, and M4, respectively). Details about the dashed lines will be discussed in the next paragraph. In this same study, increased N fertilization resulted in higher biomass yields as illustrated in Fig. 4. Biomass yields were calculated from grain yield and straw-grain ratio data given in their Tables 6 and 8. Crops with yields of the magnitudes of those presented in Fig. 4 should have had maximum LAI values in excess of 2.5, with the possible exception of the nonfertilized treatment. Extension of the lines fitted to the biomass-ET relationships to zero yield indicated that early-season *E* was about 33 cm. The slope (ET efficiency) was 1.74 and 1.96 kg m^{-3} for the lower and higher N rates, respectively. Note that the ET was quite similar for both N rates for each irrigation treatment.

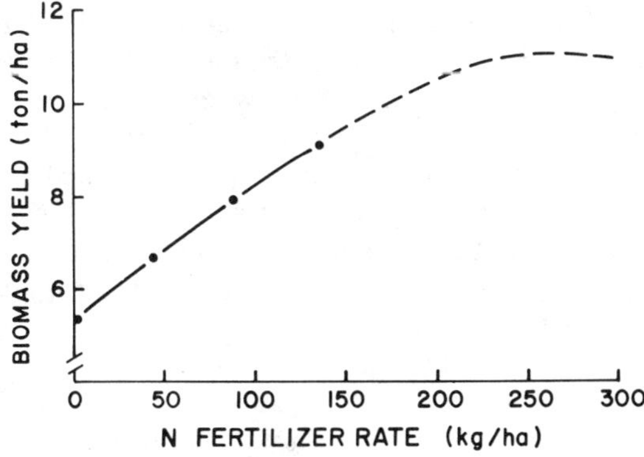

Fig. 4—Influence of N fertilization on biomass yield of wheat receiving four postemergence irrigations. The dashed line is an assumed extrapolation. Data from Jensen and Sletten (1965a).

Although ET efficiencies as influenced by fertilizer calculated from Fig. 2 and 3 are not greatly different, it is interesting to speculate about larger differences in ET efficiency. It seems highly probable that ET efficiency can be higher for wheat than the 1.96 kg m^{-3} found for the 134 kg ha^{-1} N fertilization rate shown in Fig. 3. For example, it is not uncommon to obtain biomass yields of 15 tons ha^{-1} (about 92 bu $acre^{-1}$ with a straw-grain ratio of 1.65) for modern short-stem varieties of irrigated wheat at Bushland, Tex., where the Jensen and Sletten (1965a) work was done. According to J. T. Musick (personal communication), seasonal ET has averaged about the same in recent years as it did 25 years earlier when the Jensen and Sletten work was done. A fertilizer response curve for the 1956 wheat study on plots receiving three irrigations is given in Fig. 4. It is obvious from the curve that the 134 kg ha^{-1} N rate had not caused yield to reach a maximum. The dashed curve extrapolated from the data should represent a reasonable approximation for a maximum yield of about 11 metric tons ha^{-1}. Assuming no change in ET for the higher yield, the possible biomass-ET curve is given on the upper dashed line of Fig. 3, the ET efficiency being 2.47 kg m^{-3}. The lower dashed line shows the possible biomass-ET curve for the 45 kg ha^{-1} N treatment, again assuming no change in ET. The resulting ET efficiency is 1.46 kg m^{-3}. Thus, there appears to be considerable variation in ET efficiency when management practices such as fertilizer, improved varieties, and plant population cause large differences in yield if plant water supply is either nonlimiting or available in a fixed quantity.

Evapotranspiration efficiencies calculated using the slopes of the lines fitted to data of Fig. 2 and 3 are plotted as a function of biomass yield in Fig. 5. The closed circles are taken from the Jensen and Sletten wheat study (Fig. 3) and the open circles are from the Power et al. wheat study (Fig. 2). The three sets of data points connected by the solid lines were taken from the one, two, and three irrigation treatments in the Jensen and Sletten study. The connected solid lines represent changes in ET efficiency caused by fertilizers where the water supply was fixed, and conforms to principles of Viets' (1962) Model D. The slope of each solid line is the inverse of the seasonal ET after correction for early-season *E*. The dashed lines showing no change in ET efficiency with yield increases connect data from fixed management and variable water supply treatments. The dashed curves conform to Viets' Model B at high yield levels where ET efficiency approaches an asymptote and is essentially constant. However, the constant ET efficiency value can be management specific, as illustrated by the two constant levels in ET efficiency from the Jensen and Sletten study.

When yield decreases below some critical threshold because of nutrient deficiencies, there can be a significant decrease in seasonal ET. The zero-N treatments in the 1956 wheat data of Jensen and Sletten clearly show this trend for irrigation treatments causing significant fertilizer response. For treatments with three postemergence irrigations, for example, ET was 17 cm less for the zero-N rate than for the average of the 90 and 134 kg ha^{-1} N rate. Two major reasons for such observations when top growth is greatly reduced include: (i) maximum seasonal LAI values of 2 or less cause decreased *T*, and (ii) roots of small plants explore less soil volume. A good ex-

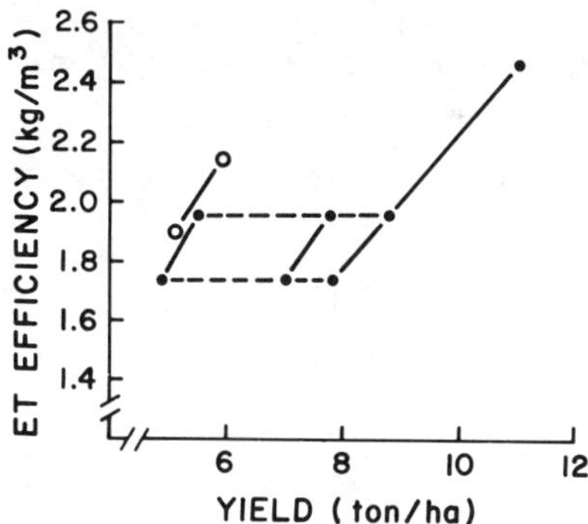

Fig. 5—The relationship between ET efficiency and biomass yield of wheat using relationships established in Fig. 2 and 3. Solid lines represent fixed ET and variable fertility; dashed lines represent fixed fertility and variable ET. Solid circles from Jensen and Sletten (1965a); open circles from Power et al. (1961).

ample of a study demonstrating a reduced soil rooting volume causing lower ET is that of Brown (1971), who reported dryland wheat grain yields of 1,610 kg ha^{-1} from an unfertilized wheat plot and 3,630 kg ha^{-1} from a plot receiving 268 kg ha^{-1} N fertilizer. Soil water supply had been depleted as much as possible before maturity in both plots as a result of little rainfall after heading. Unfertilized plants extracted about 6.0 cm less water from the soil than the high-N-treatment plants.

Because ET is less for crops where nutrient deficiencies or other non-water-related stresses are severe, ET efficiencies such as those shown in Fig. 5 do not continue to decrease linearly with lower yields. Limited evidence from the Power et al. (1961) study and the Jensen and Sletten (1965a) wheat study imply that ET efficiencies do not decrease much below levels shown in Fig. 5. I speculate that relatively fixed ET efficiencies found for container-grown plants where variable nutrient supply significantly affected yields, such as those of Kisselbach (1916), were apparently the result of low LAI. The T/ET_{max}–LAI curve for a dry soil surface in Fig. 1 could be relevant to container experiments where E is prevented. For LAI < 2, T/ET_{max} *per unit LAI* is fairly constant. For LAI > 2, T/ET_{max} per unit LAI decreases rapidly. If LAI of container grown plants is calculated relative to neighboring plants rather than container surface area, it is likely that LAI values, even for large plants, could be < 2 in most instances. I believe this phenomenon could explain much of the divergence of views about ET efficiencies as related to fertilizer or other management factors.

Most field ET efficiency data where management is a variable do not contain enough detail to make comparisons similar to those shown in Fig. 2 and 3. It is possible, however, to qualitatively test the validity of Viets' Model D by using whole-season ET and grain or dry matter yields for ET efficiency calculations. I calculated the slope of ET efficiency vs. yield for several fertilizer and population studies from the literature, using the following equation:

$$\text{Slope} = (Y_H/ET_H) - [(Y_L/ET_L)(Y_H - Y_L)], \qquad [1]$$

Table 1—The mean slope of the relationship between evapotranspiration efficiency and yield as influenced by management for several crops.

Crop	Management	Mean slope (m^{-1})	n	Reference
Wheat grain	N fertilization	2.9 (± 0.3)	5	Bond et al. (1971)
Wheat grain	N fertilization	1.2 (± 0.4)	24	Jensen & Sletten (1965a)
Sorghum grain	Population	2.7 (± 0.2)	3	Olson (1971)
Sorghum grain	N fertilization	1.7 (± 0.1)	22	Jensen & Sletten (1965b)
Maize forage	N fertilization + population	2.7 (± 0.7)	6	Carlson et al. (1959)
Sorghum biomass	Population	3.8 (± 0.6)	2	Olson (1971)
Maize biomass	Population	2.8 --	1	Olson (1971)
Forage Sorghum	Population	3.2 --	1	Olson (1971)
	Mean	2.6		

where Y is yield and the subscripts represent high (H) and low (L) yields or ET for comparing treatments. For the calculations, only paired treatments that caused at least a 25% difference in yield were used. Mean slopes for each study are given in Table 1 along with the calculated standard deviations where more than one comparison made up the mean slope. The mean value of all the slopes in Table 1 was 2.6 m^{-1}. The inverse is 0.38 m (38 cm), a reasonable value of average seasonal ET for the studies reported. These data further support Model D of Viets (1962).

On the basis of the qualitative information in Table 1 and the more precise data in Fig. 2 and 3, it seems probable that a difference in E among various management treatments for the period of the season when LAI < 1 could account for most of the differences in measured ET efficiencies given in Table 1.

III. AN ALTERNATIVE METHOD FOR ESTIMATING BIOMASS PRODUCTION

A. Light Interception and Biomass Production

Hanks (Chapter 9A) points out the practical value of using yield–ET relationships to estimate the impact of water supply on yield. There is little doubt about the importance of having a quantitative understanding of the relationships, especially as competition for limited water supplies increase. However, because yield-ET relationships can be management-specific as shown above, and climate-specific because E_{max} can be quite variable, it is desirable to have a more general technique for estimating yields as related to management, water supply, and climate.

Monteith (1977) demonstrated that cumulative seasonal light interception for several crops grown with adequate soil water supply was closely related to biomass production, as shown in Fig. 6. There was considerable similarity among the diverse crops reported, although lines for the different crops have different intercepts. Radiation values were for the photosynthetically active range, and assumed to be 50% of the solar radiation. Biomass yields included roots. Although Monteith's results from crops in

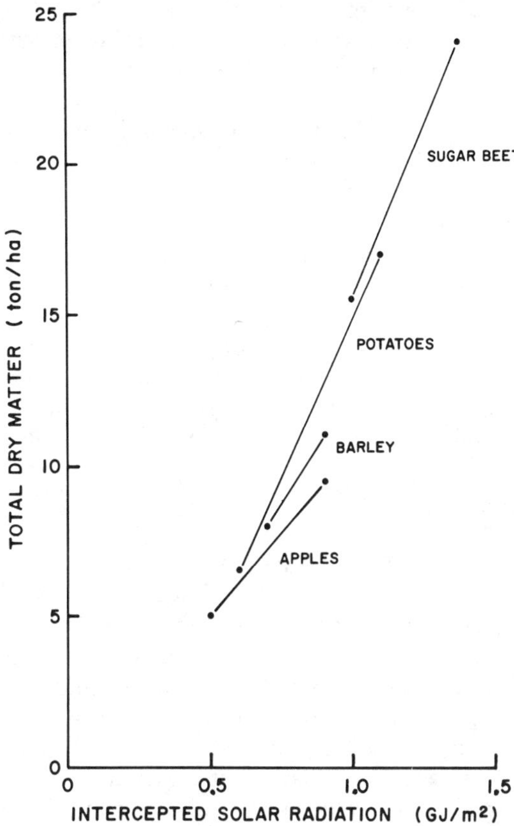

Fig. 6—Relation between total dry matter at harvest and radiation intercepted by foliage throughout growing seasons for apples, barley, potatoes, and sugar beets grown in Britain. Data from Monteith (1977).

England have not been verified closely under American conditions, there is good reason to believe in their generality. Hesketh and Baker (1967) showed that net photosynthesis rates of maize and cotton canopies measured for 15-min periods were almost linearly related to light interception.

Light interception and biomass production of maize was measured for several plant populations by Williams et al. (1965). During sustained periods when light interception was basically 100%, biomass production, including the "recoverable" roots, averaged about 40 g m^{-2} d^{-1} and the solar radiation averaged 31.4 MJ m^{-2} d^{-1}. Assuming the nonmeasurable roots amounted to 2 g m^{-2} d^{-1}, then the biomass production per unit intercepted solar radiation was about 1.34 g MJ^{-1}. From the biomass and radiation data of Stewart et al. (1977) for the 1975 maize studies at Davis, Calif., the cumulative radiation from 33 days after planting (when vegetative cover was about complete) until maturity was 2,314 MJ m^{-2}. Aboveground biomass production on the highest production plots was about 22 tons ha^{-1}. Assuming an average of 80% of the radiation was intercepted for that period and total dry matter = 1.1 × aboveground dry matter, then the biomass production per unit intercepted radiation was 1.31 g MJ^{-1}. Using similar assumptions from the 1975 maize study in Logan, Utah (Stewart et al., 1977), the seasonal intercepted radiation from 16 July to 29 September was 1,464 MJ m^{-2} and the total biomass was 19.8 tons ha^{-1}, or 1.35 g MJ^{-1}.

Relatively simple models of light interception and ET are available from the literature. Both light-interception and ET models require information about canopy characteristics such as LAI or some other estimate of crop-cover and density characteristics. Therefore, when the necessary canopy characteristics can be measured or reasonably estimated, then biomass production and ET can be reasonably estimated as independent processes.

The relationship given in Fig. 1 is an example of a relatively simple model of T as related to ET_{max} and LAI, when there is no limitation of water supply to roots. Because E is negligible for a dry soil surface, then the T model for dry soil is a good approximation of ET as related to LAI and E_{max}. The interception of radiation by crop canopies often fits the exponential equation

$$R/R_o = 1 - a \exp[-b(\text{LAI})], \quad [2]$$

where R is the radiation intercepted by the canopy, R_o is the radiation at the top of the canopy, and a and b are empirical constants. Using R/R_o measurements taken near solar noon for sorghum measured by Arkin et al. (1978, their Fig. 5), Eq. [2] fits the data well, with $a = 0.81$ and $b = 0.27$ for the 51-cm row-spacing and LAI values in the 0.5–4.5 range.

Assuming the solar noon radiation interception value is an approximation of the integrated daily value and that R_o is proportional to the maximum possible biomass production, DMP_{max}, then R/R_o is also the fraction of potential biomass production, DMP/DMP_{max}.

B. Evapotranspiration Efficiency and Soil Water Deficit

When stomata close or partially close during the day to reduce ET, interception of radiation will continue (although some reductions may occur because of leaf wilting or rolling) but biomass production per unit intercepted radiation will obviously be reduced. Since both DMP/DMP_{max} and ET/ET_{max} are affected by soil-water-induced stomata closure, models that can accurately estimate ET under limited soil water conditions should also be directly useful in estimating the reduction in DMP/DMP_{max}.

C. Relative Evapotranspiration Efficiency and Leaf Area Index

Because both ET/ET_{max} and DMP/DMP_{max} are related to LAI in the models discussed above, it is possible to calculate the relative ET efficiency as the ratio of DMP/DMP_{max} to ET/ET_{max}. Using Eq. [2] for a 51-cm row-spacing sorghum canopy to calculate DMP/DMP_{max}, and the dry-soil

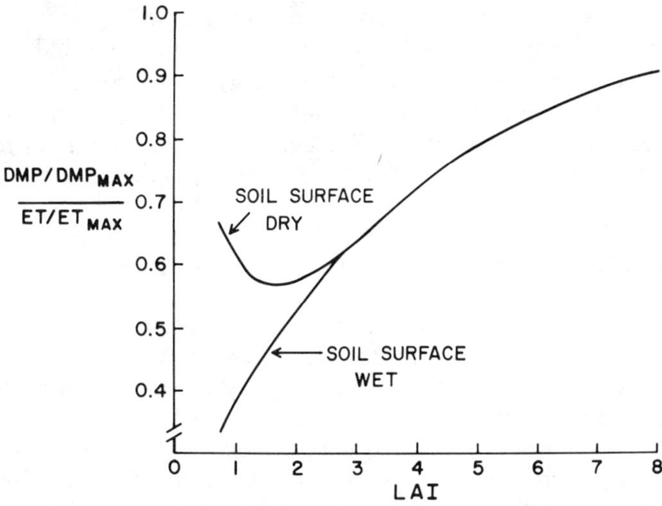

Fig. 7—The relative ET efficiency $(DMP/DMP_{max})/(ET/ET_{max})$ as related to LAI of sorghum for dry and wet soil surfaces. The subscript *max* refers to the maximum possible biomass production (DMP) and ET.

T/ET_{max} curve in Fig. 1 to calculate ET/ET_{max} for a dry soil surface when E is negligible, the relative ET efficiency as related to LAI is shown in Fig. 7. The curve for a wet soil surface results from the assumption that $E + T = ET_{max}$ regardless of LAI when the soil surface is freely evaporating. The dry soil surface curve is also the relative T efficiency curve. A relative T efficiency calculation for a wet soil surface is also possible, using the wet soil surface curve in Fig. 1. Calculations of relative T efficiency for wet soil surfaces are higher than for dry soil surface, ranging from about 0.75 to 0.90 for LAI values between 1.5 and 5.0, respectively.

Although the curves in Fig. 7 could be somewhat different for various crops and row configurations, the results indicate some general principles that can be concluded about ET efficiency:

1) There is a minimum threshold LAI value, below which ET efficiency is dependent on the soil surface wetness. This threshold LAI is about 2.5 and a significant amount of radiation is not intercepted by most crops below this threshold.

2) There is a maximum threshold LAI where ET efficiency approaches a maximum possible value. At this threshold, all radiation is intercepted if plant spacing is such that plant leaves overlap one another in all directions. Extrapolation of the curve in Fig. 7 to higher LAI suggests an LAI of about 12 for this threshold for sorghum. Williams et al. (1965) found a threshold LAI of about 12 for radiation interception to be 100% in maize.

3) For crops with LAI values between the minimum and maximum thresholds, any factor that can increase LAI will increase ET efficiency. The proportion of increase in ET efficiency is greater for low LAI values. If the soil surface is wet most of the time, this conclusion is also true for LAI values below the minimum threshold.

4) If the soil surface is dry most of the time, ET efficiency is relatively constant for crops with LAI values below the minimum threshold.

The two relatively simple models from which the above conclusions were reached depend on good definitions of DMP_{max} and ET_{max} for absolute accuracy in evaluating ET efficiency. Within the usual range of normal growing season temperatures, both DMP_{max} and ET_{max} are strongly correlated because they both are so closely coupled to the radiation level. In fact, the definition proposed for DMP_{max} is a direct linear proportionality between radiation amount and biomass production, i.e., 1.3 g MG^{-1} for maize.

Many field-tested empirical equations for calculating ET_{max} are also closely coupled to radiation. In Penman's well-known combination ET_{max} equation, for example, the major factor causing a change in the proportionality between radiation and ET_{max} is the vapor pressure deficit $(e^* - e)$. Thus, for cases where practically all radiation is intercepted (DMP $\cong$ DMP_{max}), differences in ET efficiency will primarily be caused by differences in $(e^* - e)$ for a specific crop, as Tanner and Sinclair point out (Chapter 1A). However, for cases where $DMP/DMP_{max} < 1$, ET-efficiency not only needs to be normalized by $(e^* - e)$ but also by DMP_{max}. The threshold LAI value at which DMP/DMP_{max} approaches 1 varies considerably between the Tanner and Sinclair analysis and that shown in Fig. 7. They assume, using their Eq. [9] for determining the LAI receiving direct radiation, that the maximum LAI threshold is about 3, whereas the same threshold from Fig. 7 is about 12. Equation [9] in Chapter 1A was not supported by experimental data, whereas Eq. [2] in this chapter was produced from field data of Arkin et al. (1978). Field data of Williams et al. (1965) suggest a similar radiation interception response as expressed in Eq. [2], especially for high LAI. Thus, the field data suggest the threshold LAI for DMP $\cong$ DMP_{max} is much greater than the threshold value for ET $\cong$ ET_{max}.

IV. SUMMARY AND CONCLUSIONS

In this discussion I have attempted to demonstrate some difficulties in assuming a general relationship between normalized ET or T and biomass production or yield. A relatively simple alternative was presented using cumulative canopy-intercepted radiation for estimating biomass production when no water stress is involved, and using independent E and T models for estimating ET.

The rather large difference in threshold LAI values where ET = ET_{max} and where DMP = DMP_{max} is the primary reason why crops grown on similar amounts of available water, but with variable management such as fertilizers, usually have similar seasonal ET but can have quite different yields. On the other hand, when water supply is varied through irrigation and other management is held constant, yields are usually quite proportional to the summation of seasonal ET, especially when the ET summation is started about the time plants begin to grow rapidly and cover the ground.

Much of the preceeding discussion has been based on a synthesis of several field experiments by many scientists in several locations, none of which were explicitly designed to measure ET efficiency in the manner presented herein. Thus, several assumptions were made in the attempt to demonstrate why farm management practices that influence LAI development can also influence ET efficiency. The theme of this book is efficient use of water in the broadest sense, from soil surrounding the roots to air surrounding the leaves of plants that respond to the environment at various stages of growth. I have emphasized less dynamic processes like differences in leaf area development that take periods of days and weeks to express themselves, and used daily estimates of ET_{max} and DMP_{max} for processes that are, in reality, highly variable within a day. The theoretical arguments Tanner and Sinclair use emphasize the quicker processes, like stomatal opening and leaf hydraulic resistance, which change in minute or hour time scales, to conclude that T efficiency is a stable term for LAI > 3 when normalized by $(e^* - e)$. Arguments for farm management influencing ET efficiency by Viets (1962) were based on entire season yield and ET. Thus, it is understandable that there can be a diversity of views about the very important processes of crop growth and water use.

Certainly, slower physiological processes are no more important than faster processes. It is indeed the integration of the faster processes that produce the slower ones. The way physiological processes occurring on a time scale of minutes are integrated to influence the growth of a plant over a day, week, or entire growing season is probably the least understood and most important research question in agronomic science today.

Finally, I would reinforce Tanner and Sinclair's conclusions that because ET and biomass production can be reasonably estimated using suitable models, opportunities for improving the efficient use of water become more clearly defined. Clearer quantitative definition of water use and crop production should enhance opportunities for agronomists to assist farmers in devising management strategies that minimize risks due to uncertainties in weather and water supplies.

LITERATURE CITED

Aase, J. K. 1978. Relationship between leaf area and dry matter in winter wheat. Agron. J. 70:563–565.

Adams, J. E., G. F. Arkin, and J. T. Ritchie. 1976. Influence of row spacing and straw mulch on first stage drying. Soil Sci. Soc. Am. Proc. 40:436–442.

Al-Khafaf, S., P. J. Wierenga, and B. C. Williams. 1978. Evaporative flux from irrigated cotton as related to leaf area index, soil water, and evaporative demand. Agron. J. 70: 912–917.

Arkin, G. F., J. T. Ritchie, and J. E. Adams. 1974. A method for measuring first-stage soil water evaporation in the field. Soil Sci. Soc. Am. Proc. 38:951–954.

Arkin, G. F., J. T. Ritchie, and S. J. Maas. 1978. A model for calculating light interception by a grain sorghum canopy. Trans. ASAE 21:303–308.

Arkley, R. J. 1963. Relationships between plant growth and transpiration. Hilgardia 34:559–584.

Bond, J. J., J. F. Power, and W. O. Willis. 1971. Soil water extraction by N-fertilized spring wheat. Agron. J. 63:280–283.

Brown, P. L. 1971. Water use and soil water depletion by dryland winter wheat as affected by nitrogen fertilization. Agron. J. 63:43–46.

Carlson, C. W., J. Alessi, and R. H. Mickelson. 1959. Evapotranspiration and yield of corn as influenced by moisture level, nitrogen fertilization, and plant density. Soil Sci. Soc. Am. Proc. 23:242–245.

de Wit, C. T. 1958. Transpiration and crop yields. Versl. Landbouwk. Onderz. 64.6. Inst. of Biol. and Chem. Res. on Field Crops and Herbage, Wageningen, The Netherlands.

Hanks, R. J., L. H. Allen, and H. R. Gardner. 1971. Advection and evapotranspiration of wide-row sorghum in the central Great Plains. Agron. J. 62:520–527.

Hesketh, J. D., and D. Baker. 1967. Light and carbon assimilation by plant communities. Crop Sci. 7:285–293.

Jensen, M. E., and W. H. Sletten. 1965a. Evapotranspiration and soil moisture–fertilizer interrelations with irrigated winter wheat in the southern High Plains. USDA/TAES Conserv. Res. Rep. no. 4.

Jensen, M. E., and W. H. Sletten. 1965b. Evapotranspiration and soil moisture–fertilizer interrelations with irrigated grain sorghum in the southern High Plains. USDA/TAES Conserv. Res. Rep. no. 5.

Jensen, M. E., and J. L. Wright. 1978. The role of evapotranspiration models in irrigation scheduling. Trans. ASAE 21:82–87.

Kanemasu, E. T., L. R. Stone, and W. L. Powers. 1976. Evapotranspiration model tested for soybean and sorghum. Agron. J. 68:569–572.

Kiesselbach, T. A. 1916. Transpiration as a factor in crop production. Nebraska Agric. Exp. Stn. Res. Bull. no. 6. Lincoln.

Marlatt, W. E. 1961. The interaction of microclimate, plant cover, and soil moisture content affecting evapotranspiration rates. Colorado State University Dep. Atmos. Sci. Tech. Paper 23. Fort Collins.

Monteith, J. L. 1977. Climate and efficiency of crop production in Britain. Philos. Trans. R. Soc. London, Ser. B 281:277–294.

Olson, T. C. 1971. Yield and water use by different populations of dryland corn, grain sorghum, and forage sorghum in the western corn belt. Agron. J. 63:104–106.

Power, J. F., D. L. Grunes, and G. A. Reichmann. 1961. The influence of phosphorus fertilization and moisture on growth and nutrient absorption by spring wheat: I. Plant growth, N uptake, and moisture use. Soil Sci. Soc. Am. Proc. 25:207–210.

Ritchie, J. T. 1971. Dryland evaporative flux in a subhumid climate: I. Micrometeorological influences. Agron. J. 63:51–62.

Ritchie, J. T. 1972. Model for predicting evaporation from a row crop with incomplete cover. Water Resour. Res. 8:1204–1212.

Ritchie, J. T. 1974. Evaluating irrigation needs for southeastern U.S. p. 262–279. *In* Contribution of irrigation and drainage to world food supply. Am. Soc. Civ. Eng., Biloxi, Miss.

Ritchie, J. T., and E. Burnett. 1971. Dryland evaporative flux in a subhumid climate: II. Plant influences. Agron. J. 63:56–62.

Stewart, J. I., R. E. Danielson, R. J. Hanks, E. B. Jackson, R. M. Hagan, W. O. Pruitt, W. T. Franklin, and J. P. Riley. 1977. Optimizing crop production through control of water and salinity levels in the soil. Utah Water Lab. Tech. Rep. PRWG 151-1. Logan. p. 191.

Tanner, C. B. 1957. Factors affecting evaporation from plants and soils. J. Soil Water Conserv. 12:221–227.

Tanner, C. B., and W. A. Jury. 1976. Estimating evaporation and transpiration from a row crop during incomplete cover. Agron. J. 68:239–243.

Viets, F. G., Jr. 1962. Fertilizers and the efficient use of water. Adv. Agron. 14:223–264.

Williams, W. A., R. S. Loomis, and C. R. Lepley. 1965. Vegetative growth of corn as affected by population density. I. Productivity in relation to interception of solar radiation. Crop Sci. 5:211–215.

CHAPTER 2A

Soil Properties and Efficient Water Use: An Overview[1]

W. R. GARDNER

University of Arizona
Tucson, Arizona

I. INTRODUCTION

The loss of water to the atmosphere by plants is an inevitable consequence of the necessity of a physical pathway for gaseous exchange through which CO_2 can be acquired from the atmosphere. Since few plants are capable of storing a significant fraction of the water that must be transpired daily, there must be a continuous replenishment of the plant water through the root system. The soil pores in the vicinity of the plant roots provide the reservoir by which this water is stored and through which it is transmitted to the roots. Soils differ in their ability to absorb, store, and transmit water, and these physical properties determine in part the suitability of any given environment for plant growth. This chapter deals with the principles of water retention and transmission by soils, which must be understood if optimum value is to be realized from the management of such a system. Later chapters, Chapter 10 in particular, detail the ways in which these soil properties may be managed or altered in order to achieve specific objectives.

II. THE SOIL PROFILE AS A WATER RESERVOIR

A. Soil Water Retention

1. MATRIC POTENTIAL

The total amount of water that can be stored in the soil is obviously determined by the total volume of the voids or pore spaces. However, this volume is only indirectly related to the amount of water that will be available for uptake by the roots at any given time. The two physical properties

[1] Research supported in part by the College of Agricultural and Life Sciences, University of Wisconsin-Madison (Agric. Exp. Stn. Proj. no. 2496).

Limitations to Efficient Water Use in Crop Production.

Table 1—Conversion units for water potential.†

cm H_2O	10^5	10^4	10^3	10^2	10	1
J kg^{-1} (kPa)	10^4	10^3	10^2	10	1	0.1
m H_2O	10^3	10^2	10	1	0.1	0.01
bars	10^2	10	1	0.1	0.01	0.001

† Each column represents equivalent energies.

or characteristics most useful in the quantitative description of the soil water system are the potential energy of the soil water and the soil hydraulic conductivity. The potential energy of the soil water is defined relative to pure water at a specified temperature and elevation. A saturated soil with a negligible solute concentration will have a potential energy of zero. As the soil water content is decreased, the potential energy of the remaining water decreases due to the increasing tenacity with which it is held by adsorption and capillary forces. This potential energy is defined as the energy required to transport an infinitesimally small quantity of water from a reference body of pure water at the same temperature and elevation to the point of interest in the soil. The units of this energy are most appropriately specified per unit mass, joules (J) kg^{-1} in the SI system. However, it is often convenient to give the potential energy on a unit volume or a unit weight basis. Energy per unit volume has the same dimensions as pressure so that pressure units are used to express this energy. Atmospheres (atm) and bars have been the most common units. One bar is 10^6 dyn cm^{-2} or 10^{10} dyn m^{-2}. In the SI system, the Pascal is the unit of pressure. This is such a small unit that the kiloPascal (kPa) or megaPascal (MPa) is more commonly used. Table 1 gives conversion units for the most frequently used systems of energy units.

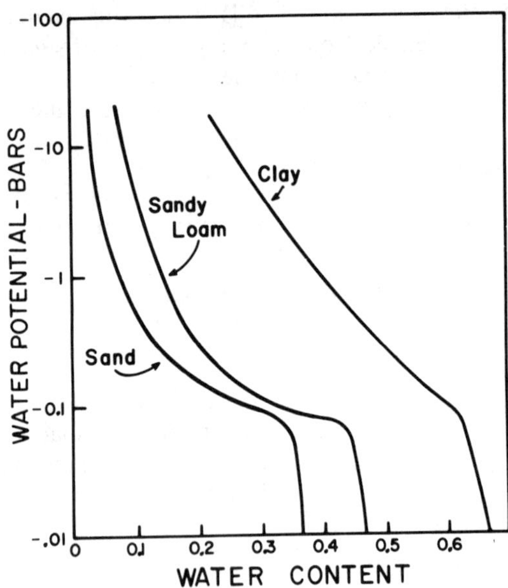

Fig. 1—Desorption curves showing matrix suction as a function of volumetric water content for three soils representing different textures.

Since the potential energy is negative, it is sometimes helpful to connote this fact in the name of the energy term. Soil suction and soil matric potential thus both refer to the potential energy of the soil water. The suction is simply the negative of the matric potential.

In Fig. 1 are shown typical curves relating soil water content to suction. At low suctions, less than about 1 bar, the shape of these curves is largely determined by the soil structure or the pore size distribution. Since the soil structure is not constant but can be a function of tillage practices, organic matter, exchangeable cations, soil solution concentration, and other factors, this part of the retention curve is, within limits, subject to manipulation. At the higher suctions, especially above 5 bars, the soil texture—particularly the clay content—determines the amount of water retained. In the range of high suctions, there is little that can be done to alter the water retention. Important exceptions to these generalizations are the swelling soils (vertisols) high in clay content, where the porosity is not a constant. Such soils exhibit considerable shrinking and swelling in the field, as evidenced by deep cracks. The change in soil density is almost directly related to the changes in soil water content, and the porosity is not constant but continuously variable.

The shape of the retention curves is very complex, but for many purposes it is useful to be able to describe them with mathematical functions. One such function that appears very general is the relation (van Genuchten, 1980)

$$\theta = [1 + (ah)^n]^{-m}; \qquad [1]$$

where θ is the volumetric water content; *a, n,* and *m* are soil parameters; and *h* is the suction head. The parameter *n* increases with increasing coarseness of texture, being on the order of 1.5–2 for clays and loams to as high as 5 or 10 for sands; *m* is usually <1.

The relation between the water content and the suction head is not necessarily unique. The curves shown in Fig. 1 and Eq. [1] are for the draining or desorption process by which the soil is initially saturated and water is continuously removed from the soil. For the opposite, or sorption, process, one may find a different curve, especially for sands that have a large proportion of large pores. At the same potential, the desorption curve shows more water in the soil than the sorption shows. While hysteretic effects are quite important in some soils, especially during infiltration and redistribution, it is beyond the scope of this chapter to deal further with the problem. Unless otherwise stated, we will consider only the desorption curve (Mualem, 1974, 1977; Reeves & Miller, 1975).

For reasons that will be explained later in this chapter, we will mainly be concerned with water contents corresponding to about 0.1 bar suction as an upper limit to about 5 bars suction as a lower limit. This covers most of the plant growth range; however, other soil organisms may have optimum growth beyond this range either at higher or lower water contents. For example, soil fungi may continue to grow at suctions as high as 80 bars with some bacteria doing well at 15 bars. At suctions < 0.1 bar, soil aeration tends to be a problem, especially in finer-textured soils.

2. GRAVITATIONAL POTENTIAL

When one is concerned with water retention and movement in the field, the effect of gravity upon the potential energy of the soil water must be taken into account. This is done through the gravitational potential. When the potential is expressed as energy per unit weight, this potential becomes simply the elevation above any suitable datum and is expressed in length units (meters in the SI system).

3. OSMOTIC POTENTIAL

Soluble salts in the soil solution also lower its potential energy. This effect is taken into account through the osmotic potential term in the total potential. It is approximately equal to RC, where R is the universal gas constant (in appropriate units) and C is the concentration of soluble salts. Such salts have little effect upon water movement in the soil and will be disregarded in the remainder of this chapter. However, they are extremely important to plant water use, and their effect cannot be neglected in considering efficient water use under irrigated conditions. This important problem is considered in detail by Hoffman and van Genuchten in Chapter 2C.

4. HYDRAULIC HEAD

The sum of the suction and the gravitational potential is defined as the hydraulic head (H). In saturated soils the potential energy may be positive due to positive hydrostatic pressure in the soil pores. In nature this is due to the weight of water above the point at which the pressure is specified. The hydraulic head in a saturated soil is the sum of the pressure head and the gravitational head

$$H = p + z \text{ saturated, and} \tag{2a}$$

$$H = -h + z \text{ unsaturated;} \tag{2b}$$

where p is the pressure head, z is the elevation, and h is the suction head, all given in meters for our purposes.

The hydraulic head determines the direction and rate of water movement in soils. Water always moves from regions of higher head to regions of lower head.

B. Soil Water Movement

The rate of water movement in soils is proportional to the gradient of the hydraulic head and to the hydraulic conductivity, k, of the soil:

$$q = -k\,(\mathrm{d}H/\mathrm{d}z) = k[(\mathrm{d}h/\mathrm{d}z) + 1] \tag{3}$$

for vertical flow with both z and q taken as positive in the positive z direction. For horizontal flow the second term in parentheses in Eq. [3], which represents the gravitational gradient, is omitted and we have

$$q = k(\mathrm{d}h/\mathrm{d}z). \quad [4]$$

The unsaturated conductivity k is a second important soil characteristic. At saturation it is equal to the hydraulic conductivity, or saturated permeability, but as the soil dries, it decreases very rapidly with decreasing water content. Typical curves of conductivity as a function of suction are shown in Fig. 2.

In addition to the two soil characteristic functions described above, it is often useful to define a third, derived function. In those cases where hysteresis can be neglected, and when the soil is sufficiently uniform so that the curve relating soil water potential to water content is single-valued, Eq. [4] can be rewritten in terms of a water content gradient

$$q = k(\mathrm{d}h/\mathrm{d}\theta)(\mathrm{d}\theta/\mathrm{d}x) = -D(\mathrm{d}\theta/\mathrm{d}x). \quad [5]$$

A new function, the soil water diffusivity (D), is defined by

$$D \equiv -\mathrm{d}h/\mathrm{d}\theta \quad [6]$$

Figure 3 shows the diffusivity for the three soils as Fig. 1 and 2.

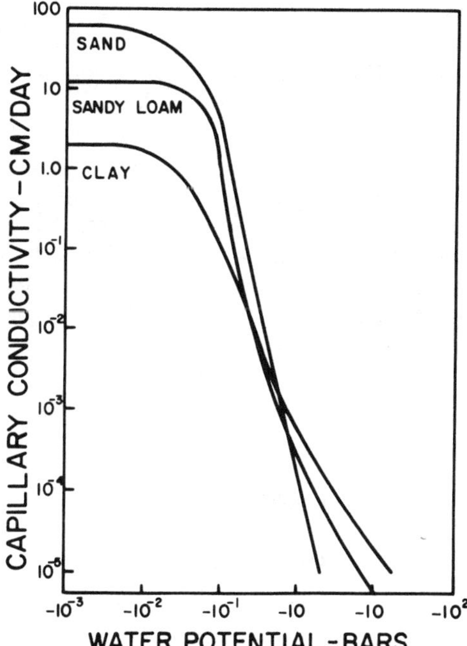

Fig. 2—Capillary conductivity as a function of matric suction for soils representing three different textures.

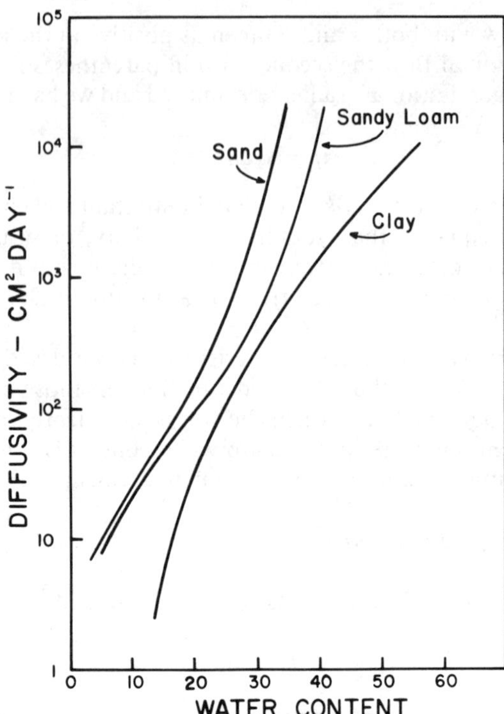

Fig. 3—Soil water diffusivity as a function of volumetric water content for three soils representing different textures.

If we combine the flow equation with an equation expressing the conservation of matter, we get the overall water flow equation

$$\partial\theta/\partial t = \nabla \cdot D\nabla\theta + \partial k/\partial z, \qquad [7]$$

where the symbol ∇ is vector notation that merely means differentiation with respect to all axes of the appropriate coordinate system and t is time. The first term on the right-hand side of Eq. [7] represents water movement due to water content gradients, and the second is that due to gravity. Since it is not possible to have large water content gradients over great distances, high rates of water movement due to water content gradients can occur only over short distances.

C. The Soil Water Budget

1. THE COMPONENTS OF THE WATER BUDGET

In considering the soil profile as a water reservoir, it is useful to separate the processes by which water enters or leaves the soil profile in terms of an overall balance. This may be written as

$$P + I - Q = \mathrm{ET} \pm D \pm \Delta W. \quad [8]$$

In this equation, which may be taken as describing the average situation over any arbitrary unit of time, P is the amount of precipitation; I is the amount of irrigation water applied, if any; and Q is the runoff over the soil surface. The left-hand side represents the inputs of water to the soil profile. On the right-hand side, the term ET represents water loss both by evaporation from the soil surface and transpiration by plants. In a later section of this chapter, the soil evaporation process will be considered in some detail. The second term represents the drainage or deep percolation out of the plant root zone. This term may be negative, representing what is known as *capillary rise.* It is included on this side of the equation since it is often difficult to determine the sign of this term, and the direction of flow may change during the course of a season or even during a drying cycle between irrigations. The final term, ΔW, represents the change in the water content of the soil profile. Equation [7] represents the mathematical rules governing Eq. [8]. This root uptake term is discussed in more detail by Taylor in Chapter 3A, but some of its broad features will be considered later in this chapter.

2. INFILTRATION

Infiltration is the process of downward entry of water into the soil at the soil surface. The flow equation has been solved for a wide range of infiltration conditions and the reader is referred to the literature for details (Philip, 1969b).

The role the soil plays in the infiltration process can most simply be considered in terms of an approximate solution to the flow equation due to Philip (1957). At short times, say, the first few minutes or hours after water is ponded in the soil surface, the cumulative amount of water infiltered into the soil is given by the equation

$$I = S\sqrt{t} + At, \quad [9]$$

where S is called the sorptivity and A is a constant with the dimensions of conductivity at about two-thirds of the hydraulic conductivity k. At larger times the equation becomes approximately

$$I = I_o + kt. \quad [10]$$

The first terms in the right-hand side of Eq. [9] and [10] represent the influence of water content gradients at the wetting front and the second terms the influence of gravity. At small times, water content gradients dominate and lateral and vertical movement are essentially the same. At long times gravity becomes dominant, with the infiltration rate equal to the conductivity. If the soil is initially saturated, the sorptivity is obviously zero. As a first approximation, sorptivity decreases linearly between a maximum for an air-dry soil and zero at soil saturation. In an actual field situation, the infiltration rate will thus be a function of the initial soil water content and will decrease with time. It eventually approaches a final infiltration rate that is a function only of the average conductivity of the wetted profile. For water entering a relatively dry soil, the depth of wetting can be estimated by dividing the cumulative infiltration by the average increase in soil water content.

Equations [9] and [10] represent idealized conditions requiring more uniformity than is often exhibited in soils. Actual field conditions may be better represented by other equations (Collis-George, 1977).

Equations [9] and [10] represent the maximum rate at which water can enter the soil, which will occur only if water is ponded continuously on the soil surface. Under natural precipitation the rate at which water is applied at the soil surface may be less than the maximum rate at which it can be accepted by the soil. If the rate of precipitation is finite but greater than the saturated conductivity of the soil, water will initially enter the soil at the rate of precipitation. However, the soil water content at the surface will gradually increase with time until it reaches saturation. At this time if the precipitation continues, water becomes ponded on the soil surface and the infiltration rate now becomes limited by the soil. The time to this ponding is proportional to the square of the sorptivity and inversely proportional to the square of the precipitation rate. After ponding, the infiltration process can be described as though ponding had existed from the beginning of the infiltration process (White et al., 1979). If the precipitation rate is always less than the hydraulic conductivity, then there will be no ponding or runoff and the soil water content will adjust to just that value which will make the unsaturated conductivity equal to the infiltration rate. This situation has been used to measure the soil water conductivity as a function of water content.

A frequent problem in many soils is the sealing or partial sealing of the soil surface due to a crust formation. Rainfall impact may disperse the surface soil particles so that even though permeability has been restored by tillage, sealing may reoccur with each new rainfall event. Unstable soils high in Na may disperse when exposed to irrigation waters of low solute content. This crusting tends to reduce both the sorptivity and the conductivity, and results in a lower degree of saturation of the soil than would otherwise occur (Hillel & Gardner, 1970). Erratic rainfall also complicates the infiltration process but in some cases this may be dealt with in a relatively simple way (Reeves & Miller, 1975). Management techniques for dealing with some of these problems are discussed by Unger and Stewart in Chapter 10A.

Another aspect of the infiltration process related to efficient water management is the problem of the variability of soils. Since the flow of water through a pore varies as the fourth power of the radius of the pore, a few large pores in sequence, root channels, and worm holes, etc., can result in a wide variation of permeabilities across a given field. This leads to very nonuniform wetting under flood or furrow irrigation (Nielsen et al., 1973).

Because the duration of ponding water on the soil surface determines the amount of infiltration, furrow irrigation or a similar situation where the upper portion of the furrow is exposed to water much longer than the lower will also result in nonuniform irrigation. Thus, if efficient water management requires uniform application of water to the soil profile, the conveyance or application system should be designed to control the rate of application rather than leaving this to the soil. Sprinkler irrigation systems and trickle or drip irrigation systems are designed in large part for this reason.

3. REDISTRIBUTION AND DRAINAGE

Following the cessation of the infiltration process when water is no longer entering at the soil surface, there continues a downward movement of water in the soil profile. This decreases the profile water content. This

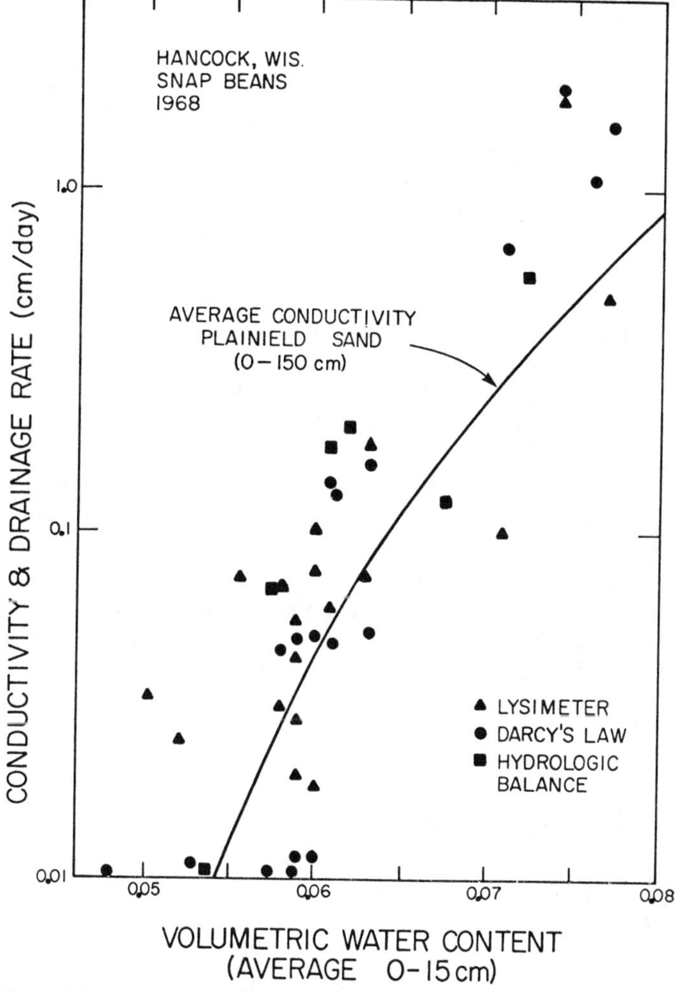

Fig. 4—Rate of drainage from the upper 150 cm of a field profile of Plainfield sand as a function of average water content of the profile. If the hydraulic gradient is taken as unity, the conductivity (represented by the smooth curve) should be equal to the drainage rate. The drainage rate was determined by three different methods.

process is largely dominated by gravity unless there exists a very shallow water table, in which case the rate of drainage is less than that which would otherwise exist. When drainage is determined by the gravitational component of the potential, the drainage rate at any depth becomes equal to the unsaturated conductivity at the depth. Thus

$$dW/dt = -k. \qquad [11]$$

If the profile is not uniform but has wide textural variations, one would have to use an average conductivity to characterize the profile in Eq. [11], but the principle is the same. Since the conductivity can be expressed as a function of the water content of the soil, it is possible in the case of a uniform profile to relate the drainage rate to the average water content of the profile (Black et al., 1969; Lambert et al., 1971). In Fig. 4 the drainage rate of a sandy profile is plotted as a function of the water content and the conductivity (Black et al., 1970). This kind of a curve or the equivalent mathematical expression is much preferred over the traditional "field capacity" concept in which it is assumed that drainage proceeds very rapidly down to a given water content, below which it is negligible. In terms of water-use efficiency, negligible usually implies a drainage rate that is small compared with the evapotranspiration rate. In practice it has sometimes been found difficult to measure drainage rates on the order of the few millimeters per day corresponding in magnitude to the transpiration rate. Such rates have sometimes been assumed to be negligible when they were in fact not so. From Fig. 4 it can easily be seen that when the water content is less than about 0.06 on a volumetric basis, the drainage rate < 1 mm day^{-1} and for many purposes can be neglected, at least over a short length of time. Davidson et al. (1970) have also shown that this simple solution of the flow equation gives a good estimate of the rate of drainage out of the profile, although at any given point it may not give a good average of the water content. Evaporation of water from the soil surface does not affect this drainage rate significantly, except to reduce the volume of soil water that will drain. On the other hand, this drainage process may reduce significantly the rate of evaporation of water from a bare soil (Gardner et al., 1970). Factors that affect this downward drainage also will affect water loss by evaporation.

4. REDISTRIBUTION PLUS WATER UPTAKE

If downward drainage is not negligible compared with plant transpiration, it is useful to be able to estimate the interaction between transpiration rate and drainage rate. This can be done easily if we assume that the water uptake pattern is approximately uniform, and if we know the profile drainage rate as a function of profile water content. A convenient mathematical expression for this latter function, valid over the relatively narrow range of water contents with which we will normally be concerned, is $k = k_o \exp(aW)$, where W is the profile water content in cm, and k_o and a are soil parameters. While (mathematically) k_o is the conductivity when $W = 0$, this extrapolates the formula beyond the range of physical reality. If we assume that the uptake is uniform, we can further assume that this uptake affects the drainage data through its influence on the water content. We can write for the rate of transpiration plus drainage

$$dW/dt = -[T + k_o \exp(aW)], \quad [12]$$

where T is the transpiration rate. If we integrate and let $W = W_i$ when $t = 0$, we obtain the following equation:

$$W = W_i - Tt - (1/a) \ln \left[\frac{T + k_o \exp(aW_i)}{T + k_o \exp(aW)}\right]. \quad [13]$$

It can be seen that the second term in Eq. [13] represents the transpiration and the last term represents the loss due to drainage. If we let W become small enough so that the drainage rate is much less than the transpiration rate, i.e., $[k_o \exp(aW) \ll T]$, Eq. [13] can be further simplified. The initial drainage rate is set equal to the infiltration rate at that moment just prior to cessation of infiltration. If we denote that rate by I and assume that $I \gg T$, Eq. [13] reduces to

$$W = W_i - Tt - (1/a) \ln (I/T). \quad [14]$$

We see from Eq. [14] that the drainage loss is a function of the initial drainage rate and the transpiration rate. Sprinkler irrigation, by reducing the initial soil water content and initial drainage rate below that which would occur under ponding, thus reduces drainage losses. Even though water uptake is not usually uniform with depth, this approach can still be used to give a surprisingly good estimate of drainage from the root zone (Lambert et al., 1971).

5. TWO- AND THREE-DIMENSIONAL INFILTRATION

Under furrow, drip, and trickle irrigation, flow during the infiltration process is horizontal as well as vertical. Since horizontal water movement from a fixed source must continuously decrease with time, the equation for horizontal flow (in one dimension) contains only the first, or square root of time, term from Eq. [9]. Where flow is vertical as well as horizontal, vertical flow must eventually dominate over horizontal. There is a limit to the horizontal distance water can be moved during infiltration (Philip, 1969a). This distance depends on the soil texture and structure. Lateral movement is much less significant in sands than in clays and clay loams. The dimensions of the furrow of the area of soil surface covers with water during a drip irrigation cycle also influence the extent of lateral movement. Both theoretical and experimental studies permit one to make estimates of this distance (Raats, 1971, 1977; Earl & Jury, 1977; Jury & Earl, 1977; Merrill et al., 1978).

6. EVAPORATION FROM THE SOIL SURFACE

An important aspect of water management is the manipulation of evaporation from the soil surface. A bare soil wetted uniformly with depth will initially lose water at a rate of determined by atmospheric factors. After this initial or constant rate period, the soil surface becomes dry and the later stages of evaporation are limited by the soil properties. Mulching to reduce this initial drying rate and to prolong it will initially result in a reduction in the loss of water from the soil. The amount to be saved by mulching or equivalent practices is reduced with increased evaporation time. Since water lost by evaporation from the soil surface is water not available from transpiration, efforts to increase water-use efficiency often concentrate on reducing evaporation losses. This important subject is dealt with in more detail in the discussion by H. R. Gardner in Chapter 2B.

III. SOIL WATER UPTAKE PATTERNS

Important to an understanding of the soil as a reservoir for the storage of water for plant use is an understanding of the transport of water to the plant roots, into and through the plant to the leaves. While considerable progress has been made on this problem, it is far from solved and well beyond the scope of this chapter. Chapter 3A by Taylor gives sufficient details that need not be repeated here. One major limitation to the application of any of the uptake models is the lack of an adequate model for describing root distribution and root uptake of water. Rather than dwell on the theory, it is deemed more useful here to examine the empirical data on water uptake and use that data as a guide to the relative availability of the water in the soil

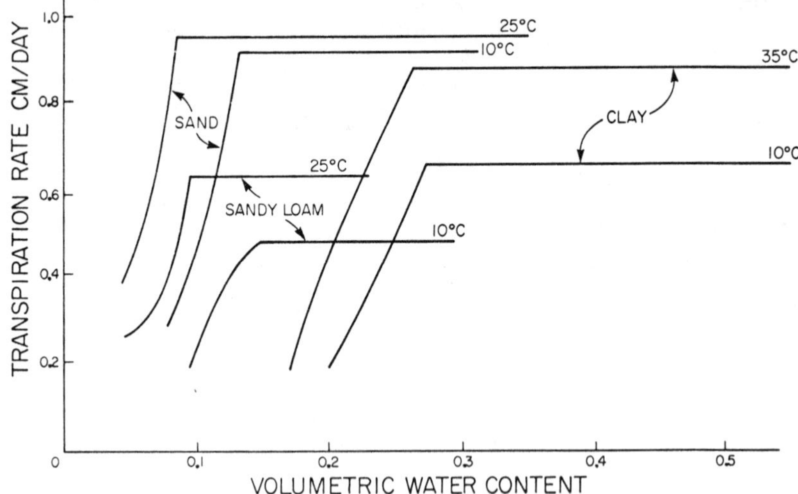

Fig. 5—Transpiration rate as a function of soil water content for birdsfoot trefoil in three different soils at two different root temperatures. Experimental procedure described in Gardner and Ehlig (1963).

profile. For this purpose, we look at the actual field data. When data are taken with sufficient care and detail, we observe a number of regularities in uptake patterns.

We observe, first, that the transpiration rate seems relatively unaffected by the soil water content or soil water potential over a considerable range, and it is only when the potential and water content fall below a given value that the transpiration rate begins to decrease. It continues to decrease with decreasing water content. This is shown in Fig. 5, in which the transpiration from birdsfoot trefoil for root temperatures of 25 and 10°C are given as a function of soil water content for soils of three different textures. The ambient temperature in the greenhouse was about 23°C for both sets of experiments. For practical purposes, the transpiration rate may be taken as a constant dictated by atmospheric conditions at high water contents, and as a linearly decreasing rate at low water contents. The breakpoint at which transpiration decreases depends upon the unsaturated conductivity of the soil (Gardner & Ehlig, 1963), but appears to fall mainly in the range of −1 to −10 bars. Similar behavior is observed for deep, nonuniform root systems as for shallow containers, i.e., a constant transpiration followed by a linear decrease in transpiration when plotted as a function of water remaining in the soil profile (Rawlins et al., 1968). This occurs when the water potential in the upper part of the root system is in this −1- to −10-bar range, according to present evidence.

As the extraction zone moves downward, the water potential in the upper part of the root zone decreases roughly exponentially with time. The plant leaf water potential must always remain below this soil water potential. Eventually the leaf water potential becomes sufficiently low to cause a significant stomatal closure and a concomitant reduction in transpiration. The depth of extraction at which this occurs will depend upon the soil water storage and the extent and resistance to water flow of the root system.

A second observation is that, starting with a uniformly wet soil profile, water is initially extracted from the region nearest the surface with the zone of extraction progressing downward through the profile. Such a pattern can be seen in Fig. 6. Whenever the soil water near the surface is replenished by precipitation or irrigation, the extraction zone goes back to the surface and moves downward again. The shape of the extraction zone, once established, seems to remain about the same as it moves downward through the root zone. Figure 7 shows the amount of water extracted as a function of depth as found by Ogata et al. (1960) for alfalfa. The first curve is for about 10 days after deep irrigation when the water potential at the 30-cm depth was about −1 bar. The second curve is for 22 days after irrigation with the alfalfa showing signs of wilting and the transpiration rate showing an abrupt decrease after this date. The water potential at the 30-cm depth was about −5 bars. The lower curve shows the water extraction at the end of a 50-day period with the plants badly wilted. Until the plants wilted, the shapes of the extraction curves were very similar.

Comparable curves for other crops are shown in Fig. 8. The alfalfa curve is the center curve from Fig. 7; the sorghum data are from Van Bavel et al. (1968) and Stone et al. (1973a, 1973b). The corn and soybean data are

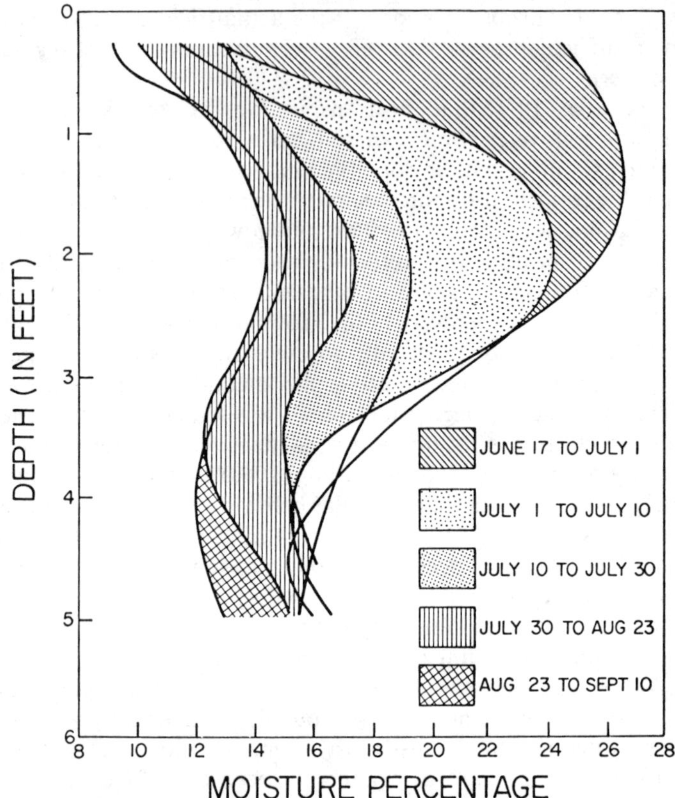

Fig. 6—Soil moisture profiles of the covered corn plots showing zones of water disappearance during the growing season (Russell & Danielson, 1956).

from Allmaras et al. (1975). The similarity of these curves can be shown by scaling both the axes appropriately. If the maximum water extracted from the upper part of the root zone is taken as 1.0, all curves can be normalized to go through the same point at the surface. If the depth for each curve is then multiplied by an appropriate scale factor, the points at which just 50% of the surface extraction is found will all coincide. This has been done for a large number of extraction patterns. Where it was possible to estimate the inflection point on the transpiration curve, the extraction pattern at this time was selected. When this was not available, the pattern when the soil water potential at aobut the 25-cm depth was about −3 bars was selected. In the cases where information was inadequate, the final extraction at the end of the season was selected. The normalized curves for many different crops are shown in Fig. 9. These include data on corn (Taylor & Klepper, 1973; Russell & Danielson, 1956; Letey & Peters, 1957; Dreibelbis & Youker, 1956; Allmaras et al., 1975; Gard et al., 1961; Corey & Blake, 1953; Allessi & Power, 1975). Data are also plotted for alfalfa (Hagan & Peterson, 1954; Taylor & Haddock, 1956; Ogata et al., 1960; Kohl & Kolar, 1976; Lucey & Tesar, 1965; Campbell et al., 1960), sorghum (Blum, 1972; Stone et al., 1973a, 1973b; Van Bavel et al., 1968; Ritchie et al., 1972),

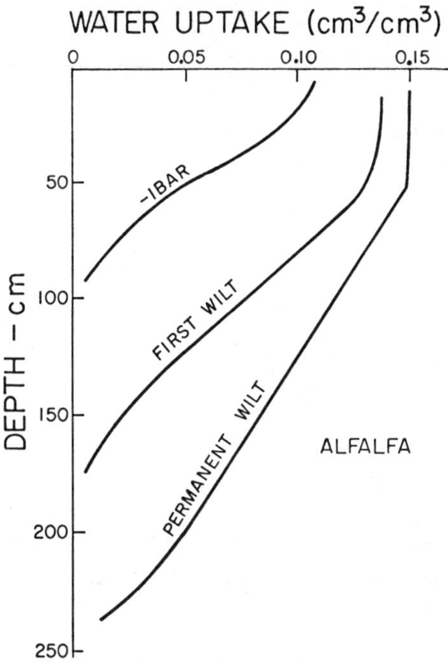

Fig. 7—Water extraction by alfalfa 10, 22, and 58 days after deep irrigation. Data from Ogata et al. (1960).

soybeans (Reicosky et al., 1972; Timmons et al., 1967), cotton (Amemiya et al., 1963; Doss et al., 1964; Taylor & Klepper, 1975; Al-Khafaf et al., 1978), wheat (Papendick et al., 1971), barley (Conrad, 1937), flax (Alessi & Power, 1970), and Douglas fir (Nnyamah & Black, 1977). A wide range of soil textures were represented, though few exhibited pronounced textural differences with a given soil profile. As explained above, the extraction

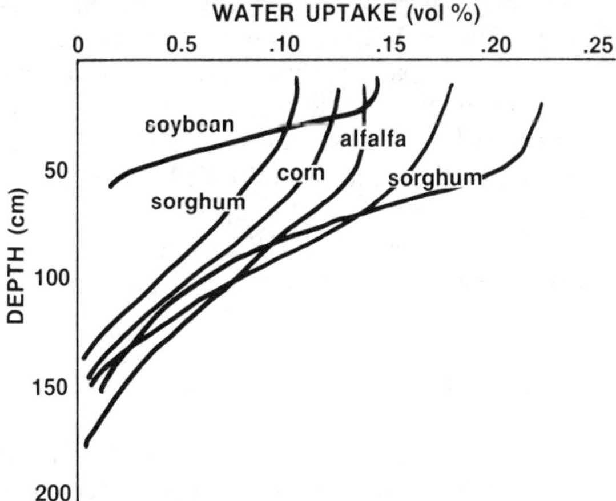

Fig. 8—Water extraction pattern for several crops at time of initial wilting. Data from Van Bavel et al. (1968), Stone et al. (1973a, 1973b), and Allmaras et al. (1975).

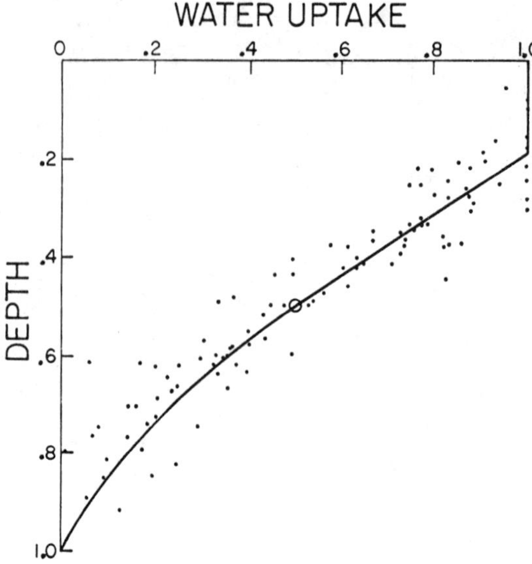

Fig. 9—Normalized water uptake curves for some 40 different experiments. Data for day of first wilt, as estimated.

from the top 25 cm was set equal to 1.0. The depth of the root zone was then multiplied by a factor such that the smooth curve through the data points (fitted by eye) all pass through the point (0.5, 0.5). Each data point was then multiplied by the scale factors and all were plotted on the same graph. The smooth cruve drawn through the points is a straight line chosen arbitrarily such that it decreases from 1.0 to 0 in going from depth 0.2 to depth 0.8. If one accepts this as a reasonable representation of all curves, 40% of the extraction is from the first 20% of the root depth, then 33%, 20%, and 7% from successive depths. This is reminiscent of the old rule of thumb that 40% of the water is extracted from the first foot, then 30, 20, and 10% from successive feet. Actually, there is a bit more extraction from the lower depths than given by this straight line, but not a great deal.

Assuming an extraction pattern that decreases linearly with depth has several advantages over other possible empirical relations. The 50% point, indicated by the open circle in Fig. 9, can be used to represent the effective depth of extraction. When this depth is multiplied by the difference between the initial and final water contents of the upper portion of the root zone, one gets the total water extraction, the ΔW term in Eq. [8]. If drainage and evaporation have been negligible, this corresponds to the transpiration.

IV. THE AVAILABLE WATER CONCEPT

Based upon the above discussions, we can arrive at a concept of available water that is an improvement over previous concepts in that it takes the principles of water retention and movement explicitly into account. It is still empirical with respect to plant response, and will remain so until the theory of water uptake is more completely worked out. This concept is best illustrated by examining Fig. 10 and 11. In Fig. 10, the hypothetical rate of water loss from the soil profile is plotted as a function of the average volumetric water content of the profile. Curves for sand and clay are shown; the intermediate soil textures will fall between these two curves. The steep portion at the right-hand end of each curve represents the rate of water loss due to drainage plus transpiration. If the profile water content is maintained at too high a level, drainage losses will tend to be excessive. The steep portion at the left-hand side of each curve represents the region where transpiration is reduced due to stomatal closure with consequent reduction in plant growth. The plateau on each curve represents the region where water is adequate and drainage is negligible. As is well-known, there is substantially more water in the clay than in the sand on this region. The same data are plotted as a function of soil water potential at the same average water content. In Fig. 11, it is to be noted that the −1/3-bar convention for field capacity is too low for these two soils, as will usually be the case. A better estimate is −1/10 bar. At the other end of the readily available range, the curves decrease linearly with the logarithm of the water potential since the water content and water potential are roughly logarithmically re-

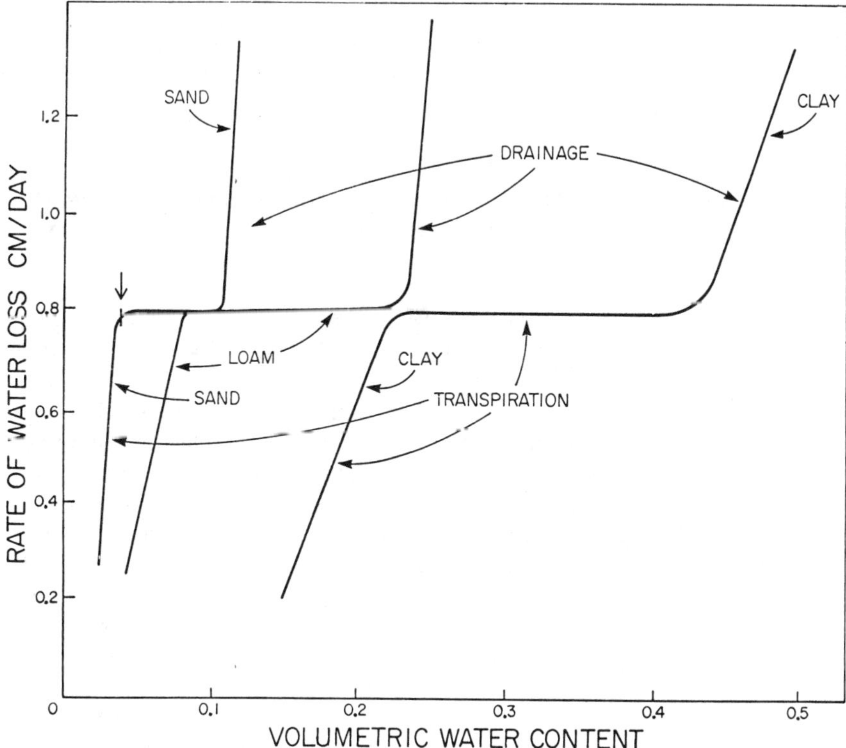

Fig. 10—Hypothetical rate of water loss from a soil profile due to drainage and transpiration as a function of soil water content.

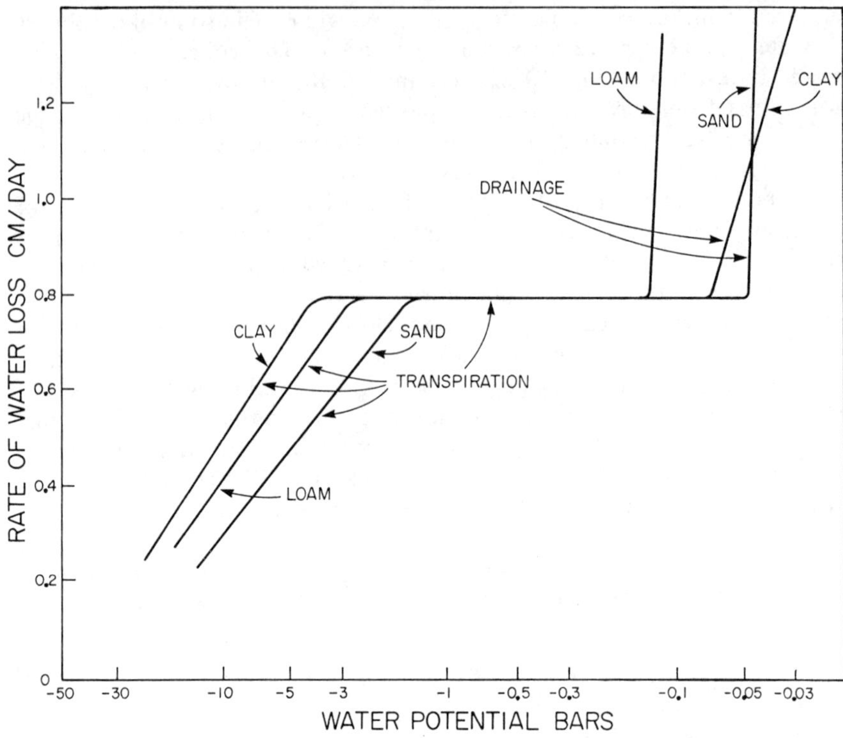

Fig. 11—Hypothetical rate of water loss from a soil profile due to drainage and transpiration as a function of soil water potential.

lated. The exact position of the breakpoint will depend upon plant and atmospheric factors, and there is no single value for the lower limit that will apply to all soils.

V. CONCLUSIONS

The principles of water movement into and through soils are becoming well-understood, and methods are now becoming available for obtaining necessary soil characteristics in the field. Greater water-use efficiency depends upon understanding and making use of these principles. Although the mathematical description of water uptake and transpiration by plants is still not completely satisfactory, the general nature of the relation between soil water potential and plant response is sufficiently well-known so that plant characteristics can be taken into account quantitatively. It is no longer adequate to assume that a single pair of water potential values will characterize the upper and lower limits of water availability for all soils and all crops. If water is to be efficiently utilized, each crop-soil combination must be considered separately. Greater water-use efficiency will only come with more careful attention to each component of the water budget and maintenance of the soil water content and water potential in the appropriate range.

LITERATURE CITED

Alessi, J., and J. F. Power. 1970. Influence of row spacing, irrigation, and weeds on dryland flax yield, quality and water use. Agron. J. 62:635–637.

Alessi, J., and J. F. Power. 1975. Response of early maturing corn hybrid to planting date and population in the Northern Plains. Agron. J. 67:762–765.

Al-Khafaf, S., P. J. Wierenga, and B. J. Williams. 1978. Evaporative flux from irrigated cotton as related to leaf area index, soil water, and evaporative demand. Agron. J. 70:912–917.

Allmaras, R. R., W. W. Nelson, and W. B. Voorhees. 1975. Soybean and corn rooting in southwestern Minnesota: I. Water uptake sink. Soil Sci. Soc. Am. Proc. 39:764–771.

Amemiya, M., L. N. Namken, and C. J. Gerard. 1963. Soil water depletion by irrigated cotton as influenced by water regime and stage of plant development. Agron. J. 55:376–379.

Black, T. A., W. R. Gardner, and C. B. Tanner. 1970. Water storage under a row crop on a sandy soil. Agron. J. 62:48–51.

Black, T. A., W. R. Gardner, and G. W. Thurtell. 1969. The prediction of evaporation, drainage, and soil water storage for a bare soil. Soil Sci. Soc. Am. Proc. 33:655–660.

Blum, A. 1972. Effect of planting date on water use and its efficiency in dryland grain sorghum. Agron. J. 64:775–778.

Campbell, R. E., W. E. Larson, W. E. Aasheim, and P. L. Brown. 1960. Alfalfa response to irrigation frequencies in the presence of a water table. Agron. J. 52:437–441.

Collis-George, N. 1977. Infiltration equations for simple soil systems. Water Resour. Res. 13: 395–403.

Conrad, J. P. 1937. The effects of variations in the yields of barley upon the amount and distribution of the residual soil moisture. Agron. J. 29:145–152.

Corey, A. T., and G. R. Blake. 1953. Moisture available to various crops in some New Jersey soils. Soil Sci. Soc. Am. Proc. 17:314–317.

Davidson, J. M., L. R. Stone, D. R. Nielsen, and M. E. LaRue. 1969. Field measurement and the use of soil water properties. Water Resour. Res. 5:1312–1321.

Doss, B. D., D. A. Ashley, and O. L. Bennett. 1964. Effect of moisture regime and stage of plant growth on moisture use by cotton. Soil Sci. 98:156–161.

Dreibelbis, F. R., and R. E. Youker. 1956. Soil moisture distribution on irrigated corn plots. Soil Sci. Soc. Am. Proc. 20:292–295.

Earl, K. D., and W. A. Jury. 1977. Water movement in bare and cropped soil under isolated trickle emitters. II. Analysis of cropped soil experiments. Soil Sci. Soc. Am. J. 41:856–861.

Gard, L. E., G. E. McKibben, and B. A. Jones, Jr. 1961. Moisture loss and corn yields on a silt-pan soil as affected by three levels of water supply. Soil Sci. Soc. Am. Proc. 25:154–157.

Gardner, W. R., and C. F. Ehlig. 1963. The influence of soil water on transpiration by plants. J. Geophys. Res. 68:5719–5724.

Gardner, W. R., D. Hillel, and Y. Benyamini. 1970. Post-irrigation movement of soil water. 2: Evaporation and redistribution. Water Resour. Res. 6:1148–1153.

Hagan, R. M., and M. L. Peterson. 1954. Soil moisture extraction by irrigated pasture mixtures as influenced by clipping frequency. Agron. J. 46:288–292.

Hillel, D., and W. R. Gardner. 1970. Transient infiltration into crust-topped profiles. Soil Sci. 109:69–76.

Jury, W. A., and K. D. Earl. 1977. Water movement in bare and cropped soil under isolated trickle emitters: I. Analysis of bare soil experiments. Soil Sci. Soc. Am. J. 41:852–856.

Kohl, R. A., and J. J. Kolar. 1976. Soil water uptake by alfalfa. Agron. J. 68:536–538.

Lambert, J. A., W. R. Gardner, and J. R. Boyle. 1971. The hydrologic response of a young pine plantation to weed removal. Water Resour. Res. 7:1013–1019.

Letey, J., and D. B. Peters. 1957. Influence of soil moisture levels and seasonal weather on efficiency of water use by corn. Agron. J. 49:362–365.

Lucey, R. F., and M. B. Tesar. 1965. Frequency and rate of irrigation as factors in forage growth and water absorption. Agron. J. 57:519–523.

Merrill, S. D., P. A. C. Raats, and C. Dirksen. 1978. Laterally confined flows from a point source at the surface of an inhomogeneous soil column. Soil Sci. Soc. Am. J. 42:851–857.

Mualem, Y. 1974. A conceptual model of hysteresis. Water Resour. Res. 10:514–520.

Mualem, Y. 1977. Extension of the similarity hypothesis used for modeling the soil water characteristics. Water Resour. Res. 13:773–780.

Nielsen, D. R., J. W. Biggar, and K. T. Erh. 1973. Spatial variability of field measured soil-water properties. Hilgardia 42(7):215–259.

Nnyamah, J. U., and T. A. Black. 1977. Rates and patterns of water uptake in a Douglas-fir forest. Soil Sci. Soc. Am. J. 41:972–979.

Ogata, G., L. A. Richards, and W. R. Gardner. 1960. Transpiration of alfalfa determined from soil water content changes. Soil Sci. 89:179–182.

Papendick, R. I., V. L. Cochran, and W. M. Woody. 1971. Soil water potential and water content profiles with wheat under low spring and summer rainfall. Agron. J. 63:731–734.

Philip, J. R. 1957. The theory of infiltration: 4. Sorptivity and algebraic infiltration equations. Soil Sci. 84:257–264.

Philip, J. R. 1969a. Absorption and infiltration in two- and three-dimensional systems. p. 503–525. *In* R. E. Rijtema and H. Wassink (ed.) Water in the unsaturated zone. Proc. IASH/UNESCO Symp., Wageningen, The Netherlands. 1966. UNESCO, Paris. Vol. 1.

Philip, J. R. 1969b. Theory of infiltration. Adv. Hydrosci. 5:215–296.

Raats, P. A. C. 1971. Steady infiltration from point sources, cavities, and basins. Soil Sci. Soc. Am. Proc. 35:689–694.

Raats, P. A. C. 1977. Laterally confined, steady flows of water from sources to sinks in unsaturated soil. Soil Sci. Soc. Am. J. 41:294–304.

Rawlins, S. L., W. R. Gardner, and F. N. Dalton. 1968. *In situ* measurement of soil and plant leaf water potential. Soil Sci. Soc. Am. Proc. 32:468–470.

Reeves, M., and E. E. Miller. 1975. Estimating infiltration for erratic rainfall. Water Resour. Res. 11:102–110.

Reicosky, D. C., R. J. Millington, A. Klute, and D. B. Peters. 1972. Patterns of water uptake and root distribution of soybeans (*Glycine max.*) in the presence of a water table. Agron. J. 64:292–297.

Ritchie, J. T., E. Burnett, and R. C. Henderson. 1972. Dryland evaporative flux in a subhumid climate. III: Soil water influence. Agron. J. 64:168–173.

Russell, M. B., and R. E. Danielson. 1956. Time and depth patterns for water use by corn. Agron. J. 48:163–166.

Stone, L. R., M. L. Horton, and T. C. Olson. 1973a. Water loss for irrigated sorghum field: I. Water flux within and below the root zone. Agron. J. 65:492–495.

Stone, L. R., M. L. Horton, and T. C. Olson. 1973b. Water loss from irrigated sorghum field: II. Evapotranspiration and root extraction. Agron. J. 65:495–497.

Taylor, H. M., and B. Klepper. 1973. Rooting density and water extraction patterns for corn (*Zea mays* L.). Agron. J. 65:965–968.

Taylor, H. M., and B. Klepper. 1975. Water uptake by cotton root systems: an examination of assumptions in the single root model. Soil Sci. 120:57–67.

Taylor, S. A., and J. L. Haddock. 1956. Soil moisture availability related to power required to remove water. Soil Sci. Soc. Am. Proc. 20:284–288.

Timmons, D. R., R. J. Holt, and R. L. Thompson. 1967. Effect of plant population and row spacing on evapotranspiration and water-use efficiency of soybeans. Agron. J. 59:262–265.

Van Bavel, C. H. M., G. B. Stirk, and K. J. Brust. 1968. Hydraulic properties of a clay loam soil and field measurements of water uptake by roots: I. Interpretation of water content and pressure profiles. Soil Sci. Soc. Am. Proc. 32:310–317.

van Genuchten, M. Th. 1980. A closed-form equation for predicting the hydraulic conductivity of unsaturated soils. Soil Sci. Soc. Am. J. 44:892–898.

White, I., D. E. Smiles, and K. M. Perroux. 1979. Absorption of water by soil: The constant flux boundary condition. Soil Sci. Soc. Am. J. 43:659–664.

CHAPTER 2B

Soil Properties and Efficient Water Use: Evaporation of Water from Bare Soil[1]

H. R. GARDNER

Agricultural Research Service, USDA
Fort Collins, Colorado

I. INTRODUCTION

Evaporation from the soil surface plays a significant role in water-use efficiency. If water is to be managed as efficiently as possible, the dynamics of the evaporation process must be understood. This process is complex but can be generalized to be understood.

Lemon (1956) described the evaporation process in his review of soil water evaporation, which included work by Kolasew (1941). Kolasew divided the soil water evaporation into three phases (Fig. 1). The first stage of the evaporation process is the constant rate or the energy-limiting phase. When the soil can no longer supply water fast enough to use all the available energy, the soil-limiting phase (phase 2) begins. The third phase occurs when the evaporation rate is very low, decreasing almost linearly with time.

II. QUANTITATIVE DESCRIPTIONS OF SOIL WATER EVAPORATION (E)

A cumulative evaporation curve for a semi-infinite wetted soil with an infinitely high evaporation demand can be described by an equation relating evaporation to the square root of time. When the depth of wetting is finite, the diffusivity equation can be used to calculate the cumulative evaporation, as was done by Gardner (1959). He solved this equation by scaling the cumulative evaporation data and by applying an appropriate relationship between the water content and the initial and air-dry diffusivities. This equation requires the assumption of homogeneous soil. Gardner (1974), using columns of undisturbed field soil, constructed scaled curves empirically of the form of those obtained from the diffusivity equation. These curves take into account the redistribution of water and hysteresis during the wetting

[1] Contribution from the Agricultural Research Service, USDA, Fort Collins, Colo.

Limitations to Efficient Water Use in Crop Production.

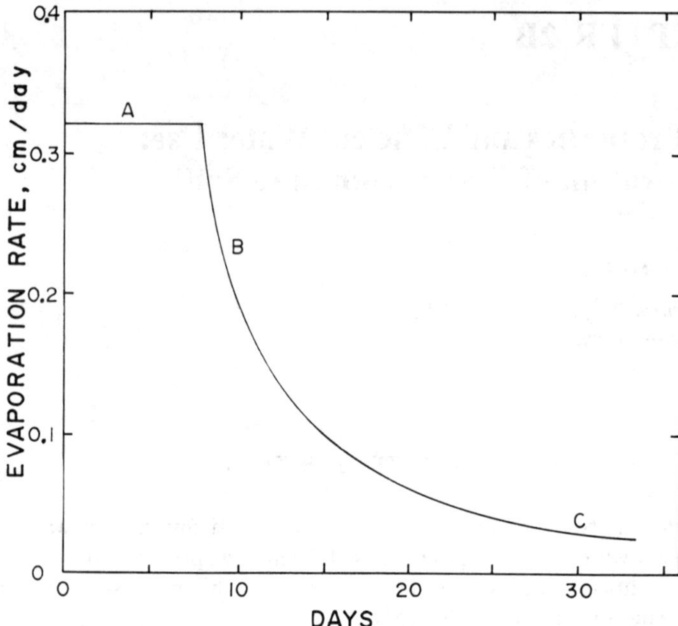

Fig. 1—Three-phase evaporation process: (*A*) Energy-limiting phase, (*B*) Rapidly falling rate, (*C*) Slowly falling rate.

and drying process. Both the calculated and the empirical curves require an infinitely high evaporation demand. Gardner and Hillel (1962) showed that when constant energy-limited evaporation takes place, the same curve that applied for the infinitely high-demand case could be considered an envelope curve. This envelope curve could be used to describe the loss after the end of the constant rate by displacing the curve to the right until it connected to the constant-rate curve. For the semi-infinite depth case, the constant rate ends at a calculated loss that is equal to $\pi/4$ times the value of the envelope curve for the same time. This same relationship is approximately true for the finite-depth case, since the decrease in evaporation rate is caused mostly by the decrease in diffusivity, rather than in water content, in both the finite and infinitely deep cases.

Figure 2 shows fractional water-loss curves for the semi-infinite-depth case with infinitely high evaporation demand, for the finite-depth case with infinitely high evaporation demand, and for the energy-limiting case with finite depth and two rates of evaporation. The envelope curve was constructed from the high-rate finite-depth data by assuming that fractional loss at the end of the constant rate phase was $\pi/4$ times the value of the envelope curve. The $\pi/4$ relationship holds well for data presented by Bond and Willis (1970) and Gardner and Hillel (1962).

Using the above relations, cumulative evaporation can be determined for a case involving constant-rate evaporation and a given time when the soil is limiting. The constant rate is plotted on a graph such as that shown in Fig. 2, and extended until it reaches $\pi/4$ times the cumulative evaporation

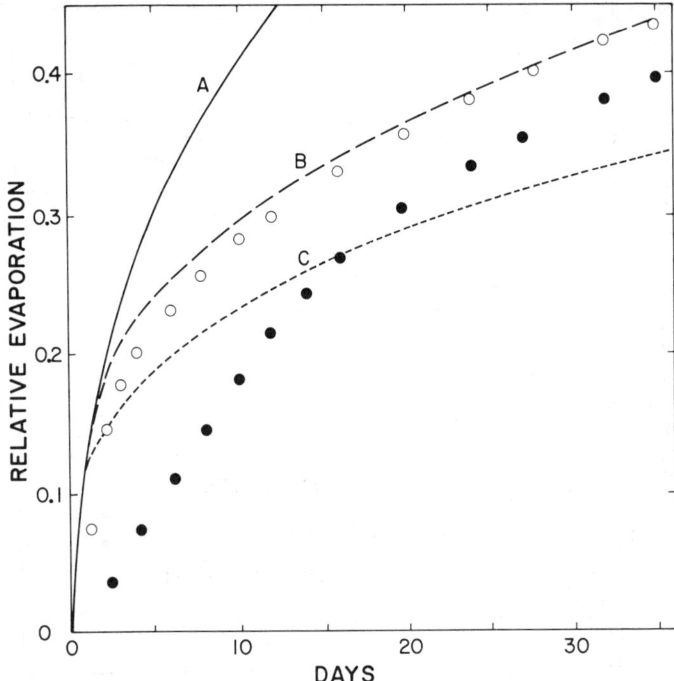

Fig. 2—Fractional loss of water by evaporation as affected by wetting depth and evaporation demand. Curve *A* is semi-infinite depth with infinitely high evaporation demand. Curve *B* is finite depth with infinitely high evaporation demand. Curve *C* is envelope Curve $A \times \pi/4$. Open circles are 9.6 cm water added with 0.72-cm day^{-1} evaporation rate. Closed circles are 9.6 cm water added with 0.18 cm day^{-1} evaporation rate.

curve to the right, and until it meets the constant-rate curves at the $\pi/4$ point. This displaced curve then becomes the falling-rate portion of the predicted curve. Because the data and calculations discussed involve constant evaporative demand but field conditions involve diurnal fluctuation of evaporative demand, it must be pointed out that the difference between the losses under the two conditions can be neglected if the average demand for the diurnal case is equal to that for the constant case. This has been shown in both laboratory (Hanks et al., 1967) and field (Van Bavel & Reginato, 1965).

III. MANAGEMENT OPTIONS FOR MODIFICATIONS OF EVAPORATION

The effects of treatments to conserve water cause a decrease in either the initial constant-rate evaporation or the rate of water movement to the soil surface. These two effects are shown in Fig. 3. The cumulative evaporation from an untreated soil would present a curve such as *A*. If a mulch is added to the soil surface, or any other treatment is imposed that only affects

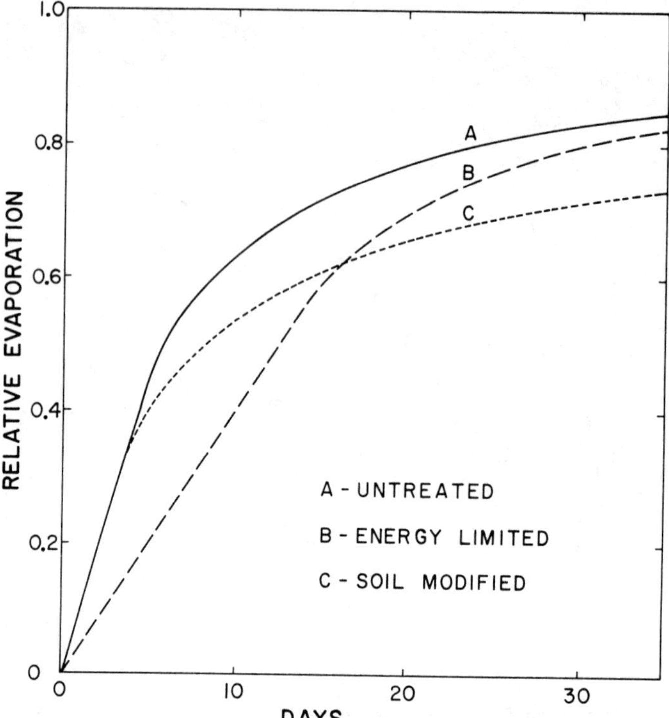

Fig. 3—Hypothetical fractional water loss as affected by treatment.

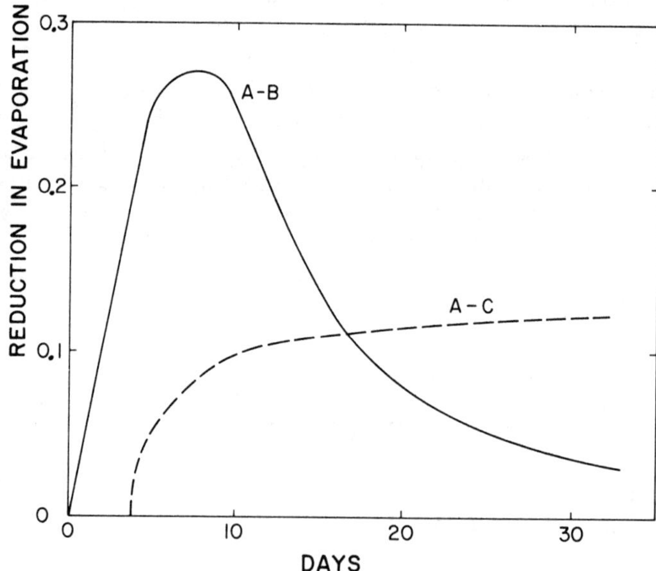

Fig. 4—Reduction in evaporative loss caused by energy-limiting (*B*) and soil-limiting (*C*) treatments from Fig. 2 compared with Curve *A*.

the energy supply or the vapor gradient above the soil surface, the slope of the energy-limiting phase of the curve will be affected, as illustrated by curve *B*. Any treatment that affects the water-transmitting properties of the soil will cause a change in the cumulative evaporation curve, as shown by curve *C*. The evaporation reductions caused by the two treatments, curve *A* minus curve *B* and curve *A* minus curve *C*, are shown in Fig. 4. The mulch, which only affects the energy supply, can have a much greater immediate effect on the cumulative evaporation, whereas the treatment affecting the soil properties can have a longer lasting effect. More thorough descriptions of the evaporation process may be found in: Gardner (1959), Gardner (1974), Hillel (1975, 1976), Hillel and Berliner (1974), Hillel and Talpaz (1977), and Hillel et al. (1975). Hillel (1977) has a good general review. The kinds of treatments that are commonly available in the field and their relative effectivenesses are reported in Chapter 10.

IV. EVAPORATION AND SEEDLING ESTABLISHMENT

Evaporative loss of water can have a great effect on crop production through its effect on plant growth during germination and seedling establishment (refer to Chapter 7A for additional discussion). The water required in the soil for success in germination and establishment of plants can be determined by using some of the available mathematical descriptions of soil-water-plant relations.

Wanjura and Buxton (1972a) presented data indicating the time required to obtain radicle emergence, as related to temperature and soil water potential. At a given temperature, there is very little difference in time required for radicle emergence, as long as the soil water potential is greater than −3 bars.

Because the soil water potential during evaporation is lowest at the soil surface and increases with depth, and because the water potential decreases with time at any given depth, seeds must be planted deep enough so that the soil water potential remains within the range necessary for germination and for a long enough time for proper germination.

If soil water content vs. depth is scaled by dividing the depth by time, t, as $t^{1/2}$, all curves of water content (θ) vs. depth for various times will fall on the same line for data that follows the $t^{1/2}$ relationship. Theoretically, a given water content will move downward with time (as $t^{1/2}$) as evaporation occurs, as long as the depth of wetting does not affect evaporation. Using this scaled plot, the water-content distribution can be predicted for any time, and if the relationship between water potential (ψ_S) and θ is known, the depth below which ψ_S will remain above the desired −3-bar value for the desired time can be determined graphically. A scaled plot of θ vs. depth/$t^{1/2}$ is shown in Fig. 5. If, after a rain, the evaporation rate is constant, the time that the soil surface remains wet may be added to the time that the soil will remain above the given water potential at the desired depth, as determined using the scaled water content data.

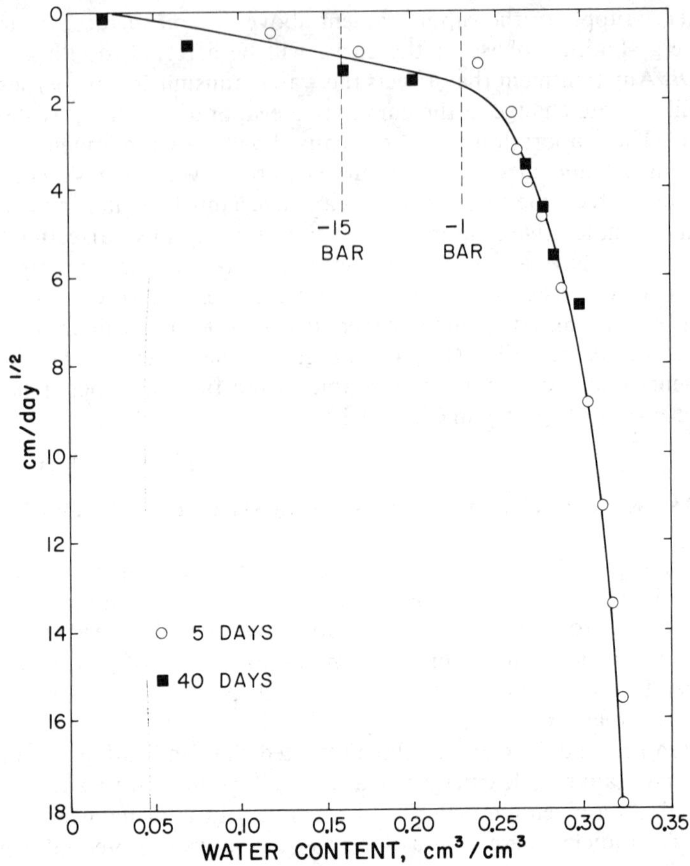

Fig. 5—Scaled plot of water content (θ) vs. $t^{1/2}$ (time). Data from Hanks et al. (1967).

An analysis of data from Wanjura and Buxton (1972b) shows that the initial plant root extends downward as $t^{1/2}$, which means that if the root moves downward in a zone wetter than -3 bars, then the root tip will remain within that zone because that zone is moving downward at the same rate. If the depth of wetting is shallow, then the drying front could move downward faster than the expected $t^{1/2}$ relationship.

V. SUMMARY

The information necessary to produce the scaled curves used to describe E in Fig. 2 are easily obtained without any knowledge of solutions to the water-flow equation. These data may be obtained from soil columns that are wetted to a range of depths and weighed for determination of E. The use of these curves is a very practical way to estimate the effects of various mulching treatments without resorting to complicated mathematical analyses and tedious laboratory procedures.

LITERATURE CITED

Bond, J. J., and W. O. Willis. 1970. Salt water evaporation: first stage of drying as influenced by surface residue and evaporative potential. Soil Sci. Soc. Am. Proc. 34:924–928.

Gardner, H. R. 1974. Prediction of water loss from a fallow field soil based on soil water flow theory. Soil Sci. Soc. Am. Proc. 38:379–382.

Gardner, W. R. 1959. Solutions of the flow equation for the drying of soils and other porous media. Soil Sci. Soc. Am. Proc. 23:183–187.

Gardner, W. R., and D. I. Hillel. 1962. The relation of external evaporative condition to the drying of soils. J. Geophys. Res. 67:4319–4325.

Hanks, R. J., H. R. Gardner, and M. L. Fairbourn. 1967. Evaporation of water from soils as influenced by drying with wind or radiation. Soil Sci. Soc. Am. Proc. 31:593–598.

Hillel, D. 1975. Simulation of evaporation from bare soil under steady and fluctuating evaporativity. Soil Sci. 120:230–237.

Hillel, D. 1976. On the role of soil moisture hysteresis in the suppression of evaporation from bare soil under diurnally cyclic evaporativity. Soil Sci. 122:309–314.

Hillel, D. 1977. Computer-simulation of soil water dynamics: a compendium of recent work. Int. Dev. Res. Ctr., Ottawa, Canada.

Hillel, D., and P. Berliner. 1974. Waterproofing surface-zone soil aggregates for water conservation. Soil Sci. 118:131–135.

Hillel, D., and H. Talpaz. 1977. Simulation of soil water dynamics in layered soils. Soil Sci. 123:54–62.

Hillel, D., C. H. M. van Bavel, and H. Talpaz. 1975. Dynamic simulation of water storage in fallow soil as affected by mulch of hydrophobic aggregates. Soil Sci. Soc. Am. Proc. 39: 826–833.

Kolasew, F. E. 1941. Ways of suppressing evaporation of soil moisture. Sb. Rab. Agr. Fix. 3:67.

Lemon, E. R. 1956. The potential for decreasing soil moisture evaporation loss. Soil Sci. Soc. Am. Proc. 20:120–125.

Van Bavel, C. H. M., and R. J. Reginato. 1965. Precision lysimetry for direct measurement of evaporative flux. p. 129–136. *In* Methodology of plant ecophysiology. Proc. of the Montpellier Symp. UNESCO, Paris. Arid Zone Res. XXV.

Wanjura, D. F., and D. R. Buxton. 1972a. Water uptake and radicle emergence of cottonseed as affected by soil moisture and temperature. Agron. J. 64:427–431.

Wanjura, D. F., and D. R. Buxton. 1972b. Hypocotyl and radicle elongation of cotton as affected by soil environment. Agron. J. 64:431–434.

CHAPTER 2C

Soil Properties and Efficient Water Use: Water Management for Salinity Control

GLENN J. HOFFMAN AND M. TH. VAN GENUCHTEN

U.S. Salinity Laboratory, USDA-ARS
Riverside, California

I. INTRODUCTION

A mixture of soluble salts is present in all soils. This mixture includes salts originally dissolved in the applied water, applied as amendments, weathered from soil minerals, or imported with subsurface water flow. If the concentration of these salts becomes excessive, crop yields will be reduced because of the decrease in osmotic potential of the soil water. To prevent harmful accumulation of salts, the soil profile must be leached periodically with an amount of water in excess of that used by evapotranspiration. Thus, where salinity is a hazard, the concept of efficient water use must be expanded to include an increment of water to meet the leaching requirement (L_r). The leaching requirement is defined as the minimum fraction of the total amount of applied water that must pass through the soil root zone to prevent a reduction in crop yield from an excess accumulation of salts (USSL Staff, 1954). Leaching occurs whenever irrigation and rainfall exceed evapotranspiration. In humid regions, rainfall is normally sufficient to periodically flush salt from the profile. In subhumid regions, additional irrigation water must frequently be applied to assure adequate leaching. Depending on the degree of salinity control required, leaching may occur continuously or intermittently at intervals of a few months to a few years.

Soils not only contain a mixture of salts, but distribution of those salts is neither uniform on an area basis or with soil depth, nor constant in time. Nonuniformity within a field is caused in large part by irrigation systems that supply water unevenly. As a consequence of this and irrigation scheduling that exceeds evapotranspiration, a large fraction of the applied water may leach through the root zone and be wasted as drainage. Present technology, including high-frequency irrigation, makes it possible to deliver water more efficiently thereby reducing excessive drainage. As we implement this technology and strive to conserve water, the minimum value to which drainage can be reduced, the leaching requirement, becomes a vital component in efficient water management.

Limitations to Efficient Water Use in Crop Production.

Two quantities establish the leaching requirement: the salt concentration of the applied water and the salt tolerance of the crop. The salt concentration of the applied water, including rainfall, is readily measured. Crop salt-tolerance, however, is more difficult to measure and has traditionally been established by measuring yields where waters of varying salt concentrations have been applied at leaching fractions approaching 0.5. Maas and Hoffman (1977) summarized salt-tolerance data for more than 60 agricultural crops. They reported salt tolerance by means of two parameters: the threshold and the rate of yield decline as salinity increases beyond this threshold. The threshold value is the maximum average salt concentration in the root zone that does not reduce yield. Where the quantity or quality of the applied water permits, efficient salinity management should not permit the average root-zone salinity to exceed the threshold value.

The leaching fraction, L, is defined as

$$L = q_D/q_I = C_I/C_D, \qquad [1]$$

where q and C represent long-term average values of the volumetric flux and the salt concentration, respectively, and the subscripts I and D refer to irrigation (including rainfall) and drainage water, respectively. Because L_r is defined as the minimum leaching fraction needed to prevent yield reduction:

$$L_r = q_D^*/q_I = C_I/C_D^*, \qquad [2]$$

where the superscript * distinguishes required from actual values (USSL Staff, 1954). The salt concentration of a relatively dilute soil solution is roughly linearly related to the electrical conductivity (σ). Because σ is easily measured, it is advantageous to express L_r as:

$$L_r \cong \sigma_I/\sigma_D^*. \qquad [3]$$

Inherent in the above equations are the assumptions that no salts precipitate, dissolve, or are removed by the crop.

Several empirical methods have been proposed to relate σ_D^* to some readily available soil salinity value. Bernstein (1964) assumed σ_D^* to be the electrical conductivity of the soil saturation extract (σ_e) at which yield in salt-tolerance experiments was reduced by 50% ($\sigma_{e_{50}}$). Later, van Schilfgaarde et al. (1974) contended that the value of σ_D^* could be increased to the σ of soil water at which roots can no longer extract water. Assuming the soil water content to be half the saturation extract water content, this value of σ is about twice σ_e extrapolated to zero yield from salt tolerance data (σ_{e_0}). Concurrently Rhoades (1974) proposed that σ_D^* could be estimated from

$$\sigma_D^* = 5\,\sigma_t - \sigma_I, \qquad [4]$$

where σ_t is taken as the salt-tolerance threshold. As will be shown later (Table 1), experimental evidence indicates that the method of Bernstein

Table 1—Comparisons of the leaching requirement (L_r) predicted by four models with experimentally determined leaching requirements for several crops.

Crop	Experimental data		Salt-tolerance data			L_r predictions				Reference
	σ_I	L_r	Threshold	$\sigma_{e_{50}}$	σ_{e_0}	$\sigma_I/\sigma_{e_{50}}$	$\sigma_I/(2\sigma_{e_0})$	$\sigma_I/(5\sigma_t-\sigma_I)$	Fig. 6	
	dS m^{-1}									
Alfalfa	2.0	0.13	2.0	8.8	15.7	0.23	0.06	0.25	0.17	Bower et al. (1969)
	4.0	0.29				0.45	0.13	0.67	0.26	
Alfalfa	1.0	0.06	2.0	8.8	15.7	0.11	0.03	0.11	0.09	Bernstein & Francois (1973)
	2.0	0.15				0.23	0.06	0.25	0.17	
Alfalfa	0.9	<0.1	2.0	8.8	15.7	0.10	0.03	0.10	0.08	Ingvalson et al. (1976)
	1.3	>0.1				0.15	0.04	0.15	0.12	
	1.6	>0.1				0.18	0.05	0.19	0.14	
	2.0	~0.2				0.23	0.06	0.25	0.17	
	3.1	>0.2				0.35	0.10	0.45	0.23	
	3.3	>0.2				0.38	0.11	0.49	0.24	
Alfalfa	1.4	<0.05	2.0	8.8	15.7	0.16	0.04	0.16	0.13	USSL Staff (1977)
Fescue, tall	2.0	0.13	3.9	13.3	22.8	0.15	0.04	0.11	0.10	Bower et al. (1970)
	4.0	0.26				0.30	0.09	0.26	0.18	
Lettuce	2.2	0.26	1.3	5.1	9.0	0.43	0.12	0.51	0.24	Hoffman et al. (1979)
Lettuce	1.4	0.25	1.3	5.1	9.0	0.27	0.08	0.27	0.18	Lonkert et al. (1976, personal communication)
Orange	1.4	<0.1	1.7	4.8	8.0	0.29	0.09	0.20	0.15	USSL Staff (1977)
Sorghum	2.2	0.08	4.0	11.0	36.0	0.20	0.03	0.12	0.10	Hoffman et al. (1979)
Sudangrass	2.0	0.14	2.8	14.4	26.1	0.14	0.04	0.19	0.13	Bower et al. (1970)
	4.0	0.24				0.28	0.08	0.40	0.22	
Tomato	2.2	>0.17	2.5	7.6	12.6	0.29	0.09	0.21	0.16	Hoffman (1980)
Wheat	2.2	0.08	6.0	13.0	20.1	0.17	0.05	0.08	0.07	Hoffman et al. (1979)
Wheat	1.4	0.07	6.0	13.0	20.1	0.11	0.03	0.05	0.04	Lonkert et al. (1976, personal communication)

(1964) overestimates, and the method of van Schilfgaarde et al. (1974) underestimates, the leaching requirement. The method of Rhoades (1974) also overestimates the leaching requirement, except at low leaching requirements.

The objectives of this discussion were to evaluate four models for predicting the mean root-zone salinity based upon experimental data, and to determine the L_r by converting the mean root-zone salinity prediction to the crop salt-tolerance threshold. The leaching requirement is presented as a function of the salt concentration of the applied water and the salt-tolerance threshold for the crop. The predicted L_r is compared with experimentally determined leaching requirements for a number of crops.

II. MATHEMATICAL ANALYSIS OF ROOT-ZONE SALINITY

Following the analysis of Gardner (1967), which was expanded by Raats (1974), the continuity equation for one-dimensional vertical flow of water through soil can be expressed as

$$\partial\theta/\partial t = -\partial q/\partial z - \lambda, \qquad [5]$$

where θ is the volumetric water content of the soil, t is time, q is the volumetric flux density of water, z is the vertical coordinate that increases positively downward and has its origin at the soil surface, and λ is the rate of water uptake by plant roots per volume of soil, per unit time. For steady flow, $\partial\theta/\partial t = 0$ and Eq. [5] reduces to

$$\mathrm{d}q/\mathrm{d}z = -\lambda. \qquad [6]$$

The steady-state mass balance for salt, neglecting the effects of chemical precipitation, dissolution, diffusion, and dispersion, is

$$\mathrm{d}(qC)/\mathrm{d}z = 0. \qquad [7]$$

Integrating Eq. [7] and noting that at $z = 0$, the salt flux equals $q_I C_I$ leads to

$$q/q_I = C_I/C. \qquad [8]$$

Evaluation of Eq. [8] at the bottom of the root zone ($q = q_D$; $C = C_D$) leads directly to the definition of the leaching fraction (Eq. [1]). This shows that the leaching fraction is based upon steady-state mass balance considerations. Expanding Eq. [7] by introducing Eq. [6] and [8] gives

$$\lambda = -q_I C_I \mathrm{d}(C^{-1})/\mathrm{d}z, \qquad [9]$$

where C^{-1} is the inverse of the concentration or alternatively the dilution

Table 2—Equations for the salt concentration (C) as a function of soil depth (z), the linearly averaged root-zone salt concentration ($\overline{C}$), and the water-uptake-weighted average root-zone salt concentration ($<C>$) for three different water uptake functions (λ). The dimensionless variable x is z/Z.

Variable	Method: Exponential uptake function (Raats, 1974)	Method: Trapezoidal uptake function (W. R. Gardner, Chapter 2A)	Method: 40–30–20–10 uptake function.
Water uptake function	$\lambda_1 = \frac{T}{\delta}\exp(-z/\delta)$	$\lambda_2 = \begin{cases} 2T/Z & 0.0 \le x < 0.2 \\ 2T(4-5x)/3Z & 0.2 \le x < 0.8 \\ 0 & 0.8 \le x \le 1.0 \end{cases}$	$\lambda_3 = \frac{T}{5Z}(9-8x)$
Concentration vs. depth	$\frac{C_1}{C_I} = [L+(1-L)\exp(-z/\delta)]^{-1}$	$\frac{C_2}{C_I} = \begin{cases} [1-2(1-L)x]^{-1} & 0.0 \le x < 0.2 \\ [1-(1-L)(40x-25x^2-1)/15]^{-1} & 0.2 \le x < 0.8 \\ L^{-1} & 0.8 \le x \le 1.0 \end{cases}$	$\frac{C_3}{C_I} = [1-(1-L)(9x-4x^2)/5]^{-1}$
Linearly averaged root-zone salt concentration	$\frac{\overline{C_1}}{C_I} = \frac{1}{L} + \frac{\delta}{ZL}\ln[L+(1-L)\exp(-Z/\delta)]$	$\frac{\overline{C_2}}{C_I} = \frac{1}{5L} - \frac{1}{2-2L}\ln[0.6-0.4L) + \frac{\tan^{-1}[3(1-L)/5L]^{1/2}}{[5L(1-L)/3]^{1/2}}$	$\frac{\overline{C_3}}{C_I} = \frac{10\alpha}{1-L}[\tan^{-1}(9\alpha) - \tan^{-1}(\alpha)]$ $\alpha = \left(\frac{1-L}{81L-1}\right)^{1/2}$ $(81L > 1)$
Uptake-weighted average root-zone salt Concentration	$\frac{<C_1>}{C_I} = \frac{\ln(L)}{L-1}$	$\frac{<C_2>}{C_I} = \frac{\ln(L)}{L-1}$	$\frac{<C_3>}{C_I} = \frac{\ln(L)}{L-1}$

(Raats, 1974). Equation [9] shows that the rate of water uptake may be calculated as the product of $q_I C_I$ and the negative of the slope of the dilution profile, $d(C^{-1})/dz$. Integration of Eq. [9] between $z = 0$ and any arbitrary depth z within the soil root zone gives the cumulative water uptake (w) between the soil surface and z and can be expressed as

$$w(z) = \int_0^z \lambda \, dz = q_I(1 - C_I/C). \qquad [10]$$

At the bottom of the root zone ($z = Z$), w becomes equal to the transpiration rate (T), and C equals C_D. Equation [10] then reduces to

$$T = q_I(1 - L). \qquad [11]$$

The water uptake function can be calculated from a measured steady-state salt profile using Eq. [9]. Alternatively, the steady-state salt distribution can be calculated from Eq. [10], provided the water uptake function (λ) is known. Table 2 gives three models for λ. The exponential uptake function (λ_1 in Table 2) was first proposed by Raats (1974). The parameter δ in this equation is an empirical constant. The trapezoidal uptake function (λ_2 in Table 2) is identical to the normalized uptake function described by W. R. Gardner (Chapter 2A). The third uptake function (λ_3) assumes that water uptake from successively deeper quarter-fractions of the root zone is proportioned as 40, 30, 20, and 10% of the transpiration rate. This uptake pattern is reminiscent of the old "40–30–20–10" rule used by Rhoades and Merrill (1976), among others. The three water uptake functions are plotted in Fig. 1. For the exponential uptake function, the parameter δ was set to $0.2\ Z$. In contrast to the trapezoidal and "40–30–20–10" uptake functions, the area under the exponential function is not exactly equal to 1.0, but to 0.993. This deviation from 1.0 is a consequence of the exponential nature of this uptake equation, and is considered insignificant.

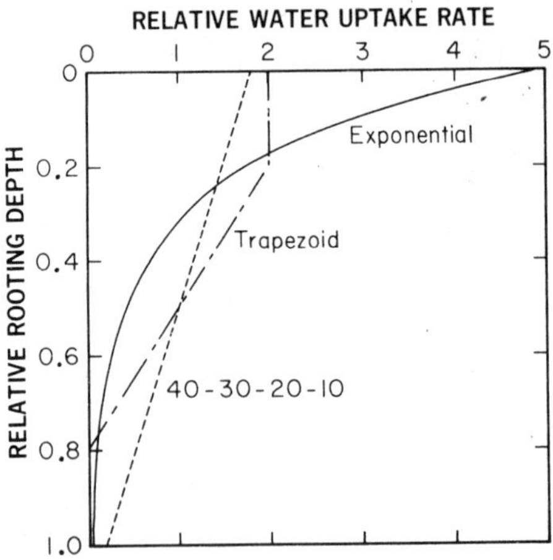

Fig. 1—Water uptake patterns as predicted by the three models presented in Table 2, relative to an uptake rate uniform throughout the rooting depth.

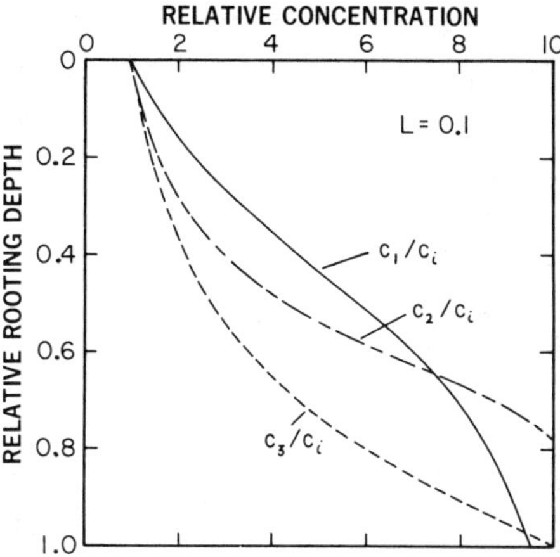

Fig. 2—Soil salinity profiles throughout the root zone at a leaching fraction (L) of 0.1; C_I, C_1, C_2, and C_3 are the salt concentrations in the irrigation water and those predicted using the exponential, trapezoidal, and 40–30–20–10 water uptake models, respectively.

Substitution of the three uptake functions of Table 2 into Eq. [10], solving the indicated integral, and using Eq. [11] results in the salt-concentration distribution equations given in Table 2. A graphical illustration of these distributions for $L = 0.1$ is given in Fig. 2. The shapes of these curves are similar to measured salt distributions (e.g., Bower et al., 1969, 1970).

There are at least two possible approaches for calculating the average salt concentration in the root zone. The linearly averaged salt concentration $\overline{C}$, can be calculated as

$$\overline{C} = \frac{1}{Z} \int_0^Z C\, dz, \qquad [12]$$

where Z is the depth of the root zone. Application of Eq. [12] to the equations for the salt concentration as a function of depth in Table 2 leads to expressions of $\overline{C}$ as a function of L for each of the uptake functions (see Table 2). Likewise, one can calculate the average root-zone salt concentration by means of a water-uptake weighted function ($<C>$),

$$<C> = \int_0^\infty \lambda C\, dz \Big/ \int_0^\infty \lambda\, dz, \qquad [13]$$

where the upper integration limit has been extended to infinity to accommodate the exponential uptake function (λ_2 and λ_3 vanish below the root zone). Substitution of the expressions for λ and C in Table 2 into Eq. [13] results in the water-uptake-weighted average root-zone salt concentrations as given in Table 2. Note that the resulting expression for $<C>$ is the same for all three water uptake functions. Bernstein and Francois (1973) showed earlier that $<C>$ is independent of the assumed water uptake distribution.

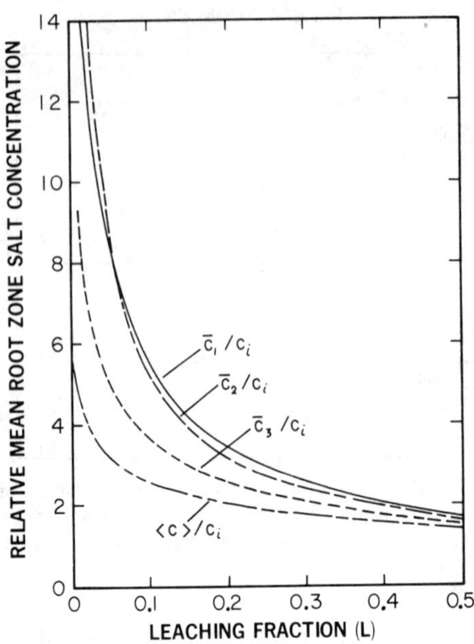

Fig. 3—Relative mean root-zone salt concentrations as a function of leaching fraction (L); $\overline{C}_1$, $\overline{C}_2$, and $\overline{C}_3$ are for the exponential, trapezoidal, and 40–30–20–10 water uptake models, respectively, of Table 2, and $<C>$ is the uptake-weighted average salt concentration for each of the three uptake models.

For comparison, the relationships between the relative mean root-zone salt concentrations and L for all four models are given in Fig. 3. The parameter δ for the exponential uptake function was taken as $0.2Z$; the reason for this choice will be obvious in the following section. Of interest in Fig. 3 is the lack of sensitivity of $<C>/C_I$ to L as compared with the other models.

III. MEAN ROOT-ZONE SALINITY

To estimate the leaching requirement based on the prediction of the average root-zone salinity, it is first necessary to verify the accuracy of the models for predicting leaching fraction. For that purpose, we took soil salinity data from the experiments of Bower et al. (1969, 1970), Bernstein and Francois (1973), Bernstein et al. (1975), Ingvalson et al. (1976), and Hoffman et al. (1979). In most, but not all, of these experiments chloride irrigation waters were used so that precipitation at low leaching fractions should not cause an error. For waters high in sulfate or carbonate, precipitation could be significant at low leaching fractions and the computer model proposed by Rhoades and Merrill (1976) may be more accurate.

For vertification of the linearly averaged expressions, the mean root-zone salinity value was calculated as the linear average of the experimentally determined salt concentration values. Each mean salinity value was sub-

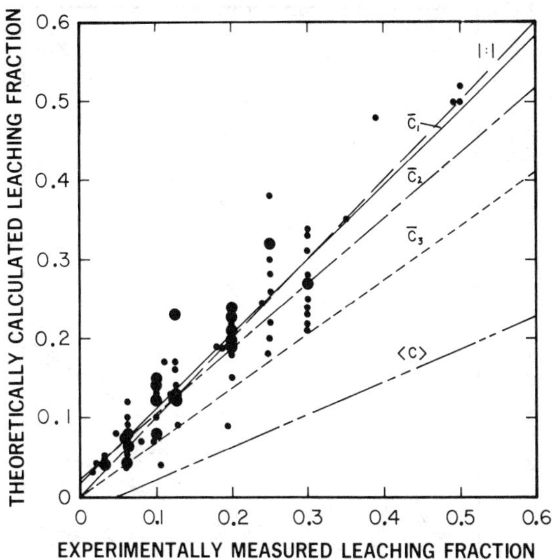

Fig. 4—Comparison of experimentally measured leaching fractions from seven experiments with theoretically calculated leaching fractions from $\overline{C}_1$, where $\delta = 0.2Z$. Also shown are the linear regressions for $\overline{C}_2$, $\overline{C}_3$, and $<C>$ for comparison. Data points are for $\overline{C}_1$ only. Larger data points indicate more than one observation.

stituted into each of the linear-average models and the equations were solved graphically for the "theoretical" leaching fraction. For vertification of the uptake-weighted expression, the experimental salt concentration values were weighted using the 40–30–20–10 water uptake function. This procedure is valid because, as shown earlier, the uptake-weighted average salt concentration is independent of the particular uptake function. The "theoretical" leaching fraction was then calculated by inserting the resultant $<C>$ value into the uptake-weighted expression. Figure 4 compares these theoretical L-values with the experimentally measured values. With the selection of $\delta = 0.2Z$, $\overline{C}_1$ best predicts the mean root-zone salinity. The other expressions all substantially underestimate L. The lines in Fig. 4 are statistically fit linear regressions through the data for each model, although only the data for $\overline{C}_1$ are shown. The correlation coefficient of the linear regression for $\overline{C}_1$ is 0.94.

The relatively good agreement of $\overline{C}_1$ and $\overline{C}_2$ with the experimental data lends credibility to their application. A disadvantage of $\overline{C}_1$, as compared with $\overline{C}_2$, is that a value must be assigned to the empirical parameter δ. However, we propose the use of $\overline{C}_1$ because of its better correlation with the experimental data.

The mean root-zone salinity, based on $\overline{C}_1$, is shown in Fig. 5 as a function of the salinity of the applied water and the leaching fraction. Note that the salinity is expressed in terms of the electrical conductivity of the soil saturation extract (σ_e) assuming $\sigma = 2\sigma_e$. Thus, the curve for $L = 1.0$ in Fig. 5 does not fall on the 1:1 line. As an example, if soil samples revealed that the mean soil salinity was 5 dS m^{-1} and the salinity of the applied water

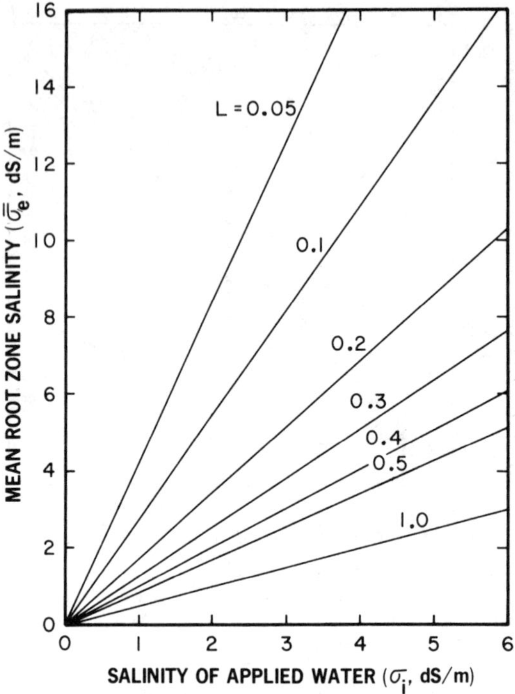

Fig. 5—The mean root-zone soil salinity as a function of the salinity of the applied water and leaching fraction (L). Curves are based upon the linearly averaged root-zone salt concentration for the exponential water uptake function ($\overline{C}_1/C_I$ in Table 2).

(σ_I) was 3 dS m^{-1}, the leaching fraction would be estimated to be 0.21 under steady-state conditions.

The large differences between experiment and theory for $\overline{C}_1$ and $\overline{C}_3$ indicate that, at least under saline conditions, the 40–30–20–10 rule of thumb often quoted as the distribution of water uptake through the root zone underestimates the water lost from the upper portion of the root zone. The water uptake distribution is approximately 71–21–6–1 based on $\overline{C}_1$, and 50–35–14–1 based on $\overline{C}_2$. Although evaporation was ignored in the theoretical development, it may account in large part for the high proportion of water lost near the soil surface.

IV. LEACHING REQUIREMENT PREDICTION

From the definition of leaching requirement as stated in Eq. [3], the unknown term in predicting L_r is σ_D^*. Unfortunately, this definition is not suited for use of the large number of experiments designed to establish the salt tolerance of agricultural crops. A more convenient quantity for evaluating leaching requirement is the crop salinity threshold value. If mean soil salinity is replaced by the threshold, the result is a graph similar to Fig. 5,

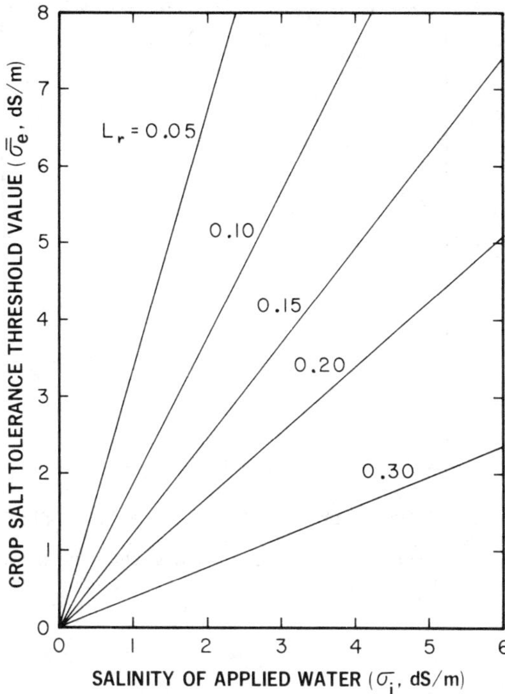

Fig. 6—Graphical solution for the leaching requirement (L_r) as a function of the salinity of the applied water and the salt-tolerance threshold value for the crop.

with L replaced by L_r. The resulting relationship, however, can not be applied directly, because the mean root-zone salinity for a given salinity of the applied water must be reduced if L is lower than L_r to prevent yield reduction or, conversely, L could be increased if the crop could tolerate more salinity. Unfortunately, no mathematical relationship has been developed to span this gap.

Plants adjust osmotically as soil salinity increases (Maas & Nieman, 1978). In salt-tolerance trials, this adjustment is sufficient to prevent yield loss until soil salinity surpasses the salt-tolerance threshold. Because salt-tolerance trials usually are designed to maintain leaching fractions approaching 0.5, the osmotic adjustment consistent with no yield loss can be estimated by the mean soil salinity at 50% leaching, as given in Fig. 5. As a first approximation, the leaching requirement can be expressed as a function of the threshold value by reducing $\overline{C_1}$ at any given L by $\overline{C_1}$ at 50% leaching. This relationship among salinity of the irrigation water, crop salt-tolerance threshold, and leaching requirement is illustrated in Fig. 6.

Comparisons among calculated leaching requirements and those measured experimentally are given in Table 1. Linear regressions correlating these calculated and measured values are given in Fig. 7 for each of the predictive models listed in Table 2. Predicting L_r by $\sigma_I/\sigma_{e_{50}}$ consistently overestimates, and by $\sigma_I/2\sigma_{e_0}$ consistently underestimates, the leaching require-

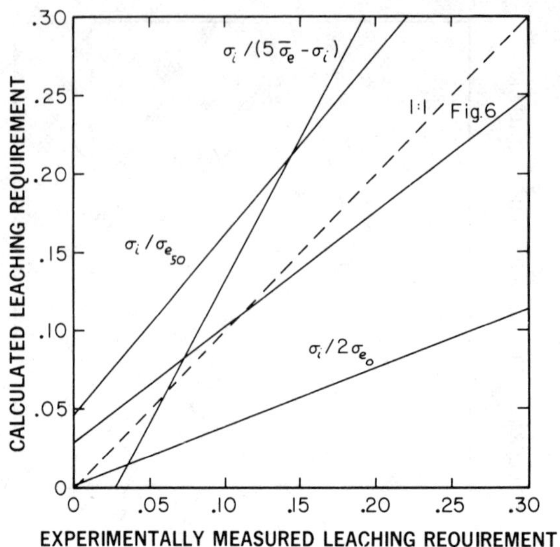

Fig. 7—Comparison of experimentally measured leaching requirements for six crops with empirically calculated leaching requirements. Data for establishing the least-squares linear regressions are taken from Table 1 for those experiments where a precise leaching requirement was reported.

ment. Predicting L_r from $\sigma_I/(5\sigma_t - \sigma_I)$ also overestimates the L_r, except at low L_r. Although not perfect, the L_r predictions from Fig. 6 agree well with the measured values throughout the range of L_r's of agricultural interest. In view of our ability to measure the leaching fraction in the field, Fig. 6 is a suitable estimate of leaching requirement.

LITERATURE CITED

Bernstein, L. 1964. Salt tolerance of plants. USDA Info. Bull. 283. U.S. Government Printing Office, Washington, D.C.

Bernstein, L., and L. E. Francois. 1973. Leaching requirement studies: sensitivity of alfalfa to salinity of irrigation and drainage waters. Soil Sci. Soc. Am. Proc. 37:931–943.

Bernstein, L., L. E. Francois, and R. A. Clark. 1975. Minimal leaching with varying root depths of alfalfa. Soil Sci. Soc. Am. Proc. 39:112–115.

Bower, C. A., G. Ogata, and J. M. Tucker. 1969. Rootzone salt profiles and alfalfa growth as influenced by irrigation water salinity and leaching fraction. Agron. J. 61:783–785.

Bower, C. A., G. Ogata, and J. M. Tucker. 1970. Growth of sudan and tall fescue grasses as influenced by irrigation water salinity and leaching fraction. Agron. J. 62:793–794.

Gardner, W. R. 1967. Water uptake and salt-distribution patterns in saline soils. p. 335–341. *In* Isotope and radiation techniques in soil physics and irrigation studies. Int. Atomic Energy Agency Proc. Ser.

Hoffman, G. J. 1980. Leaching requirement for salinity control in agriculture. p. 277–283. *In* Proc. Int. Symp. on Salt Affected Soils, Karnal, India. Feb. 1980.

Hoffman, G. J., S. L. Rawlins, J. D. Oster, J. A. Jobes, and S. D. Merrill. 1979. Leaching requirement for salinity control: I. Wheat, sorghum, and lettuce. Agric. Water Manage. 2:177–192.

Ingvalson, R. D., J. D. Rhoades, and A. L. Page. 1976. Correlation of alfalfa yield with various index of salinity. Soil Sci. 122:145–153.

Maas, E. V., and G. J. Hoffman. 1977. Crop salt tolerance—current assessment. J. Irr. Drain. Div. Am. Soc. Civil Eng. 103(IR2):115–134.

Maas, E. V., and R. H. Nieman. 1978. Physiology of plant tolerance to salinity. Chapt. 13. *In* G. A. Jung (ed.) Crop tolerance to suboptimal land conditions. ASA Spec. Pub. no. 32. Am. Soc. of Agron., Madison, Wis.

Raats, P. A. C. 1974. Steady flows of water and salt in uniform soil profiles with plant roots. Soil Sci. Soc. Am. Proc. 38:717–722.

Rhoades, J. D. 1974. Drainage for salinity control. *In* J. van Schilfgaarde (ed.) Drainage for agriculture. Agronomy 17:433–461. Am. Soc. of Agron., Madison, Wis.

Rhoades, J. D., and S. D. Merrill. 1976. Assessing the suitability of water for irrigation: theoretical and empirical approaches. FAO Soils Bull. 31:69–109.

U.S. Salinity Laboratory Staff. 1954. Diagnosis and improvement of saline and alkali soils. Agric. Handb. no. 60, USDA. U.S. Government Printing Office, Washington, D.C.

U.S. Salinity Laboratory Staff. 1977. Interim Report—Minimizing salt in return flow through irrigation management. USEPA Interagency Project IAG-D4-0370. Environ. Prot. Agency Rep. no. EPA-6000/2-77-134. U.S. Government Printing Office, Washington, D.C.

van Schilfgaarde, J., L. Bernstein, J. D. Rhoades, and S. L. Rawlins. 1974. Irrigation management for salt control. J. Irr. Drain. Div. Am. Soc. Civil Eng. 100(IR3):321–338.

CHAPTER 3A

Managing Root Systems for Efficient Water Use: An Overview[1]

HOWARD M. TAYLOR

Texas Tech University
Lubbock, Texas

I. INTRODUCTION

When aquatic plants migrated to land eons ago, roots developed as specialized structures to absorb mineral nutrients and water stored in the soil profile. This stored water then allowed the plants to survive on land and grow between rainstorms.

Soil water uptake, U, is usually assumed to be proportional to the product of the rooting activity, hydraulic conductivity, and potential energy terms for any specific soil volume element of uniform properties. Numerous models are described in the literature. They differ principally in their assumptions about the appropriate measurements needed to define root activity, hydraulic conductivity, and potential energy. Once U is determined for a specific soil volume and time increment, it is then summed over all soil volumes occupied by roots to provide a value for transpiration.

All the current water uptake models contain coefficients that are site-, time-, and crop-specific. Some published models have not even been verified for the specific conditions that they purport to model. For these reasons, no "best" model can be selected at the present time. This chapter uses the model of Taylor and Klepper (1978) as a convenient framework to subdivide the discussion. This model requires an independent measurement or estimate of transpiration as an input. This transpiration equals water uptake by the entire root system plus water withdrawn from storage in plant tissue (usually, but not always, negligibly small). The root system water uptake is sectioned into water uptake from a number of soil volumes, each with uniform soil and root system properties. Water uptake rates (U_i, cm^3 day^{-1}) from any specific soil volume V_i (cm^3) are governed by the equation

$$U_i = V_i L_{vi} q_i (\psi_{si} - \psi_p + \psi_{zi} + \psi_{fi}), \quad [1]$$

[1] Contribution from the Agricultural Research Service, USDA, and Journal Paper no. J-9468 of the Iowa Agriculture and Home Economics Experiment Station. Ames, Iowa (Project no. 2083).

Limitations to Efficient Water Use in Crop Production.

where

L_{vi} = root length density (cm roots cm^{-3} soil in V_i),

q_i = specific root uptake rate [cm^3 H_2O cm^{-1} root $(\psi_{si} - \psi_{rxi})^{-1}$ day^{-1}],

ψ_{si} = total soil water potential at the midpoint of V_i (bars),

ψ_p = plant xylem water potential at the ground surface (bars),

ψ_{zi} = the positional water potential in the xylem due to gravity (bars),

ψ_{fi} = the loss of potential due to friction as water flows axially along xylem elements (bars), and

ψ_{rxi} = the water potential in the xylem at the midpoint of volume V_i (bars).

The factors that affect transpiration rate, tissue water storage, and plant water potential at the ground surface are discussed in other chapters. This chapter covers effects on efficient use of water of (i) root growth rates and duration (which affect root length density [L_{vi}] and rooting volume), (ii) specific root water uptake efficiency (q_i), which includes unsaturated hydraulic conductivity of soil, and (iii) axial resistance to water flow in root xylem.

II. RATE AND DURATION OF ROOT GROWTH

Roots grow because new cells are formed and then expand in volume. This expansion occurs because of a net influx of water into the newly formed cells. The solute potential in the cells, ψ_π, creates a tendency for water to flow inward. The cell pressure potential (often called turgor pressure), ψ_{pc}, onto the cell walls and associated membranes causes a tendency for water to flow outward. Cells continue to grow only when the pressure potential is sufficient to cause cell walls to continue to strain. For roots growing in soil, there is an additional complicating factor. The force created by pressure potential in the root cells must not only be greater than the constraints imposed by the cell wall and its associated membranes, but also sufficient to overcome the additional constraint imposed by the surrounding soil matrix. A slow decrease in water content of soil containing the roots (the normal field condition) will cause: (i) the solute potential (ψ_π) in the roots to decrease (they will osmotically adjust; Sharp & Davies, 1979), (ii) cell wall extensibility to be reduced (Hsiao & Bradford, Chapter 6A, (iii) water uptake rate per unit length of root to be reduced (Taylor & Klepper, 1975), and (iv) mechanical constraint of the root by the surrounding soil matrix to be increased (Taylor & Ratliff, 1969a). The osmotic adjustment phenomenon will tend to cause the cells to continue to grow at their formerly well-watered rate but the other three phenomena will tend to cause reduced root growth rates.

Shoot and root development are closely coordinated as growth proceeds (Brouwer, 1965). If the environment is constant, a logarithmically linear relationship usually is found between the weights of shoots and roots during vegetative growth (Russell, 1977). The root system of a small young plant obviously occupies less soil volume than that of a large older plant if both are growing in uniform soil. However, environmental factors can cause fourfold differences in the ratio of leaf area (a transpiration demand factor) to root length (a water supply factor) (Eavis & Taylor, 1979). Shoot weight/root weight and leaf area/root length ratios also change as plants shift from vegetative to reproductive development (Russell, 1977).

Crop dry matter yield at a specific site usually increases linearly with total transpiration (de Wit, 1958; van Keulen, 1975; Tanner and Sinclair, Chapter 1A, this volume). This fact suggests that increasing the quantity of water available for transpiration will increase yield if water supply is limiting yield.

One method for increasing the supply of available water is through increasing the total soil volume occupied by the plant's roots. In terms of the model, this increased soil volume can be described either by an increased size of each V_i or by an increased number of soil volumes. With isolated plants—or with plants undergoing interspecific competition—the increased volume of soil occupied by roots can occur as a result of either a greater lateral spread or a greater rooting depth. Unfavorable soil conditions reduce the rate of root elongation in both directions and thus often reduce the amount of water available for transpiration. In field crops, however, the rate of lateral spread is often unimportant because the roots from one row of plants will reach those of the adjacent rows quite early in the growing season. All of the available water is absorbed from between crop rows unless soil conditions are very unfavorable. Taylor (1980) has discussed the progress of a research program to increase soybean rooting depth in a deep fertile soil of Iowa.

A second method for increasing available water supply is through increasing root length density (L_v) in soil zones from which water is lost by evaporation or deep drainage or where water remains in the profile at the plant's physiological maturity. The increased rooting density sometimes occurs as a result of increased initiation of secondary roots, of increased elongation rate of the secondary roots, or reduced root shedding.

A third method for increasing total available water is through increasing the specific radial root uptake rate, q_i, in soil zones where water must be used rapidly by the plant or become lost. Soil management techniques that increase soil water content increase q_i (Taylor & Klepper, 1975). These soil management techniques are discussed by Unger and Stewart (Chapter 10A). In addition, q_i is altered by many plant-related mechanisms. These mechanisms are discussed more fully by Fiscus (Chapter 4C).

A fourth method for increasing total available water is through reducing axial resistance to water flow in the root xylem. This reduction in resistance allows more of the diurnal depression in leaf water potential to be transmitted to xylem of roots located deep within the profile. Thus, the difference in potential between soil and root xylem is increased. Because total uptake rate increases as this potential difference increases, less water will remain at physiological maturity when roots have continually extended into moist soil (Taylor & Klepper, 1978).

If the plant must grow and reach maturity on a limited water supply, low axial resistance may be a disadvantage. Passioura (1972) has postulated (and provided some evidence to support the hypothesis) that high axial resistance will ration a limited water supply and under some circumstances will allow cereal crops to produce a greater grain yield than if axial resistance was lower. The factors affecting axial resistance and methods for modifying it are discussed more fully by Klepper (Chapter 3B).

Rates of root initiation, growth, and shedding, as well as specific uptake rates of root segments, are all affected by soil environments. The effects of some of these variables on root characteristics and functions are discussed below.

III. SOIL ENVIRONMENTAL EFFECTS ON ROOT GROWTH AND FUNCTION

A. Aluminum Effects on Root Growth

Root growth is reduced in many crop plants when the roots encounter subsoil with a low pH. As soil pH decreases, both Al ion concentration and ion activity increase. This increase in Al reduces root elongation rates (Adams & Lund, 1966). For example, cotton root elongation rates decreased about 60% as the soil-solution Al activity increased from $0.10 \times 10^{-5} M$ to $1.0 \times 10^{-5} M$ (Fig. 1).

Aluminum toxicity also affects root morphology and function. Aluminum-injured roots are characteristically stubby and distorted with patches of brown cortical tissue. The roots generally lack fine branching

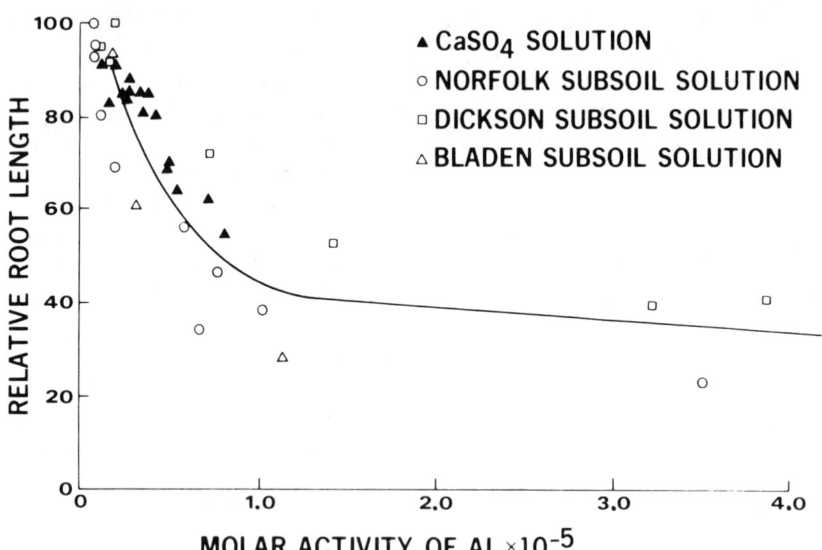

Fig. 1—Effect of molar activity of Al in subsoil solutions in situ and in subsurface nutrient solutions on cotton radicle growth. Redrawn from Adams and Lund (1966).

(Foy, 1974). Binucleate cells occur more frequently in Al-injured roots than in normal ones. Cotton root cells of the pericycle are damaged by exposure for 12 hours to 1 ppm Al concentration (Huck, 1972). Ion and water uptake rates (per unit of root tissue) are lower in Al-damaged than in normal cotton roots (Lance & Pearson, 1969). More complete discussions of Al toxicity effects on root growth are given in Adams (1974), Foy (1974), and Blue and Dantzman (1977).

Roots of some crop species are more tolerant than others to high concentrations of Al. Adams and Pearson (1970) found that an Al activity of 33.3 μM reduced length of cotton roots grown for 48 hours by 66%, whereas it reduced the length of peanut roots by 11%. Brenes and Pearson (1973) showed that the acidity of an Oxisol and an Ultisol reduced root growth of corn and sorghum but that a soil pH as low as 4.0 did not restrict root growth of stargrass. Brenes and Pearson (1973) showed that Al toxicity occurred at a much greater concentration of Al ions in the Oxisols and Ultisols of Puerto Rico than in the soils of the southeastern United States.

One method for increasing the utilization of water stored in a subsoil with a high Al activity is to increase subsoil pH. Subsoil pH can be increased by mechanically mixing lime into the subsoil (Pearson et al., 1973), by applying high rates of a neutral N salt (such as $Ca(NO_3)_2$) to an acid-tolerant crop such as coastal bermudagrass) (Adams & Pearson, 1969), and by applying high rates of animal manure (Doss et al., 1976; Lund, 1978). Soil management techniques to increase subsoil pH are covered more extensively by Reicosky (Chapter 10C).

A second method that may increase utilization of water from low-pH soils is switching to crops that are more tolerant to Al. Examples are changing from cotton to peanuts (Adams & Pearson, 1970) or from sorghum to corn (Brenes & Pearson, 1973).

A third method for increasing the utilization of water in acid subsoils (without switching crops) is to grow Al-tolerant cultivars. Differential tolerance of cultivars to Al toxicity has been shown for barley (Reid, 1976), wheat (Neenan, 1960), rye (Mugwira et al., 1976), soybeans (Devine, 1976), alfalfa (Devine et al., 1976), rice (Howeler & Cadavid, 1976), sweet potato (Munn & McCollum, 1976), and corn (Lutz et al., 1971).

Several different crop species thus exhibit differential tolerances to Al toxicity. We should be able to increase yields through breeding and selection of Al-tolerant cultivars for most crops when they are grown on soils where deep percolation occurs annually and where the subsoil is sufficiently acid to cause Al toxicity problems.

B. Root Growth and Function as Affected by Toxic Concentrations of Other Metals

Manganese, iron, copper, nickel, and zinc ions also are phytotoxic, but their effects on root growth of field crops are either less than on top growth or not well-catalogued (Foy et al., 1978). Toxic concentrations of these ions reduce general growth of the plants and thus reduce leaf area development and water use. This reduction in water use in turn increases the probability that a greater proportion of the annual precipitation will percolate below the root zone, which also may be less deep because of the toxic metal concentrations.

Differential tolerances among plant ecotypes are known for each of these ions (Foy et al., 1978). This fact increases the likelihood that a plant selection program can enhance efficiency of water use on soils that contain phytotoxic quantities of these elements; however, much more information is needed about the physiology of metal toxicity in order to determine effective and rapid selection criteria.

Changes in soil pH usually alter toxicities of these metals. The chemistry causing these altered toxicities is complex and outside the scope of this chapter.

C. Soil Salinity Effects on Root Growth and Function

Saline soils exist in many parts of the world. They cause both direct and indirect effects on root growth and function. First, salts reduce the total water potential of the soil solution, which is one of the factors that control water uptake. If plant roots do not absorb sufficient salt to adjust their osmotic potentials to that of the soil solution, water availability is reduced.

Second, increases in soil salinity decrease the rate of root elongation. This tendency is illustrated by the unpublished cotton radicle elongation data of J. E. Box, Jr. and H. M. Taylor (Fig. 2). These data were obtained using glass-fronted boxes (Pearson et al., 1970) placed on an 18° tilt from the vertical in a lighted growth room maintained at 28 ± 1°C. Bulk oven-dry soil from the plow layer of a Dothan loamy sand (Plinthic Paleudults) was treated with 7% by weight NaCl solutions of 0, −5, −10, −15, −20, and −25 bars osmotic potential. This created total water potentials of −1.0, −6.8, −12.1, −17.4, −22.8, and −28.4 bars, respectively, as measured with a thermocouple psychrometer. A 36-cm layer of treated soil was placed in a compartment of the glass-fronted box and covered by a 1-cm layer of the −1.0-bar total potential soil. Cotton ('Empire') seeds were pregerminated for 10 hours in distilled water. Five seeds were inserted, with the radicle downward, into the upper 1-cm soil layer. The compartment was then covered with clear polyethylene film. Radicle tip position was marked as a function of time. By the end of 5 days, the radicles averaged 29.5 cm deep in the −1.0-bar soil and 8.0 cm in the −28.4-bar soil. In all cases, the radicles continued to extend downward into the treated soil throughout the 5-day period, but they grew fastest in the highest potential soil.

Third, increases in salt concentration reduce permeability for water and ions (Waisel, 1972) and these in turn affect the specific root water uptake rate, q_i. Details are provided by Fiscus (Chapter 4C this volume).

Fourth, ionic composition of the saline solution affects the general nutritional status of the plant. Nutritional status affects leaf area development, stomatal density and responsiveness, thickness of cuticular

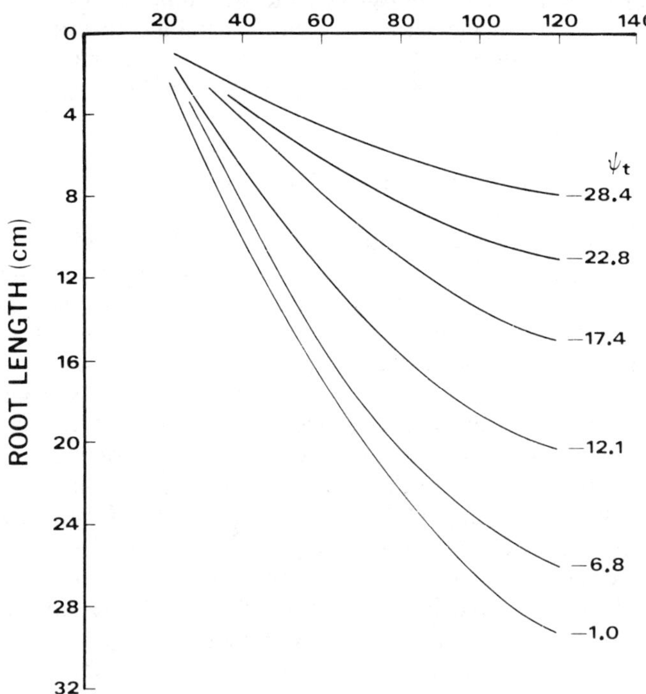

Fig. 2—Effects of various total soil water potentials (ψ_t) on cotton radicle lengths during a 120-h growth period. Different NaCl solutions caused the different total soil water potentials. Root lengths were measured from the interface of a nonsaline upper soil and a saline subsoil. (J. E. Box, Jr., and H. M. Taylor, unpublished data.)

wax, leaf thickness, and other aboveground plant characteristics that alter transpiration rates and tissue water storage (Poljakoff-Mayber, 1975).

Fifth, salinity levels affect the width of Casparian strips (Ginzburg, 1964,[2] as quoted by Waisel, 1972). In halophytes, the Casparian strips cover almost all of the endodermal end walls but cover a much smaller fraction of the end walls in nonhalophytes. Presumably, this difference in endodermal anatomy affects specific root water uptake rate, q_i.

Sixth, increases in salinity may upset hormonal balances, which control many plant responses such as root growth rate, membrane permeability to water and ions, and transpiration rates.

Finally, increases in soil pH due to alkalinity may affect soil structure, altering strength, aeration, and hydraulic and ionic conductivities of soils. Methods of managing saline soils for most efficient use of water are covered by Hoffman and van Genuchten (Chapter 2C).

Plant tolerance to high levels of salinity is reduced by high temperature, strong light, low atmospheric humidity, and low amounts of O_2 in the soil solution (Nieman & Shannon, 1976). There is substantial

[2] C. Ginzburg, 1964. Ph.D. Thesis. Hebrew University, Jerusalem.

genetic variability in whole plant tolerances, and presumably also in root tolerances, to high levels of salinity (Venables & Wilkins, 1978). Some progress has been made in increasing tolerance to salinity through selection programs (Epstein, 1976). The exact mechanisms through which this additional tolerance was achieved are still somewhat in doubt.

D. Phytotoxins and Root Growth

Many biochemical compounds produced by higher plants and by microorganisms affect root development, including root elongation. Scopoletin, an alkaloid obtained from various members of the nightshade family, inhibits root growth (Avers & Goodwin, 1956) and *Agropyron repens* excretes a compound that is toxic to corn roots (Bandeen & Buckholtz, 1967). A compound in extracts of alfalfa, probably saponin, reduces elongation of cotton radicles (Megie et al., 1967). Several other examples of phytotoxins, either exuded from living roots or present in decomposing plant residues, are discussed by Patrick and Toussoun (1965) and by McCalla and Norstadt (1974). In addition, phytotoxins produced by plant parasites are reviewed by Strobel (1974).

Root exudates sometimes benefit root growth, either as direct stimulants or as protectants against pest attack. Flax exudes HCN, which protects its roots against some pathogenic fungi (Timonin, 1941). African marigolds exude a polythienyl that is toxic to several nematode species (Slootweg, 1956) and asparagus exudes a compound toxic to the stubby root nematode (Rohde & Jenkins, 1958).

This brief discussion points out that many phytotoxins exist in soil. Some of their effects on root growth and function are documented, but great gaps exist in our knowledge about interactions among phytotoxins and soil physical and chemical effects on root function (Newman, 1978). At the present time, it is impossible to consistently control the toxins in such a way as to improve the efficiency of water use by crop plants. There is hope, however, that genotypes can be found that will be tolerant to specific phytotoxins. McCalla and Norstadt (1974) discuss such a possibility for patulin, a phytotoxin prevalent in mulch-tilled wheat lands.

E. Pesticide Effects on Root Growth and Function

Ideally, any pesticide controls its intended target without affecting nonpests. Practically, pesticides usually have some effect on nontargeted species. Ashton and Crafts (1973) discussed 16 classes of herbicides and indicated that 10 of them contained one or more compounds that affected plant roots (Table 1). Some herbicides affected geotropic response, whereas others affected lateral root initiation, cell wall formation, mitosis, protein synthesis, cell wall lignification, and integrity of cellular membranes.

Trifluralin (α,α,α-trifluro-2,6-dinitro-*N,N*-dipropyl-*p*-toluidine), an herbicide, also reduces nodulation of soybeans. Kust and Struckmeyer

Table 1—Classes of herbicide chemistry and those known to directly affect plant roots (from Ashton & Crafts, 1973).

1. Aliphatics	5. Bipyridiliums	9. Phenols	13. Triazoles
2. Amides†	6. Carbamates†	10. Phenoxys†	14. Uracils†
3. Arsenicals	7. Dinitroanilines†	11. Thiocarbamates†	15. Ureas
4. Benzoics†	8. Nitriles†	12. Triazines†	16. Unclassified†

† Herbicides known to directly affect plant roots.

Table 2—Effect of trifluralin on dry weight of roots and nodulation of soybeans (from Kust & Struckmeyer, 1971).

Trifluralin rate	Root weight	Number of nodules
kg ha^{-1}	g	
0	0.19 a*	25.0 a
0.36	0.20 a	10.0 b
0.74	0.14 b	0.5 c
1.10	0.13 b	0.0 c

* Column values followed by the same letter do not differ significantly at the 5% level according to Duncan's Multiple Range Test.

(1971) showed that a trifluralin application rate of 0.74 kg ha^{-1} to the upper 5 cm of soil reduced root weight from 0.19 g in the control to 0.14 g for the treated plants and reduced the number of soybean nodules from 25 per plant in the control to 0.5 per treated plant (Table 2). A more detailed description of herbicide effects on roots is given by Currey and Teem (1977).

Fungicides seem to affect root growth and function much less than herbicides but may combine with herbicides in synergistic effects (Kappelman et al., 1971).

It seems evident that, at the present time, pesticides must be individually evaluated for their direct effects on root elongation and water absorption by root systems.

F. Effects of Aqueous Ammonia and Ammonium Ion on Root Growth

Much of the N fertilizer is applied in the form of ammonical N. The phytotoxicity of this ammonical-N has been known for many years (Bennett, 1974), but sometimes the direct effects of this N form on root growth have not been understood.

Bennett and Adams (1970) reported that cotton radicle elongation rates were reduced when aqueous NH_3 concentrations were >0.17–0.2 mM in soil solutions and 0.16–0.22 mM in nutrient solutions. Bennett (1974) suggested that lack of charge and the reaction of alkali-like ions with fat would facilitate NH_3 entry through cell membranes and thus would affect many biochemical reactions in plants.

Aqueous NH_3 not only is present when ammonical-N fertilizers are used but also may occur when fresh organic residues containing considerable protein are added to the soil. Megie et al. (1967) found that un-ionized NH_3 concentrations >10 ppm were lethal to cotton roots (Table 3). The

Table 3—Effect of a layer of decomposing alfalfa residue on soil pH, NH_4-N content, NH_3 content, and cotton seedling root growth at different distances from the residue (from Megie et al., 1967).

Distance from residue layer	Soil reaction	NH_4-N	Calculated NH_3	Root growth
cm	pH	ppm	ppm	
0	9.1	470	470	Seed did not germinate.
3	8.8	321	123	Seed cracked and radicle died.
6	8.2	212	29	radicle emerged but died.
9	6.1	85	0.4	Radicle and secondary roots appeared normal.
12	6.0	28	0.1	Radicle and secondary roots appeared normal.

NH_3 concentrations increased linearly with the N content of organic material added to the test soil (Megie et al., 1967).

Phytotoxic effects of aqueous NH_3 and NH_4 ions can best be alleviated by judicious management of fertilizer and organic matter applications. Usually, toxic concentrations of NH_3 will be reduced below toxic levels within 2 or 3 weeks after application.

G. Soil Strength and Root Growth

Excessive soil strength, sometimes called physical impedance or mechanical impedance, can severely distort normal root growth patterns. This excessive soil strength can be caused by a high soil bulk density, high soil cohesion, and drying of a consolidated or cohesive soil mass. The excessive soil strength sometimes occurs as a result of natural conditions but often is accelerated by human activities, such as by plowing or soil compaction with tractor tires.

Excessive soil strength reduces the likelihood that plant roots will enter a specific soil mass. Ninety-five percent of cotton radicles entered and penetrated through a 2.5-cm layer of soil at 5 bars but none penetrated at 30 bars soil strength, as measured with a specific static-tip penetrometer (Taylor & Gardner, 1963; Taylor et al., 1966). Even where soil strengths are great, roots can enter the soil volume if soil pores are larger in diameter than the root tips (Wiersum, 1957). Root tips differ considerably in diameter (Cannell, 1977). If the pores are too small to be entered by the main root axes but laterals are not restricted, the laterals proliferate and a much-branched root system is the result (Russell & Goss, 1974). When the pore size prevents laterals from entering, their growth also is restricted and the entire root system is stunted.

Root elongation rates, of both main axes and laterals, decrease as soil strengths increase. One indication of this fact is shown by the data of Russell and Goss (1974). At a low applied pressure (low-strength) barley roots were long and slender after 6 days growth, but at a 0.5-bar applied pressure (high-strength) the roots were short and stubby (Fig. 3).

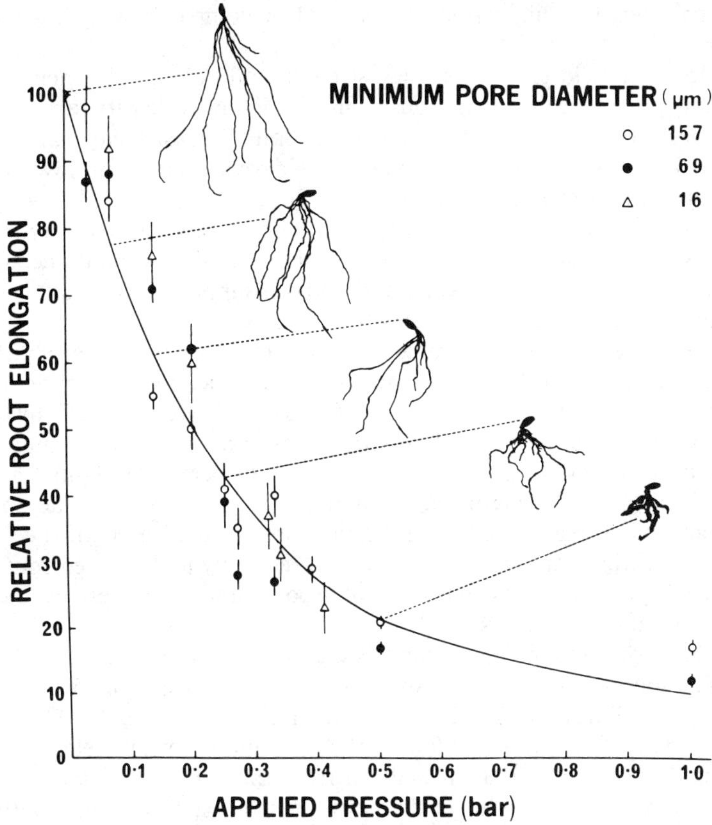

Fig. 3—Effect of applied pressure on rate of elongation of seminal axes of barley plants grown for 6 days in ballotini beads. Redrawn from Russell and Goss (1974).

Elongation rates of cotton and peanut main root axes progressively decreased as soil strength increased. At any specific quantitative value, peanut root elongation rates were decreased proportionately less than those of cotton (Taylor & Ratliff, 1969a). Russell and Goss (1974) also showed that crop species reacted differently in their elongation rates at specific soil strengths.

Greacen and Oh (1972) hypothesized that osmotic regulation in the elongating cells of the roots they studied could account for the differences among plants in their reactions to mechanical constraint. If this postulate is correct, we should expect different species to exert different maximum root growth pressures. Maximum root growth pressures indeed are somewhat different among crop species (Pfeffer, 1893; Stolzy & Barley, 1968; Eavis et al., 1969; Taylor & Ratliff, 1969b). Russell and Goss (1974) have questioned the Greacen and Oh postulate by pointing out that root elongation rates at specific soil constraints could differ among plant species if root hormones altered the cell wall constraints upon cell enlargement. It seems likely that both osmotic regulation and hormonal relaxation of cell wall constraints will be altered in high-strength soils, when compared with low-strength ones.

In most field situations, high-strength soil volumes are neither horizontally nor vertically continuous. Pans of high bulk density (and strength) may occur between layers of low-strength soil. Cracks, shear planes, worm holes, and root channels in some cases provide paths of low strength through otherwise high-strength layers (Sutton, 1969; Russell, 1977). The mere presence of a root-restricting pan will not necessarily affect crop yield. Effects of soil pans and other high-strength zones always should be analyzed in terms of their effects on water and nutrient supply, anchorage, and other factors that directly affect top growth.

Perennial plants act similarly to annual ones during establishment (Barton et al., 1966). However, once established, perennial crops are less affected than annual crops by high-strength layers (Fryrear & McCully, 1972). Part of the reason for the different reactions of annual and perennial crops to high soil strengths may be that some roots of perennial plants finally do penetrate the high-strength soil volumes that annual plant roots do not have enough time to penetrate before their life span is over. Part of the reason, however, is that perennial crop roots can reoccupy undestroyed root channels from previous years, but these old root channels are destroyed during tillage for annual crops (Russell, 1977).

Some of these old root channels are quite persistent. Lund and Elkins (1978) showed that cotton yields were increased for at least 5 years when bahiagrass ('Pensacola') sod was plowed after 3 years' growth. The average cotton yield after sod was 179% of the continuous cotton yield. Lund and Elkins (1978) showed that the subsoil from the bahiagrass plots had eight times more pores >1 mm in diameter than did subsoil from continuous cotton plots.

Several soil management techniques are available to reduce the strength of a soil layer and to reduce the effects of high-strength layers on crop yield. These are discussed by Unger and Stewart in Chapter 10A. Also, it may be possible to select for cultivars with greater than normal ability to penetrate high-strength layers. Charles B. Elkins (Auburn, Ala., personal communication) has shown that Auburn 7 genotype of *Festuca arundinacea* Schreb. has a greater proportion of large roots (> 3.4 mm in diameter) than Auburn 81 genotype, and that these large roots were more likely than small roots to penetrate "compacted" soil. In cases where high Al activity, excessive CO_2, low soil O_2, or low soil temperatures interact with excessive soil strength to reduce root penetration (Tackett & Pearson, 1964a, 1964b; Pearson et al., 1970), genetic manipulation to increase root elongation rates seems more profitable than where high soil strength alone is the controlling variable.

H. Soil Aeration and Root Growth

Optimum root growth occurs only when the soil is well-aerated so that the supply of O_2 is sufficient to meet the demands of not only the plant roots but also the soil organisms. In well-aerated soils, O_2 used by roots and soil organisms is replaced by diffusion through continuous air-filled pores (Grable, 1971). Mechanisms for gaseous transport have been reviewed by Grable (1971), Currie (1970), and Greenwood (1975).

Anaerobic conditions occur when the demand for O_2 exceeds the rate at which it is supplied to a specific soil volume. Anaerobic conditions can occur within soil structural units even though much of the soil profile is well-aerated (Currie, 1961). Anaerobiosis is more likely to occur when air-filled porosity is less than about 10%, but this value depends on soil pore arrangement and continuity (Cannell, 1977).

Root growth requires a continuous supply of O_2 for respiration. If the supply is cut off, cotton radicle elongation stops within a few minutes (Huck, 1970). If very low O_2 partial pressures (0.01 bar or less) are continued for a few hours, the root system will die (Williamson & Splinter, 1969), but Greenwood (1971), who surveyed 51 data sets, concluded that root elongation was 70% of maximum even when O_2 partial pressures were as low as 0.06 bar.

Anaerobic conditions around a root also decrease its permeability to water (Glinka & Reinhold, 1962), affecting the specific root water uptake rate, q_i.

The anaerobiosis literature must be interpreted very carefully. Root elongation is reduced much more by a specific low-O_2 treatment if the entire plant is treated than if only the root system is treated (Eavis et al., 1971) (Fig. 4). This difference occurs between O_2 moves from leaves to roots through specialized structures called aerenchyma if the leaves are located in ambient air. Roots of most crop plants normally do not grow much below the water table; therefore, it is self-evident that internal transport of O_2 usually does not meet the entire root system demand. Waterlogging effects on root growth and function have been discussed in greater detail by Allen (1977).

Soil volumes that are low in O_2 may contain concentrations of ethylene that inhibit root elongation (Smith & Russell, 1969; Smith & Robertson, 1971), but root growth is later resumed in an ethylene-free environment although root branching may be modified (Crossett & Campbell, 1975). The relative importance of modified ethylene and hormonal concentrations on root growth in field soils has not been clarified sufficiently. A further discussion of effects of root hormones, including ethylene, on root growth is given by Torrey (1976).

Draining excessively wet soils and tilling excessively compact soil layers seem to offer the easiest solutions to low-O_2 and high-ethylene levels. These soil management techniques are discussed by Unger and Stewart (Chapter 10A). In some cases, one can shift from a sensitive crop to one that tolerates flooded soil conditions. Selecting genotypes of a species for their tolerances to low O_2 levels by obtaining energy in anaerobic respiration or by evading the problem through development of aerenchyma also may prove successful, but such efforts are not yet widespread enough to evaluate for worldwide significance.

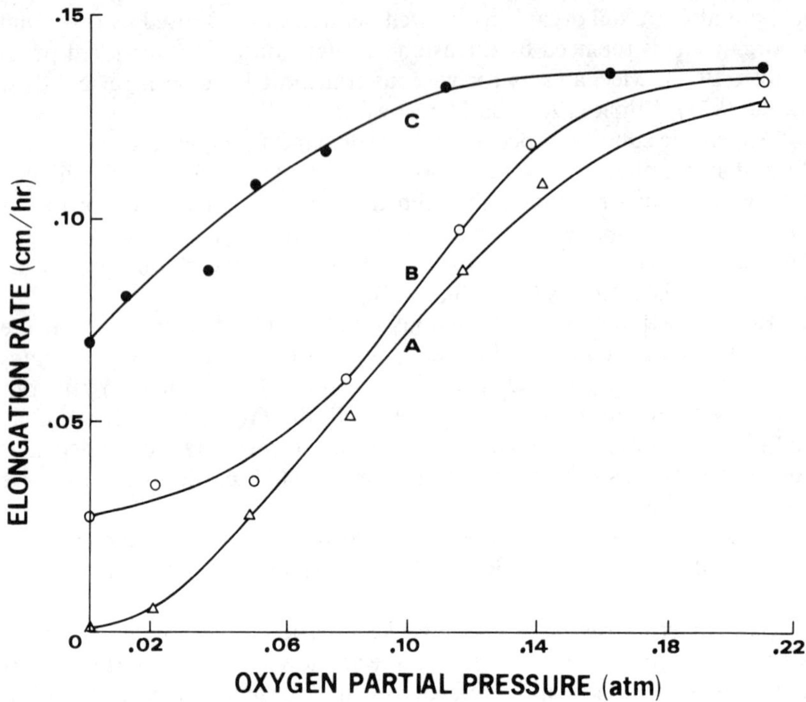

Fig. 4—Effect of varying O_2 partial pressure on pea seedling elongation rates when roots and shoots were treated separately: (*A*) the entire seedling was enclosed in the same partial pressure; (*B*) cotyledon and shoots were at 0.21 atm. partial pressure O_2 while roots were at different partial pressures; and (*C*) roots were in soil air with 0.21 atm partial pressure O_2 while cotyledons and shoots were at different partial pressures. Redrawn from Eavis et al., 1971.

I. Soil Temperature and Root Growth

Although not subject to such extreme diurnal fluctuations, root-zone soil temperatures tend to follow seasonal trends in air temperatures. As a result, the integrated effects of temperature on plant performance are complex. Temperature maxima for root growth and shoot growth differ (Russell, 1977), so root-shoot ratios are affected by temperature, with the lowest root-shoot ratios occurring near the temperature for maximum shoot growth. Temperature maxima differ among plant species (Richards et al., 1952).

Cotton radicles sometimes can grow as deep as 20 cm before the cotyledons emerge; therefore cotton radicle growth sometimes is used as a simplified system hopefully to predict effects of soil factors on root growth in general. Arndt (1945) showed that cotton radicle elongation increased linearly with constant temperatures between 18 and 33°C but then decreased rapidly, with elongation rates at 18 and 39°C being about equal. Temperature and time interact as they affect cotton radicle elongation. Larry F. Ratliff (quoted by Taylor et al., 1972) pregerminated cotton seeds until the radicle had just emerged from the seedcoat, then transplanted the seeds into soil at

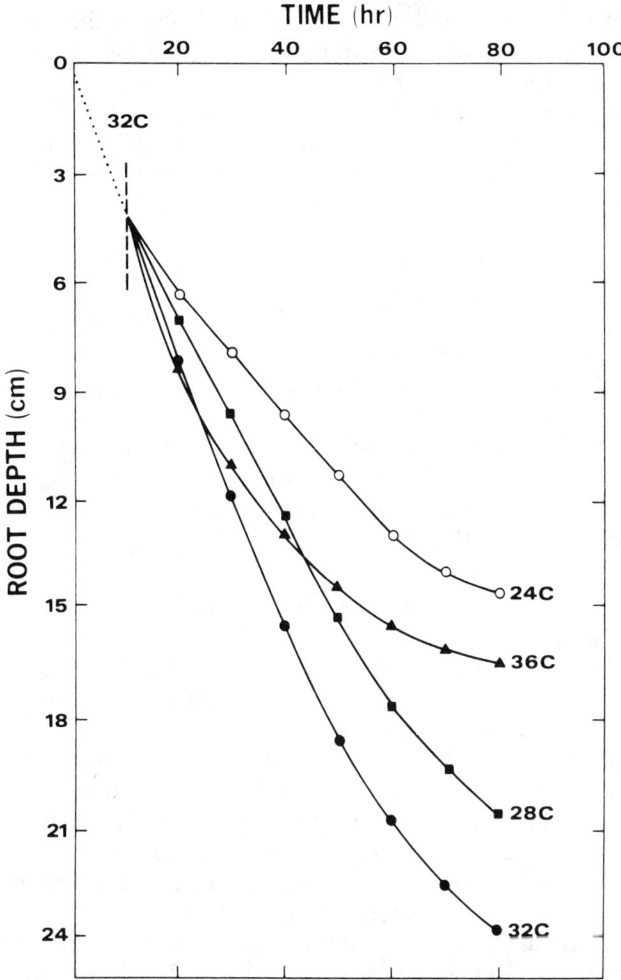

Fig. 5—Cotton radicle tip location during the first 70 h after differential soil temperatures were applied at 10 h. Each data point represents an arithmetic mean of 16 plants. Redrawn from Taylor et al. (1972).

32°C. After 10 h at 32°C, soil temperatures of 24°, 28°, 32°, or 36°C were imposed for an additional 70 h. Radicles at 36°C initially grew fastest but by 40 h those at 32° were elongating the most rapidly. During the 70-hour treatment period radicle growth was 10, 17, 20, and 12 cm at 24, 28, 32, and 36°C, respectively (Fig. 5). These different elongation rates should affect seasonal rooting depth, but little is known about rates of rooting-depth increases as functions of temperature deep within the soil profile.

Soil temperatures also affect anatomical and morphological characteristics of root systems. Brouwer and Hoogland (1964) found temperature effects on (i) the distance from the root apex to the first microscopically visible branch roots, (ii) the branching pattern of lateral roots, and (iii) the number of suberized endodermal cells within 20 cm of the root apex. The first two chracteristics affect rooting density (L_v), while the third probably affects specific root uptake rate (q_i).

The angle at which roots of some species grow from the horizontal differs with soil temperature (Mosher & Miller, 1972). Onderdonk and Ketcheson (1973) showed that corn roots grew most horizontally (10° from horizontal) at a constant 17°C. Sheppard and Miller (1977) showed that the geotropic response of corn radicles was not dictated by cyclic temperatures per se but was regulated by the time distribution of cyclic temperature changes. Sheppard and Miller (1977) also showed that the geotropic response varied with distance from the seed, probably as a result of differences in auxin transport from the seed (Greenwood et al., 1973). When all other conditions are the same, differences in geotropic response affect both rooting depth and rooting density (L_v).

The uptake rate of water per unit root length (q_i) increases with soil temperature. This increase not only occurs when soil temperature is increased for a short time during an experiment, but q_i also is affected by the temperature history of the roots (Cooper, 1973).

J. Soil Water and Root Growth

This section primarily concerns the effects of deficient soil water contents on root growth and function. Excessive soil water contents are covered in the section on soil aeration and root growth.

Much of the weight of a living root is water. Gardner and Danielson (1964) showed that roots grown at -0.05 bars soil water potential contained about 15.5 g water g^{-1} dry weight, whereas at -10 bars the roots contained about 6 g water g^{-1} dry weight. Roots must absorb much more water than they retain because some water is required in formation of new plant mass and large quantities are required to offset transpiration. Because water is required for maintenance of turgor (which in turn is required for cell elongation), one immediately suspects that root elongation rates are reduced with reduced plant water potentials. In fact, this conclusion often if drawn from experiments that show reduced elongation rates with lowered soil water contents. There is some evidence, however, that neither plant water potential nor soil water potential per se affects taproot elongation very much as long as both are above -10 to -15 bars. Considerable controversy still exists in the literature about the direct role of soil water at potentials > -10 bars on root growth.

Part of this controversy arises because changes in soil water content affect so many other soil and plant factors. Examples are how water affects soil strength, aeration, temperature, salinity, and toxic concentrations of Al. These factors in turn affect root growth.

Part of the controversy also exists because investigators have used different measures of "root growth." The data of Gardner and Danielson (1964) and Cole and Alston (1974) show that wet weights are affected much more than root lengths as soil water contents decrease. In fact, Taylor and

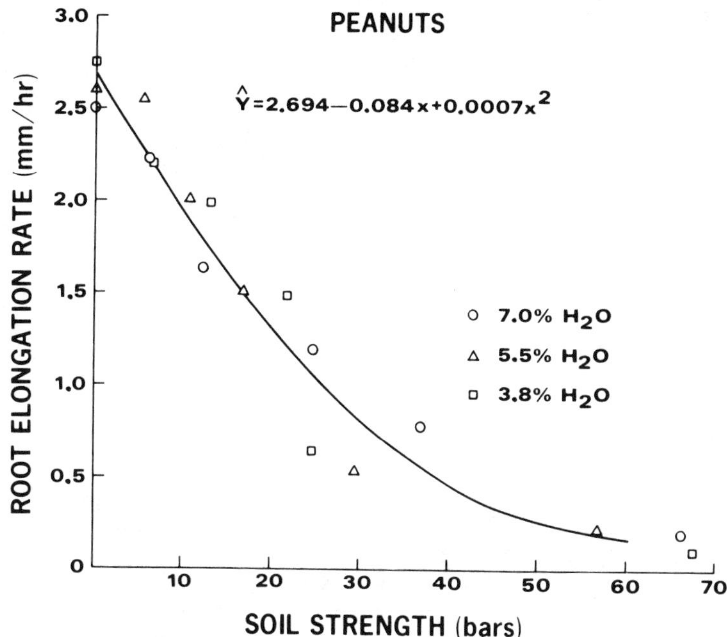

Fig. 6—Effect of soil strength (penetrometer resistance) and soil water content on peanut root elongation for the period 40–80 h after transplanting into soil. At 7.0% H_2O, total potential was −0.19 bars and at 3.8% H_2O, total potential was −12.5 bars. Redrawn from Taylor and Ratliff (1969b).

Ratliff (1969b) found that peanut root volumes or root weights progressively decreased with soil water potentials between −0.19 and −12.5 bars but that elongation rates were not affected in this range of water potentials as long as soil strength effects were considered (Fig. 6).

Soils obviously can become too dry for root elongation to be maintained at its maximum rate, but some root growth of corn and tomatoes can occur at −40 bars total soil water potential and root tips can remain alive even at much lower potentials than that (Portas & Taylor, 1976). Roots extending into soils at potentials below −5 bars usually intercept only slight amounts of water. However, worm holes, root channels, and soil cracks where soil strength is low may provide pathways for roots to grow through soil layers with low potentials into wetter layers where root interception again becomes important.

Low soil water contents not only reduce root elongation rates but also may accelerate root shedding. Some root shedding occurs at all times during the plant's growth cycle, but shedding usually accelerates during periods of rapid reproductive development. Root shedding in a specific soil volume also accelerates when soil water contents decrease there. As an example, rooting density of cotton decreased whenever soil water content decreased to about 7% by volume, or to a total soil water potential of between −1 and −2 bars. Roots at each of four depths, from 30–150 cm, followed this same pattern in this loose sandy soil (Fig. 7).

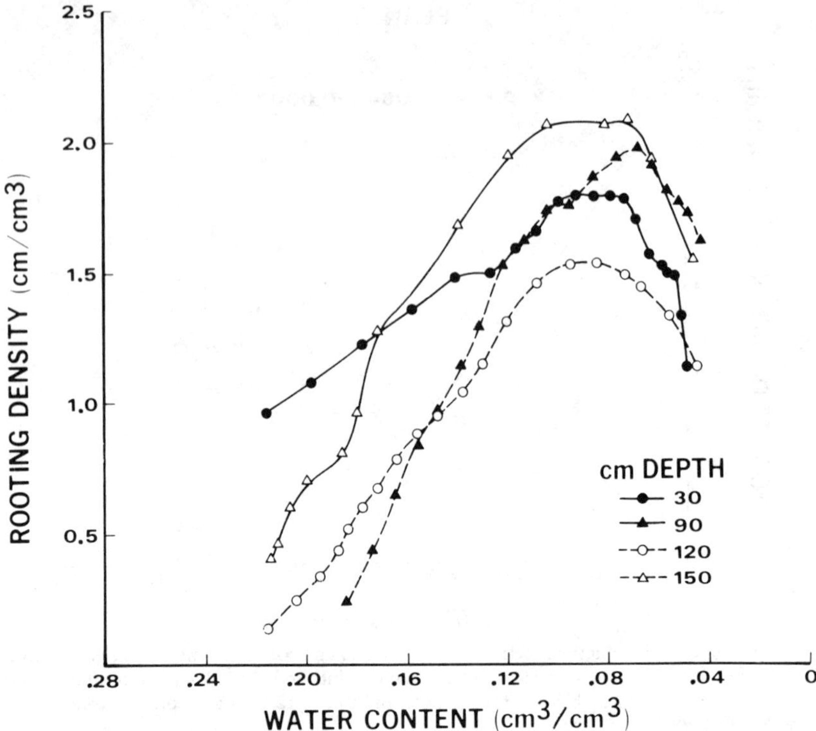

Fig. 7—Cotton rooting density as a function of soil water content at four depths in a drying profile treatment at the Auburn, Ala., rhizotron. Redrawn from Taylor and Klepper (1974).

Roots grow fastest where water contents are high (without causing inadequate aeration) and soil strengths are lowest. Root shedding also is lowest when water contents are highest. Because of these two facts, plant roots are concentrated in soil zones wetted by high-frequency irrigations, either from tricklers (Bar-Yosef, 1977) or from being furrow-irrigated on one side of the plants (Portas, 1970,[3] 1973).

Lowered soil water contents reduce root elongation rates (which may reduce rooting depths), increase root shedding (which reduces L_v), decrease soil hydraulic conductivity (which reduces q_i), reduce root diameter (Cole & Alston, 1974) thus reducing root-soil contact (which reduces q_i), reduce the water potential of the soil surrounding a specific root (which decreases the difference in water potential between soil and root xylem), and probably increase root suberization (which reduces q_i).

The most obvious palliatives for these conditions are either to increase soil water contents through soil management techniques outlined by Unger and Stewart (Chapter 10A) or to reduce water uptake rates by reducing transpiration rates through crop management techniques outlined by Loomis (Chapter 8A).

[3] C. A. M. Portas, 1970. Ph.D. Dissertation. University of Luanda, Angola (in Portuguese, with English and French summaries).

K. Nutrient Supply and Root Growth

Nutrients supplied to deficient plants increase growth of both tops and roots if other growth conditions are adequate. The nutrient supply affects the form of the root system which in turn affects the nutrient supply. Extension rate of the seminal axis of barley was not affected by the supplies of N, P, and K. The total length of laterals per centimeter of axis increased when levels of N and P increased around sections of the seminal roots but not when K increased (Drew, 1975). Increased root length density increases water uptake rates of that specific soil volume. The increased uptake rate will soon dry the soil sufficiently to increase root death in that soil volume (Taylor & Klepper, 1974); therefore practical significance of the increased root length density due to localized concentrations of N and P has not been clarified.

There are specific ion effects on root growth per se. For example, a deficiency of Ca ions reduces the growth of root tips, root laterals, and root hairs (Jackson, 1967). Soybean taproot elongation rates were 0.57, 2.29, and 2.90 mm $hour^{-1}$ at Ca concentrations of 0, 0.05, and 0.25 ppm, respectively, around the root tips (Lund, 1970). Lateral roots were more sensitive than taproots to lack of Ca (Lund, 1970).

Boron also is required around the root tip for elongation to proceed at normal rates in some crops. Elongation rates of tomato (Albert & Wilson, 1961) and field bean (Neales, 1960) roots are sensitive to B deficiency, but maize is relatively insensitive. Boron deficiency seems to cause supraoptimal levels of indoleacetic acid (Bohnsack & Albert, 1977).

Soluble compounds containing Ca or B are readily available. In most cases, fertilizing the soil is the easiest and most economical way of increasing root growth in subsoils low in Ca or B. However, the possibility also exists of selecting plants that are more efficient in extracting Ca or B in subsoils with supplies of these ions.

L. Interactions Among Soil Factors and Root Growth

There are major interactions among the various soil factors in their effects both on elongation rates of main roots (and therefore rooting depth) and on growth and development of laterals (and therefore rooting density). If an increase in soil strength or a decrease in soil temperature each causes a 50% reduction in root elongation from that of the maximum, some elongation still occurs. Pearson et al. (1970) found that cotton taproot length 80 h after planting was about 24 cm when soil strength was 0.05 bar, soil temperature was 32°C, and soil pH was 6.2 (zero Al activity). The root length was 12 cm at a soil temperature slightly below 24°C at a soil strength of about 10 bars or a pH slightly above 5.2 (Al activity about 2.5 μM). Root length was only slightly affected, however, when soil temperatures were changed from 24 to 28 to 32°C at soil strengths of 10 bars or soil pH of 5.2. Many other such interactions occur.

These types of interactions are especially important for water, the subject of this book. Water interacts in one way or another with every soil factor discussed in this chapter. These interactions make our task of predicting the effects on crop yield of various soil and plant modifications exceedingly difficult, if not impossible, at the present time.

IV. MANAGEMENT STRATEGIES FOR ROOT SYSTEMS

There are essentially three strategies for managing root systems to increase the efficient use of water. First, the total annual supply of water can be increased by manipulating the rooting pattern. This strategy will be most effective when the soil profile is fully recharged before each growing season. Second, management systems can be devised for rationing the available water supply during the growing season. This strategy is most likely to benefit yield where crops currently extract the total annual water supply before crop maturity. Third, uptake of nutrients can be increased, thus shifting the relationships between evaporation of water from soil and from leaf surfaces, especially in dryland areas.

Each local situation must be examined in detail to determine the root system modification strategy that has the greatest probability of success or, indeed, to determine if some management strategy not involving roots per se might be a more logical strategy.

A. Manipulating the Rooting Volume

1. INCREASING TOTAL ROOTED VOLUME

One way that the annual plant water supply can be increased is to increase the volume of soil occupied by roots. There are three methods of doing this. First, the soil profile can be modified to increase root extension rates. This chapter has discussed several soil factors—biological, chemical, and physical—that affect root elongation and thus the soil volume explored by roots. This volume can be increased if the soil factor dominating root extension is modified to increase the soil's desirability as a rooting medium. The amount of increase in the total annual supply of water depends on many soil, plant, and climatic factors, but can be substantial. In the southeastern United States, for example, cotton roots sometimes do not penetrate compacted layers that occur 20–25 cm below the soil surface. The cotton plants thus can extract only the water stored in the upper 20- to 25-cm layer through the periodic rainfall plus a small amount of water moved upward by capillarity during dry periods. Cotton plants often start midday wilting 4 or 5 days after a rain. In contrast, tilling the soil to disrupt the compact layer allows cotton rooting to a meter or more. This increased rooting depth increases the extractable water by about 6 cm, enough water to delay onset of water stress for another 10 days at an evapotranspiration rate of 0.6 cm day^{-1}. Increases of this amount probably could be attained on 5% of the cultivated area of Alabama, as one example.

Second, the plant species can be modified through breeding or selecting individuals that are more tolerant of the specific soil condition that limits root extension. The most notable example is the success in screening plants for Al tolerance (Devine, 1976). In soils where toxicity levels are not lethal but still reduce root growth greatly, it seems likely that the extractable water supply can be increased by amounts up to 10 cm of water. Areal extent of soils that fall in the exact category postulated here is unknown.

Third, the species to be planted can be shifted from one with roots that are particularly sensitive to a root-inhibiting soil condition to one with roots that are tolerant. The shift from cotton to peanuts as subsoil acidity increases (Adams & Pearson, 1970) takes advantage of the greater tolerance of peanut roots to high levels of Al activity. Historically large land areas already have made this shift when pine trees replaced cotton in the southeastern United States.

2. INCREASING WATER EXTRACTION FROM SOIL DEPTHS USUALLY PERMEATED BY ROOTS

Water evaporates from the upper few centimeters of soil quite rapidly after a rain when the crop canopy is incomplete. Under these conditions, water used from this layer for transpiration reduces losses from direct evaporation. Water can be used more rapidly from this layer as rooting density increases, as radial root resistance decreases, or as the roots located there are more evenly scattered through the layer rather than concentrated along ped surfaces. Any of the transpiration that can be satisfied by extraction from this upper soil layer reduces extraction from deeper layers and thus extends the water supply.

In many soils, much of the profile explored by roots will be at or near the wilting point by crop maturity, regardless of the rooting density there. No advantage would be gained there by manipulating the water extraction rate, either by slowing or speeding it. In many soils, however, some water remains in the soil layers explored by roots. It seems likely that 5–10 cm more water could be extracted from some soil profiles and used for crop production if rooting density were increased, radial resistance reduced, or the roots more evenly distributed, or if the roots had arrived there sooner. The three strategies—soil modification, genetic selection within a species, or shifts among crops—again are applicable to this situation.

B. Rationing the Water Supply

In many areas of the world, crops must grow to maturity on a finite, limited water supply. It may be desirable in those circumstances to retard the use of water during early growth in order to retain available water for use during reproductive growth. Loomis (Chapter 8A) discussed shoot management techniques to accomplish this end. There are also root function management techniques that may be appropriate. Water can conceivably be rationed through a mechanism of increased axial resistance of the root tissue (Klepper, Chapter 3B). The possibility that water can be rationed by increased radial root resistance is discussed by Fiscus (Chapter 4C).

Water use also can conceivably be rationed by reducing the total root length that supplies water to a unit leaf area, but this technique rests on the assumption that water uptake rate per unit length of root does not vary with transpiration rate. Research by Bunce (1978) and Eavis and Taylor (1979) shows, however, that radial resistance probably does vary with transpiration rate. The effect of reducing the ratio between total root length and total leaf area is not yet clarified.

Water use may also be rationed by allowing the soil to retain root-restricting characteristics that reduce rates of rooting-depth increases or those that allow roots to remain concentrated along surfaces of large peds. The latter technique causes greater soil resistance in the process of water transport to the root surface.

C. Increasing the Nutrient Supply

Increases in supplies of yield-limiting nutrients will increase yield in many semi-arid areas where all of the rainfall is evaporated, either from the plant leaves or from the nonvegetated bare areas between plants. If we use the definition of aboveground biomass divided by water transpired from plant surfaces for water-use efficiency, it can be increased by fertilization. This increase can occur either by (i) increasing the concentration of the yield-limiting ion(s) at the root surface, or (ii) increasing the root surface area (root length) in the soil layer (usually the plow layer) that contains the higher concentrations of nutrients. Increased ionic concentrations at the root surface likely will be more effective than increased root surface, which will tend to increase the rate of water extraction from that specific layer and thus reduce ionic diffusion to the root surface.

V. SUMMARY AND CONCLUSIONS

Numerous soil factors affect rate of root elongation, longevity of roots, and effectiveness of unit root length. The rate of root elongation is a major factor determining the volume of soil occupied by roots and thus of the total annual water supply in many localities. Elongation rate and longevity of laterals are major factors, along with effectiveness of unit root length, that control water and ion uptake from a particular soil volume.

The total annual supply of water to crops can be increased in many localities by increasing rooting depth. For some situations, soil modification to reduce the intensity of the root-limiting factor is the easiest and most economical route to more efficient use of water. In other situations, the most economical route is to change to a species with roots that tolerate the adverse soil condition while continuing to function. In still other situations, crop cultivars must be developed that are insensitive to the root-limiting condition.

Some hope also exists that a finite, limited water supply can be rationed by increasing the longitudinal resistance within the root xylem, increasing the resistance to water movement between root epidermis and xylem lumen, or reducing the root length relative to the leaf area. These water rationing schemes are less well-documented than the rooting volume techniques.

The efficient use of water for crop production can also be enhanced in semi-arid areas by adding fertilizer elements to soils deficient in them. This technique usually increases total above-ground mass without increasing water use because all water was destined to be evaporated anyway, either from crop or soil surfaces. This technique shifts the balance between soil and plant evaporation more toward plant evaporation.

Tanner and Sinclair (Chapter 1A) indicated that plants grown in humid environments, such as the eastern United States, suffer yield-depressing levels of water stress during part of most growing seasons. They asked that a greater research effort be made in humid areas to pinpoint geographical areas, crop species, and climatic conditions where irrigation will increase yield. While that research definitely is needed, attention also should be given to the distinct possibility that plant root systems can be managed to drastically reduce or eliminate the need for irrigation in many of these situations. Some sacrifice in yield often can be tolerated because root system modification may be a cheap alternative to the capital investments required for irrigation water delivery.

Much of the material discussed in this article is speculative and may be refuted when adequate field tests are conducted. The recent articles comparing the agronomic and physiological responses of soybeans and grain sorghum to water deficits in Australia (Constable & Hearn, 1978; Burch et al., 1978; Turner et al., 1978; Rawson et al., 1978) or soybeans to water deficits in Iowa (Mason et al., 1980) are examples of coordinated research to fill such a need.

LITERATURE CITED

Adams, F. 1974. Soil solution. p. 441–481. *In* E. W. Carson (ed.) The plant root and its environment. University of Virginia Press, Charlottesville.

Adams, F., and Z. F. Lund. 1966. Effect of chemical activity of soil solution aluminum on cotton root penetration of acid subsoils. Soil Sci. 101:193–198.

Adams, F., and R. W. Pearson. 1969. Neutralizing soil acidity under bermudagrass sod. Soil Sci. Soc. Am. Proc. 33:737–742.

Adams, F., and R. W. Pearson. 1970. Differential response of cotton and peanuts to subsoil acidity. Agron. J. 62:9–12.

Albert, L. S., and C. M. Wilson. 1961. Effect of boron on elongation of tomato root tips. Plant Physiol. 36:244–251.

Allen, L. H., Jr. 1977. Soil water and root development. Soil Crop Sci. Soc. Fla. Proc. 36:4–9.

Arndt, C. H. 1945. Temperature-growth relations of the roots and hypocotyls of cotton seedlings. Plant Physiol. 20:200–220.

Ashton, F. M., and A. S. Crafts. 1973. Mode of action of herbicides. John Wilay & Sons. New York.

Avers, C. J., and R. H. Goodwin. 1956. Scopoletin inhibits root growth. Am. J. Bot. 43: 612–620.

Bandeen, J. D., and K. P. Buckholtz. 1967. Competitive effects of quackgrass upon corn as modified by fertilization. Weeds 15:220–224.

Barton, H., W. G. McCully, H. M. Taylor, and J. E. Box. 1966. Influence of soil compaction on emergence and first-year growth of seeded grasses. J. Range Manage. 19:118–121.

Bar-Yosef, B. 1977. Trickle irrigation and fertilization of tomatoes in sand dunes: water, N, and P distributions in the soil and uptake by plants. Agron. J. 69:486–491.

Bennett, A. C. 1974. Toxic effects of aqueous ammonia, copper, zinc, lead, boron, and manganese on root growth. p. 669–683. *In* E. W. Carson (ed.) The plant root and its environment. University of Virginia Press, Charlottesville.

Bennett, A. C., and F. Adams. 1970. Concentration of NH_3 (aq) required for incipient NH_3 toxicity to seedlings. Soil Sci. Soc. Am. Proc. 34:259–263.

Blue, W. G., and C. L. Dantzman. 1977. Soil chemistry and root development in acid soils. Soil Crop Sci. Soc. Fla. Proc. 36:9–15.

Bohnsack, C. W., and L. S. Albert. 1977. Early effects of boron deficiency on indolacetic acid oxidase levels of squash root tips. Plant Physiol. 59:1047–1050.

Brenes, E., and R. W. Pearson. 1973. Root responses of three gramineae species to soil acidity in an oxisol and an ultisol. Soil Sci. 116:295–302.

Brouwer, R. 1965. Root growth of cereals and grasses. p. 153–166. *In* F. L. Milthorpe and J. D. Ivins (ed.) The growth of cereals and grasses. Butterworths, London.

Brouwer, R., and A. Hoogland. 1964. Responses of bean plants to root temperatures: II. Anatomical aspects. Jaarb. I.B.S. (Wageningen):23–31.

Bunce, J. A. 1978. Effects of shoot environment on apparent root resistance to water flow in whole soybean and cotton plants. J. Exp. Bot. 29:595–601.

Burch, G. J., R. C. G. Smith, and W. K. Mason. 1978. Agronomic and physiological responses of soybean and sorghum crops to water deficits: II. Crop evaporation, soil water depletion and root distribution. Aust. J. Plant Physiol. 5:169–177.

Cannell, R. Q. 1977. Soil aeration and compaction in relation to root growth and soil management. Appl. Biol. 2:1–86.

Cole, P. J., and A. M. Alston. 1974. Effect of transient dehydration on absorption of chloride by wheat roots. Plant Soil 40:243–247.

Constable, G. A., and A. B. Hearn. 1978. Agronomic and physiological responses of soybean and sorghum crops to water deficits: I. Growth, development and yield. Aust. J. Plant Physiol. 5:150–167.

Cooper, A. J. 1973. Root temperature and plant growth. Res. Rev. 4. Commonw. Bur. Hort. Plant. Crops. Ser. Commonw. Agric. Bur., Farnham Royal.

Crossett, R. N., and D. J. Campbell. 1975. The effects of ethylene in the root environment upon the development of barley. Plant Soil 42:453–464.

Currey, W. L., and D. H. Teem. 1977. Herbicides and root development. Soil Crop Sci. Soc. Fla. Proc. 36:23–28.

Currie, J. A. 1961. Gaseous diffusion in the aeration of aggregated soils. Soil Sci. 92:40–45.

Currie, J. A. 1970. Movement of gases in soil respiration. p. 152–171. *In* J. G. Gregory (ed.) Sorption and transport processes in soils. Soc. Chem. Inc. Monogr. no. 37. London.

Devine, T. E. 1976. Genetic potentials for solving problems of soil mineral stress: aluminum and manganese toxicities in legumes. p. 65–72. *In* M. J. Wright (ed.) Plant adaptation to mineral stress in problem soils. Cornell Univ. Agr. Exp. Stn., Spec. Pub. Ithaca, N.Y.

Devine, T. E., C. D. Foy, A. L. Fleming, C. H. Hanson, T. A. Campbell, J. E. McMurtrey, III, and J. W. Schwartz. 1976. Development of alfalfa strains with differential tolerance to aluminum toxicity. Plant Soil 44:43–79.

De Wit, C. T. 1958. Transpiration and crop yields. Versl. Landbouwk. Onderz. Agric. Res. Rep. no. 663. Pudoc, Wageningen, The Netherlands.

Doss, D. B., Z. F. Lund, F. L. Long, and L. Mugwira. 1976. Dairy cattle waste management: its effect on forage production and runoff water quality. Alabama Agric. Exp. Stn. Bull. 485.

Drew, M. C. 1975. Comparison of the effects of a localized supply of phosphate, nitrate, ammonium, and potassium on the growth of the seminal root system, and the shoot, in barley. New Phytol. 75:479–490.

Eavis, B. W., L. F. Ratliff, and H. M. Taylor. 1969. Use of a dead-load technique to determine axial root growth pressure. Agron. J. 61:640–643.

Eavis, B. W., and H. M. Taylor. 1979. Transpiration of soybeans as related to leaf area, root length and soil water content. Agron. J. 71:441–445.

Eavis, B. W., H. M. Taylor, and M. G. Huck. 1971. Radicle elongation of pea seedlings as affected by oxygen concentration and gradients between shoot and root. Agron. J. 63:770–772.

Epstein, E. 1976. Adaptation of crops to salinity. p. 73–82. *In* M. J. Wright (ed.) Plant adaptation to mineral stress in problem soils. Cornell Univ. Agric. Exp. Stn., Spec. Pub. Ithaca, N.Y.

Foy, C. D. 1974. Effects of aluminum on plant growth. p. 601–642. *In* E. W. Carson (ed.) The plant root and its environment. The University of Virginia Press, Charlottesville.

Foy, C. D., R. L. Chaney, and M. C. White. 1978. The physiology of metal toxicity in plants. Annu. Rev. Plant Physiol. 29:511–566.

Fryrear, D. W., and W. G. McCully. 1972. Development of grass root systems as influenced by soil compaction. J. Range Mange. 25:254–257.

Gardner, H. R., and R. E. Danielson. 1964. Penetration of wax layers by cotton roots as affected by some soil physical conditions. Soil Sci. Soc. Am. Proc. 28:457–460.

Glinka, Z., and L. Reinhold. 1962. Rapid changes in permeability of cell membranes to water brought about by carbon dioxide and oxygen. Plant Physiol. 37:481–486.

Grable, A. R. 1971. Effects of compaction on content and transmission of air in soils. p. 154–164. *In* K. K. Barnes et al. (ed.) Compaction of agricultural soils. Am. Soc. Agric. Eng. Monogr., St. Joseph, Mich.

Greacen, E. L., and J. S. Oh. 1972. Physics of root growth. Nature (London) New Biol. 235: 24–25.

Greenwood, D. J. 1971. Soil aeration and plant growth. Rep. Prog. Appl. Chem. 55:423–431.

Greenwood, D. J. 1975. Measurement of soil aeration. p. 261–272. *In* Soil physical conditions and crop production. Tech. Bull. 29. Ministry Agric. Fish Food, Her Majesty's Stationary Office, London.

Greenwood, M. S., J. R. Hillman, S. Shaw, and M. B. Wilkins. 1973. Location and identification of auxin in roots of *Zea mays.* Planta 109:369–374.

Howeler, R. H., and L. F. Cadavid. 1976. Screening of rice cultivars for tolerance to Al toxicity in nutrient solutions as compared with a field screening method. Agron. J. 68:551–555.

Huck, M. G. 1970. Variation in taproot elongation rate as influenced by composition of the soil air. Agron. J. 62:815–818.

Huck, M. G. 1972. Impairment of sucrose utilization for cell wall formation in the roots of aluminum damaged cotton seedlings. Plant Cell Physiol. 13:7–14.

Jackson, W. A. 1967. Physiological effects of soil acidity. *In* R. W. Pearson and F. Adams (ed.) Soil acidity and liming. Agronomy 12:43–124. Am. Soc. of Agron., Madison, Wis.

Kappelman, A. J., Jr., G. A. Buchanan, and Z. F. Lund. 1971. Effect of fungicides, herbicides and combinations on root growth of cotton. Agron. J. 63:3–5.

Kust, C. A., and B. E. Struckmeyer. 1971. Effects of trifluralin on growth, nodulation and anatomy of soybeans. Weed Sci. 19:147–152.

Lance, J. C., and R. W. Pearson. 1969. Effect of low concentration of aluminum on growth and water and nutrient uptake by cotton roots. Soil Sci. Soc. Am. Proc. 33:95–98.

Lund, Z. F. 1970. The effect of calcium and its relation to several cations. Soil Sci. Soc. Am Proc. 34:456–459

Lund, Z. F. 1978. Root development in dairy manure-treated soil. J. Environ. Qual. 7:473–477.

Lund, Z. F., and C. B. Elkins. 1978. Creating soil conditions favorable to rooting. *In* J. M. Brown (ed.) Proc. Beltwide Cotton Production–Mechanization Conf., Dallas, Tex. 9–12 Jan. 1978. Natl. Cotton Coun., Memphis, Tenn.

Lutz, J. A., Jr., G. W. Hawkins, and C. F. Genter. 1971. Differential response of corn inbreds and single crosses to certain properties of an acid soil. Agron. J. 63:803–805.

Mason, W. K., H. M. Taylor, A. T. P. Bennie, H. R. Rowse, D. C. Reicosky, Y. S. Jung, A. A. Righes, R. L. Yang, T. C. Kaspar, and J. A. Stone. 1980. Soybean row spacing and soil water supply: their effect on growth, development, water relation and mineral uptake. Adv. in Agric. Tech. NC-5, Sci. and Educ. Agency, USDA, Peoria, Ill.

McCalla, T. M., and F. A. Norstadt. 1974. Toxicity problems in mulch tillage. Agric. Environ. 1:153–174.

Megie, C. A., R. W. Pearson, and A. E. Hiltbold. 1967. Toxicity of decomposing crop residues to cotton germination and seedling growth. Agron. J. 59:197–199.

Mosher, P. N., and M. H. Miller. 1972. Influence of temperature on the geotropic response of corn roots. Agron. J. 64:459–462.

Mugwira, L. M., S. M. Elgawhary, and K. I. Patel. 1976. Differential tolerances of triticale, wheat, rye, and barley to aluminum in nutrient solution. Agron. J. 68:782–787.

Munn, D. A., and R. E. McCollum. 1976. Solution culture evaluation of sweet potato cultivar tolerance to aluminum. Agron. J. 68:989–991.

Neales, T. F. 1960. Some effects of boron on root growth. Aust. J. Biol. Sci. 13:232–248.

Neenan, M. 1960. The effects of soil acidity on growth of cereals with particular reference to the differential reaction of varieties thereto. Plant Soil 12:324–328.

Newman, E. I. 1978. Root microorganisms: their significance in the ecosystem. Biol. Rev. 53: 511–554.

Nieman, R. H., and M. C. Shannon. 1976. Screening plants for salinity tolerance. p. 359–367. *In* M. J. Wright (ed.) Plant adaptation to mineral stress in problem soils. Cornell Univ. Agric. Exp. Stn., Ithaca, N.Y.

Onderdonk, J. J., and J. W. Ketcheson. 1973. Effect of soil temperature on direction of corn root growth. Plant Soil 39:177–186.

Passioura, J. B. 1972. The effect of root geometry on the yield of wheat growing on stored water. Aust. J. Agric. Res. 23:745–752.

Patrick, Z. A., and T. A. Toussoun. 1965. p. 440–459. *In* K. F. Baker and W. C. Snyder (ed.) Ecology of soil-borne plant pathogens. University of California Press, Berkeley.

Pearson, R. W., J. Childs, and Z. F. Lund. 1973. Uniformity of limestone mixing in acid subsoil as a factor in cotton root penetration. Soil Sci. Soc. Am. Proc. 37:727–732.

Pearson, R. W., L. F. Ratliff, and H. M. Taylor. 1970. Effects of soil temperature, strength and pH on cotton seedling root elongation. Agron. J. 62:243–246.

Pfeffer, W. 1893. Druck und Arbeitsleistung durch Wachsende Pflanzen. Abh. Sachs. Ges. (Adad.) Wiss. 33:235–474.

Poljakoff-Mayber, A. 1975. Morphological and anatomical changes in plants as a response to salinity stress. p. 97–117. *In* A. Poljakoff-Mayber and J. Gale (ed.) Plants in saline environments. Springer-Verlag, New York.

Portas, C. A. M. 1973. Development of root systems during the growth of some vegetable crops. Plant Soil 39:507–518.

Portas, C. A. M., and H. M. Taylor. 1976. Growth and survival of young plant roots in dry soil. Soil Sci. 121:170–175.

Rawson, H. M., N. C. Turner, and J. E. Begg. 1978. Agronomic and physiological responses of soybean and sorghum crops to water deficits: IV. Photosynthesis, transpiration and water use efficiency of leaves. Aust. J. Plant Physiol. 5:195–209.

Reid, D. A. 1976. Screening barley for aluminum tolerance. p. 269–275. *In* M. J. Wright (ed.) Plant adaptation to mineral stress in problem soils. Cornell Univ. Agric. Exp. Stn., Ithaca, N.Y.

Richards, S. J., R. M. Hagan, and T. M. McCalla. 1952. Soil temperature and plant growth. p. 303–480. *In* B. T. Shaw (ed.) Soil physical conditions and plant growth. Academic Press, New York.

Rohde, R. A., and W. R. Jenkins. 1958. Basis for resistance of *Asparagus officinalis* var. altilis L. to stubby-root nematode *Trichodorus christiei* Allen. Maryland Agric. Exp. Stn. Bull. A-97. College Park.

Russell, R. S. 1977. Plant root systems: their function and interaction with the soil. McGraw-Hill Book Co. (U.K.) Ltd., London.

Russell, R. S., and M. J. Goss. 1974. Physical aspects of soil fertility—the response of roots to mechanical impedance. Neth. J. Agric. Sci. 22:305–318.

Sharp, R. E., and W. J. Davies. 1979. Solute regulation and growth by roots and shoots of water-stressed maize plants. Planta 147:43–49.

Sheppard, S. C., and M. H. Miller. 1977. Temperature changes and the geotropic reaction of the radicle of *Zea mays* L. Plant Soil 47:631–644.

Slootweg, A. F. G. 1956. Rootrot of bulbs caused by *Pratylenchus* and *Hoplolaimus* Spp. Nematologica 1:192–201.

Smith, K. A., and P. D. Robertson. 1971. Effect of ethylene on root extension of cereals. Nature (London) 234:148–149.

Smith, K. A., and R. S. Russell. 1969. Occurrence of ethylene and its significance in anaerobic soil. Nature (London) 222:769–771.

Stolzy, L. H., and K. P. Barley. 1968. Mechanical resistance encountered by roots entering compact soils. Soil Sci. 105:297–301.

Strobel, G. A. 1974. Phytotoxins produced by plant parasites. Annu. Rev. Plant Physiol. 25: 541–566.

Sutton, R. F. 1969. Form and development of conifer root systems. Commonw. Forestry Bur. Tech. Commun. no. 7. Oxford, England.

Tackett, J. L., and R. W. Pearson. 1964a. Oxygen requirements for cotton seedling root penetration of compacted soil cores. Soil Sci. Soc. Am. Proc. 28:600–605.

Tackett, J. L., and R. W. Pearson. 1964b. Effect of carbon dioxide on cotton seedling root penetration of compacted soil cores. Soil Sci. Soc. Am. Proc. 28:741–743.

Taylor, H. M. 1980. Postponement of severe water stress in soybeans by rooting modifications: a progress report. p. 161–178. *In* F. T. Corbin (ed.) Proc. World Soybean Res. Conf. II, Raleigh, N.C. 26–29 Mar. 1979. Westview Press, Boulder, Colo.

Taylor, H. M., and H. R. Gardner. 1963. Penetration of cotton seedling taproots as influenced by bulk density, moisture content, and strength of soil. Soil Sci. 96:153–156.

Taylor, H. M., M. G. Huck, and B. Klepper. 1972. Root development in relation to soil physical conditions. p. 57–77. *In* D. I. Hillel (ed.) Optimizing the soil physical environment toward greater crop yields. Academic Press, New York.

Taylor, H. M., and B. Klepper. 1974. Water relations of cotton: I. Root growth and water use as related to top growth and soil water content. Agron. J. 66:584–588.

Taylor, H. M., and B. Klepper. 1975. Water uptake by cotton root systems: an examination of assumptions in the single root model. Soil Sci. 120:57–67.

Taylor, H. M., and B. Klepper. 1978. The role of rooting characteristics in the supply of water to plants. Adv. Agron. 30:99–128.

Taylor, H. M., and L. F. Ratliff. 1969a. Root elongation rates of cotton and peanuts as a function of soil strength and soil water content. Soil Sci. 108:113–119.

Taylor, H. M., and L. F. Ratliff. 1969b. Root growth pressures of cotton, peas and peanuts. Agron. J. 61:398–402.

Taylor, H. M., G. M. Roberson, and J. J. Parker, Jr. 1966. Soil strength—root penetration relations for medium- to coarse-textured soil materials. Soil Sci. 102:18–22.

Timonin, M. I. 1941. The interaction of higher plants and soil microorganisms: III. Effect of by-products of plant growth on activity of fungi and actinomycetes. Soil Sci. 52:395–413.

Torrey, J. G. 1976. Root hormones and plant growth. Annu. Rev. Plant Physiol. 27:435–459.

Turner, N. C., J. E. Begg, H. M. Rawson, S. D. English, and A. B. Hearn. 1978. Agronomic and physiological responses of soybean and sorghum crops to water deficits: III. Components of leaf water potential, leaf conductance, $^{14}CO_2$ photosynthesis and adaptation to water deficits. Aust. J. Plant Physiol. 5:179–194.

van Keulen, H. 1975. Simulation of water use and herbage growth in arid regions. Pudoc, Wageningen, The Netherlands.

Venables, A. V., and D. A. Wilkins. 1978. Salt tolerance in pasture grasses. New Phytol. 80: 613–622.

Waisel, Y. 1972. Biology of halophytes. Academic Press, New York.

Wiersum, L. K. 1957. The relationship of the size and structural stability of pores to their penetration by roots. Plant Soil 9:75–85.

Williamson, R. E., and W. E. Splinter. 1969. Effects of light intensity, temperature and root gaseous environment on growth of *Nicotiana tabacum* L. Agron. J. 61:285–288.

CHAPTER 3B

Managing Root Systems for Efficient Water Use: Axial Resistances to Flow in Root Systems—Anatomical Considerations[1]

BETTY KLEPPER

Agricultural Research Service, USDA
Pendleton, Oregon

I. INTRODUCTION

Root systems anchor plants against dislodgement by wind and other forces, provide an early sink for carbohydrates prior to seed production, have selective accumulation mechanisms for mineral uptake from the soil, and provide a low-resistance pathway in xylem vessels for long-distance transport of water, minerals, and a few organic compounds from deep in the soil up to the shoots. This low-resistance xylem pathway sets vascular plants apart from other plants. Evolution of terrestrial plants of stature depended on development of these tissues that are capable of maintaining water supplies for shoots far above underground sources.

We are accustomed to the idea that shoots must have low-resistance delivery systems to replace water lost daily in transpiration, but we sometimes fail to extend our thinking to the root system. In fact, xylem tissues in roots are more conductive than those in stems (Baxter & West, 1977), and there is evidence that evolution of conductive tissue components has proceeded at different rates in response to different selection pressures in the two organs (Wagner, 1977). Supported by soil and not subjected to winds, roots have a lighter, weaker wood with larger-diameter and thinner-walled xylem elements than are found in stems (Fayle, 1968). Thus they appear well-suited to rapid conduction.

Roots of crop plants must deliver to the shoots water that has been gathered simultaneously from both deep and shallow soils, from both moist and partially dry soils, and from soils at different temperatures, often with different physical and chemical properties. Uptake in one part of the system influences uptake elsewhere, and we would like to describe this interaction quantitatively. In order to do so we must be able to interrelate the different parts of the root system through mathematical simulations or models that

[1] Published as Technical Paper no. 5965, Oregon State University, Corvallis.

Limitations to Efficient Water Use in Crop Production.

specify such interaction. Such models merely help us think, picture, and calculate; the ultimate test is how well a model describes measured water-extraction patterns of growing root systems under field conditions.

Primitive forms of such models exist. They use an Ohm's Law Analogue, where flows are related directly to water potential differences and inversely to resistances to flow. These models partition the resistances in various ways among soil, rhizosphere (a root-affected soil layer surrounding each root), rhizoplane (the soil-root contact area), root tissues themselves, and, occasionally, xylem elements of the root. These resistances are partitioned to accommodate whatever is known or assumed about flows and potentials in the system. Since resistances are not measured directly and must be deduced from decreases in water potential under a given set of flow conditions, it is not surprising that a number of points in the system have been designated as the ''primary'' resistance. Most common are the rhizosphere, rhizoplane, and root endodermis.

In this article, I will discuss resistances in the soil-root-xylem pathway and the low-resistance interconnections among various parts of the root system by xylem elements. Since resistances are not observable in the absence of flow, I shall necessarily consider flows and potential differences. Although primary emphasis will be on the axial resistances interconnecting different parts of the root system, I shall have to consider radial resistances (across root tissues) because axial resistances cannot be discussed without considering interactions between radial and axial water-flow rates. In general, resistances to water flow within tissues of living cells of plants are several orders of magnitude greater than resistances to flow through a comparable distance in dead xylem elements (Tyree et al., 1974; Taylor & Klepper, 1978; Burch, 1979). Although mineral flows occur concurrently with water flows most of the time, I shall neglect this important area of interaction, which is discussed by Fiscus (Chapter 4C).

II. ROOT SYSTEM FORM AND DEVELOPMENT

Roots differ morphologically from shoots and yet in the various plant families have evolved conservatively to be remarkably similar to one another. Almost all roots are cylindrical. This shape facilitates root extension into soil. The more or less bullet-shaped tip is well-adapted to penetrating soil and is pressed forward by growth (osmotic) forces originating in the elongation zone behind the tip. It is only after roots have elongated and tissues have differentiated that they produce branch roots, which originate in the pericycle (outermost) layer of the stele. Consider the difficulties if lateral roots originated in the apical meristem as do shoot branches. Much greater forces would be required to move the bulky tip forward and young branch roots would be sheared off in the process. The delay in branch-root initiation is necessitated by the semi-solid nature of the rooting medium.

Because of this delay in formation, the branch-root xylem must establish a connection with the parent xylem *after parent tissues have differenti-*

ated. In monocots, which have no secondary growth, and in dicots before secondary xylem is formed, this "T-junction" probably impedes rapid water flow. In woody dicots, the connection is eventually made continuous by a cambium that lays down xylem elements in an "elbow" that carries water smoothly around the bend. When either monocot or dicot stems develop adventitious roots, they generally arise in the immediate vicinity of vascular tissues of the stem (Esau, 1967). Here again junction resistances probably exist, at least for a time, because the new root xylem must be joined to already matured xylem tissues in the parent stem. In any case, there is resistance to movement of water between branch roots and their parent roots or stems at least for a period of time during their development. These junctions and the "junction resistances" to water movement across them have not yet been studied sufficiently to assess their importance to the water economy of plants.

In young monocot roots water moves radially across the epidermis and across several concentric layers of cortical cells, including the endodermis, before reaching the stele. The stele comprises the pericycle and parenchyma, through which pass several peripheral vascular strands with or without a central pith. Generally, vessels nearer the center have larger diameters than peripheral ones, and, in the extreme case, the center may be occupied by a single large vessel. Water in the xylem of one bundle is carried longitudinally without being mixed in significant quantities with water from other bundles except as it is slowly exchanged with water in surrounding tissues. Although anastomoses occur (Esau, 1965), their extent and flow characteristics have not been determined.

Mixing does occur when xylem of branch roots joins the parent-root xylem. The water-conducting system in main roots of most monocots can be thought of as a collection of cables with associated branch lines. Each cable is a vascular bundle. Each main root contains several cables. Secondary roots and their branches belong to the vascular bundle(s) nearest to their junction with the main root. Thus a vascular bundle collects water from its associated branches in a pattern shown in Fig. 1. A number of these cables are associated with each main root. There is little interchange of water among these cables until the roots join the stem where the xylem tissues are anastomosed (Bull et al., 1972).

Water crosses the same tissues in young dicot roots to reach the xylem in the center of the stele. There is no central pith. There are two to several arms of xylem tissue radiating from the center of the stele. Phloem tissues are found between these radiating arms of xylem. In most species, branch roots are initiated in the pericycle adjacent to the xylem arms. In these cases, branch-root xylem elements form a T-junction with main-root xylem and resistances to flow are probably minimal. However, those species that form branch roots between xylem and phloem poles will have "junction resistances" if water is forced to flow through one or more layers of living cells before reaching parent-root xylem. Further research is needed to determine the anatomical features of these junctions and to assess the resistances to fluid flow through them.

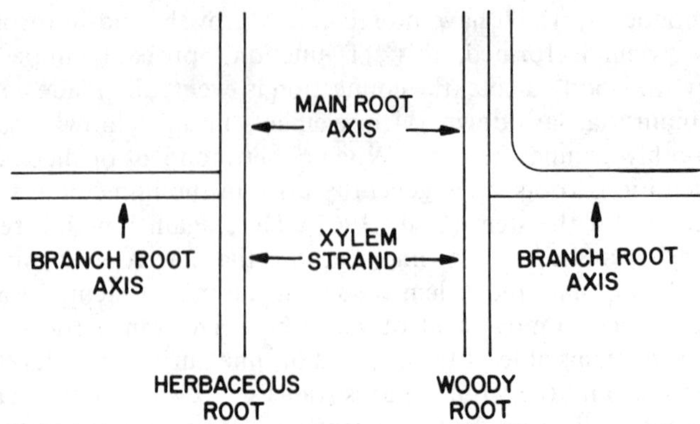

Fig. 1—Relationships of xylem strands of lateral and main roots for monocots, herbaceous dicots, and woody dicots. Each line represents a number of associated xylem vessels.

Because xylem strands of most young dicot roots gather water from branch roots adjacent to them, water absorbed on one side of the root tends to be collected into the same vessels and does not readily mix with water collected on the other side. Of course, if the root axis has spiralled during development, the water may move to the other side, but it does so within a discrete strand of xylem tissues. Thus, water in one xylem strand tends to be carried toward the top of the plant in a cable very like that in monocot roots.

When dicot roots undergo secondary growth, branch-root xylem joins the main-root xylem through smooth elbows in the upward (proximal) direction, but distal connections are much less distinct (Fig. 1). The vascular cambium forms after branch root development and lays down continuous xylem between the main root and the upward side of the lateral root so that the new xylem forms a continuous conductive tissue where water can be carried inward and upward in a single vessel made up of a number of vessel members. On the lower side, the vascular cambium of the lateral root builds concentric layers of xylem tissues, but these vessels appear to make a "T-junction" as they meet the xylem of the main root (unpublished observations). Water from vessels on the lower side of the lateral root must pass through pitted and thickened vessel sidewalls, thus increasing resistance to upward flow. There appears to be less resistance to upward than to downward passage of water where a woody branch root joins the main root. Water in a woody root, then, moves in a discrete and definite group of xylem tissues from branch to parent root and upward in a "cablelike" arrangement similar to that previously discussed for monocot roots.

Thus roots, whether primary or woody, monocot or dicot, do not readily mix water as it moves longitudinally, and flows can be pictured to a first approximation to occur as shown in Fig. 1. Anastomosis can and often does occur especially in stems, some of which have elaborate mixing systems that even-out water supplies to various parts of shoots (Esau, 1965; Bull et al., 1972). In most trees, however, water taken up by a root system on one side of the tree primarily supplies one side of the canopy (Kolesnikov,

1971). This occurs because the resistance to longitudinal water transport in wood is much less than the resistance to lateral transfer, leading to measurable differences between the water status of east and west sides of a plant during the morning and afternoon (Klepper, 1968).

Quantitative estimates of resistances to flow from one part of the water-conducting system to another are not available. If they are as great as has been assumed in the previous paragraphs, then it will be important to understand which part of the root system supplies which part of the shoot. In dicots, the lower leaves and older branches are directly "plumbed" into xylem in the center of the root; more distal leaves and branches are connected to outer layers of wood. As a tree grows, it self-prunes lower branches as heartwood is formed in the center of the tree. Water and mineral supplies pass to the crowns in sapwood. Interrelationships among different parts of crop root and shoot systems are poorly understood and need to be clarified.

III. MICROSCOPIC STRUCTURE AND CONDUCTION IN XYLEM

Xylem comprises three kinds of cells—conducting cells that are devoid of living contents at maturity; empty, dead fibers; and living parenchyma cells. The living cells are always present and serve a function or functions as yet incompletely understood. We do know that they can contain stored starch or fat, tannins, and crystals, and that the cell contents can push through the pits in adjoining walls and fill conducting elements with tyloses that block conduction (Esau, 1965). Tyloses are especially common in roots (Burstrom & Odhnoff, 1963). The fibers are apparently strengthening cells and are especially numerous in species that have evolved large-diameter, short vessels.

Although it is convenient to think of the conducting elements of xylem as pipes that run the entire length of the plant with a constant diameter and without any cross-walls or obstructions, this picture is too simplified. The conducting cells are of two kinds: tracheids and vessel elements. Tracheids pass water from one cell to another through paired pits in the secondary walls, and cells are tapered and lie alongside one another for considerable distances. They are the sole type of conducting cell in gymnosperms that exhibit measurable axial resistances (Nnyamah et al., 1978; Hellkvist et al., 1974).

Crop plants have both tracheids and vessels. Since tracheids are thought to play a minor role in water transport when the larger-diameter and more conductive vessels are present, no further discussion of the mechanisms of tracheid transport of water will be presented here. Suffice it to say that considerable axial resistance can be expected from the many cross-walls that must be traversed for water to move up the plant in tracheids.

Vessel members are laid end-to-end with perforated end-walls separating one member or cell from the other along the vessel, which consists of a series of such cells. These end-wall perforations may be scalariform, in which case the element-to-element interface is steeply tapered, permitting a large surface area for exchange. At the other extreme, the end-walls may be reduced to an annular ring or ridge, in which case the interface is more or less perpendicular to the direction of flow and vessel members are short. In either case, fluids must pass through constrictions at the end of each vessel member and will presumably be impeded by these obstructions.

These obstructions help explain the fact that measured axial resistances in xylem are often greater than those calculated from theory (Tyree & Zimmermann, 1971; Passioura, 1972; Denmead & Millar, 1976). For example, birch wood conductivity was only about one-third of that predicted from use of measured vessel diameters in Poiseuille's equation (Petty, 1978). In this case, vessel radii were 29.5 ± 2.6 μm and there were 73.3 ± 4.0 vessels mm^{-2} of stem cross section. Individual vessel members were 0.5 mm long with scalariform plates containing about 15 bars and having gaps of 5–6 μm between adjacent bars. Petty (1978) concluded that the comparatively low measured conductivities could be accounted for by the resistance to flow of scalariform end-walls and by the fact that some vessels terminated within the sample so that the actual number of conducting vessels was less than observed. In addition, presence of boundary conditions different from those specified for the Poiseuille equation (Giodano et al., 1978), especially lateral exchange of water (van Booren, 1976) or solutes (Klepper & Taylor, 1979) through porous vessel walls, may contribute to such differences. Allowance must also be made for changes in vessel diameters caused by pressure differences across xylem walls (Greacen et al., 1976) and from point to point in the system. For example, metaxylem vessel radii increase with depth in seminal roots of wheat (Meyer & Alston, 1978). Finally, presence of tyloses to block vessels and cavitation of water in some vessels (Byrne et al., 1977) may preclude their participation in conduction.

In summary, then, conductivity of xylem tissue depends on the number, radii, and structures of conducting cells comprising the tissue, and flow depends on the hydraulic potential differences that drive flow. Anatomical obstruction is sufficient in xylem elements and vessels are sufficiently narrow to allow development of axial resistance, especially if some restriction is placed on the number of conducting vessels (Passioura, 1972).

IV. PATTERNS OF WATER EXTRACTION

Effects of axial resistances on water-extraction patterns have been discussed by Passioura (1972) and by Taylor and Klepper (1978). The primary effect is the slow use of water deep in the soil profile. When water is limiting, this water-use pattern may conserve deep water for late in the growing season when water is critical. Plants are grown under sufficient stress to reduce plant size and water-use rate. Final yield, too, is lower than the potential yield for the crop, but production of some yield is more nearly assured.

Consider an example. One square meter of land surface is underlain by 1 m of profile divided into 25-cm-depth increments (each with a volume of 2.5×10^5 cm^3). Root length density is uniform with depth at 1 cm root cm^{-3} of soil (2.5×10^5 cm of root per depth increment). The average uptake rate per centimeter of root length is assumed to be 3.0×10^{-5} cm^3 h^{-1} bar^{-1}, and axial resistance is taken as either zero or constant with depth so that root xylem water potential changes at 0.07 bar cm^{-1} of depth (Taylor & Klepper, 1978). Soil water potential is assumed to be either uniform at −0.1 bar or nonuniform with upper layers drier than lower layers. Values for soil and plant-xylem water potential at midday when xylem potential at the soil surface is −25 bars are shown in Table 1. The calculated water-extraction rate at midday for the crop in wet soil is 0.064 cm h^{-1} with axial resistance and 0.075 cm h^{-1} without; in dry soil, rates are 0.052 and 0.063 cm h^{-1}, with and without axial resistance, respectively. The greatest difference in use is in the deepest layer where the plant with axial resistance absorbed water at three-quarters the rate of the plant without axial resistance.

Data from Table 1 can also be used to calculate the shoot water potential required to extract water at the maximum rate shown in the table (747 cm^3 h^{-1}). A plant with axial resistance in the uniformly wet profile would need to decrease xylem water potential at the soil surface to −28.5 bars to provide water to the shoot at the rate provided at −25 bars without axial resistance. For the drying profile, the potentials would have to be −28.9 bars with no axial resistance and −32.4 bars with axial resistance.

A more detailed analysis of diurnal patterns of uptake with axial resistance shows that deep water is tapped only when shoot water potentials are below some threshold value. In the case shown in Table 1, xylem water potentials at the soil surface would have to be < -6.2 bars before uptake from the deepest layer could occur. For example, if it were −7 bars, uptake from the deepest layer would be 6.6 cm^3 h^{-1}.

Table 1—Predicted uptake rates from 1 m of soil in 1 m^2 of land surface with and without axial resistance to water flow in root xylem. Root length density is assumed constant at 1 cm of root cm^{-3} of soil.

		No axial resistance		With axial resistance	
Depth increment	Soil water potential	Root xylem Ψ	Flow rate	Root xylem Ψ	Flow rate
cm	bars	bars	cm^3 $hour^{-1}$	bars	cm^3 $hour^{-1}$
		Uniformly wet profile			
0–25	−0.1	−25	186.75	−24.1	180.0
25–50	−0.1	−25	186.75	−22.4	167.25
50–75	−0.1	−25	186.75	−20.6	153.75
75–100	−0.1	−25	186.75	−18.9	141.0
		Drying profile			
0–25	−10	−25	112.5	−24.1	105.75
25–50	−5	−25	150	−22.4	130.50
50–75	−1	−25	180	−20.6	147.0
75–100	−0.1	−25	186.75	−18.9	141.0

Models must be more complex than that used for Table 1 because resistance to flow in small pipes depends on flow rate. Assumption of a uniform constant axial resistance does not permit resistance to change according to the Poiseuille equation. A more complex model is under development (Klepper et al., 1983).

Consider again the example in Table 1. Assume predawn xylem water potentials at the soil surface to be -5 bars. With no axial resistance, uptake from the profile would be 0.015 cm h^{-1} with 36.8 cm^3 h^{-1} from each layer in the case with wet soil. For the drier case, the net uptake rate would be 0.0029 cm h^{-1} with 36.8 cm^3 h^{-1} from the bottom layer, 30 cm^3 h^{-1} from the third layer, no extraction from the second layer, and a prediction of 37.5 cm^3 h^{-1} delivered to the top layer (sum: 29.3 cm^3 h^{-1}). Apparently part of the water taken up in the lower half of the profile would be delivered to the top layer of soil at the end of the night unless some resistance prevented it.

Most researchers assume that this apparent reverse flow of water to dry soil predicted by models is spurious, and find ways to change their models to prevent it. If such backflows occur, these quantities of water (Molz & Peterson, 1976) would not have been detected in published field experiments since soil water contents are usually measured only every few days to determine long-term changes. The overall effect of nighttime backflow would be to depress average measured water uptake rates in the upper (dry) part of the profile below that which actually occurs. Such backflow could be at least a partial explanation for measured decreases in uptake in drier soils (Taylor & Klepper, 1975).

For crop plants, potential differences that cause backflow will occur for shorter periods than those that cause uptake, since photoperiods are generally long at times when crop plants are sufficiently large to have created drying profiles such as that shown for dry soil in Table 1. Furthermore, the potential differences causing backflow are generally not as great as those that cause uptake during the day. In this example, the potential difference at dawn would be about one-half of that at midday in the top layer (10 vs. 5 bars). Thus the quantities of water that flow back into dry soil are probably small relative to those being extracted, even if resistances and capacitances are neglected.

Capacitances in the system could cause the relationship between flow and potential difference to take the form of a hysteresis loop rather than a straight line. For the example shown in Table 1, each depth increment contains 2.5×10^5 cm of root, which has to be rehydrated at the end of the day as water potentials in the plant rise to predawn values. If all roots in a depth increment are assumed to be 0.05 cm in diameter, there are about 500 cm^3 of root in each layer. To change the water potential of this volume of plant tissue by 1 bar would take about 5 cm^3 of water. If average uptake rate in a layer over a 24-h period is 80 cm^3 h^{-1}, changing root tissue by 1 bar during that hour would require a 6% of the average uptake rate just to account for the plant-tissue water changes.

During rehydration the average water uptake rate is greater than the rate at which fluid is delivered to xylem by that amount of water required to rehydrate tissues. During rehydration, the apparent resistance is greater than actual; during dehydration it is less than actual.

In relatively wet soil, root tissues will rehydrate progressively from outer layers inward and flow will continue to be in the direction of daytime uptake. As soils dry, the source of water for root-tissue rehydration will shift so that the xylem serves to rehydrate roots in dry soil from the inside out using water derived from roots in wetter soils. This is especially true if root shrinkage during the middle of the day has caused a vapor gap to occur between the soil and the root surface. Such shrinkage, of course, causes a large rhizoplane resistance and may prevent transfer of water from roots to dry soil for most, if not all, of the night.

In dry soil, roots will begin to rehydrate when the xylem water potential exceeds that of surrounding root tissues and water begins to flow in the direction opposite to normal. Since roots are cylindrical, the flow rates are dissipated as water moves outward. Outward flow is slowed by the dissipation of water over larger and larger concentric rings of cells. Transfer to soil will be delayed until the outermost layer of cells reaches a water potential above that of the soil. If a substantial root-soil vapor gap has developed during the day, this delay will be considerable. Location of high-resistance layers becomes especially important when analyzing this problem. If the endodermis impedes water transfer, then it will delay rehydration of cortical cells and this will in turn delay water transfer into surrounding dry soil.

So far we have assumed that the xylem vessels of upper roots in dry soil are in good communication with the xylem of the rest of the root system and that water potentials of xylem contents are relatively high at night. Neither of these situations is necessarily true. Recall that earlier in this chapter I proposed that branch roots of woody plants do not transfer water downward as readily as upward because xylem vessels are laid down in a smooth elbow only in the proximal direction. Thus rehydration of upper dry-root xylem may be delayed at night by resistances to transfers of water from main-root xylem to that of upper lateral roots because of lack of direct communication between distal main roots and lateral roots (Fig. 1).

Finally it is possible that xylem vessels of lateral roots in dry soil contain dissolved solutes that raise the osmotic pressure to several bars. Such a rise in xylem osmotic pressure does occur and causes guttation and "bleeding" from cut stems when roots of young seedlings are in fertile, well-watered soil. Although high osmotic pressures in older roots cannot be ruled out without experimental data, they appear unlikely since ion accumulation is a phenomenon most associated with young, active root tips, which are not a significant fraction of the root material present on older root systems in dry soil.

V. SUMMARY

Resistance to longitudinal (axial) flow of water in xylem elements, while small in comparison to the resistance encountered along comparable lengths of living plant tissues, is nevertheless of potential importance in explaining patterns of water extraction by root systems. Axial resistances arise both from structural features of the conducting members (presence of endwalls, vessels of small diameter, tyloses) and from the way xylem tissues of branch roots are related to parent-root xylem. A "junction resistance" between branch- and parent-root xylem results from the fact that branch roots must connect to parent tissues which have already matured. With secondary growth, the branch root delivers water in the upward direction along a smooth elbow, but the connection on the lower (distal) side is a "T-junction" where xylem tissues abruptly end at the outer wall of the parent-root xylem. Thus resistance to flow of water collected by lateral roots is less in the upward than in the downward direction. This difference may partially explain the apparent absence of significant deliveries of water from deep wet soils to drier soils nearer the surface by the root system.

Axial resistance changes soil water-use patterns by conserving water deep in the profile. Lower-shoot water potentials are required to provide water at a rate sufficient to replace a given amount of transpired water. Plant stress resulting from axial resistance to water flow may reduce plant size and subsequent water-use rates. Thus, by conserving deep water and by decreasing plant water requirements, axial resistance to water flow may provide benefits when water supplies are limited.

LITERATURE CITED

Baxter, P., and D. West. 1977. The flow of water into fruit trees: I. Resistances to water flow through roots and stems. Ann. Appl. Biol. 87:95–101.

Bull, T. A., K. R. Gayler, and K. T. Glasziou. 1972. Lateral movement of water and sugar across xylem in sugarcane stalles. Plant Physiol. 49:1007–1011.

Burch, G. J. 1979. Soil and plant resistances to water absorption by plant root systems. Aust. J. Agric. Res. 30:279–292.

Burstrom, H. G., and C. Odhnoff. 1963. Vegetative anatomy of plants. Svenska Bokforlaget, Stockholm.

Byrne, G. F., J. E. Begg, and G. K. Hansen. 1977. Cavitation and resistance to water flow in plant roots. Agric. Meteorol. 18:21–25.

Denmead, O. T., and B. D. Millar. 1976. Water transport in wheat plants in the field. Agron. J. 68:297–303.

Esau, K. 1965. Plant anatomy. 2nd ed. John Wiley & Sons, New York.

Fayle, D. C. F. 1968. Radial growth in tree roots—distribution, timing, anatomy. Tech. Reps. no. 9. Faculty of Forestry, University of Toronto.

Giodano, R., A. Salleo, S. Salleo, and F. Wanderlingh. 1978. Flow in xylem vessels and Poiseuille's law. Can. J. Bot. 56:333–338.

Greacen, E. L., P. Ponsana, and K. P. Barley. 1976. Resistance to water flow in the roots of cereals. p. 86–100. *In* O. L. Lange et al. (ed.) Water and plant life. Springer-Verlag, Berlin.

Hellkvist, J., G. P. Richards, and P. G. Jarvis. 1974. Vertical gradients of water potential and tissue water relations in Sitka spruce trees measured with the pressure chamber. J. Appl. Ecol. 11:637–667.

Klepper, B. 1968. Diurnal patterns of water potential in woody plants. Plant Physiol. 43:1931–1934.

Klepper, B., and H. M. Taylor. 1979. Limitations to current models describing wateruptake by plant root systems. p. 51–65. *In* J. L. Harley and R. S. Russell (ed.) The soil-root interface. Academic Press, New York.

Klepper, B., R. W. Rickman, and H. M. Taylor. 1983. Farm management and the function of field crop root systems. *In* J. Stone and W. Willis (ed.) Plant production and management under drought conditions. Elsevier Scientific Publishing Co., Ansterdam.

Kolesnikov, V. 1971. L Aksenova (transl.) The root system of fruit plants. Mir Publishers, Moscow.

Meyer, W. S., and A. M. Alston. 1978. Wheat responses to seminal root geometry and subsoil water. Agron. J. 70:981–986.

Molz, F. J., and C. M. Peterson. 1976. Water transport from root to soil. Agron. J. 68:901–904.

Nnyamah, J. U., T. A. Black, and C. S. Tan. 1978. Resistance to water uptake in a Douglas fir forest. Soil Sci. 126:63–76.

Passioura, J. B. 1972. The effect of root geometry on the yield of wheat growing on stored water. Aust. J. Agric. Res. 23:745–752.

Petty, J. A. 1978. Fluid flow through the vessels of birch wood. J. Exp. Bot. 29:1463–1469.

Taylor, H. M., and B. Klepper. 1975. Water uptake by cotton root systems: an examination of assumptions in the single root model. Soil Sci. 120:57–67.

Taylor, H. M., and B. Klepper. 1978. The role of rooting characteristics in the supply of water to plants. Adv. Agron. 30:99–128.

Tyree, M. T., C. Caldwell, and J. Dainty. 1974. The location and measurement of the resistances to bulk water movement in the shoots of hemlock (*Tsuga canadensis*). p. 84–89. *In* U. Zimmerman and J. Dainty (ed.) Membrane transport in plants. Springer-Verlag, New York.

Tyree, M. T., and M. H. Zimmerman. 1971. Theory and practice of measuring transport coefficients and sap flow in the xylem of red maple stems (*Acer rubrum*). J. Exp. Bot. 22: 1–18.

van Booren, R. 1976. Porous pipe flow. Phys. Fluids 19:481–482.

Wagner, P. 1977. Vessel types of the monocotyledons: a survey. Botan. Not. 130:383–402.

CHAPTER 3C

Managing Root Systems for Efficient Water Use: Breeding Plants for Efficient Water Use

D. K. BARNES

Agricultural Research Service, USDA, and University of Minnesota
St. Paul, Minnesota

I. INTRODUCTION

Insufficient water during some period of plant growth limits crop production on most of the world's arable land. Rainfall patterns, availability of irrigation water, temperature, and photoperiod generally have determined which crop species are adapted to a given geographical area. Microclimate differences influence the annual yields of adapted species. In this chapter I have limited my discussion to increasing the efficiency with which available water supplies are used by adapted species. Modifying plant species to change their general areas of adaption would be an extremely long-range approach to achieve more efficient use of water.

The goal of most plant breeders is to develop cultivars that produce the highest marketable yield per unit land area over a range of environments. During the selection and evaluation phases of the breeding program, various degrees of water stress are usually encountered. Unfortunately, the stress periods often occur at random, so that different growth stages are affected in the different environments. This type of stress pattern increases genotype × environment interactions without providing plant breeders with the controlled selection intensity needed to make rapid and significant changes in efficiency of water use. In cases where this efficiency may have been changed by selecting for increased marketable yield, the factor(s) responsible for the improvement have often not been identified. An increase in efficiency of water use is frequently not even considered as a possible cause for increased yielding ability.

When addressing the problems and opportunities involved with breeding to increase the efficiency with which available water supplies are used it is necessary to analyze the constraints on maximizing crop production. Estimates are not available on the potential yields of our present crop species in the absence of all environmental limitations. However, we can assume that under most environments some level of water stress will reduce the crop

Limitations to Efficient Water Use in Crop Production.

yield potential. Harvested yield will reflect composite effects of all types of stresses including water stress. The difference between the harvested yield and potential yield for the available water supply represents the amount of yield improvement that could be achieved by maximizing the efficient use of water. The efficiency with which water is used can be increased by (i) improving soil and water management practices, (ii) improving crop management practices, and (iii) planting stress-tolerant cultivars. Improved soil and water management refers to utilization of the most effective technology for fertilization, seedbed preparation, moisture conservation, and supplemental irrigation as discussed in Chapters 9, 10, and 11. Improved crop management refers to timely and proper seeding practices, pest control, weed control and timely crop harvesting as discussed in Chapter 8. The use of stress-tolerant cultivars refers to cultivars having (i) economic levels of disease, insect, and nematode resistances; (ii) tolerances to soil, water, and air problems (i.e., salinity and air pollution); and (iii) insensitivities to normal temperature and photoperiod fluctuations. I consider cultivars with improved water-use efficiencies as those with inherent physiological characteristics that will allow the increased production of dry matter per unit of water supplied by rainfall or irrigation (refer to discussions in Chapters 1 and 11).

The first priority in a plant breeding program for increasing the efficient use of water should be to determine what factors are limiting the crop from achieving its highest yield considering the available water supply. Some of these limiting factors could be a less-than-optimum root system, leaf area, and/or vascular system. In many instances breeding for improved levels of pest resistance would be required. Other situations might require breeding for either increased rate of seed germination or increased seedling vigor under unfavorable conditions to provide more uniform stands. Increased rate of seedling establishment under unfavorable conditions could produce deeper and more extensive root systems to permit more uptake of the available water in the soil profile. Breeding for traits that enable plants to obtain more of the available soil moisture without changing the yield per unit of water used can be considered a defensive approach to increasing the efficient use of water. An offensive approach would be one where changes in plant anatomy, morphology, and/or physiology would increase marketable yield per unit of water used. The offensive approach may often be considered more glamorous and challenging to scientists. However, the greatest short-term benefits will probably result from defensive breeding approaches because it is difficult to recognize, much less to utilize, many offensive types of traits until crop yield is not limited by factors other than water supply.

A survey of crop water-use (including drought-tolerance) literature from the United States during the last 10 years revealed an increase in activity during the late 1970's. Differences among cultivars have been identified for traits affecting efficient water use in many crop species, including sorghum (Henzell et al., 1975; Blum, 1974a, 1974b; Blum et al., 1977), cotton (Quisenberry & Roark, 1976), wheat (Keim & Kronstad, 1979; Hurd, 1971; Gul & Allan, 1976), pearl millet (Burton et al., 1977), soybeans (Sammons et al., 1978), alfalfa (Cole et al., 1970; Dobrenz et al., 1971), blue panicgrass (Dobrenz et al., 1969a, 1969b), Russian wildrye and intermediate wheatgrass (Hunt, 1962), Boer lovegrass (Wright & Dobrenz, 1970; Wright & Jordan, 1970), Lehmann lovegrass (Hull et al., 1978), and several cool-season forage grasses (Bleak & Keller, 1973). Based on these reports, genetic variability for traits affecting efficient water use probably can be identified in essentially all species of higher plants.

II. SOME IMPORTANT PHYSIOLOGICAL TRAITS

Several physiological characteristics have been associated with either increased drought tolerance or increased water-use efficiency. Blum (1974a) compared the responses of sorghum cultivars under soil moisture stress. Cultivars differed for leaf water potential, leaf diffusion resistance, and soil water extraction capacities. Some cultivars had lower water use prior to anthesis, relative to the total water used at maturity, than other varieties.

Blum (1974b) reported that some sorghum cultivars showed variability in cell-wall elasticity which affected the amount of water loss prior to complete wilting. Type of growth habit can be an important factor under water-stress conditions. Quisenberry and Roark (1976) compared determinate and indeterminate cultivars in upland cotton. Indeterminate types were the most productive under water stress, but the advantage decreased with increased water levels. Stebbins (1974) concluded that the indeterminate growth habit had an inherent flexibility that enabled plants to form as many flowers as the environment allowed.

Vigorous rooting characteristics have been associated with drought tolerance. Gul and Allan (1976) reported that rapid emergence was associated with a faster developing root system. Blum et al. (1977) reported heterosis in the root systems of sorghum similar to that in the top of the plant. Primary expressions of root heterosis were faster growth rates, increased root length, and increased root volumes.

Leaf morphology has received attention in water-stress research. Dobrenz et al. (1969a) reported that drought-tolerant clones of blue panicgrass had fewer stomata per unit leaf area than susceptible clones. This relationship was similar to one in barley where Miskin and Rasmusson (1970) reported that stomatal frequency did not influence rate of photosynthesis, but influenced transpiration and stomatal diffusive resistance. Burton et al. (1977) reported that when near-isogenic smooth-leaf and hairy-leaf lines of pearl millet were compared, the smooth-leaf plants had reduced transpiration (8.9–35.0%) as compared with the hairy-leaf plants. Some of the differences were attributed to the fact that the smooth plants accumulated more dew, which kept the leaves wet longer. They suggested that the smooth-leaf trait had the potential for increasing water-use efficiency and drought tolerance. The advantage for smooth leaves reported by Burton et al. is in contrast to a report by Ghorashy et al. (1971) in which they reported that soybean isolines with dense leaf pubescence had significantly lower transpiration rates than either normal or glabrous isolines. Wright and Dobrenz (1973) showed an association between water-use efficiency and seedling drought tolerance with the quality of cuticular wax on the leaves of Lehmann lovegrass. A scanning electron microscope analysis of the epicuticular wax indicated that all lines had a basic type of wax, but drought-resistant lines had an additional layer of large wax plates (Hull et al., 1978). They suggested that wax morphology could serve as a selection tool for increasing drought tolerance in Lehmann lovegrass.

III. BREEDING TO ALTER SPECIFIC TRAITS

Based on research reports and the knowledge that most sincere attempts to breed for specific plant characteristics have been successful, I have concluded that many types of traits can be modified to improve the efficiency with which the available water supply is used for essentially any plant species. However, significant progress will require an increased commitment to breeding for more efficient use of water, including improved water-use efficiency. Along with this commitment there will be a need for increased funding for research and the formation of interdisciplinary research teams.

Measuring and understanding plant-soil-water interactions is probably the most difficult area of plant-improvement research yet attempted. Much of the previous research in the area has involved either a plant breeder looking for genotypic variability that was related to yield genes, or a plant physiologist looking for physiological or morphological traits that were associated with plant-water relations. These types of research approaches have provided important background information. However, the development of improved cultivars will require the synthesis and integration of many types of information by scientists from several disciplines. It will be important that all of the scientists are equally dedicated to achieving the goal of more efficient water use in crop production. Some of the primary types of expertise that will be required include: plant physiology, soil science, plant genetics, plant anatomy, plant morphology, agricultural engineering, and electron microscopy. Plant pathology and entomology will also be required when disease and insect problems are influencing the efficient use of water in the cropping system being studied.

Improving any plant species for either the efficient use of water or water-use efficiency will require a long-term breeding program. The most effective approach would be to establish the program at a location that already has a viable breeding project on the target crop spcies. This will provide a source of information on local crop culture and on biological and environmental factors affecting crop production. It will also provide a source of locally adapted germplasm.

A. Survey of Available Germplasm

The first phase of a breeding program for more efficient water use in plants will be to survey available germplasm and to identify both tolerant (resistant) and susceptible genotypes. It is very important that a large number of sources representing a broad range of germplasm be evaluated. First priority should be placed on adapted cultivars and advanced breeding populations, because they will contain most of the pest resistances, quality traits, and adaptation characteristics needed for maximizing agronomic yields. Second priority should be given to unadapted germplasms, such as Plant Introductions, and a third priority should be given to related species. It is often difficult to identify desirable traits in poorly adapted germplasm because the desirable traits are masked by other factors. The incorporation of useful traits from poorly adapted germplasm into adapted germplasm will usually increase the time and effort required to develop a cultivar. This is because the many undesirable traits introduced from the unadapted parent must be bred out of the population while maintaining the good traits. Development of a new cultivars can require from 5 to 20 years, depending on the crop species, type of available germplasm, traits to be selected, and level of program support.

Whenever possible I favor conducting the germplasm-evaluation phase of the breeding program under field conditions. Use of normal water-stress conditions should allow for the evaluation of many thousands of plants and increase the chances of identifying traits that will be beneficial in farm environments. Periodic observations of type of plant growth, growth rate, and yield will likely be the primary data collected during germplasm evaluation. Genotypes that produce extremely tolerant and extremely susceptible reactions will be selected, and others eliminated. I believe that it is important for essentially all members of the research team (i.e., plant breeder, plant physiologist, and soil scientist) to cooperate in these evaluations because each scientist will be observing different aspects of the plant-soil-water ecosystem. It is not critical to identify the mechanisms responsible for observed differences among genotypes at this time. However, it is important to monitor environmental conditions associated with each stage of plant growth and each type of plant response in order to repeat the same types of stress in future tests. Multidisciplinary cooperation during the germplasm-evaluation phase can pay large dividends in the success of a research team. It provides a valuable opportunity for scientists to share impressions and philosophies and to discuss ideas and plans for future research.

B. Development of Selection Procedures

The second phase of a breeding program includes determining how the tolerant and susceptible genotypes selected in the first phase differ, and then developing selection procedures for the trait(s). An excellent example of this was research by Wright and Dobrenz (1973) and Hull et al. (1978) where they determined that drought tolerance in two Lehmann lovegrass breeding populations was associated with a specific type of cuticular wax on the leaf. That provided a selection criterion for use in their breeding program. Unfortunately, many traits that may be associated with efficiency of water use are complex and will require more difficult selection procedures. The most efficient selection procedures will be those where large numbers (at least several thousand per year) of plants can be rapidly evaluated at minimal cost. When possible, the use of seedlings in selection procedures can be useful.

C. Selection from Segregating Parents

It is often difficult to determine if a trait(s) identified in phase two is responsible for the improved efficiency of water use or if it is due to associated factors. Therefore, I recommend a third phase, which would be to conduct a selection program for the identified trait(s) in a different germplasm source or in segregating F_2 progenies from crosses between tolerant × susceptible parents. It is useful to use a bidirectional selection program and select for both very tolerant and very susceptible plants (Viands et al., 1981). A comparison of the selected tolerant and susceptible populations with the original populations can be used to determine if either efficient water use or water-use efficiency were significantly affected. I am not aware of reports where all three phases were carried out. I favor this approach because it precludes overlooking unusual types and combinations of plant characteristics that might either directly or indirectly improve the efficiency of water use. It also will indicate if the original selection criteria were artifacts.

D. Example of an Integrated Program

When examining drought-tolerant and -susceptible plants we need to look for differences in the obvious traits, such as stomatal number, water potential, leaf waxes, etc. However, we must also be alert for new and unique traits that could be useful. Plant species have thousands of genes and every gene offers the possibility for genetic variability. Often it is incorrectly assumed that one or two plants can represent an entire species. We must be prepared to look at very large numbers of plants and also be prepared to be surprised. As an illustration: Fig. 1A represents a classical textbook cross section of an alfalfa root, Fig. 1A and 1B show how plants can vary in the proportion of root bark and wood, and Fig. 1C shows a lobed cambium trait that changes the root anatomy. Bark width is controlled by two genes (Brick & Barnes, 1982) and lobed cambium is controlled by only one gene (Brick & Barnes, 1981). These types of traits can be readily crossed into any alfalfa germplasm.

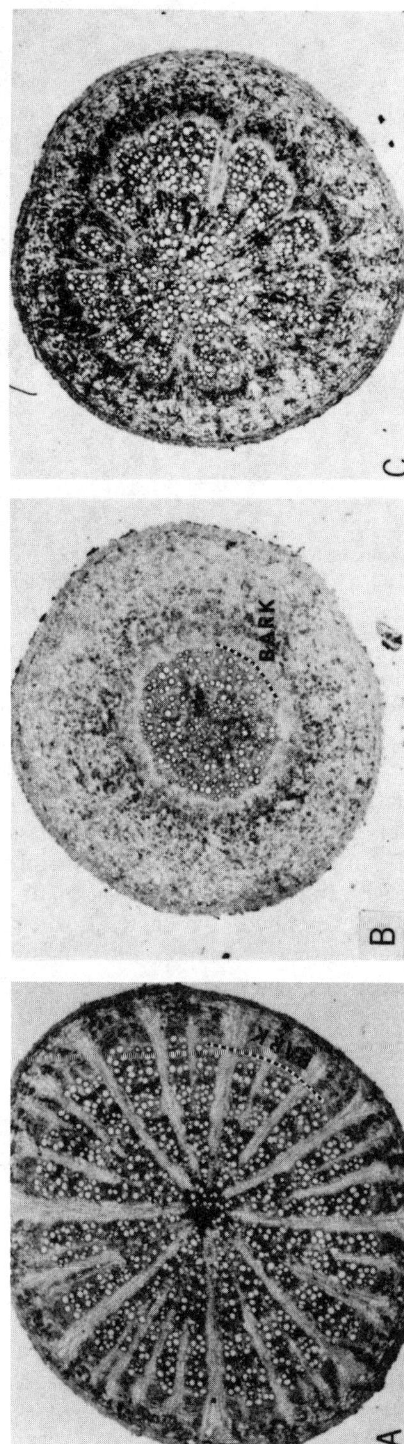

Fig. 1—Photomicrographs of alfalfa root cross sections illustrating differences occurring among plants for amount of root bark (A = small, B = large), and lobed cambium (C) rather than normal circular cambium layers (A and B).

Fig. 2—Variability observed among 12-week-old alfalfa seedlings for amount of fibrous roots.

Figure 2 represents a range in levels of fibrous root production in 12-week-old alfalfa seedlings (Viands et al., 1981). Amount of fibrous root growth can be classified as a quantitative trait that probably is controlled by several genes. Most alfalfa plants would be classified in the range of 2 or 3 on this scale. Two cycles of phenotype-recurrent selection (Viands et al., 1981) increased the mean rooting score of two germplasm pools from 2.7 to 3.8 and from 3.1 to 4.0, with many plants classified as 5. When accurate selection criteria are applied, it is usually possible to gradually increase the frequency of desirable genes in a population of plants. As an example, Viands et al. (1979) described how bacterial wilt resistance was increased in a disease-susceptible germplasm pool to a highly resistant level in nine generations of selection. In each generation the most resistant plants were selected and intercrossed. In the first few generations the selected plants were relatively susceptible. In the later generations only very resistant plants were selected. The same principles will apply to breeding for traits that can improve the efficient use of water.

IV. SUMMARY

It has been nearly a decade since Hurd (1971) expressed optimism at the possibility for breeding for drought tolerance in crop species. Moss et al. (1974) expressed the same optimism and philosophy. Both expressed the need for increased cooperation between plant breeders and plant physiologists. Few drought-tolerant cultivars have been released. However, increased recognition of the importance of improving efficient use of water (including increased water-use efficiency), improvement in instrumentation, and increased cooperation among scientists will increase the opportunity to breed plants for more efficient water use. I share the optimism expressed by Hurd and Moss et al.

LITERATURE CITED

Bleak, A. T., and W. Keller. 1973. Water requirement, yield, and tolerance to clipping of some cool-season, semiarid range grasses. Crop Sci. 13:367–370.

Blum, A. 1974a. Genotypic responses in sorghum to drought stress: I. Response to soil moisture stress. Crop Sci. 14:361–364.

Blum, A. 1974b. Genotypic responses in sorghum to drought stress. II. Leaf tissue water relations. Crop Sci. 14:691–692.

Blum, A., W. R. Jordan, and G. F. Arkin. 1977. Sorghum root morphogenesis and growth: II. Manifestation of heterosis. Crop Sci. 17:153–157.

Brick, M. A., and D. K. Barnes. 1981. Inheritance and ontogeny of a lobed-cambium trait in alfalfa roots. J. Hered. 72:419–422.

Brick, M. A., and D. K. Barnes. 1982. Inheritance and anatomy of root bark area in alfalfa. Crop Sci. 22:747–752.

Burton, G. W., W. W. Hanna, J. C. Johnson, Jr., D. B. Leuck, W. C. Monson, J. B. Powell, H. D. Wells, and N. W. Widstrom. 1977. Pleiotropic effects of the *tr* trichomeless gene in pearl millet on transpiration, forage quality, and pest resistance. Crop Sci. 17:613–616.

Cole, D. F., A. K. Dobrenz, M. A. Massengale, and L. Neal Wright. 1970. Water requirement and its association with growth components and protein content of alfalfa (*Medicago sativa* L.). Crop Sci. 10:237–240.

Dobrenz, A. K., D. F. Cole, and M. A. Massengale. 1971. Yield components and leaf characteristics associated with the water requirement of alfalfa. Crop Sci. 11:124–125.

Dobrenz, A. K., L. N. Wright, A. B. Humphrey, M. A. Massengale, and W. R. Kneebone. 1969a. Stomate density and its relationship to water-use efficiency of the blue panicgrass (*Panicum antidotale* Retz.). Crop Sci. 9:354–357.

Dobrenz, A. K., L. N. Wright, M. A. Massengale, and W. R. Kneebone. 1969b. Water-use efficiency and its association with several characteristics of blue panicgrass (*Panicum antidotale* Retz.) clones. Crop Sci. 9:213–216.

Ghorashy, S. R., J. W. Pendleton, R. L. Bernard, and M. E. Bauer. 1971. Effect of leaf pubescence on transpiration, photosynthetic rate and seed yield of three near-isogenic lines of soybeans. Crop Sci. 11:426–427.

Gul, A., and R. E. Allan. 1976. Interrelationships of seedling vigor criteria of wheat under different field situations and soil water potentials. Crop Sci. 16:615–618.

Henzell, R. G., K. J. McCree, C. H. M. van Bavel, and K. F. Schentz. 1975. Method for screening sorghum genotypes for stomatal sensitivity to water deficits. Crop Sci. 15:516–518.

Hull, H. M., L. N. Wright, and C. A. Bleckmann. 1978. Epicuticular wax ultrastructure among lines of *Eragrostis lehmanniana* Nees developed for seedling drouth tolerance. Crop Sci. 18:699–704.

Hunt, O. J. 1962. Water requirement of selected genotype of *Elymus junceus* Fisch. and *Agropyron intermedium* (Host) Beauv. and their parent-progeny relationships. Crop Sci. 2:97–99.

Hurd, E. A. 1971. Can we breed for drought resistance? p. 77–88. *In* K. L. Larson and J. D. Eastin (ed.) Drought injury and resistance in crops. Crop Sci. Soc. Am. Spec. Pub. no. 2, Madison, Wis.

Keim, D. L., and W. E. Kronstad. 1979. Drought resistance and dryland adaptation in winter wheat. Crop Sci. 19:574–576.

Miskin, K. E., and D. C. Rasmusson. 1970. Frequency and distribution of stomata in barley. Crop Sci. 10:575–578.

Moss, D. N., J. T. Woolley, and J. F. Stone. 1974. Plant modification for more efficient water use: the challenge. Agric. Meteorol. 14:311–320.

Quisenberry, J. E., and B. Roark. 1976. Influence of indeterminate growth habit on yield and irrigation water-use efficiency in upland cotton. Crop Sci. 16:762–765.

Sammons, D. J., D. B. Peters, and T. Hymowitz. 1978. Screening soybeans for drought resistance: I. Growth chamber procedure. Crop Sci. 18:1050–1055.

Stebbins, G. L. 1974. Flowering plants. Evolution above the species level. Harvard University Press, Cambridge, Mass. p. 278–279.

Viands, D. R., D. K. Barnes, and G. H. Heichel. 1981. Nitrogen fixation in alfalfa-responses to bidirectional selection for associated characteristics. USDA Tech. Bull. 1643. U.S. Government Printing Office, Washington, D.C.

Viands, D. R., D. K. Barnes, R. E. Stucker, and F. I. Frosheiser. 1979. Inheritance of resistance to bacterial wilt in two alfalfa gene pools: response to selection and quantitative analysis. Crop Sci. 19:711–714.

Wright, L. N., and A. K. Dobrenz. 1970. Efficiency of water-use and seedling drought to tolerance of Boer Lovegrass *Eragrostis curvula* Nees. Crop Sci. 10:1–2.

Wright, L. N., and A. K. Dobrenz. 1973. Efficiency of water-use and associated characteristics of Lehmann lovegrass. J. Range Manage. 26:210–212.

Wright, L. N., and G. L. Jordan. 1970. Artificial selection for seedling drought tolerance in Boer lovegrass (*Eragrostis curvula* Nees). Crop Sci. 10:99–102.

CHAPTER 4A

Water Transport and Balance Within the Plant: An Overview

WILLIAM WENKERT

Agricultural Research Service, USDA
Columbus, Ohio

I. INTRODUCTION

Transport of water in plants is of wide interest. It is the major pathway for the tremendous quantity of water that is moved from land to atmosphere. It also is essential in the normal functioning of the plant. That functioning, including transpiration, depends fundamentally on plant water status, which in turn depends on the precarious balance between water loss to the atmosphere and water uptake from the soil. If the leaf water content decreases by only a few percent, for example, the stomata may close, causing a virtual halt in both transpiration and plant production.

When transpiration starts from a leaf, the initial rate of water loss exceeds the rate of gain, and the water content and potential of the tissue falls. It does so until the gradient that drives liquid water uptake is high enough to balance water loss and the tissue comes to a steady state. This chapter is concerned with the development of these water deficits over a day, or a longer drying cycle. The treatment is necessarily broad since controls are located throughout the soil-plant-atmosphere system. In fact, some of the most interesting flow processes occur at the immediate boundaries of root and soil, leaf and air. For more detailed treatments the reader is directed to other chapters in this volume, and to earlier reviews of transport in the soil and plant (e.g., Gardner, 1960, 1968; Cowan, 1965; Cowan & Milthorpe, 1968; Newman, 1974, 1976; Jarvis, 1975; Tinker, 1976; Taylor & Klepper, 1978) and in the atmosphere above plants (Lemon, 1965; Monteith, 1964, 1973; Tanner, 1968; Thom, 1975).

A. Electrical Analogs

Throughout this discussion we will be describing the relationship between the steady flow rate of water and the force assumed to be driving it. The hope is that the two will be proportional. This approach has been used successfully to describe electrical current (Ohm's Law), diffusion (Fick's

Limitations to Efficient Water Use in Crop Production.

Law), and saturated water flow in soils (Darcy's Law). In each case the observed behavior can be expressed as

$$\delta\Phi/Q = R, \quad [1]$$

or

$$\delta\Phi/q = r, \quad [2]$$

where Q is the amount of substance moving per unit time (called "flow" here) and q is flow per unit cross-sectional area ("flux"). The potential difference that drives this transport is $\delta\Phi$ and the proportionality factors are the resistances R and r. These resistances are rarely constant but are still convenient if they can be described by other independent variables. The description of flow in any system, then, will depend on finding alternate expressions describing the behavior of R and r. These should be as general as possible and grounded in basic transport theory. Nevertheless, the ultimate definition of any resistance, i.e., the aim of our description, does not arise from theory but from phenomenological equations such as Eq. [1] and [2].

Although both R and r are called resistances, the former treats the system as a black box. It is applied, for example, to a specific electrical circuit or root system. The inverse of R has been called conductance. The other resistance has an area term within it (r = AR) and describes the flux across some discrete element, e.g., a membrane or a given thickness of air. The inverse of r is commonly called a permeability coefficient or a transfer coefficient in the respective examples. Either resistance can be expressed per unit length and the inverse of this has the same dimensions as a diffusion coefficient.

Another useful electrical analog is relative capacitance:

$$C' = (\partial V/\partial \Phi)(1/V) \quad [3]$$

This describes the relative change in water volume $\partial V/V$ with potential. While R is defined by Eq. [1] and [2] only for steady conditions, i.e., constant flow throughout the resistance element, the speed at which many systems follow a change to a new steady state is directly proportional to C', i.e., to how much water needs to be lost or gained to reach the new potential level. Capacitance will be useful, then, in describing the response to varying conditions and in evaluating when R can be measured.

B. Liquid vs. Vapor Transport

Although the continuity of water movement has been emphasized, it will be most convenient to handle the flow as two coupled but separate systems: the loss of water vapor from plant to atmosphere, and the gain and redistribution of liquid water from the soil. Not only are these different types of transport, but their effective resistances are different orders of magnitude.

There is a natural discontinuity at the leaf surface between the physics of liquid and vapor transfer. The movement of liquid water in many parts of the plant, as well as in saturated soil, is approximately proportional to gradients of water potential, $\delta\Psi$. Vapor movement, however, is proportional to the difference in vapor pressure, or approximately so to vapor concentration, δX. Rawlins (1963) demonstrated that vapor flow can be described in terms of water potential only if the resistance is made a function of vapor concentration. This is very inconvenient, however.

It should be noted here that considering water potential as the appropriate driving force for liquid flow implies that the gradients of component potentials are equally effective. If, however, there is an osmotic difference, $\delta\psi_\pi$, between the two ends of a long vascular bundle, or across a membrane that is permeable to those ions, then the osmotic gradient will not have any effect. Membrane impermeability to those ions is expressed by a reflection coefficient, σ, which is 1 or less, such that the effective osmotic difference is $\sigma\delta\psi_\pi$. The flux across a membrane would be described by

$$q = (\delta\psi_p + \sigma\delta\psi_\pi)/r, \qquad [4]$$

and is not proportional to $\delta\Psi$ when $\delta\psi_\pi$ is large and $\sigma < 1$.

There is a second type of effective discontinuity at the leaf surface. Over the normal range of leaf water potentials, the vapor concentration within is very nearly constant at the saturation value. This means that the vapor gradient, and therefore transpiration rate, is not substantially affected by changes in Ψ. For example, as leaf tissue dries from 0 to -27 bars the vapor concentration at 20°C falls from 100 to only 98% of saturation. This is equivalent to saying that the changes in the vapor phase dominate the flow. Another example is evaporation from moist soil. As long as the surface is visually moist its vapor concentration is close to saturation, and despite substantial changes in surface water potential, the evaporation rate is controlled by the vapor phase alone. Of course, this changes sharply when the liquid phase resistance begins to increase rapidly and the surface dries severely, but in the plant the stomata close long before this point.

It will hold generally then, that leaf water status affects transpiration rate only if it changes the resistance to vapor diffusion through the stomata. It follows that we could initially consider this resistance a given, derive from it a transpiration rate, and then consider how this flow generates a water potential drop in the soil and plant. The system is complete if we can describe the feedback from leaf water potential to stomatal resistance.

II. LOSS OF WATER VAPOR

A. The Potential Difference

Transpiration from a leaf will be considered as vapor movement from the internal cell wall surfaces to some plane z in the bulk air above (Fig. 1). That movement is proportional to the difference in vapor pressure between

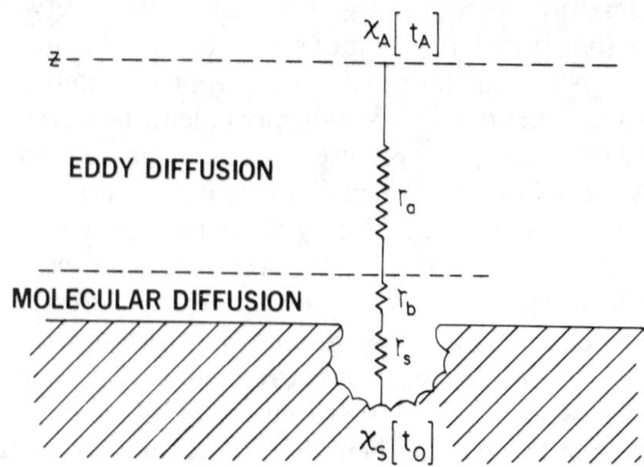

Fig. 1—Transpiration from a single leaf surface. The vapor gradient is from the cell-wall surfaces, where the concentration is saturated at the surface temperature, $\chi_S[t_O]$, to plane z, where the ambient concentration is χ_A. The stomatal resistance is r_s, the resistance to vapor diffusion through the leaf boundary layer is r_b, and resistance in the bulk air is r_a.

the two surfaces. If temperature differences are not very large, then vapor movement is also approximately proportional to vapor concentration. At height z the ambient concentration is X_A. From what was said earlier, the vapor concentration within even fairly dry tissue can be taken as the saturation concentration at leaf temperature $\Sigma_S(t_O)$.

It should be emphasized that the vapor gradient is highly dependent on leaf temperature and that this may vary quite a few degrees above, and sometimes below, air temperature. This can be made explicit by describing the gradient in two parts: (i) a gradient of saturated vapor concentration due to the temperature difference between surface and air, and (ii) the difference between saturated and actual vapor concentration in the air. Thus,

$$X_S[t_O] - X_A = (X_S[t_O] - X_S[t_A]) + (X_S[t_A] - X_A) \quad [5]$$

and

$$\cong [\Delta(t_O - t_A)] + (X_S[t_A] - X_A), \quad [6]$$

where Δ is the average change in saturated vapor concentration with temperature between t_O and t_A, as shown in Fig. 2 (Δ at 20°C is 1.07×10^{-6} g cm^{-3} °C^{-1}). The second term in brackets is analogous to the vapor pressure deficit and is solely an environmental factor. In the example shown in Fig. 2, for an air temperature of 26°C and a relative humidity of 50%, the vapor gradient (vertical arrows) changes by a factor of about 2.5 in going from a surface 4°C below air temperature to one 4°C above.

It is imperative, then, to know how leaf temperature is fixed. Intuitively, one of the main determinants of t_O should be the cooling effect of transpiration itself. This type of closed control can be understood more fully if we take the vapor gradient as a given for now and consider first the other main determinant of evaporation, the resistance to vapor transport.

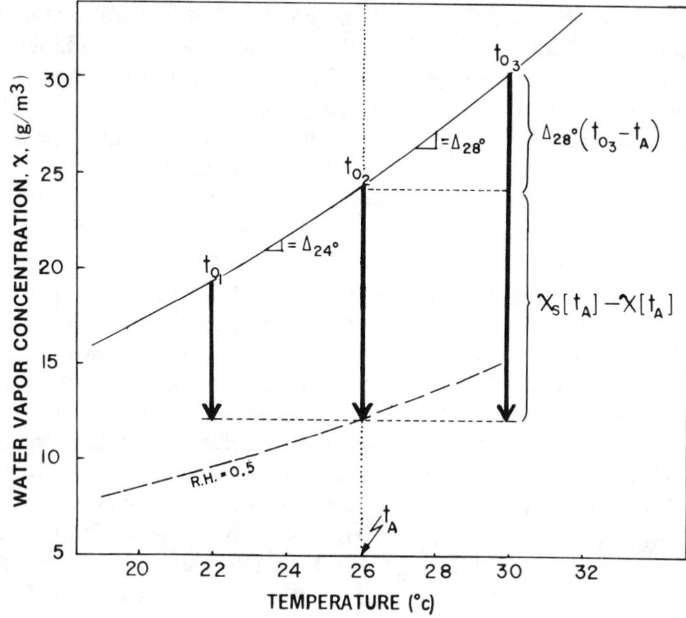

Fig. 2—The vapor gradient from leaf to air for three different surface temperatures, t_O. The air is at 26°C and 50% RH throughout. The upper line is the saturated vapor concentration and the vapor gradient is divided into the two parts described by Eq. [5].

B. Resistances

1. STOMATAL RESISTANCE: r_s

As water evaporates from cell walls within the leaf, the vapor diffuses out through both the cuticle and the stomatal pores. The former is an important pathway only when the stomata are nearly closed. It will be ignored in this treatment so that the total resistance from cell surfaces to the exterior leaf surface is called *stomatal resistance,* r_s (Fig. 1). The pivotal role this factor has in water movement cannot be overemphasized: it very often is the major resistance in the vapor pathway, it can vary over an order of magnitude, and this variation is under the control of the plant. The mechanics and behavior of stomata have been reviewed recently by Raschke (1975, 1976) and Cowan (1977). We will consider here only the two most basic concerns: sensitivity to light and sensitivity to leaf water status.

Stomata close in low light to the degree that most lower leaves in normal canopy shade have high r_s. Leaves in moderate and full sun, either in sunspots or at the top of the canopy, often have sufficient light for minimum stomatal resistance. This may be about 1.0 s cm^{-1}, probably an order of magnitude less than the resistance of shade leaves. Not all plants have exhibited this on-off behavior with light, but it does approximate many results, particularly for field plants. It will be generally true in any event that the major leaf-to-leaf variability in transpiration rate within a plant or canopy will be due to light, and this has important implications for the distribution of liquid water flow.

Stomatal resistance is generally independent of leaf water status over a wide range. It usually increases rapidly at some critical water deficit, apparently related to leaf turgor pressure. The observation of a range of resistances at approximately similar turgor pressures does not contradict this idea, as Turner (1975) has suggested, but may instead indicate a narrow critical range. The importance of leaf turgor as opposed to other aspects of water deficit is suggested first, because it is physically most likely, and second, because the critical ψ_L for closure varies with $\psi_{\pi L}$ (Turner, 1974; Brown et al., 1976; Cutler & Rains, 1977; Fereres et al., 1978; Beadle et al., 1978). The same thing is implied by studies that have shown that the critical ψ_L is lower later in the season and for field vs. growth-chamber plants, since both of these factors are generally associated with lower osmotic potential.

The critical leaf turgor for stomatal closure need not be zero or, in fact, any fixed value. Its apparent value in maize was 2–4 bars in Wenkert (1981). Critical turgors based on extracted bulk sap were about 2, 0, and 2 bars for maize, sorghum, and tobacco, respectively, (Turner, 1974), −2 to −9 bars for both sunflower and sorghum (Turner et al., 1978), and 0 to −5 bars for Sitka spruce (Beadle et al., 1978). However, the same technique used with wheat yielded critical turgors of 6–8 bars (Biscoe et al., 1976; Millar & Denmead, 1976). Correcting for the dilution error in the extracted-sap technique (Wenkert, 1980a) would probably remove the negative values in some of these references, but could only increase the high values of critical turgor in Biscoe et al. (1976) and Millar and Denmead (1976).

The view that stomata only respond to maintain a minimum leaf turgor is contradicted by studies showing a direct sensitivity to ambient humidity. These have been reviewed and discussed recently by Cowan (1977), Rawson et al. (1977), Farquhar (1978), and Edwards and Meidner (1978). It is becoming clear that this may be an important mechanism for maximizing water-use efficiency. Defined as the ratio of assimilation to transpiration, the denominator increases with decreasing ambient humidity, while the numerator does not. It is reasonable that xerophytes, with a limited water supply, may have evolved a mechanism to restrict photosynthesis to those parts of the day or season when efficiency is highest by closing their stomata when humidity is low.

Despite the potential importance of this mechanism, our concern here will be with the more basic feedback control. Sensitivity to humidity is not always significant in mesophytes, while the well-defined stomatal closure with leaf water stress is. Although the mechanics of stomata depend on the turgor of guard and subsidary cells, as discussed in Chapter 4B, and these differ from that measured in the leaf mesophyll, there is still good evidence for mesophyll control. Whatever the intermediates in the process, there is some aspect of a water deficit that is harmful to the leaf cells, such that stomatal control of water loss is required. The best candidate is cell turgor or some closely related parameter.

2. ATMOSPHERIC RESISTANCE TO WATER VAPOR AT THE LEAF SURFACE: r_b

Once at the leaf surface, the movement of vapor through the atmosphere is actually composed of two distinct processes: (i) molecular diffusion very close to the leaf, and (ii) bulk flow, or "eddy diffusion," in the air further out. Very close to the surface, plus or minus an order around 1 mm, frictional forces have slowed down any air movement enough so that its structure is laminar. If the temperature difference between surface and air is not large, then there is not any bulk vertical movement in this zone and the water molecules move only by diffusion. This resistance to vapor diffusion across the laminar boundary layer is denoted as r_b (Fig. 1).

Moving away from the surface the air flow gradually loses its laminar structure. The frictional forces associated with the roughness of the surface and the buoyant forces associated with temperature gradients both cause net vertical motion in the form of swirling air parcels or eddies. The resistance in this turbulent zone to the height z is denoted r_a in Fig. 1. Although this is bulk movement, which is orders of magnitude more rapid than molecular diffusion, the distances covered by the latter are much smaller. The diffusive resistance r_b governs transport from the leaf surface to the air a few millimeters away, while the turbulent resistance r_a governs transport within and out of the plant canopy. The former is a function of leaf characteristics while the latter depends mainly on the type of canopy. Canopy transport is discussed in Chapters 5A and 5B, and we consider here the diffusive transport at the leaf surface only.

The resistance to diffusion of water vapor through a laminar boundary layer should depend on the thickness of that region. If the temperature gradient from leaf to air is large, some air parcels will be carried in bulk by buoyant forces or free convection. This may need to be considered when a leaf is much above air temperature and the wind velocity very low (see Parkhurst et al., 1968; Vogel, 1970; Monteith, 1973). It will be ignored here except to note that it effectively puts an upper limit on r_b.

For flat plates parallel to the wind stream, and the same order of size as a leaf, this laminar-zone thickness depends on two factors: (i) it decreases with bulk wind speed, and (ii) it increases with distance traveled over the surface as the air is increasingly slowed and the laminar layer builds up. Empirically, the mean transfer resistances for one surface of smooth plates in a laminar wind stream can be summarized as

$$r_b \cong \alpha\,(d/U)^{1/2}, \qquad [7]$$

where r_b is the boundary-layer resistance to heat transfer, $\alpha = 3.2$, d is the leaf dimension in the direction of the wind (cm), and U is wind speed (cm s^{-1}). The resistance to vapor is implicitly assumed equal to the resistance for heat since the two differ by only about 10%. Values for r_b derived from heat

Table 1—Measured $r_b = \alpha(d/U)^{1/2}$ for single leaf surface in natural wind.

Reference	Leaf	α
Raschke (1956)	*Alocasia*	1.3
Gates (1964)	Oak	2.5
Knoerr & Gay (1965)	Hardwoods	3.0
Kanemasu & Tanner (1969)†	Bean	1.4
Linacre, 1972‡	Grape	2.0
Parlange et al. (1972)	Reed	1.3
Denmead & Bradley (1973)	Wheat	1.3–3.2
Stanhill & Moreshet (1966)	Model	2.7
Pearman et al. (1972)	Model	1.8

† As modified by Pearman et al. (1972) and including only measurements for which $t_o - t_A > 1.0°C$ as suggested by these authors.
‡ Modified for single surface.

transfer experiments are more commonly measured, and therefore is convenient to deal with these reported values. In all cases, Eq. [7] refers to the mean transfer over the surface, although actual leaf surfaces have large local differences in laminar-layer thickness and hence heat and vapor flux (Rashke, 1956; Parlange et al., 1971; Clark & Wigley, 1975).

Transfer from leaves in natural wind has generally been consistent with the form of Eq. [7], but there are important differences. The exponent of (d/U) will change if the laminar layer becomes unstable and breaks down at very large U or d. The main problem, however, is the coefficient α, which has been found to vary, somewhat unpredictably, over twofold. A number of studies that solved for the resistance using an independent measurement of flux and potential difference are summarized in Table 1. These measured values of r_b are all between 50 and 100% of the results for smooth plates in laminar wind. The reasons often supposed for this include any factor that will increase vertical eddy movement and erosion into the laminar layer: roughness of the leaf surface, leaf movement, wind attack angle, and high levels of turbulence in natural wind.

The difficulty involved in predicting r_b is emphasized by data from Denmead and Bradley (1973), shown in Fig. 3. For these wheat leaves in the field, the measured α changed from 1.3 to almost 3.2 over a narrow range of low wind speeds. These upper and lower values are the theoretical limits for fully laminar and turbulent winds, respectively, as presented by Parlange et al. (1971). A shift in wind structure at low velocities may be supposed, but this is not a general result. Other studies have shown a constant or inconsistently varying α over a range of wind speeds.

The most comprehensive study in natural wind may be Pearman et al. (1972), which considered different model leaf shapes and sizes over different crop surfaces and three different techniques of solving for α. They found the coefficient varied between 1.3 and 2.7, with a mean value of 1.8, which is probably close to the mean for all the studies summarized in Table 1. The scatter in their results was not associated with any particular model or wind speed. Although some of it may have been due to variations in wind attack angle (Parkhurst et al., 1968), the major conclusion was that r_b will

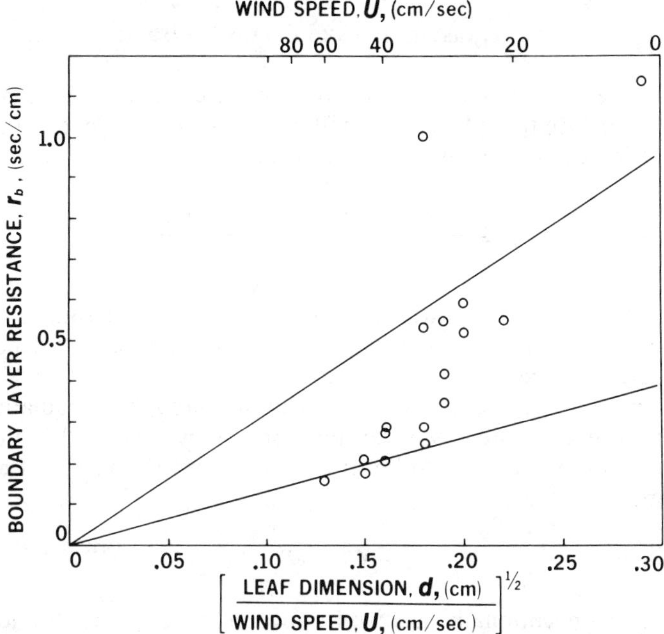

Fig. 3—The boundary layer resistance, r_b, for wheat leaves at low wind speed. The upper and lower values are from Parlange et al. (1971) for fully laminar and turbulent flow, respectively. Figure redrawn from Denmead and Bradley (1973).

vary rapidly with natural turbulence levels. This implies that we need to describe the turbulence generated upstream, and this will be a difficult effect to predict.

The significance of r_b, and that of its expected variability, depends on two factors: (i) the magnitude of the other resistances in series with it, and (ii) the feedback effect of r_b and the other resistances on leaf temperature and hence the vapor gradient.

The other aerodynamic term, r_a, when considered over distances on the order of a canopy, will often be comparable with r_b. Both resistances decrease with wind speed: r_a with U, and r_b with $U^{1/2}$. The surface resistance r_b increases with leaf size (d), while r_a decreases with plant height. Considering transfer to a plane just above the canopy: for low canopies with narrow foliage, r_a will dominate, particularly at low wind velocity. For the opposite extreme (an open forest with broad leaves) r_b will dominate. In all cases the total aerodynamic resistance is in series with the stomatal resistance and these are comparable only when the stomata are open. The maximum r_b for a single leaf surface is on the order of 1.0 s cm^{-1}. Minimum stomatal resistances are similar but can increase as much as two orders of magnitude with low light or water stress.

A change in any of the resistance terms changes the heat loss, and therefore temperature and vapor concentration, within the leaf. This complicates the effect of these resistances and can be understood only by considering how leaf temperature is fixed.

C. Transpiration and Energy Restraint

From the two preceding sections we can describe transpiration from a single leaf surface to a plane a few millimeters above the leaf as the vapor gradient (Eq. [5]) divided by the total resistance, $r_s + r_b$:

$$T = \frac{\Delta (t_O - t_A) + (X_s [t_A] - X_A)}{r_s + r_b}. \qquad [8]$$

As noted earlier, $t_O - t_A$ varies greatly in nature. The restraint on surface temperature is the requirement that energy gains and losses balance when the temperature is constant.

The major energy exchanges for a leaf are through net radiation, R_N; the latent heat associated with transpiration, λT (where λ is the latent heat of evaporation of water); and sensible heat or convection, C. At constant leaf temperature,

$$R_N - C - \lambda T = 0, \qquad [9]$$

where R_N is conventionally positive for incoming energy and C and λT are positive for net loss. The other connection between λT and C is that both increase with the surface-to-air temperature difference. Convection is directly proportional to the difference of heat per unit volume:

$$C = C_p \varrho \frac{t_O - t_A}{r_b}, \qquad [10]$$

where $C_p \varrho$ is the volumetric heat capacity of air (1.2×10^{-3} J cm^{-3} °C^{-1} at 20°C). The leaf-to-air temperature gradient cancels out when C (Eq. [10]) is divided by λT (from Eq. [8]) to yield the Bowen ratio. If this expression is solved for C and substituted into the energy balance condition (Eq. [9]), the result is:

$$\lambda T = \frac{\Delta R_N + \dfrac{C_p \varrho (X_s [t_A] - X_A)}{r_b}}{\Delta + \dfrac{C_p \varrho (r_s + r_b)}{r_b}}. \qquad [11]$$

In this form the temperature gradient is implicit in the other variables. This equation has been called the Modified Penman or Penman-Monteith equation and has been widely applied both in practical measurements and in questions of the theoretical sensitivity of λT.

The energy restraint described by Eq. [11] is essential for understanding the relative importance of the aerodynamic and stomatal resistance terms. Any change in the boundary-layer thickness changes the heat balance of the leaf through both λT and C. These terms have a different degree of coupling with leaf temperature, however (Eq. [8] and [10]), so the effect of a change in r_b on transpiration rate is not direct if the surface temperature is free to vary. For a leaf with high stomatal resistance, in high ambient humidity λT is small and most of the energy exchange is through C. A decrease in r_b, through increasing wind speed, for example, will lower t_o enough through convection to cause a decrease in transpiration despite the lower vapor resistance. For a leaf with low r_s, in low ambient humidity most of the energy loss is through λT and the leaf will be close to air temperature. A decrease in r_b will have less effect on t_o in this case since λT is not directly related to $t_O - t_A$, whereas C is. Therefore, the effect of decreased resistance will dominate and λT will increase.

The overriding effect of energy restraint explains an important and surprising observation: when stomatal resistance is minimum, say 0.5–2.0 sec cm^{-1}, very different leaf types, in very different canopies and wind velocities, have roughly similar transpiration rates not far from the net radiation value. At low humidities this maximum rate will be equal to or greater than the net radiation and the leaves will be at or below air temperature. At high humidities the maximum λT will be less than R_N and the leaves will be above air temperature.

If energy restraint governs how changes in resistance affect transpiration, then it also has application to water-use efficiency. From the definition given earlier,

$$\frac{A}{T} = \frac{\delta CO_2}{X_s[t_O] - X_A} \frac{r_b + r_s}{r_b' + r_s' + r_m'} \cong \frac{k}{X_s[t_O] - X_A} \frac{r_b + r_s}{1.4r_b + r_s' + r_m'} . \qquad [12]$$

Here r_m is mesophyll resistance to CO_2 from cell wall to carboxylation site; the difference in CO_2 concentration from atmosphere to chloroplast, δCO_2, is assumed constant; and the resistance to diffusion of CO_2 through the leaf boundary layer, r_b', is 1.4 times that of water vapor. It appears that A/T would always increase with increases in r_s or r_b, since these represent a larger fraction of the total resistance for water vapor than for CO_2. However, Cowan and Troughton (1971), Parkhurst and Loucks (1972), Jones (1976), and Cowan (1977) have considered the associated changes in leaf temperature, in analyses similar to Eq. [11], and how this affects photosynthesis. These demonstrate that the effect of an increase in resistance is variable. Under those conditions associated with leaf temperatures below ambient—low net radiation and humidity and high air temperature—A/T tends to increase with increases in r_b. Unless r_m is very low (C_4 plants?) and temperature is high, A/T generally increases with r_s.

The ideas reviewed this far suggest that upper leaves in a closed canopy contribute the bulk of the total transpiration. These have the highest net radiation and the lowest stomatal and aerodynamic resistances. Lower leaves in sunspots may also have high R_N and low r_s. Due to these two terms, transpiration rates will generally be similar to the upper canopy leaves, despite higher r_b and ambient humidity within the canopy. Lower shaded leaves will have low R_N and high r_s, and their λT may be an order of magnitude less than adjacent or upper exposed leaves. The local variability in T within a plant will be important to the next section in discussing the transport of liquid water to the sites of evaporation.

III. GAIN OF LIQUID WATER

A. Capacitance

How low water potentials in the soil and plant have to fall in order to maintain a certain steady uptake rate depends on the flow resistance across each section (Fig. 4). How rapidly these steady-state values are reached depends to a large extent on how much water needs to be removed from the system to realize that potential change. It is this change in water content that requires time, and that needs to be considered whenever the flow is not steady-state. This in fact is always more or less the case in the varying field environment.

There is another basic consideration here, in that resistances along the flow path are not independent of potentials. As far as we know, there is nothing in water potential itself which causes this. In the soil it is the loss of water volume that causes large decreases in flow for the cross-sectional

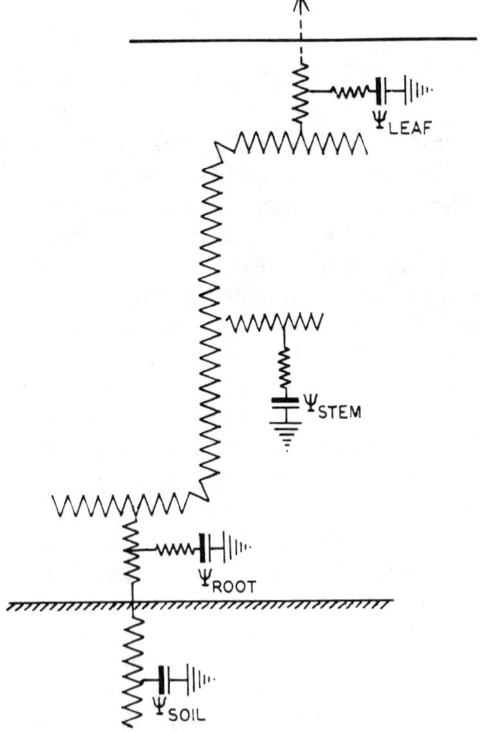

Fig. 4—Resistance analog for liquid water movement.

area, and hence large increases in resistances. In the plant there may be changes in basic membrane or stomatal resistances with water deficit. This is probably not related to water potential per se, but instead to changes in water volume and the consequent changes in turgor and osmotic potential. It will be one of the main conclusions of this chapter that the restraint on plant water potentials imposed by the component potentials has an overriding role in understanding plant water deficit.

1. CONTROLLING VARIABLES

In soils the water potential is generally a matric potential associated with surface forces. There are always some large pores present, which make little difference in potential as they fill or drain, so that soil capacitance is very large in the wet range. This is shown in Fig. 5, with data taken from Gardner (1960), and the water content expressed relative to saturation at V_o. With water loss, C becomes increasingly smaller and is very low for dry soils, say -5 to -15 bars. The absorbed water films are very thin in this range and their matric potentials change rapidly with water content, but unfortunately not in any way predictable from theory.

Capacitance of plant tissue is more complex but perhaps more amenable to theoretical treatment. The simplifying assumption is that the bulk of the water exchanged is from free solution within the symplast. Here there are two factors: changes in turgor due to the elastic behavior of the restraining cell wall, and changes in osmotic potential due to concentration. The change in matric potential of water associated with cell walls would raise the same difficulty found in soils, but we can probably ignore this in the plant since both the relative volume and relative capacitance of matric water is small at normal water content ranges.

With these assumptions, we can describe relative capacitance by

$$C' = 1/(\epsilon_t - \psi_\pi), \qquad [13]$$

where ϵ_t is the volumetric modulus of elasticity. This describes the change in turgor pressure with relative volume, which for fully turgid leaf tissue is typically 100–300 bars. The change is osmotic potential with V is described by $-\psi_\pi$, typically 10–20 bars.

There are important consequences of the relatively high elastic modulus. First, it must dominate C' in relation to ψ_π, i.e., the cells behave as if they had rigid walls so that initial changes in turgor are associated with small changes in water volume. The parallel changes in solution concentration are then relatively insignificant. We do not have enough information about whether this high modulus remains constant over a range of turgor, or decreases with drying (Wenkert et al., 1978). In any event, the relative capacitance of vacuolated tissue, which behaves as Eq. [13], is certainly very low initially compared with soil. This is shown in the example in Fig. 5, which was calculated for an initial ψ_π and ϵ_t of 10 and 200 bars, respectively, and a linear decrease in ϵ_t with decreasing turgor.

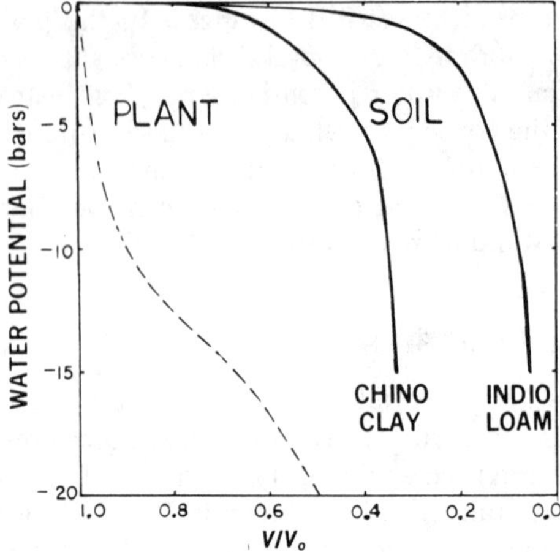

Fig. 5—Water content-potential relationships for typical soils (Gardner, 1960) and vacuolated plant tissue. The latter was calculated from Eq. [13] using $\psi_\pi = -10$ bars and an initial $\epsilon_t = 200$ bars, which decreases linearly with turgor pressure. Water contents are expressed relative to saturation at V_o.

Although Eq. [13] describes the dependence of vacuolated tissue water potential on volume, there are also important parallel changes in potential not related to water volume per se: changes in osmotic potential due to solute level, denoted $\delta\psi_{\pi o}$, often amount to roughly half of the diurnal change in total water potential (Wenkert, 1980, 1981). If loss of cell turgor represents a problem for the plant, then it is apparent that addition of solutes while Ψ is falling has important consequences in maintaining turgor. This capacity may govern the effective lower limit of leaf water potential as discussed later. Solute changes can be very significant to the observed capacitance, since they represent changes in water potential independent of changes in water volume.

2. LOCATION OF CAPACITANCE WITHIN THE TRANSPIRATION STREAM

The soil storage volume is coincident with its flow pathway. Within the plant this is often not the case and we have to recognize two different types of flow paths and resistances: first, those associated with steady-state transpiration, and second, those associated with changes in cell hydration.

The flow analog shown in Fig. 4 depicts the water-storage volumes in root, stem, and leaf as capacitors that are off of the main transpiration stream. The major axial flow, for example, is in the xylem, parallel to water in the adjacent vacuolated cells. Radial movement to or from the xylem is to some extent through cell walls, which are again parallel to the cell vacuoles. The latter case, transport through vacuolated tissue, will come up repeatedly and is central to determining how closely coupled the capacitors are with the transpiration stream.

There is a related question in this area of how we define water potential in tissue with diverse components and large possible gradients. This will come up later in discussing the leaf, for example. The most realistic solution, and the one used here, is to define tissue water potential as the mean value within the vacuoles. This is roughly what is measured any time tissue is excised, using the pressure chamber for example, since the xylem and cell walls contain water with a small relative capacitance. Hence these adjust to the vacuole potential without greatly changing the water balance of the cells. When measuring vapor pressure of intact leaves, however, such as with a psychrometer or dew-point hygrometer, there may be some discrepancy. The measured potential in this case is presumably that of the xylem immediately adjacent to the covered area.

3. TIME RESPONSES

The time it takes for the soil and plant to reach steady state following some change in transpiration rate is simply the time required for the system to lose or gain the necessary amount of water. That amount is determined by capacitance. The rate at which it moves in and out of storage is determined by the relevant resistance, R_c. If the flux at any time is proportional to the difference between actual and final water potentials, and the only changes in Ψ are described by capacitance, then the steady state is approached exponentially. This is described by the time constant

$$\tau = R_c C' V, \quad [14]$$

where τ is the time for $(1 - 1/e \sim 0.63)$ of the completed response and R_c is the resistance that the storage water encountered.

We have seen that C' in soil is very large. For the Indio loam shown in Fig. 5, the time constant per unit soil volume is about 5 min cm^{-3}. Although its capacitance decreases with drying, the resistance increases even faster. The result is that τ increases with drying, although less than an order of magnitude down to -15 bars. It is apparent that the rooted soil volume would have a very large and fairly constant time response.

In vacuolated plant cells, C' is fairly constant with species but may increase with drying. The resistance of these cells to water loss or gain, probably a major part of which resides in the plasmolemma and tonoplast, may or may not be constant with species and water status. The lack of information in this area will be considered later. The evidence suggests that if R_c does change with drying, it is probably a decrease, and the effect of this on τ might be counter-acted by an increase in C'.

The factor that certainly causes large variation in lag times in plants, and that in fact will probably turn out to be the most significant, is the volume of water available. The large differences in water storage between different plant parts and species have been discussed by Jarvis (1975). Observed time constants for leaf tissue, for which V is small, are on the order of a minute. For whole herbaceous plants t is roughly 10–100 min, while for trees it is probably around 100–1,000 min. This means that nonsteady-state conditions will always be significant following a rapid change in T, e.g., in alternating sun and shade, but significant over a diurnal cycle only for whole plants and species with larger storage volumes.

The effect of changes in solute level is to decrease the apparent capacitance so that for a given increment in Ψ, less water is exchanged and the time response is shorter. Whether the changes in $\psi_{\pi o}$ are always slow, or whether they are rapid enough so that they are significant to all transients, remains uninvestigated.

B. Soil Pathway

We are interested in water movement in the soil only to the extent that it influences plant water potential. If the gradients in potential within the soil volume are small, then that influence is a simple and direct one: the bulk ψ_{soil} is the potential at the root surface, and we need only change this boundary condition as the soil dries slowly over days. Whether we can get away with this depends on the one overriding behavior of soil water movement: the resistance defined as $\delta\Psi/q$ is highly dependent on soil water content, and hence potential (Fig. 6). There is no question that the effect of soil resistance is insignificant, relative to that within the plant, when the soil is very moist. In drier soils, however, large gradients of water potential can

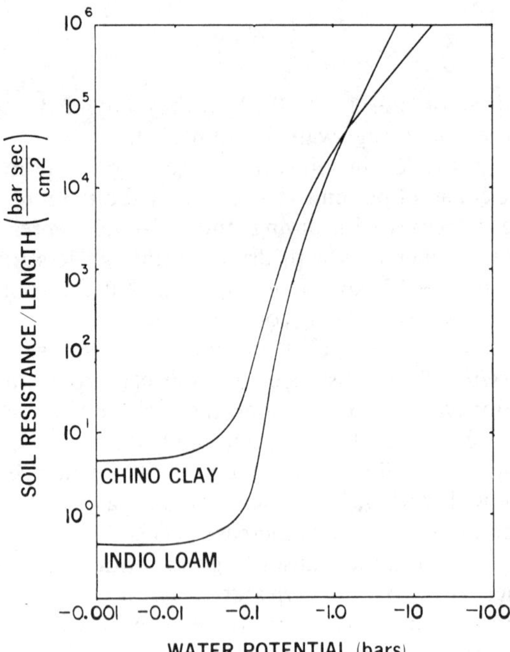

Fig. 6—Resistance per unit length for two soils as a function of water potential. Data from Gardner (1960).

develop, particularly in the soil adjacent to the root where flow converges to the root surface.

The potential difference across this so-called perirhizal zone increases with flow rate and bulk soil resistance and decreases with distance from the root and root density. Analyses by Gardner (1960) and others have all concluded that perirhizal gradients are not significant until the soil is fairly dry. Depending on the root density chosen, this may be about −15 bars (Newman, 1969a, 1969b, 1974; Williams, 1974; Nnyammah et al., 1978) or −20 to −30 bars (Gardner, 1960; Cowan, 1965). There is also uncertainty in the bulk soil resistance values at lower water potentials (Fig. 6). Reicosky and Ritchie (1976) have used an analysis similar to Newman's, and agree with his critical bulk soil resistance, but concluded that this occurred at −1 to −8 bars for sandy and clay soils, respectively. Although there are some differences here, no study using reasonable values of root density and total flow rate has predicted significant perirhizal water potential gradients in soils wetter than −1 bar. This means that we should be able to ignore soil resistance for plants with well-distributed root systems in this range.

Most field data supports the above conclusion (this will be discussed later), but there have also been observations of plant water stress and stomatal closure following decreases in soil water potential of only a few 10ths of a bar near field capacity. The best explanation for this effect seems to be that much of the water uptake is localized with very high fluxes in certain regions of the root. This may be due to differences in root permeability with age or length, or to poor root-soil contact (Herkelrath et al., 1977). This possibility was recognized in all the theoretical studies cited and has been specifically discussed by Newman (1974), Caldwell (1976), and Tinker (1976). The often-cited 40% root shrinkage measured by Huck et al. (1970) would certainly be effective in causing poor root-soil contact, but this is a very large volume change compared to other plant parts, and the root is a particularly maladaptive place to have such shrinkage. We need more information about the generality of this effect and other mechanisms that relate to root-soil contact, such as the presence of root hairs, lenticels, and mucigel. Localized uptake has been pinpointed as the critical problem by a very effective comparison of models and experiments in this area.

The general conclusion is that we can always ignore soil water movement when the soil is very moist, and probably often do so even down to −5 to −10 bars, depending on root-soil contact. This would encompass the bulk of the available soil water and have very important consequences for plant water deficit. The effect of ψ_{soil} on ψ_{plant} throughout this range would only be additive. Since we are considering a decrease of 5–10 bars at the most, and since this normally takes place gradually over days and weeks, the plant may be able to adjust. The second suggestion is that we are not interested in developing a greater root distribution within the explored soil volume. The plant has a surplus of this area (at least for water uptake) as long as the perirhizal resistance is negligible, and what should instead be encouraged is new root exploration into untapped soil water volumes, as discussed by Taylor in Chapter 3A.

C. Plant Pathways

A major problem in describing flow in whole plants is the heterogeneity of the pathways. The confluence of flow in the roots, and its bifurcation again at branches, leaves, and veins, makes it very difficult to describe the appropriate gradients and fluxes. Richter (1973) and others have emphasized how ambiguous many reported resistances are, and the usual recommendation has been for more realistic and complex models. Instead of this a much-simplified model that emphasizes the major sites and qualitative behavior of plant resistance will be discussed.

Three subsystems are considered: the radial path across a single section of root, the axial path through xylem, and the radial path out of the xylem in a cross section of a leaf. For our purpose, and for crop plants of moderate height, these three systems can be taken to represent about equal parts of the total water potential gradient from soil to leaf. As far as absolute magnitudes of resistance, we will consider only that the total resistances at moderate and high flow rates are very similar for all plants when expressed per unit ground area: almost all mesophytes have maximum transpiration rates near the maximum net radiation value, say 500 W m^{-2}, and minimum leaf water potentials of -15 bars, within a factor of 2. In Chapter 4C by Fiscus water flow in roots is discussed in detail. Also, the complexity of real plant geometry, and many reported values of plant resistances, are discussed in recent reviews by Jarvis (1975), Newman (1976), and Taylor and Klepper (1978).

1. RADIAL PATHWAY ACROSS THE ROOT

The path from the external root surface to the xylem generally comprises the major resistance in the root, always represents a significant fraction of the total plant resistance, and is in a critical location affecting all parts downstream. Despite the considerable number of studies in this area, the location and behavior of the major resistance is still debated.

There are excellent discussions by Newman (1974, 1976) on possible pathways in the root; three of these are summarized in Fig. 7. The vacuole path (pathway 1) requires crossing four membranes per cell, or two membranes and plasmodesmata. The cell-wall path (pathway 3) probably entails a jog across a membrane at the endodermis in order to move around the impermeable casparian strip. The symplasm path (pathway 2) postulates movement through the cytoplasm, either crossing two membranes per cell or traveling through plasmodesmata. All that is known for certain is that some part of the entering water crosses some part of the symplasm. This is inferred from the sensitivity of root resistance to temperature, anoxia, or metabolic inhibitors.

The central problem here, and it comes up again in all parts of the plant, is the partitioning of flow in vacuolar tissue between cell and cell-wall paths. It is an important question since we expect different behaviors for the two. Transfer studies in roots and leaves have accumulated indirect

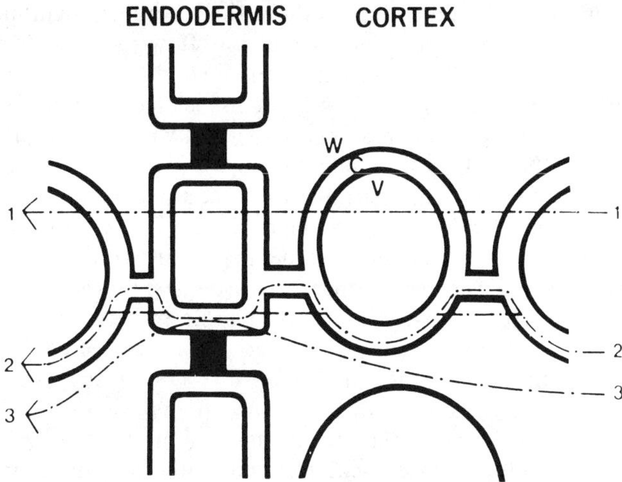

Fig. 7—Three possible pathways for radial movement across a root, where *W* is cell wall, *C* is cytoplasm, and *V* is vacuole. (Adapted from Newman, 1974.)

evidence for different pathways. These have generally emphasized the cell wall, although there are important exceptions that will be discussed. The question has also been addressed directly by Molz and Ikenberry (1974) and Molz (1976), with a model of flow in parallel cell and cell-wall pathways. This model, and an application of it by Ferrier and Dainty (1977), concluded that both pathways may be important. All these studies have been inconclusive for the same reason: there is great uncertainty—at least one or two orders of magnitude—in the specific resistances for higher plant membranes and cell walls.

Optimistically, this may be due to experimental difficulties that will be resolved. It may also be due, however, to real variation in specific resistances, even within a species. Most of this variation should be at the membranes. Spaces within the wall are small enough that they should not drain at normal tensions, although some intercellular spaces may (Meiri & Anderson, 1970). Membrane resistance, however, varies with temperature and metabolic factors, and possibly with water status.

Returning to the specific problem of the root, we do not know which pathway in Fig. 7 is the most important. Newman (1976b) concluded that the symplasm (pathway 2) is the major one. He discounted the more traditional cell-wall path based on the following reasoning: because of the metabolic sensitivity of root water uptake, the resistance within the cell walls themselves is assumed less than that at the endodermis. The computed total resistance, however, using reported values for cell-wall resistance, would then be too high to agree with observed values. The fallacy in this is the assumption that some degree of metabolic sensitivity implies a normally dominant symplasm component. When metabolic effects are large, e.g., during restricted uptake following inhibitors, the symplasm resistance may certainly dominate, but this is no longer the normal root resistance used to make the above conclusions.

An additional problem is that many of the laboratory demonstrations of metabolic sensitivity have been done at low flow rates, with relatively large external salt concentrations. Under these conditions, a large proportion of the water potential gradient may be associated with active salt uptake rather than the hydraulic gradient associated with transpiration. That is, we may often be looking at a metabolic sensitivity of ion uptake or exclusion, and not of membrane resistance to water movement.

The actual water pathway across the root must be a combination of all three pathways, which varies with relative membrane resistances. The importance of the symplasm path, especially through plasmodesmata, is still open to question. Although continuity of cytoplasm between cells is well-established, substantial water movement here has never been demonstrated (Spanswick, 1976). The model used by Molz (1976) suggests that the diameter of plasmodesmata is critical. We are in any event certain that the radial root pathway is very heterogeneous, that is includes some living tissue, and that it will be very difficult to predict flow behavior a priori. The problems of estimating the resistances to water flow in roots are discussed further by Fiscus in Chapter 4C.

2. AXIAL PATH ALONG THE ROOT, STEM, AND LEAF

Flow within the xylem has often been compared with pipe flow. There are some problems with calculating accurate values, due to uncertain geometry and wall properties, and these are summarized by Taylor and Klepper (1978). Petty (1978), for example, has found the resistance of birch vessels to be substantially higher than that of equivalent capillaries, largely due to vessel end walls. Nevertheless, it does seem well-established that the flow is slow enough to be laminar, and, from the Hagen-Poiseuille equation at least approximately proportional to the fourth power of the lumen radii. The significance of this is that the larger vessels carry the bulk of the flow and these should be the vessels that conform best to the pipe flow model.

The only possibility for large changes in resistance here is if vessels block due to cavitation or disease. Both of these, however, are probably distinguishable from normal plant resistance behavior. Disease will generally be gradual and irreversible. If cavitation affects large vessels, the effect will be rapid and irreversible, although the vessels might refill at night. Although there is evidence that cavitation can occur in drying plants (Byrne et al., 1977), it is not clear that it is a common response. The fact that it would be most prevalent in large vessels, at higher flow rates and tensions, would make it very maladaptive.

We can tentatively conclude that the normal resistance within the xylem varies only with viscosity. This would show some temperature dependence, but this is not a large effect and can be easily predicted. This contrasts with the radial transport to and from the xylem, where the flow path and membrane behavior are uncertain.

3. RADIAL MOVEMENT OUT OF THE LEAF VEINS

We are concerned here with movement from vascular bundles both to the evaporating surfaces of cell walls and to the mesophyll cells where average leaf water potential is defined. How much of the latter is in the pathway of the former is, of course, the same problem encountered in the root. The answer must be even more qualitative here since the geometry is so complex. It is noted that cell-to-cell contact in leaf tissue is very irregular, and that there may or may not be bundle sheath extension cells acting as connectors between xylem and epidermis. As water moves outward to the two leaf surfaces, there are parallel pathways for liquid water movement in the cell walls and symplast, as well as an additional one for movement of vapor in the intercellular spaces.

The first concern here is the direction and terminus of transpirational water; i.e., what are the major sites of evaporation within the leaf? The traditional view, that water evaporates at equal rates from all exposed cell walls, has been contested from two different sides.

Boyer (1977) has concluded that evaporation rates are relatively higher next to the vascular bundle, i.e., that substantial water movement across the leaf is in the vapor rather than liquid phase. The argument for this seems doubtful. He demonstrated that the resistance associated with changes in leaf hydration—in this case pushing water out of the leaf with a pressure chamber—is much higher than that associated with steady-state transpiration flow. The same has been demonstrated by Tyree and Cheung (1977), and, as discussed earlier, is expected whenever part of the transpirational flow is not through the vacuoles. Boyer rejected the idea that the cell walls represent significant pathways, however, based on his finding that the time response for water exudation from the whole leaf was an order of magnitude greater than that from dispersed single cells. He concluded that cell-wall resistances were high, that most of the flow was cell-to-cell, and that lower resistance associated with transpiration was due to a shorter path length.

It is true that similar kinetics for single cell and tissue would imply that the wall is a low-resistance pathway. The opposite, however, is not true. During exudation from applied pressure, movement is laterally along the leaf from cell vacoules to the veins, which in this case averaged 17 cells apart. A major proportion of that lateral flow could be through cell walls and still have a time constant that is much larger than that for individual cells. The type of kinetics can be similar whatever the pathway (Molz & Ikenberry, 1974). It seems that during transpiration an even larger proportion of water would flow in the walls, as these are sinks for water lost to the intercellular spaces. Another difficulty with the Boyer model is that it requires a cutin layer to prevent evaporation from the cell walls adjacent to the stomatal pore.

The opposite view—that the major proportion of transpiration occurs from cells immediately adjacent to the stomatal cavity, especially the epidermal and guard cells—has been put forth by Meidner (1975) and Sheriff and Meidner (1974). Close to the vapor exit there is necessarily a shorter path, and therefore lower resistance, for diffusion. A second piece of evidence is the very low osmotic potentials and inferred water potentials of guard cells (Edwards & Meidner, 1978), implying that these are major sites of water loss. The third type of evidence is tracer studies, since then supported by Byott and Sheriff (1976) and Burbano et al. (1976), which suggest substantial water loss from the inner walls of guard, subsidiary, and epidermal cells.

All these studies have also suggested that the major pathway to the evaporation sites in mesophytic leaves is vertically from the veins, through the bundle sheath extension cells, and then laterally through the epidermis. The generality of this role for the sheath extension cells is questionable, even in mesophytes, since many species have them only on the largest veins (Esau, 1965). However, the fact that species with the most extension cells have widely spaced veins does seem to support the role of the epidermis as an important lateral conductor. In any event, this proposed pathway would be very significant since the mesophyll would be largely bypassed.

Cowan (1977) discussed this point and demonstrated that the distribution of evaporation within the leaf may vary depending on small temperature gradients. For a leaf that is below air temperature, the epidermis and guard cells would be warmer than the mesophyll; this, combined with their proximity to the stomatal pore, would cause them to be the major site of evaporation. For a leaf above air temperature, evaporation from the mesophyll would increase. A temperature gradient from mesophyll to epidermis of only 0.1°C gave significant changes in Cowan's analog model. We can also expect that large vapor gradients would accentuate epidermal evaporation independent of local temperature differences. A quantitative understanding of these effects would require extending the energy balance and resistance analog, presented earlier, below the leaf surface. We will consider only that the fraction of transpirational flow moving in the mesophyll, where leaf water potential is defined, may vary both with species and environmental conditions.

D. ψ_L Behavior—Moist Soil

Having considered the possible locations and natures of resistances, we can now discuss the relationship between plant water potential and transpiration. This is begun most directly for the moist soil case. We have seen earlier that water potential at the root surface should be nearly zero, and that any potential drops will be due to resistance within the plant only. Nevertheless, due to the complexity that has been demonstrated for the plant pathway, there is very little that can be said a priori. We will begin by considering reported relationships between leaf water potential and T for moist soil or solution culture, and then discuss the generality and possible reasons for each. This extends the reviews in this area by Hailey et al. (1973), Newman (1974), Kaufmann (1976), Jarvis (1975), and Cowan (1977).

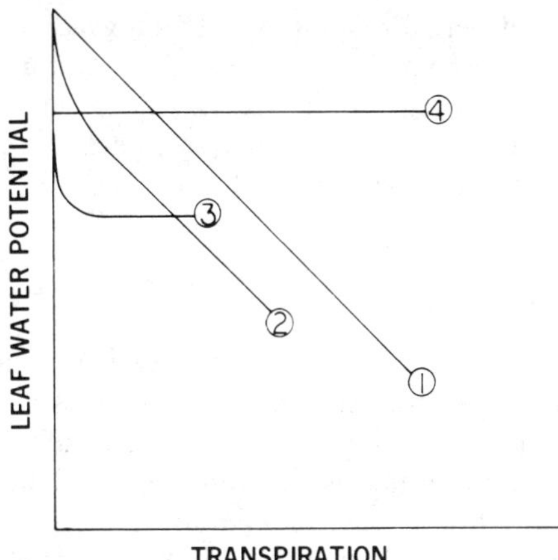

Fig. 8—Graphical representation of four reported relationships between leaf water potential and transpiration rate at high soil water potential. See text for explanation of individual curves.

1. REPORTED ψ_L VS. T

Most reported results of ψ_L vs. T for moist soil fall into the four rough categories, depicted in Fig. 8. These are:

1) A linear decrease in ψ_L with T—if the difference between leaf and soil water potentials at zero flow is small enough to be ignored or accounted for by some of the factors to be discussed, then this response can be interpreted as a constant plant resistance (Barrs, 1973; Hailey et al., 1973; Neumann et al., 1974; Camacho et al., 1974; Dube et al., 1975; Denmead & Millar, 1976; Ackerson & Krieg, 1977; Bunce, 1978; Ike et al., 1978a).
2) An apparent high resistance at low flow, but approaching a constant value at higher T (Janes, 1970; Boyer, 1974, 1977; Biscoe et al., 1976; Ackerson & Krieg, 1977; Sterne et al., 1977; Roberts & Knoerr, 1978; Jones, 1978).
3) Initial high resistance followed by a decreasing apparent resistance with flow rate such that ψ_L is held constant at a plateau value (Macklon & Weatherley, 1965; Tinklin & Weatherley, 1966; Stoker & Weatherley, 1971; Faiz & Weatherley, 1977; Bunce, 1978).
4) Apparent decreases in resistance with flow such that ψ_L is held constant not far from ψ_{soil} (Barrs, 1973; Troughton et al., 1974; Camacho et al., 1974).

Although these are all reports in which T was actually measured, there are also a preponderance of studies showing large diurnal ψ_L changes that are consistent with responses 1 or 2 but definitely not with 3 or 4.

To generalize, most of the results showing constant resistance (1), or approaching constant resistance at higher T (2), were with field plants. All the results showing a constant water potential over a range of T (3 and 4) were with plants grown in pots or solution. In all cases the apparent resistances were either constant or decreased with flow. This supports our earlier conclusion that perirhizal water potential gradients in moist soil are insignificant, since these would increase with flow rate. Also, cavitation within the xylem at higher tensions is evidently not a normal response.

A simple resistance analog for predicting ψ_L seems appropriate only for response 1. Some of the flow characteristics that could give rise to other responses are discussed below.

2. EFFECTS OF SALT UPTAKE

There is a possibility for nonlinearity between T and ψ_L anytime the water potential difference from soil to xylem includes a significant osmotic component and the reflection coefficient between the two is <1 (Eq. [4]). As long as the osmotic gradient is constant, ψ_L vs. T could still be linear, although shifted to a lower water potential at all T. Nonlinearity arises when the osmotic difference changes with flux. This has been made explicit in models by Fiscus (1975) and Dalton et al. (1975), and its application to explaining some specific experimental results has been discussed by Newman (1974, 1976a) and Fiscus (1977a, 1977b).

A plant that is actively taking up ions from moist soil or dilute solution can have a significant osmotic potential within its xylem. If the effective reflection coefficient between xylem and soil is <1, then there is a water potential gradient even at zero flow. When transpiration begins the internal concentration is diluted with flow. The pressure gradient then gradually replaces the relatively less effective osmotic gradient—if exactly so, then the xylem water potential, ψ_x, remains constant while flow is increasing (horizontal middle line in Fig. 9). If σ is low enough, the increase in $\psi_{\pi x}$ can exceed the decrease in ψ_{px} so that ψ_x increases even while the pressure-driven flow is increasing (lower line in Fig. 9).

It is evident that this effect will be significant only when T and σ are small and the osmotic gradient is large. The latter depends on a significant

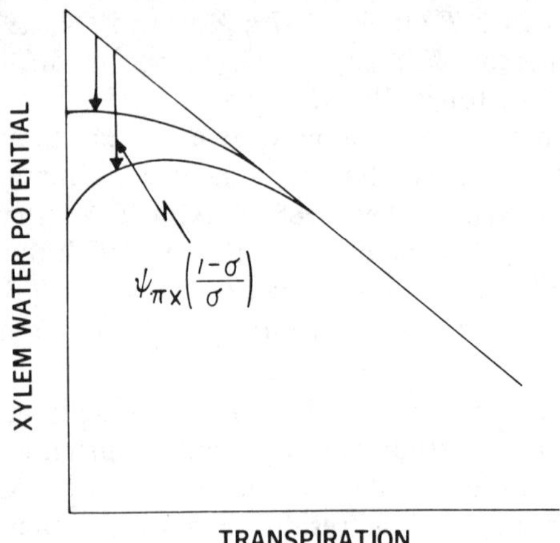

Fig. 9—Possible relationships between xylem water potential in the root and transpiration rate. These depend on xylem osmotic potential, ψ_π, and the effective reflection coefficient, σ, from Eq. [4].

rate of active salt uptake, which may increase with external salt levels. Both this factor, and the required low flow rate, suggest that this free-space salt effect will be more evident in potted or solution culture plants.

Although it is clear that salt accumulation in the xylem can cause large apparent changes in resistance at low flow, its application to the reported responses shown in Fig. 8 seems limited. Nonlinearity between T and water potential, such as shown in Fig. 9, would be measured at the leaf only if the high xylem salt levels are distributed throughout the leaf cell walls, and not if the salts were removed from free space in the upper plant parts. The magnitude of the effect is of course limited by xylem osmotic potential, which may rarely be < -1 or -2 bars. If the reflection coefficient is constant, the expected response with increasing flow near flow $= 0$ is a low or negative resistance (Fig. 9), not the high resistance characteristic of (2) and (3) in Fig. 8. At higher flow there is nothing in the proposed mechanism that tends toward a constant leaf water potential over a range of T, the typical responses 3 and 4 in Fig. 8.

There is also a possibility of solute accumulation outside the xylem due to the filtering effect of the root barrier. Nulsen and Thurtell (1978) have shown an apparent increase in plant resistance with time, presumably due to salt buildup outside root. However, this effect cannot explain a decreasing resistance with flow unless higher flow rates actually decrease external salt levels.

3. CAPACITANCE EFFECTS

A second possibility for apparent changes in resistance is associated with capacitance. If the transpiration rate is changing, and the plant capacitance is large, then the associated time responses can lead to hysteresis in ψ_L vs. T as shown in Fig. 10. When T starts from zero, the rate

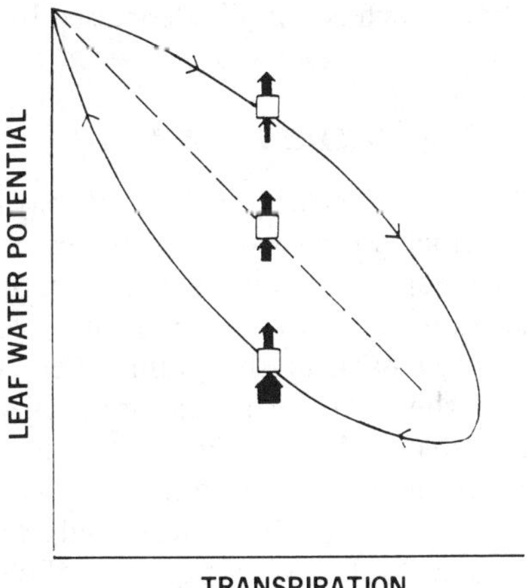

Fig. 10—Hysteresis in the relationship between leaf water potential and transpiration rate caused by the time lag associated with changes in tissue hydration. Plant tissues are represented by the open square, and the relative magnitudes of liquid water flux into and out of the tissue are represented as solid arrows.

of water loss from the leaves initially exceeds uptake. Leaf water potential requires time to fall to the value that maintains steady-state flow, and until that time the plant resistances downstream are underestimated. When T is decreasing the opposite occurs, and resistances are overestimated. This type of behavior always occurs diurnally to some extent and has been demonstrated to a very significant degree for grapevines (Liu et al., 1978) and apple trees (Powell & Thorpe, 1977). It is difficult to evaluate how significant this has been in other studies with herbaceous plants, but resistances calculated early or late in the day are especially suspect. The root resistance in Douglas fir calculated by Nnyammah et al. (1978), for example, increased about twofold over the day, but midday values were constant over all transpiration rates.

There may, of course, be other changes with time, either real changes in membrane resistance or slow adjustments to salt balance that produce apparent changes at low flow. The very long times (> 1 h) cited by Bunce (1978) required for steady behavior in potted soybean may be related to the latter and does not follow the expected capacitance behavior shown in Fig. 10.

Another type of nonsteady flow occurs in tissue that is growing. Boyer (1974, 1977) has suggested this as the reason for high apparent resistance at low flow rates (response 2 in Fig. 8). However, as Cowan (1977) has pointed out, a significant decrease in ψ_L due to growth should only decrease the apparent resistance. If the flow paths for transpiration and growth were identical, then the effect of each on ψ_L would be additive: ψ_L vs. T would have the same slope but be shifted downward by the amount of the water potential gradient associated with growth. If the flow paths differ, which seems more probable, then some of the transpiration stream bypasses the mesophyll vacuoles, either in the cell walls or the bundle sheath-extension path. The apparent resistance at low flow is less in this case since the vacuole water potential, which is initially depressed by growth, is less sensitive to T.

4. THE LOCATION OF ψ_L IN THE FLOW PATH

A third possibility of nonlinearity in ψ_L vs. T may be a variable relation between the point at which we measure leaf water potential and the flow path. Within the whole plant, for example, the potential at each leaf is partly influenced by the transpiration rates from other leaves, since the total flow affects the potential back at the point where the pathways are in common. It was seen earlier that leaf-to-leaf variability in T can be large. It was implied by Denmead and Millar (1976) that many of the results demonstrating variable resistance might be explained by a variation in the proportion of the total flow moving through the measured leaf.

In the example in Fig. 11, the water potential of a monitored upper leaf is greater for a given transpiration rate when fewer other leaves are transpiring. The dashed line would result, for example, if there were increasing stomatal closure of lower (unmeasured) leaves as water potential decreased. The same pattern would be followed if the different $T - \psi_L$ points repre-

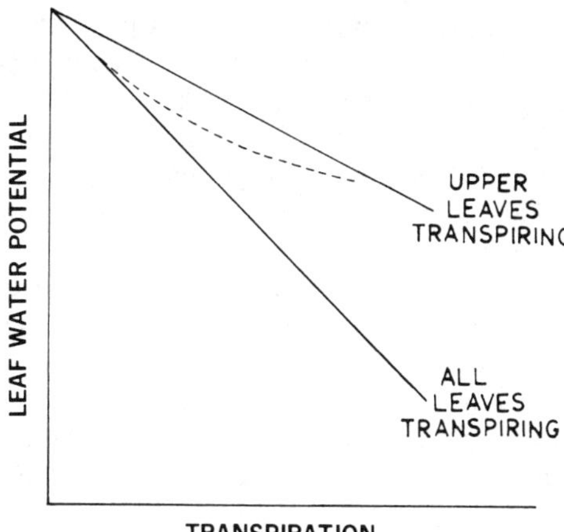

Fig. 11—Dependence of water potential at a specific leaf on the transpiration rate from other leaves.

sented different leaves across a single plant: the leaves at low T (in shade for example) would have a lower than average proportion of the total flow through the plant, while leaves at high T would have a higher proportion. Both of these errors result in a response that matches type 2 in Fig. 8. Although they may have been present to some degree in all studies, it would be difficult to conclude that they are the sole cause of the consistent type 2 responses.

There may also be flow-distribution changes within the leaf. The suggestion by Cowan (1977) that this may vary with environmental conditions was discussed earlier. He also pointed out that ψ_L, as measured in the meso-

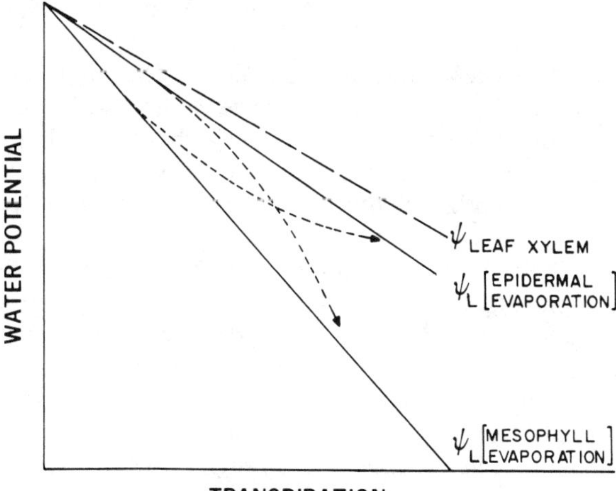

Fig. 12—Dependence of the leaf water potential—transpiration rate relationship on pathway of liquid water flow between leaf xylem and evaporation sites. Water potential in the leaf xylem is the upper line with large dashes. Leaf water potential (solid lines), defined as the average value within the mesophyll vacuoles, is less than that in the xylem, depending on the degree to which the mesophyll is in the flow pathway.

phyll, will be less tightly coupled to the transpiration rate as less of the evaporation takes place there. For a leaf below air temperature, for example, most of the evaporation may be from the epidermal and guard cells. The mesophyll would then remain closer to the potential in the vascular bundle, especially so if the pathway to the epidermis bypasses the mesophyll to a large extent. This is illustrated by the middle line in Fig. 12.

There seems to be no reason, however, for this type of nonlinearity to consistently be the response 2 type (decreasing resistance with flow). Also, there is certainly no mechanism for maintaining a constant ψ_L over a range of flow rates (responses 3 and 4). Moreover, the magnitude of any effect is limited to the drop in water potential from xylem to mesophyll within the leaf, and this is probably exaggerated in Fig. 12. It was argued earlier that the in situ hygrometer or psychrometer measures a potential close to that of the adjacent xylem, while the pressure chamber measures the mean mesophyll potential. If the effect suggested by Fig. 12 is significant, then the difference between the two measurements would be largest for conditions favoring mesophyll evaporation, and at low water potentials. The pressure chamber should read lower than the hygrometer in the dry range. However, results have either shown just the opposite (Baughn & Tanner, 1976), or a good agreement between the two instruments over all potentials (Campbell & Campbell, 1974; Ike et al., 1978b). This implies that the xylem-mesophyll gradient in leaf tissue is not large.

Many of the reported ψ_L vs. T results summarized in Fig. 8 cannot be convincingly explained. The attempt at least makes explicit which variables may be involved and should be controlled in future experiments. The most troublesome results are those showing constant ψ_L with T, especially at higher flow rates. It is particularly difficult to explain some of the associated low water potentials at $T = 0$, and the insensitivity of this intercept to nutrient solution potential (Macklon & Weatherly, 1965) or death of the roots (Stoker & Weatherly, 1971). These results are not typical, however, and have never been shown for field plants. The implication that there is a mechanism for adjusting liquid phase resistance, such that ψ_L remains constant, seems doubtful. It seems more likely that if there were regulation it would be based on the component potentials. It would be interesting to know how these potentials change in experiments reporting constant leaf water potential over a range of flow rates.

There has, however, been a commonly measured change in resistance even with field plants: high apparent resistance at low flow rates, decreasing to a constant value at high T (response 2). All of the factors discussed in this section could have some application here, but none can explain the consistency with which this particular curve is measured. It is difficult to avoid the conclusion that some basic change in flow path or membrane permeability may often occur at low flow rates or high water status. Which of these two factors is the important one is not clear.

Powell (1978) has suggested that root resistance is controlled by root turgor—that resistance of endodermal cell membranes decreases with the loss of turgor associated with transpirational flow. However, Newman

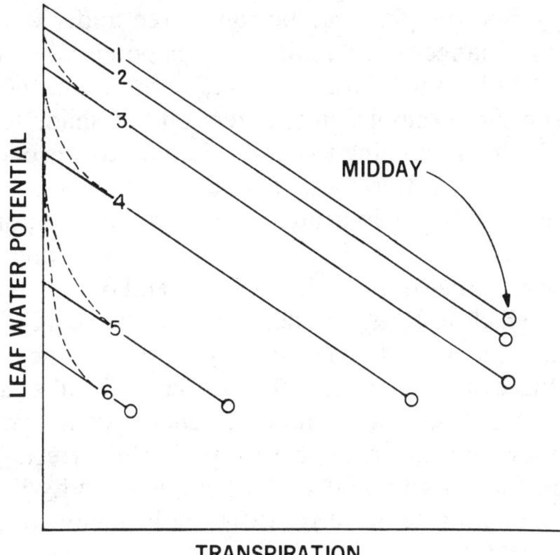

Fig. 13—Changes in leaf water potential with transpiration as soil dries. Individual curves represent increasing time from an initially wet soil condition. The stomata close, for progressively longer periods, from time 4 onward. The dashed lines show probable behavior when a few roots are in wetter soils.

(1974) pointed out that this type of control predicts different resistances depending on whether the flow is driven by transpiration/suction or by pressure applied to the roots, while in fact these two give generally similar results. Also, Green et al. (1979) showed that membrane resistance may actually increase with turgor loss, and this is the wrong direction to explain the results in Fig. 8.

E. ψ_L Behavior—Drying Soil

Despite the uncertainties considered this far, we need to go on and ask how leaf water potential changes as the soil dries. We will assume the simplest plant model: the liquid phase resistances and the steady-state pathway are both constant, salt transport and capacitance are ignored, and the potential drop within the soil is always negligible. These are reasonable assumptions for field plants, particularly at the high flow rates associated with minimum potentials during the day. Under these restrictions, ψ_L should decrease linearly with increasing T, and extrapolate to ψ_{soil} at zero flow. We will assume, further, that stomatal resistance increases sharply at a fixed critical turgor pressure.

With these assumptions, the behavior of leaf water potential over a drying cycle can be expected to look like Fig. 13. As soil water potential falls with time, the drop in ψ_L parallels it over the range in which the stomates, and hence maximum T, are unaffected (up to t = 3 in Fig. 13). As drying progresses, and leaf turgor approaches zero, the stomata of transpiring leaves close at some point. Associated with this there is an important change in the water potential gradients within the plant. Although the soil continues to dry, and the plant can be considered under water stress, there may be very little change in the midday water potential of exposed leaves (circles). There will be much larger changes in minimum potential in non-transpiring tissue, for example, in the stem and in shade leaves. This is a result of the whole plant coming to equilibrium, and approaching equilibrium with the soil, as flow decreases. It is obvious that stomatal closure and plant water stress are better indicated by changes in the potential of non-transpiring tissue, e.g., ψ_{stem} or $\psi_{stem} - \psi_L$, than by ψ_L alone.

The ψ_L at which the stomata close will depend on the osmotic potential in those leaves, which in turn may depend on osmotic adjustment and the rate of drying. If the leaf has the capacity, and the time, to decrease the minimum $\psi_{\pi L}$ successively each day, then the critical soil and leaf water potentials for stomatal closure are similarly extended. In soils with low capacitance, drying may be rapid enough to restrict this effect. This would be especially severe for potted plants and, as emphasized by Begg and Turner (1977), may be a major reason for the different behaviors of potted vs. field plants.

The continuously linear $\psi_L - T$ behavior described in Fig. 13 actually ignores two important difficulties in dry soil. First, soil water potentials are rarely uniform. A few deep roots into moist soil, for example, may allow the plant to wet up slowly at night, as indicated by the dashed lines in Fig. 13. There would actually be all degrees of this effect, and hence nonlinearity to some extent throughout the range of T.

The second difficulty with drying soil is that at some point the rhizosphere water potential gradients and soil resistances become a significant part of the total. The slope of the lines in Fig. 13 would then increase. Additionally, since the water content at the root surface and surrounding soil now undergoes significant diurnal changes, there may be large time lags present. This is evident, for example, in the $\psi_L - T$ data of Sterne et al. (1977) and Jones (1978), but only when soil was dry. Plant capacitance may be involved here also, since this is most significant in the dry range, but the involvement of soil is suggested by the necessary apparent change in resistance.

Although there may be increasingly nonlinear behavior with soil drying, the model in Fig. 13 still has relevance for minimum midday potentials, and for understanding how these change with drying. It implies that throughout the range in which soil resistance can be ignored, we are limited only by the capacity of osmotic adjustment to keep up with soil drying. How low can leaf osmotic potential fall to maintain turgor? There is a possibility of toxicity at low ψ_π, as indicated by salt-stress studies. On the other hand, some xerophytes have survived at extremely low water potentials.

There may be a much more basic restriction to the range of leaf osmotic potential. For adapted plants, leaf osmotic potential may be as low as is necessary to allow extraction of the bulk of the soil water. It is evident in Fig. 5 that there simply is not much water left below soil water potentials of -5 bars. Almost all the water is held within -15 bars, and for these two soils, over 60% of this has been lost by −1 bar and over 90% by −5 bars. Maintaining positive turgor and low stomatal resistance down to −5 bars requires a minimum ψ_L of perhaps −20 bars, depending on plant resistance. An investment by the plant in a much lower osmotic potential, and a root system that could maintain low perirhizal gradients in dry soil, would allow the stomates to remain open only a short time longer in a drying period. Looked at in this way, the striking uniformity in minimum leaf water potential for all mesophytes is simply a reflection of the uniformity in the capacitance distribution of most soils.

This model proposes that plants can maintain open stomates, and probably maximum photosynthesis, as long as the soil water potential is above a critical value. This value is low enough to encompass almost all the available soil water as long as (i) rhizosphere gradients remain small (root-soil contact is adequate), (ii) plant resistance does not increase appreciably, and (iii) drying is slow enough to allow osmotic adjustment within the plant. All these assumptions may often hold, and to this extent plants are well-adapted to soil drying.

This has relevance to many of the proposals for increasing drought resistance. Any of these that operate independently of soil water level must be very limited. For example, one could lower the critical soil water potential for stomatal closure, either by increasing the plant water potential (lowering resistances), or by increasing plant tolerance to low water potential (perhaps through changes in ϵ_T and ψ_π). Either would have little effect in maintaining production below soil water potentials of −5 bars since there is very little additional water to be gained. Some of these strategies may, however, have application to decreasing the critical soil potential for survival. There would be little flow in this case, so that the resistances again are irrelevant, but increasing plant tolerance to low water potentials could be important.

Production, nevertheless, requires water. When this is limited the successful strategies must be based on increasing the efficiency with which it is used. This can be the basic instantaneous water-use efficiency, defined earlier, or long-term measures that consider water conservation and different plant sensitivities at different stages. Some of the possible strategies are reviewed by Fischer and Turner (1978), and Loomis in Chapters 8A of this book. The problem is that any mechanism which reduces water loss through the stomata always tends to reduce production, and is therefore unwanted when water is not limited. The optimum behavior will always be a compromise as long as there is uncertainty in the future soil water balance.

IV. SUMMARY

A comprehensive model of water transport in plants can be imagined, based on the water status of soil, plant, and atmosphere, and on the functions describing resistances to flow in each. Unfortunately, any such theory that was physically basic and general would require far more information than we now have.

Transport at the leaf surface is difficult due to variable turbulence, and in the soil due to large changes in resistance with water content. In both areas, there may be large gradients at the immediate boundaries of the plant where the geometry is complex. The best that can be done in many cases is simply to delimit the conditions for which simpler, less structural, models are adequate. By considering the energy restraint on evaporation, we can greatly decrease the uncertainty from errors in aerodynamic resistance, and even ignore the latter for certain combinations of humidity and stomatal resistance. At high moisture levels we can ignore soil resistance, although we need more information on how this changes with drying.

It seems that the same tactic might be used in the plant where the flow processes are more complex. We know so little about the relative permeabilities of different materials, and their geometry within the plant, that the main pathways for water movement outside the xylem are uncertain. Most importantly, within the plant we are confronted with an additional dimension: a directed set of mechanisms that influence water movement to the plant's benefit. It has been noted that midday water potentials of illuminated leaves are very similar for most species, with perhaps twofold differences. Many of these species operate close to their critical turgor pressures, so that a decrease in midday water potentials of just a few bars may be harmful. This implies very precise control, yet resistance analog models have order-of-magnitude uncertainties in many required variables. The hope of accurately predicting water potentials, then, using a detailed model of tissue components, seems far off. It seems that in parallel with these structural models, we need more information about the restraints on water status that are governed by the plant.

LITERATURE CITED

Ackerson, R. C., and D. R. Krieg. 1977. Stomatal and nonstomatal regulation of water use in cotton, corn, and sorghum. Plant Physiol. 60:850–853.

Barrs, H. D. 1973. Controlled environment studies of the effects of variable atmospheric water on photosynthesis, transpiration and water status of *Zea mays* L. and other species. p. 249–258. *In* Plant response to climatic factors. Proc. UNESCO Uppsala Symp. UNESCO, Paris.

Baughn, J. W., and C. B. Tanner. 1976. Excision effects on leaf water potential of five herbaceous species. Crop Sci. 16:184–190.

Beadle, C. L., N. C. Turner, and P. G. Jarvis. 1978. Critical water potential for stomatal closure in Sitka spruce. Physiol. Plant. 43:160–165.

Begg, J. E., and N. C. Turner. 1977. Crop water deficits. Adv. Agron. 28:161–217.

Biscoe, P. V., Y. Cohen, and J. S. Wallace. 1976. Community water relations: diurnal and seasonal changes of water potential in cereals. Philos. Trans. R. Soc. London Ser. B. 273:565–580.

Boyer, J. S. 1974. Water transport in plants: mechanism of apparent changes in resistance during absorption. Planta 117:187–207.

Boyer, J. S. 1977. Regulation of water movement in whole plants. *In* Integration of activity in the higher plant. Symp. Soc. Exp. Biol. 31:455–470.

Brown, K. W., W. R. Jordan, and J. C. Thomas. 1976. Water stress induced alterations of the stomatal response to decreases in leaf water potential. Physiol. Plant. 37:1–5.

Bunce, J. A. 1978. Effects of shoot environment on apparent root resistance to water flow in whole soybean and cotton plants. J. Exp. Bot. 29:595–601.

Burbano, J. L., T. D. Pizzolato, P. R. Morey, and J. D. Berlin. 1976. An application of the Prussian Blue technique to a light microscope study of water movement in transpiring leaves of cotton. J. Exp. Bot. 27:134–144.

Byott, G. S., and D. W. Sheriff. 1976. Water movement into and through *Tradescanti virginiana* (L.) leaves: II. Liquid flow pathways and evaporative sites. J. Exp. Bot. 27:634–637.

Byrne, G. F., J. E. Begg, and G. K. Hansen. 1977. Cavitation and resistance to water flow in plant roots. Agric. Meteorol. 18:21–35.

Caldwell, M. M. 1976. Root extension and water absorption. p. 63–85. *In* O. L. Lange et al. (ed.) Water and plant life. Springer-Verlag, Berlin.

Camacho-B., S. E., A. E. Hall, and M. R. Kaufmann. 1974. Efficiency and regulation of water transport in some woody and herbacious species. Plant Physiol. 54:169–172.

Campbell, G. S., and M. D. Campbell. 1974. Evaluation of a thermocouple hygrometer for measuring leaf water potential in situ. Agron. J. 66:24–27.

Clark, J. A., and G. Wigley. 1975. Heat and mass transfer from real and model leaves. p. 413–422. *In* D. A. deVries and N. H. Afgan (ed.) Heat and mass transfer in the biosphere. Part I. Scripta, Washington, D.C.

Cowan, I. R. 1965. Transport of water in the soil-plant-atmosphere system. J. Appl. Ecol. 2: 221–239.

Cowan, I. R. 1977. Stomatal behavior and environment. Adv. Bot. Res. 4:117–228.

Cowan, I. R., and F. L. Milthorpe. 1968. Plant factors influencing the water status of plant tissues. p. 137–193. *In* T. T. Kozlowski (ed.) Water deficits and plant growth. Vol. 1. Academic Press, New York.

Cowan, I. R., and J. A. Troughton. 1971. The relative role of stomata in transpiration and assimilation. Planta 97:325–336.

Cutler, J. M., and D. W. Rains. 1977. Effects of irrigation history on responses of cotton to subsequent water stress. Crop Sci. 17:329–335.

Dalton, F. N., P. A. C. Raats, and W. R. Gardner. 1975. Simultaneous uptake of water and solutes by plant roots. Agron. J. 67:334–339.

Denmead, O. T., and E. F. Bradley. 1973. Proc. 1st Australian Conf. on Heat and Mass Transfer, Monash University, Melbourne, Australia.

Denmead, O. T., and B. D. Millar. 1976. Water transport in wheat plants in the field. Agron. J. 68:297–303.

Dube, P. A., K. R. Stevenson, G. W. Thurtell, and H. H. Neumann. 1975. Steady state resistance to water flow in corn under well watered conditions. Can. J. Plant Sci. 55:941–948.

Edwards, M., and H. Meidner. 1978. Stomatal responses to humidity and the water potentials of epidermal and mesophyll tissue. J. Exp. Bot. 29:771–780.

Esau, K. 1965. Plant anatomy. John Wiley & Sons, New York.

Faiz, S. M. A., and P. E. Weatherley. 1977. The location of the resistance to water movement in the soil supplying the roots of transpiring plants. New Phytol. 78:337–347.

Farquahar, G. D. 1978. Feed forward responses of stomata to humidity. Aust. J. Plant Physiol. 5:787–800.

Fereres, E., E. Acevedo, D. W. Henderson, and T. C. Hsiao. 1978. Seasonal changes in water potential and turgor maintenance in sorghum and maize under water stress. Physiol. Plant. 44:261–267.

Ferrier, J. M., and J. Danity. 1977. Water flow in *Beta-vulgaris* storage tissue. Plant Physiol. 60:662–665.

Fischer, R.A., and N. C. Turner. 1978. Plant productivity in the arid and semiarid zones. Annu. Rev. Plant Physiol. 29:277–317.

Fiscus, E. L. 1975. The interaction between osmotic- and pressure-induced water flow in plant roots. Plant Physiol. 55:917–922.

Fiscus, E. L. 1977a. Determination of hydraulic and osmotic properties of soybean root systems. Plant Physiol. 59:1013–1020.

Fiscus, E. L. 1977b. Effects of coupled solute and water flow in plant roots with special reference to Brouwers' experiment. J. Exp. Bot. 28:71–77.

Gardner, W. R. 1960. Dynamic aspects of water availability to plants. Soil Sci. 89:63–73.

Gardner, W. R. 1968. Availability and measurement of soil water. p. 107–135. *In* T. T. Kozlowski (ed.) Water deficits and plant growth. Vol. I. Academic Press, New York.

Gates, D. M. 1964. Leaf temperature and transpiration. Agron. J. 56:273–277.

Green, W. N., J. M. Ferrier, and J. Dainty. 1979. Direct measurement of water capacity of *Beta-vulgaris* storage tissue and resulting values for cell membrane hydraulic conductivity. Can. J. Bot. 57:981-985.

Hailey, J. L., E. A. Hiler, W. R. Jordan, and C. H. M. van Bavel. 1973. Resistance to water flow in *Vigna sinensis* L. (Endl.) at high rates of transpiration. Crop Sci. 13:264-267.

Herkelrath, W. N., E. E. Miller, and W.R. Gardner. 1977. Water uptake by plants: I. Divided root experiments. Soil Sci. Soc. Am. J. 41:1033-1038.

Huck, M. G., B. Klepper, and H. M. Taylor. 1970. Diurnal variations in root diameter. Plant Physiol. 45:529-530.

Ike, I. F., G. W. Thurtell, and K. R. Stevenson. 1978a. The relationship between steady-state transpiration rates and leaf water potential in cassava (*Manihot esculenta* cv. Llamera). Can. J. Bot. 56:1537-1539.

Ike, I. F., G. W. Thurtell, and K. R. Stevenson. 1978b. Evaluation of the pressure chamber technique for measurement of leaf water potential in cassava (*Manihot* species). Can. J. Bot. 56:1638-1641.

Janes, B. E. 1970. Effect of carbon dioxide, osmotic potential of nutrient solution and light intensity on transpiration and resistance to flow of water in pepper plants. Plant Physiol. 45:95-103.

Jarvis, P. G. 1975. Water transfer in plants. p. 369-394. *In* D. A. deVries and N. H. Afgan (ed.) Heat and mass transfer in the biosphere. Part 1. Scripta, Washington, D.C.

Jones, H. G. 1976. Crop characteristics and the ratio between assimilation and transpiration. J. Appl. Ecol. 13:605-622.

Jones, H. G. 1978. Modelling diurnal trends of leaf water potential in transpiring wheat. J. Appl. Ecol. 15:613-626.

Kanemasu, E. T., and C. B. Tanner. 1969. Stomatal diffusion resistance of snap beans: I. Influence of leaf-water potential. Plant Physiol. 44:1547-1552.

Kaufmann, M. R. 1976. Water transport in plants: current perspectives. p. 313-327. *In* I. F. Wardlaw and J. B. Passioura (ed.) Transport and transfer processes in plants. Academic Press, New York.

Knoerr, K. R. and L. W. Gay. 1965. Tree leaf energy balance. Ecology 45:17-24.

Lemon, E. R. 1965. Micrometerorology and the physiology of plants in their natural environment. p. 203-227. *In* F. C. Steward (ed.) Plant physiology, a treatise. Vol. 4A. Academic Press, New York.

Linacre, E. T. 1972. Leaf temperatures, diffusion resistances, and transpiration. Agric. Meteorol. 10:365-382.

Liu, W. T., W. Wenkert, L. H. Allen, Jr., and E. R. Lemon. 1978. Soil-plant water relations in a New York vineyard: resistances to water movement. J. Am. Soc. Hortic. Sci. 103: 226-230.

Macklon, A. E. S., and P. E. Weatherley. 1965. Controlled environment studies of the nature and origin of water deficits in plants. New Phytol. 64:414-427.

Meidner, M. 1975. Water supply, evaporation and vapor diffusion in leaves. J. Exp. Bot. 26: 666-673.

Meiri, A., and W. P. Anderson. 1970. Observations on the effects of pressure difference between the bathing media and exudates of excised maize roots. J. Exp. Bot. 21:899-907.

Millar, B. D., and O. T. Denmead. 1976. Water relations of wheat leaves in the field. Agron. J. 68:303-307.

Molz, F. J. 1976. Water transport through plant tissue: the apoplasm and symplasm pathways. J. Theor. Biol. 59:277-292.

Molz, F. J., and E. Ikenberry. 1974. Water transport through plant cells and cell walls: theoretical development. Soil Sci. Soc. Am. Proc. 38:699-704.

Monteith, J. L. 1964. Evaporation and environment. *In* G. E. Goff (ed.) The state and movement of water in living organisms. Symp. Soc. Exp. Biol. 19:205-234.

Monteith, J. L. 1973. Principles of environmental physics. Edward Arnold, London.

Neumann, H. H., G. W. Thurtell, and K. R. Stevenson. 1974. In situ measurements of leaf water potential and resistance to water flow in corn, soybean and sunflower at several transpiration rates. Can. J. Plant Sci. 54:175-184.

Newman, E. I. 1969a. Resistance to water flow in soil and plant: I. Soil resistance in relation to amounts of root: theoretical estimate. J. Appl. Ecol. 6:1-12.

Newman, E. I. 1969b. Resistance to water flow in soil and plant: II. A review of experiment and evidence on the rhizosphere resistance. J. Appl. Ecol. 6:261-272.

Newman, E. I. 1974. Root and soil water relations. p. 363–440. *In* E. W. Carson (ed.) The plant root and its environment. University Press, Charlottesville, Va.

Newman, E. I. 1976a. Interaction between osmotic and pressure-induced water flow in plant roots. Plant Physiol. 57:738–739.

Newman, E. I. 1976b. Water movement through root systems. Philos. Trans. R. Soc. London, Ser. B. 273:463–478.

Nnyammah, J. W., T. A. Black, and C. S. Tan. 1978. Resistance to water uptake in a Douglas fir forest. Soil Sci. 126:63–76.

Nulsen, R. A., and G. W. Thurtell. 1978. Osmotically induced changes in the pressure-flow relationship of maize root systems. Aust. J. Plant Physiol. 5:469–476.

Parkhurst, D. F., P. R. Duncan, D. M. Gates, and F. Kreith. 1968. Wind-tunnel modeling of convection of heat between air and broad leaves of plants. Agric. Meteorol. 5:33–47.

Parkhurst, D. F., and O. L. Loucks. 1972. Optimal leaf size in relation to environment. J. Ecol. 60:505–537.

Parlange, J.-Y. P. E. Waggoner, and G. H. Heichel. 1971. Boundary layer resistance and temperature distribution on still and flapping leaves: I. Theory and laboratory experiments. Plant Physiol. 48:437–442.

Parlange, J.-Y., P. E. Waggoner, and G. H. Heichel. 1972. Boundary layer resistance and temperature distribution on still and flapping leaves: II. Field experiments. Plant Physiol. 50:60–63.

Pearman, G. I., H. L. Weaver, and C. B. Tanner. 1972. Boundary layer heat transfer coefficients. Agric. Meteorol. 10:83–92.

Petty, J. A. 1978. Fluid flow through the vessels of birch wood. J. Exp. Bot. 29:1463–1469.

Powell, D. B. B. 1978. Regulation of plant water potential by membranes of the endodermis in young roots. Plant, Cell and Environment 1:69–76.

Powell, D. B. B., and M. R. Thorpe. 1977. Dynamic aspects of plant-water relations. p. 259–279. *In* J. J. Landsberg and C. V. Cutting (ed.) Environmental effects on crop physiology. Academic Press, London.

Raschke, K. 1956. Uber die physikalischen beziehungen zwischen warmeubergangszahl, strahlungsaustausch, temperatur and transpiration einesblattes. (In German.) Planta 48:200.

Raschke, K. 1975. Stomatal action. Annu. Rev. Plant Physiol. 26:309–340.

Raschke, K. 1976. How stomata resolve the dilemma of opposing priorities. Philos. Trans. R. Soc. London Ser. B. 273:551–560.

Rawlins, S. L. 1963. Resistance to water flow in the transpiration stream. *In* I. Zelitch (ed.) Stomata and water relations in plants. Conn. Agric. Exp. Stn. New Haven Bull. 664:69–85.

Rawson, H. M., J. E. Begg, and R. G. Woodward. 1977. The effect of atmospheric humidity on photosynthesis, transpiration and water use efficiency of leaves of several plant species. Planta 134:5–10.

Reicosky, D. C., and J. T. Ritchie. 1976. Relative importance of soil resistance and plant resistance in root water absorption. Soil Sci. Soc. Am. J. 40:293–297.

Richter, H. 1973. Frictional potential losses and total water potential in plants: a re-evaluation. J. Exp. Bot. 24:983–994.

Roberts, S. W., and K. R. Knoerr. 1978. In situ estimates of variable plant resistance to water flow in *Ilex opaca* Ait. Plant Physiol. 61:311–313.

Sheriff, D. W., and H. Meidner. 1974. Water pathways in leaves of *Hedera helix* L. and *Tradescantia virginiana* L. J. Exp. Bot. 25:1147–1156.

Spanswick, R. M. 1976. Symplasmic transport in tissues. p. 35–53. *In* U. Luttge and M. G. Pitman (ed.) Encyclopedia of plant physiology. New Ser., Vol. 2: Transport in plants II: Part B. Tissues and organs, Springer-Verlag, Berlin.

Stanhill, G., and S. Moreshet. 1966. p. 52–75. *In* 2nd Annu. Rep. to USDA Proj. AIO-SWC-29. U.S. Government Printing Office, Washington, D.C.

Sterne, R. E., M. R. Kaufmann, and G. A. Zentmyer. 1977. Environmental effects on transpiration and leaf water potential in avocado. Physiol. Plant. 41:1–6.

Stoker, R., and P. E. Weatherley. 1971. The influence of the root system on the relationship between the rate of transpiration and depression of leaf water potential. New Phytol. 70: 547–554.

Tanner, C. B. 1968. Evaporation of water from plant and soil. p. 73–106. *In* T. T. Kozlowski (ed.) Water deficits and plant growth. Academic Press, New York.

Taylor, H. M., and B. Klepper. 1978. The role of rooting characteristics in the supply of water to plants. Adv. Agron. 30:99–128.

Thom, A. S. 1975. Momentum, mass and heat exchange of plant communities. p. 57–109. *In* J. L. Monteith (ed.) Vegetation and the atmosphere. Vol. 1. Academic Press, London.

Tinker, P. B. 1976. Roots and water: transport of water to plant roots in soil. Philos. Trans. R. Soc. London, Ser. B. 273:445–461.

Tinklin, R., and P. E. Weatherley. 1966. On the relationship between transpiration rate and leaf water potential. New Phytol. 65:509–517.

Troughton, J. H., S. E. Camacho, and A. E. Hall. 1974. Transpiration rate, plant water status, and resistance to water flow in *Tidestromia oblongifoliz.* Carnegie Inst. Washington Yearb. 73:830–835.

Turner, N. C. 1974. Stomatal behavior and water status of maize, sorghum and tobacco under field conditions: II. At low soil water potential. Plant Physiol. 53:360–365.

Turner, N. C. 1975. Concurrent comparisons of stomatal behavior, water status, and evaporation of maize in soil at high and low water potential. Plant Physiol. 55:932–936.

Turner, N. E., J. E. Begg, and M. L. Tonnet. 1978. Osmotic adjustment of sorghum and sunflower crops in response to water deficits and its influence on the water potential at which stomata close. Aust. J. Plant Physiol. 5:597–608.

Tyree, M. T., and Y. N. S. Cheung. 1977. Resistance to water flow in *Fagus grandifolia* leaves. Can. J. Bot. 55:2591–2599.

Vogel, S. 1970. Convective cooling at low airspeeds and the shape of broad leaves. J. Exp. Bot. 21:91.

Wenkert, W. 1980. Measurement of tissue osmotic potential. Plant Physiol. 65:614–617.

Wenkert, W. 1981. The behavior of osmotic potential in leaves of maize. Environ. Exp. Bot. 21:231–239.

Wenkert, W., E. R. Lemon, and T. R. Sinclair. 1978. Water content-potential relationship in soya bean: changes in component potentials for mature and immature leaves under field conditions. Ann. Bot. (London) 42:295–307.

Williams, J. 1974. Root density and water potential gradients near the plant root. J. Exp. Bot. 25:669–674.

CHAPTER 4B

Water Transport and Balance Within the Plant: Stomatal Mechanics and Gas Exchange

J. ROBERT COOKE

Cornell University
Ithaca, New York

I. INTRODUCTION

Water transport within plants occupies a place of prominence in crop production—in both an economic and a physiological sense. It is this latter aspect that is the focus of Chapter 4A by Wenkert. Actually the scope is slightly broader, since an understanding of water transport within the plant depends in a fundamental manner upon the role of the two interfaces: plant-atmosphere and plant-soil.

The discussion that follows may be grouped into three categories: (i) general comments on mathematics in plant-water relations, (ii) specific comments on the mechanics of stomatal aperture, and (iii) some consideration of gaseous transport.

II. MATHEMATICS IN PLANT-WATER RELATIONS

Wenkert concentrates predominantly upon the physical aspects of water transport and succeeds in giving some useful order-of-magnitude characterizations of that process. Mathematical models occupy a prominent supporting role in that discussion. Indeed, the study of plant-water relations has benefited more than most aspects of the study of crop production from the application of mathematics. That mathematics will play an even more decisive role in crop physiology in the future is clear and is illustrated throughout this book.

Two rather different approaches are manifest in the literature. One may consider water transport on a whole-plant basis. On the other hand, one may consider more limited and specialized studies, e.g., stomatal diffusion, water movement through sieve plates, the role of the Casparian strip, etc. Crop physiologists tend to emphasize the former, although both approaches are productive.

Limitations to Efficient Water Use in Crop Production.

The "macro view" tends to emphasize a more empirical, statistically oriented view, while the "micro view" is probably more amenable to a rational, first-principle approach. Both approaches should be encouraged and, if successful, should lead to complementary points of view. In any case, a more incisive use of a quantitative approach is likely to result.

A rather heavy dependence upon the diffusion resistance concept (Cooke & Rand, 1980) is evident in the extensive literature stretching back to the turn of the century when Brown and Escombe introduced the concept. The utility of this approach in the study of diffusion of water vapor and CO_2 between the plant and its environment has been well-established. Its utility is greatest for the study of steady-state diffusion (Holcomb & Cooke, 1977). However, the characterization of the transient response of plants is also important, so the usefulness of the concept has been stretched as capacitors and time-dependent resistances are introduced to deal with the more general situation.

The concept of water potential, which evolved from classical thermostatics (or thermodynamics as it is commonly called), has gained widespread usage in the literature and is freely applied to nonequilibrium conditions. While passive transport of water in plants depends upon gradients in water potential, we should become more interested in the individual components of water potential. For example, the stomatal pore width, which regulates the plant-atmosphere exchange of gases, depends upon the turgor pressures of both the guard cell and the adjacent cells. A knowledge of water potential alone does not suffice to describe their fundamental role. In fact, the two turgor pressures—and not just their differences—must be known. More about this later.

Another example of the importance of studying the individual components of water potential is the relation of cell turgor to plant growth. While extensively studied in the plant physiology literature, only modest allusion has been made to this in the crop physiology literature. Some limited attention is being given by agricultural engineers to the mechanical aspects of plant tissue viewed as a liquid- and gas-filled porous medium. However, these studies should be extended to deal with the relationship of water transport to the mechanical aspects of cell enlargement and growth. Another equally important direction is the investigation of water movement in a leaf when subjected to a pressure chamber environment (Stroshine et al., 1979).

III. MECHANICAL ASPECTS OF STOMATAL APERTURE

Because of the crucial role played by stomata in regulating the exchange of gases between a plant and its environment, stomata have been examined with great intensity for many years. While most of the current attention in the literature has concentrated upon gaining an understanding of the biochemical aspects of stomatal response, the historical interest in the mechanical aspects continues. In a series of recent papers additional insight has been presented.

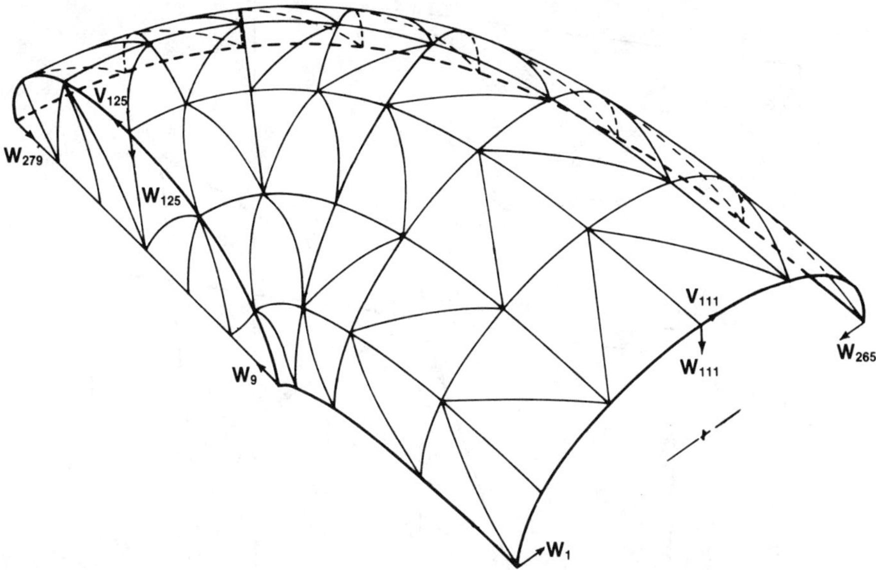

Fig. 1—A finite element discretization of a quadrant of a guard cell (reproduced with permission; Cooke et al. [1976]).

Cooke et al. (1976) analyzed the behavior of elliptical stomata using a computer-based technique—the finite-element method—which is especially well-suited for dealing with the complex geometries frequently associated with biological problems. The finite-element method is rapidly gaining in popularity as a result of decreasing cost and increasing availability of computing facilities, and the flexibility of the method in dealing with complex geometries and directional and space-dependent properties.

Figure 1 depicts the application of a structural engineering shell analysis program to a study of guard cell mechanics. By symmetry arguments one need only consider one quadrant of a doubly-elliptical shell, i.e., a torus with an elliptical cross section and an elliptical top view. This particular program is based upon the use of curvilinear triangles and a suitable discretizing mesh was constructed. Then continuity considerations were applied to contiguous triangular elements. Likewise, boundary conditions were applied to enforce continuity at the boundaries of the quadrant. The elastic properties of the shell were taken to be different in the directions parallel to and perpendicular to the microfibrils in the guard cell. Variable guard cell wall thickness such as occurs at the stomatal pore was also considered. A problem of this complexity would be completely intractable without the use of the finite element method and a computer. I believe this will become an increasingly attractive technique for studies of efficient water utilization in plants.

Figure 2 depicts the multilinear dependence of pore width of elliptical stomata upon the opposing pressures in the guard and subsidiary cells of an open stomata. When the pressure inside the guard cell increases (with constant subsidiary cell pressure), the pore width increases. On the other hand,

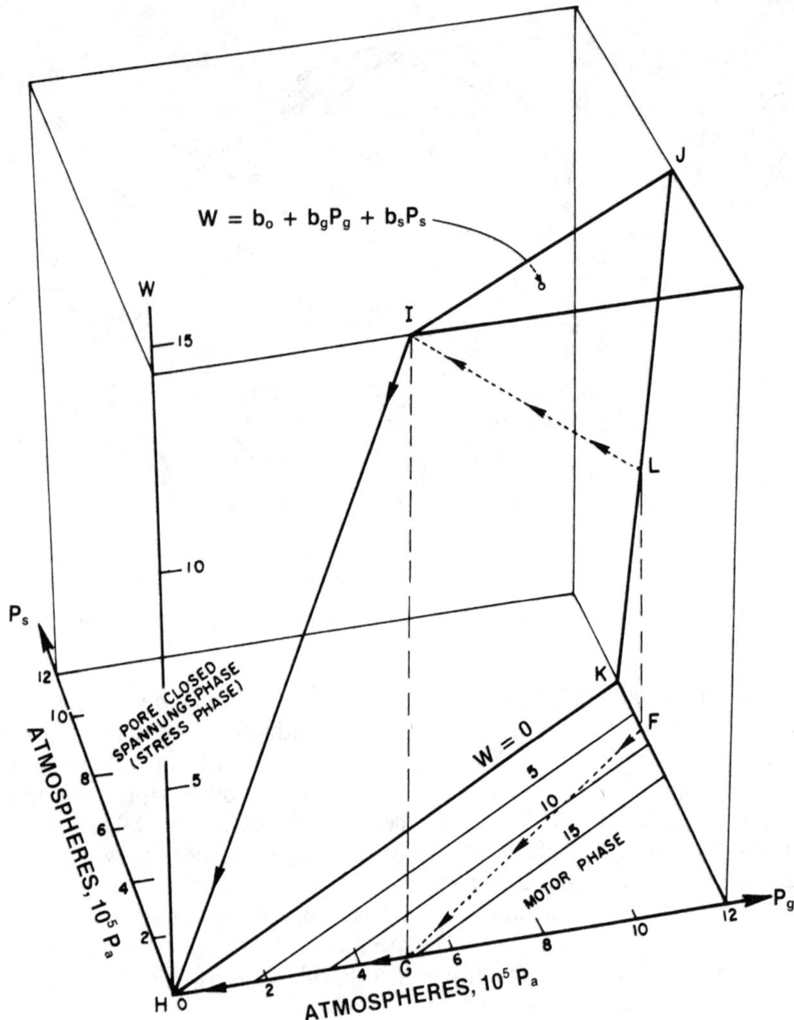

Fig. 2—Stomatal pore width as a function of guard cell and subsidiary cell turgor pressures (reproduced with permission, Cooke et al. [1976]).

for a constant guard-cell pressure, an increase in the subsidiary cell pressure decreases the pore width. These competing or "antagonistic" effects determine the pore width and hence the exchange rates for water vapor and CO_2 for given concentration differences in the gases. In short, these pressures, and not water potential per se, regulate the transport rate for water vapor.

Another interesting property results from this essential "antagonism" between the guard and subsidiary cells. As shown in Fig. 2, the pore width can actually increase with a decreasing guard-cell pressure (P_g), provided the subsidiary cell pressure (P_s) is decreasing sufficiently rapidly. Consider the path FG in the P_g, P_s pressure plane. At point F the pore width is given by locating the projection of point F (i.e., L.) on the HIJK plane, which specifies the width, w. In moving from point F to G the constant width contours projected from the HIJK plane onto the P_g, P_s plane indicate an increasing pore width, despite a decreasing guard cell pressure P_g. This is also apparent from the change from point L to point I. Then with a zero subsidiary cell pressure and a decreasing guard cell pressure, the pore closes. Note that because $b_g \neq b_s$ in the width equation (cf. Fig. 2), the pore width is not simply a function of the pressure difference, as many have supposed.

Once closed, however, the pressures do not affect the (zero width) pore in the region identified as the "stress phase." Notice that this is a well-defined region of the P_g, P_s plane.

The relative magnitudes of the three coefficients (b_o, b_g, b_s in Fig. 2) are determined by the geometrical and mechanical properties of the stomate. This sensitivity may be subject to genetic manipulation. The classical "thickened wall" hypothesis has been shown to not be a crucial factor in causing the opening of stomata. The radial cellulose microfibrils in the guard cell have been shown to enhance the antagonistic response to the guard and subsidiary cell pressures. Nevertheless, pore opening can occur without this radial stiffening. On the other hand, the elliptical shape of the guard cell plays a decisive role. A circular torus of reasonable dimensions, for example, would not exhibit the proper response. Inflation of a circular torus of reasonable dimensions results in pore closure, while deflation produces the opposite behavior, which would be catastrophic for a drought-stressed plant.

The multilinear relationship depicted in Fig. 2 does not apply for arbitrarily large pressures, of course. When the shell analysis takes geometric nonlinearities into consideration (Cooke et al., 1977) a maximum pore opening is predicted and the cross section of the guard cell does not decrease without limit. The pore length, however, does remain essentially constant, as predicted by the linear model.

IV. MECHANICS OF GASEOUS TRANSPORT

In addition to establishing the steady state pore width for prescribed pressures, this antagonism relationship also influences the transient response of the plant to environmental changes.

Consider the scenario depicted (Delwiche & Cooke, 1977) in Fig. 3 for the opening of a stomate. Suppose the osmotic pressure in the guard cell increases as in Fig. 3a. The hydrostatic (or turgor) pressure initiates pore opening. This permits a greater flux of water vapor from the substomatal cavity through the pore and into the atmosphere. The water loss from the cavity walls then decreases the turgor pressure, P_s, of the subsidiary cell. At this point the "antagonism" effect becomes apparent. Because the pore width is more sensitive to a unit drop in P_s than to a unit change in P_g, and because the pressure drop in P_s is greater than the drop in P_g, the pore opens rapidly.

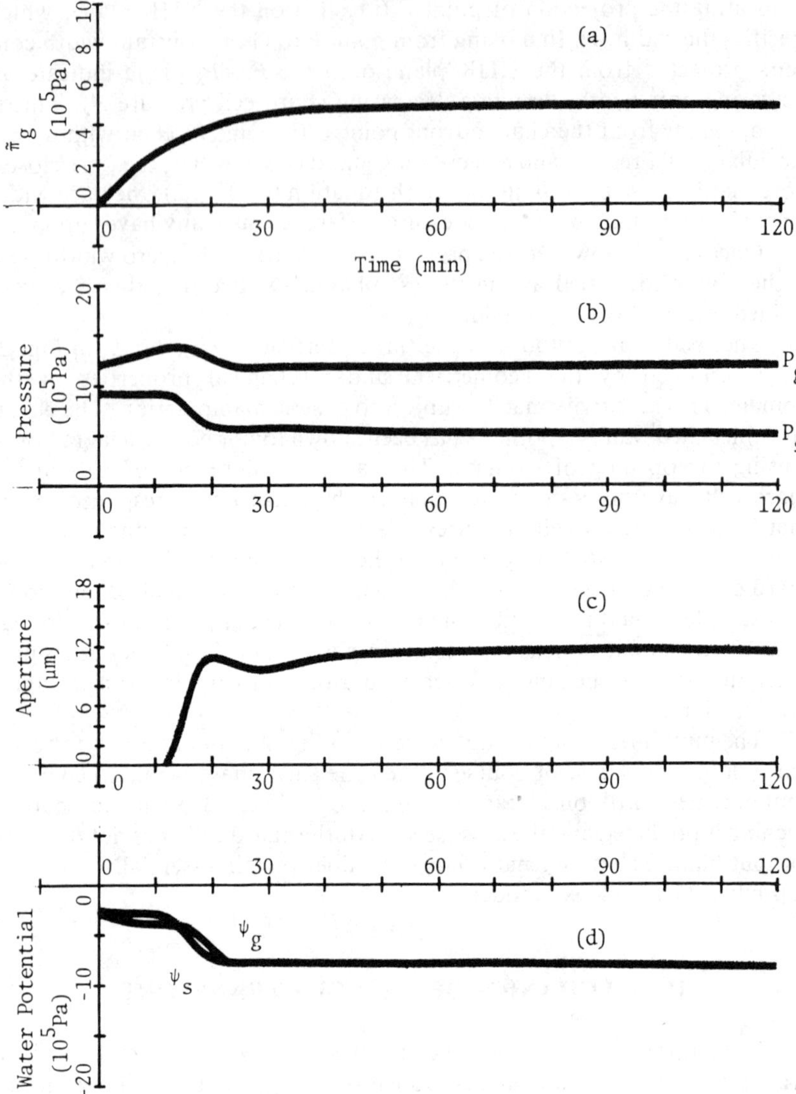

Fig. 3—Simulation of transient stomatal response to an increase in osmotic potential: (*a*) applied osmotic potential, (*b*) guard-cell and subsidiary-cell turgor pressures, (*c*) width of stomatal aperture, and (*d*) water potential for guard cells and subsidiary cells (reproduced with permission of J. Theor. Biol.; Delwiche and Cooke [1977]).

The behavior suggests that the rapidity of the pore opening is more sensitive to subsidiary-cell pressure change than to the guard-cell pressure change. Perhaps a lack of appreciation of the major role played by the subsidiary cell has led to an excessive concern with the limitation on the speed with which active transport into the guard cell and changing osmotic conditions can alter pore width.

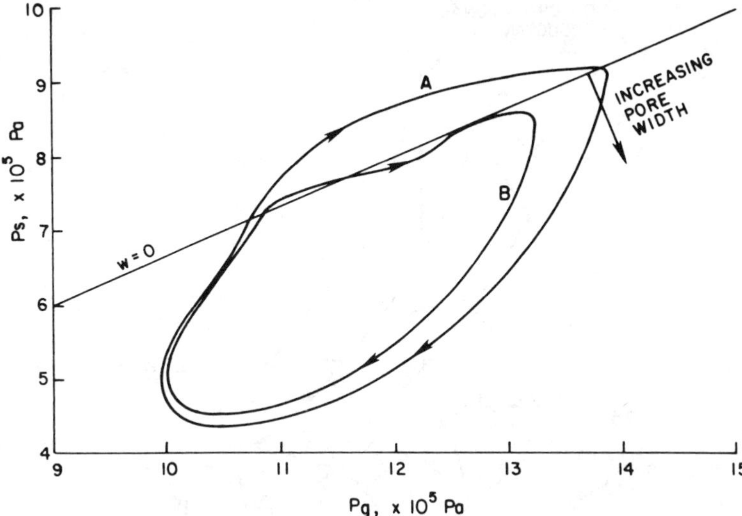

Fig. 4—(*A*) Limit cycle oscillations exhibited by the hydropassive feedback loop (Delwiche & Cooke, 1977). (*B*) Effect of CO_2 feedback loop on stomatal oscillations (Upadhyaya et al., 1980).

Endogenous rhythms in stomatal pore width are known to exist for many plants under certain *constant* environmental conditions. These oscillations can be shown to emerge naturally as a stable limit cycle in a simplified model (Delwiche & Cooke, 1977) consisting of a system of non-linear ordinary differential equations. Upadhyaya et al. (1981) have extended the model to include CO_2 effects and have shown that water-use efficiency under drought conditions is increased by the presence of these oscillations. Experimental verification is needed for this important calculation. Also, the mathematical model of the stomatal oscillations provides a basis for the explication of a generic biological oscillation in the absence of a varying excitation.

The outer curve in Fig. 4 shows the stable limit cycle that results when only the hydropassive, hydraulic aspects are considered. The oscillation proceeds as follows: water evaporates from the mesophyll and diffuses through the stomatal pore to the leaf exterior. This lost water is replenished by water moving from the roots and from the guard cells into the mesophyll cells. The resulting guard cell turgor drop decreases the pore width, which in turn slows the rate of evaporation and causes water to accumulate again in the mesophyll cells. This leads to a greater guard-cell turgor and an increased pore width. Then the process begins another cycle with a period on the order of 10–50 min.

Rand et al. (1981) have explored in more detail the mathematical properties of this model and have established the existence of an unstable limit cycle by using the theory of the Hopf bifurcation.

When the role of CO_2 is considered, an additional oscillation having a period of 2–5 min becomes apparent, as shown by the inner curve of Fig. 4

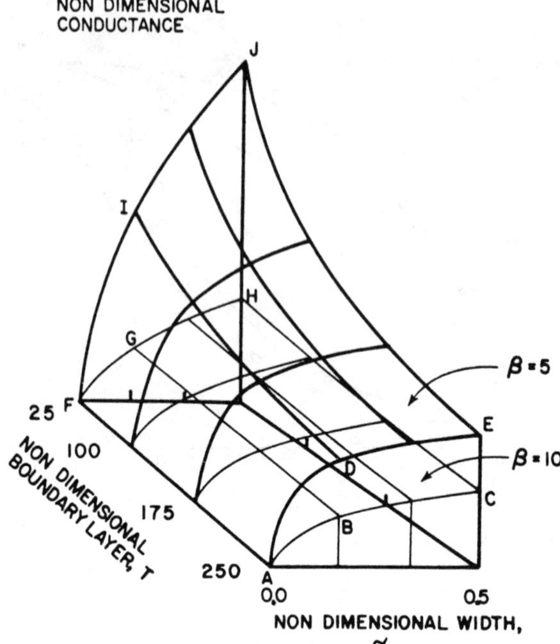

Fig. 5—A nondimensionalized display of stomatal diffusion rate as a function of pore width (α), boundary-layer thickness (T), and stomatal spacing (β) (reproduced with permission; Cooke and Rand [1980]).

(Upadhyaya et al., 1980). The CO_2-based oscillation is superimposed on the water-based oscillation and occurs only when the stomatal pore is nearly closed, i.e., in the vicinity of the $w = 0$ line, representing a closed stomate.

Let's turn now to the relationship between pore size and diffusion rate. The diffusion rate from a leaf is known to *not* be linearly proportional to the area of the pores and to depend upon pore spacing. Is there an optimal spacing for stomata that will maximize gaseous exchange? A scale model electrolytic tank analog of diffusion from a stomate can be used conveniently (Holcomb & Cooke, 1977). This modeling technique is especially convenient for studying diffusion when complex geometries are involved. For example, the technique is especially well-suited for the study of scale models of various stomatal configurations and stomatal distribution patterns.

As shown in Fig. 5 (Holcomb & Cooke, 1977), the diffusion rate J (on a unit leaf area basis) increases monotonically with increasing pore width, α, with decreasing pore spacing, β, and with decreasing boundary-layer thickness, T, so the answer to the question regarding an optimum spacing of stomata is "no." Figure 5 can also be used to predict the (limited) potential for altering gas-exchange properties of a leaf by genetic selection. Specifically, for a leaf with a "thick" boundary layer (i.e., large T), the stomate behaves essentially as an off-on valve. Said differently, little increase in gas exchange occurs (see curves BC and DE) even if the pore width increases substantially. On the other hand, the change (AB and AD) is dramatic for small changes when the pore is nearly closed.

But with a "thin" boundary layer the change is more gradual and loses this off-on quality (see FGH and FIJ). The diffusion rate increases substantially as the relative pore spacing (β) decreases—i.e., FIJ increases more markedly than does FGH.

V. SUMMARY

Let me conclude with an expression of optimism about the future of research dealing with the topics of this chapter (and, indeed, of this book). I believe that progress will be even more rapid than implied by Wenkert in Chapter 4A. The ability to meaningfully incorporate greater physiological detail in crop production studies will expand rapidly as the increasing availability of computers and as the utility of mathematical models becomes more apparent. The finite element method is one such computer-based tool which has great promise and deserves greater attention. The international energy crisis will surely focus more attention upon the importance of efficient water utilization in crop production.

LITERATURE CITED

Cooke, J. R., J. G. DeBaerdemaeker, R. H. Rand, and H. A. Mang. 1976. A finite element shell analysis of guard cell deformations. Trans. ASAE 19:1107–1121.

Cooke, J. R., and R. H. Rand. 1980. Diffusion resistance models. p. 93–121. *In* J. D. Hesketh and J. W. Jones (ed.) Predicting photosynthesis for ecosystem models. Vol. 1. CRC Press, Boca Raton, Fla.

Cooke, J. R., R. H. Rand, H. A. Mang, and Josse DeBaerdemaeker. 1977. A nonlinear finite element analysis of stomatal guard cells. Am. Soc. Agric. Eng. Paper no. 77-5511. ASAE, St. Joseph, Mich.

Delwiche, M. J., and J. R. Cooke. 1977. An analytical model for the hydraulic aspects of stomatal dynamics. J. Theor. Biol. 69:113–141.

Holcomb, D. P., and J. R. Cooke. 1977. An electrolytic tank analog determination of stomatal resistance. Am. Soc. Agric. Eng. Paper no. 77-5510. ASAE, St. Joseph, Mich.

Rand, R. H., S. K. Upadhyaya, J. R. Cooke, D. W. Storti. 1981. Hopf bifurcation in a stomatal oscillator. J. Math. Biol. 12:1–11.

Stroshine, R. L., J. R. Cooke, R. H. Rand, J. M. Cutler, and J. F. Chabot. 1979. Mathematical analysis of pressure chamber efflux curves. Am. Soc. Agric. Eng. Paper no. 79-4585. ASAE, St. Joseph, Mich.

Upadhyaya, S. K., R. H. Rand, and J. R. Cooke. 1980. A mathematical model of the effects of CO_2 on stomatal dynamics. Am. Soc. Agric. Eng. Paper no. 80-5517. ASAE, St. Joseph, Mich.

Upadhyaya, S. K., R. H. Rand, and J. R. Cooke. 1981. The role of stomatal oscillations on plant productivity and water-use efficiency. Am. Soc. Agric. Eng. Paper no. 81-4017. ASAE, St. Joseph, Mich.

CHAPTER 4C

Water Transport and Balance Within the Plant: Resistance to Water Flow in Roots

EDWIN L. FISCUS

Agricultural Research Service, USDA
Fort Collins, Colorado

I. INTRODUCTION

Water transport through root systems or segments of roots is frequently expressed in the plant physiological literature by the relationship

$$J_v = L_p(\Delta P_h - \sigma\Delta\pi), \qquad [1]$$

which is the equation presented for irreversible volume flow through semipermeable membranes (see Katchalsky & Curran, 1965). Here J_v is the flow of volume per unit area of root surface ($cm^3\ cm^{-2}\ s^{-1}$), ΔP_h is the hydrostatic pressure difference between the outside and the inside (xylem elements) of the root (bars), σ is the dimensionless reflection coefficient or osmotic efficiency, and $\Delta\pi$ is the osmotic pressure difference between the inside and outside of the root (bars). The term L_p is the so-called hydraulic conductivity coefficient ($cm^3\ cm^{-2}\ s^{-1}\ bar^{-1}$), which, unlike many expressions used in soils terminology, has the distance or length term absorbed into itself. Owing to the nature of the development of this relationship, the driving forces are expressed only as finite differences and not spatial derivatives (cf. Katchalsky & Curran, 1965; Chapter 10). In fact, the specific manner and anatomical structures over which these potential differences are dissipated is still a matter of controversy. Thus, Eq. [1] might be generally less useful if it were necessary to rigorously define the gradient. This distinction between a difference and a gradient is carried through to L_p, which, because the geometry of the system is not completely known, should properly be termed a conductance. In trying to establish some uniformity of terminology between plant physiology and other branches of science and engineering, we will refer to L_p as a conductance. One must bear in mind that other types of conductances are possible, depending on what parts of the system are considered unknown. We will use the symbol L_p for systems defined by Eq. [1], where the cross-sectional flux area is known, but not the gradient. The term conductivity will be confined to cases where the driving force is expressed as

Limitations to Efficient Water Use in Crop Production.

a spatial derivative. The same considerations apply to the additional inverse coefficients, the resistivity, and the resistance. A flux density ($V\,l^{-2}\,t^{-1}$) may be defined by a conductivity or conductance, but a flux ($V\,t^{-1}$) only by a conductance. Further, we could make use of the hydraulic head concept, which is more widely used in soil sciences, and write Eq. [1] as

$$J_v = L_p(\Delta H - \sigma\Delta\pi), \qquad [2]$$

where ΔH is the difference in hydraulic head ($\Delta H = \Delta h_p + \Delta h_g$), Δh_p is the difference in pressure head, and Δh_g is the difference in gravitational head. The notation used will not alter the sense of the following discussion.

It is generally considered to have been through the influence of the papers of Gradmann (1928) and van den Honert (1948) that the concept of resistance became part of the language of plant-water relations. However, Richter (1973) pointed out that it was really an earlier paper by Huber (1924) that was deserving of this credit. In any case, it was proposed that the flow of water through the various parts of the soil-plant-atmosphere continuum could be treated as a catena, analogous to the flow of current in an electrical circuit. This concept became established in the literature and came to be known as the *Ohm's law analogy.* The concept may be credited with stimulating much useful research in plant-water relations since its adoption. However, it did suffer from a serious shortcoming in that it was essentially an "instantaneous" concept, or was interpreted that way by many investigators. By "instantaneous" we mean that the value of the resistance to flow was calculated simply as the ratio of the instantaneous applied force to the instantaneous flux. Frequently implicit in the interpretation of such calculations are the assumptions that the relationship is linear and that it passes through the origin. So much confusion has been generated because of the difficulties inherent in the concept of the instantaneous resistance that we feel it is desirable to spend some time discussing two of the fundamental resistance concepts. The discussion will by no means be comprehensive.

II. A CONCEPTUAL MODEL OF FLOW RESISTANCE

As an illustration of the resistance concepts, we will make use of the system in Fig. 1. Throughout, however, one must bear in mind that this system is only for illustration of the resistance concepts and may or may not bear any resemblence, real or analogous, to any particular part of the soil-plant-atmosphere system. This simple system consists of a vertical pipe with a nonlimiting inlet, four exit ports, and a pressure transducer at the bottom. At any time during the following discussion any combination of the exit ports may be closed to illustrate a particular point. The flow of water will occur from left to right and the efflux from the system, J_w, will be determined as a function of the pressure, P_h, at the base of the pipe. The measured pressure will, of course, depend only on the height, Z, of the water column, with the reference elevation, Z_o, being at the level of the

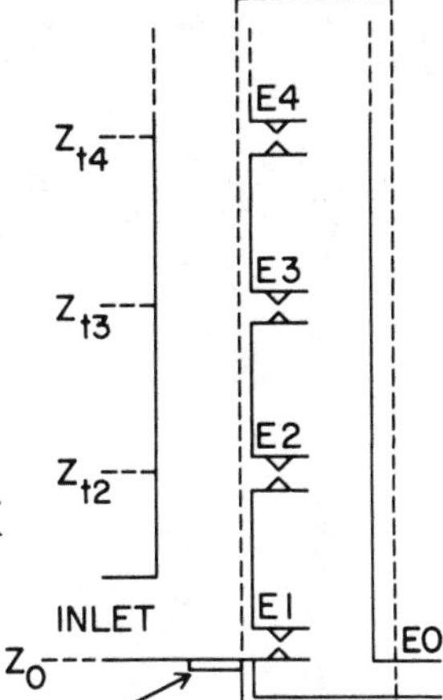

Fig. 1—Vertical pipe system with water outlet ports E1, E2, E3, and E4, and collective outlet at E0. The region within the dashed line is inaccessible.

transducer. The apertures of the exit ports may differ and only their relative sizes will be specified for each case considered. All efflux from the system will be collected at the outlet, EO, which will also be considered to be at Z_o. Further, the area enclosed by the dashed lines will be inaccessible to measurement. As the simplest case, consider that all exit ports except E1 are closed. In this situation, the flow of water from the system, J_w, will be described by a straight line passing through the origin, line I in Fig. 2A. This line may be defined as

$$P_h = J_w r. \quad [3]$$

Obviously, the proportionality factor, r, may be calculated as P_h/J_w and is everywhere the same. The ratio P_h/J_w is also the tangent of the angle α formed by the horizontal axis and the force/flux line. For future reference, we will define tan α (= P_h/J_w) as the instantaneous resistance. As it happens, tan α is also the slope of the force/flux curve in this instance, and it is this slope, at any point, that we will define as the "differential" resistance. In this case, line I of Fig. 2B describes both resistance functions. Throughout the rest of this paper we will use r_i and r_d to mean the instantaneous and differential resistances, respectively.

Next, consider the situation where only port E2 is open and that it has the same aperature as E1. In this case, no water may exit the system until the threshold height (Z_{t2}), therefore threshold pressure (P_{t2}), has been reached (line II, Fig. 2A). Further increases in pressure beyond P_{t2} result in the same proportional increases in flow as in line I. Line II may be described as

$$P_h = J_w r + P_{t2}. \quad [4]$$

The differential resistance, $\mathrm{d}P_h/\mathrm{d}J_w$ ($= r_d$), is the same for lines I and II, and one would not conclude from this that there was any difference in the apertures of ports E1 and E2. However, if one calculates the instantaneous

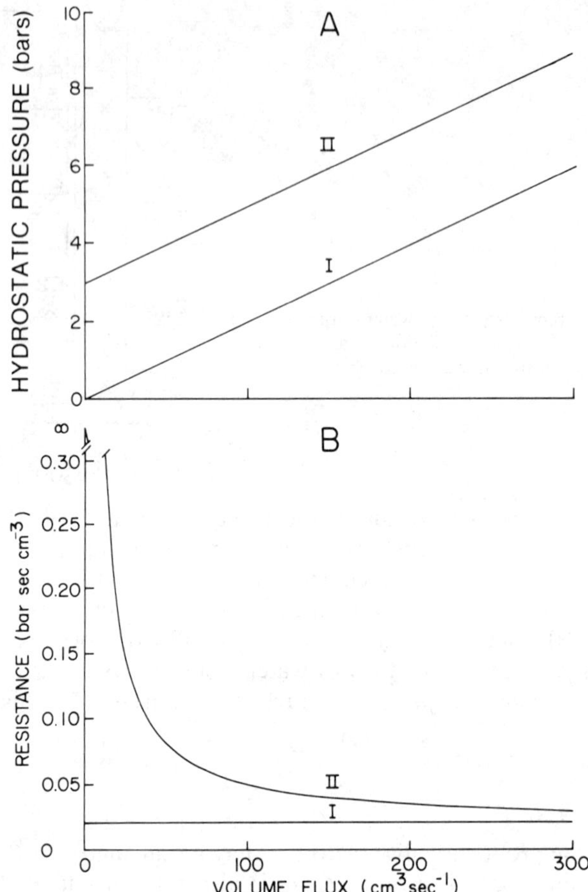

Fig. 2—Plot of pressure at base of pipe against volume flux (*A*), and system resistances against volume flux (*B*). In each plot line I is for the case when only pore E1 is open, and line II is when only pore E2 is open.

resistance for line II of Fig. 2A, quite a different picture emerges. The instantaneous resistance (line II, Fig. 2B) may take on any value between infinity as $J_w \rightarrow 0$ and the differential resistance value as J_w becomes infinite. Such a resistance function may clearly be erroneously interpreted as showing nonlinear flow characteristics. As a result, much time and effort may be expended searching for the causes of this nonlinearity when in fact it does not exist.

To further illustrate the pitfalls of the instantaneous resistance concept, consider one further example. In this example, E1 is closed and E3 and E4 have identical characteristics, which include much larger apertures than E2. In the first instance, E4 is closed and the flow curve will look like line I in Fig. 3A.

In the second instance, E3 is closed and E4 is opened so that flow will resemble line II of Fig. 3A. The corresponding differential resistance func-

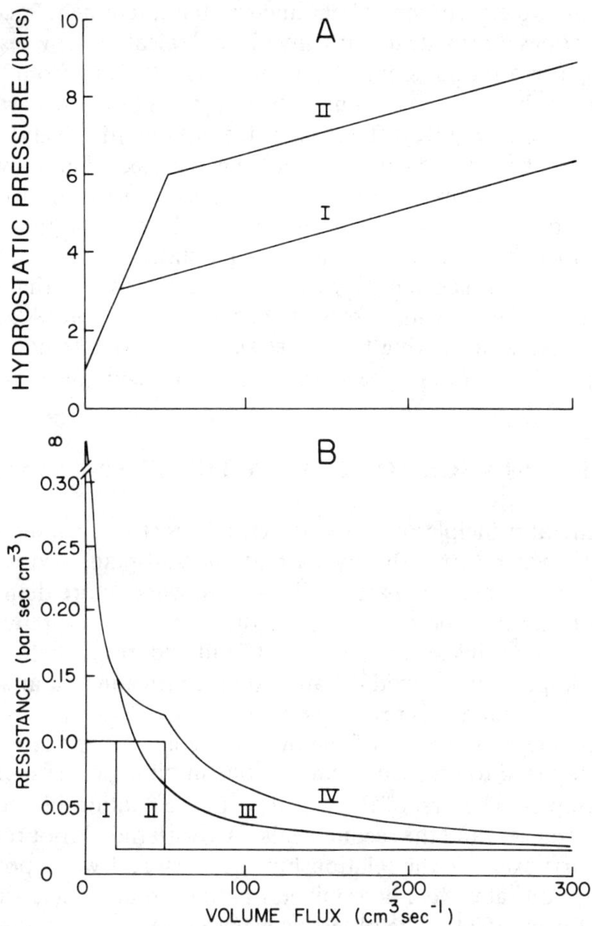

Fig. 3—Plot of pressure at base of pipe against volume flux (A), and system resistances against volume flux (B). Lines I & III are for cases when E2 and E3 are open, and lines II & IV when E2 and E4 are open. The relative apertures are E3 and E4 > E2.

tions will look like lines I and II of Fig. 3B. These lines clearly indicate that the only difference between the two systems is in the height of the ports. The instantaneous resistance functions, lines III and IV of Fig. 3B, not only suggest two highly nonlinear systems, but substantial differences between them. Unfortunately, the instantaneous resistance functions also tend to mask the essential difference between the two cases. In short, the differential resistance function clearly and concisely conveys the behavior of the system, whereas the instantaneous concept tends to mask the particularly relevant features.

If all that we desire is to reproduce the force/flux curve, then calculating the instantaneous resistance will serve, and so will a commercial copy machine. In neither case will the procedure tell us anything about what makes the plant system operate as it does. The concept of instantaneous resistance does not provide us with any insights concerning how the system should respond to environmental parameters that might be related to the resistance, nor does it provide us with any physiological hypotheses to test. At best one might reason out similar qualitative mechanisms from both resistance functions, however, assigning values to physiological parameters on that basis may lead to quite different conclusions quantitatively.

To be sure, a more logical choice in the preceding examples would have been to measure the flow through each exit port as a function of the pressure drop across that port. The total flow would then be the simple sum over all the operating ports. However, we have stated that these areas are inaccessible. So, too, it is often the case in biological systems that we are not able to measure the relevant forces at their sites of action. Also, it is frequently not clear what driving forces are acting in a system nor even where they are acting. This, too, has been a source of confusion over the years.

III. RESISTANCE TO FLOW IN THE PLANT SYSTEM

The general principles of the differential resistance should be equally applicable to water fluxes in any part of the soil-plant-atmosphere continuum. However, there are parts of the system where more than one force, flux, or pathway may be operating. In these cases, interpretation of the force/flux relationships becomes more difficult and it is in these cases that it is especially important to avoid the additional confusion that arises through the use of the instantaneous resistance concept.

To illustrate how the use of the instantaneous rather than differential resistance may lead to erroneous conclusions in biological systems, we will cite two examples. The first of these deals with the relationship between leaf and root resistances, and the second deals only with radial root resistance.

In the first example the relationships between leaf water potential and water absorption rates were determined in attached and detached leaves of sunflower (Boyer, 1974). The objective was to determine whether the large variable plant resistance resided in the leaves or the roots. In the case of attached leaves the investigator observed a force/flux relationship similar to our Fig. 3A, line I or II. The pattern observed for detached leaves was similar to our line II of Fig. 2A.

Calculations based on these data [(Boyer, 1974) Fig. 2A and 2B] showed large changes in r_i in both the attached and detached leaves. Because of the large change in r_i of the detached leaves, the investigator concluded that the major variable resistance to flow within the plant resided in the leaves. However, examination of the differential resistance characteristics of his data reveals that (i) there is an approximate sixfold change in r_d in the attached leaves over the range of water absorption rates examined, (ii) over the same approximate range of flux rates there is practically no change in r_d for the detached leaves, and (iii) for the higher absorption rates encountered, r_d for the attached leaves is about 2.6 times larger than r_d for the detached leaves. Therefore, on the basis of r_d, one must conclude that the major resistance lies not in the leaves but in some other part of the plant, probably the roots. In fact, the data under discussion show that at high flux rates the leaf r_d is approximately 38% of the total system r_d and only 6% of the total r_d at low flux rates. In this instance the use of r_i was totally misleading.

The second example deals with the volume flow through detopped root systems in response to an applied hydrostatic pressure. The details of the experimental technique may be found in Lopushinsky (1964) and Mees and Weatherley (1957). Basically, the technique consists of severing the shoot from the root system and sealing the roots in a chamber with the stump protruding through the lid. A source of water and nutrient ions is made available and the rate of solute and water flow from the stump measured under various levels of hydrostatic pressure in the chamber. The technique was developed to simulate the high rates of volume flow through the root system that the intact plant may generate during conditions of rapid transpiration. Even though we can succeed in producing high flow rates in this manner, the crucial question remains as to whether or not the root system conductances measured under these conditions are comparable to those that exist in the intact plant. We will approach this question later and will, for now, accept that the technique is valid.

The nonlinear nature of the pressure-induced flow through detopped root systems led some workers to conclude that L_p (eq. [1]) was changing. We were able to show (Fiscus, 1975, 1977) that even though the differential resistance was changing, it was possible for this to occur while L_p remained constant. Because we were able to express r_d in terms of physiological parameters, we were able to demonstrate that the nonlinear nature of the flow could be accounted for on the basis of a variable osmotic difference term, while L_p could remain constant. That is to say that the differential resistance was shown to be a function of interacting forces and fluxes. It is clear that in such a case calculation of r_i was a meaningless exercise. In addition, we were able to show that L_p could be calculated as the reciprocal of the limiting r_d at high flow rates. This last point brings us to another area of confusion regarding resistance terminology, and that arises because of the penchant of some workers for freely using a reciprocal relationship between conductance and resistance. There is a natural proclivity to do so when the concept of instantaneous resistance is invoked. However, in many cases, the use of conductance and resistance in this manner can be very misleading [cf. Fiscus & Kramer (1975)]. Where the concept of differential resistance is used, it is frequently quite clear that such reciprocal usage may cause great problems. Here again, frequently implicit in the use of this reciprocal relationship is the assumption that the force/flux relationship is linear and passes through the origin, so that the instantaneous conductance can therefore be just as meaningless as the instantaneous resistance.

Criticism of our original nonlinear root-water flow model by Newman (1976b) is valid and, we think, will lead to modifications, although not exactly as he suggests. We feel now, as we stated then (Fiscus, 1975), that the basic effects we elucidated regarding coupled solvent/solute flow in root systems and the differential resistance characteristics will remain valid, even though the details of the system may change with regard to the number of compartments necessary. In the original model (Fiscus, 1975), we used the term "apparent" resistance but have since come to believe that the term "differential" resistance is more descriptive and far preferable.

In addition to the confusion created by the use of the instantaneous resistance, there are several other somewhat controversial problem areas in plant-water transport research that we feel are worthy of brief mention. Most of the following remarks will be confined to root-water transport and will be presented as they occur to us. The list will not be comprehensive.

A. Radial vs. Axial Resistance

We have previously assumed that it is the resistance to radial water flow that controls the overall flux patterns in our root systems. By radial water flow we mean only that which occurs between the root surface and the xylem elements. That the radial resistance is limiting in our systems is strongly supported by the temperature response of root water flow under a constant pressure. The early work of Kramer (1942, 1948), Kuiper (1964), and, more recently, Markhart et al. (1979) generally show a temperature response that is much greater than can be accounted for on the basis of viscous flow phenomena. More important, the sharp breaks in the Arrhenius plots of the responses in many of these instances are characteristic of membrane phase transitions or some form of metabolic mediation. So far as we know, there is no evidence to indicate that membrane properties control the axial flow of water along significant portions of a root. Also, the many examples of metabolic-inhibitor treatments slowing root water transport suggest that something other than the axial resistance is limiting in those instances. This is not to suggest that axial resistance cannot be limiting, especially in species with smaller vessels and longer roots. None of the roots in any of our studies on beans and soybeans was longer than 3 m.

B. Driving Force for Radial Flow

Failure to distinguish between systems driven by an external pressure excess and those driven by an internal pressure reduction (tension) can be misleading. In an intact transpiring plant the xylem tension (ψ_{px}), as well as the leaf water potential (ψ_L), is not merely a result of the atmospheric demand and the root conductance. It is also a function of the osmotic phenomena opposing (or aiding) flux at the root-water flow barrier. One therefore would expect the relationship between ψ_L (or ψ_{px}) and the root water flux to be nonlinear in a manner similar to the relationship between J_v and ΔP_h (Eq. [1]; Fiscus, 1975, 1977; Fiscus & Kramer, 1975), assuming that the root medium is unchanged and ignoring any longitudinal losses. This nonlinearity does not require "leaky" membranes (Aston & Lawlor, 1979) or changes in L_p (Michel, 1977), but is due in general to the coupling of water and solute transport in the roots.

C. Radial Pathway(s)

Just as there can be interaction between different forces and fluxes in root systems, it is also possible for interactions to occur between alternate pathways for water movement in a system. The pathway of radial water movement has been the subject of controversy for years. The reviews by Newman (1974, 1976a) present as many of the alternate views as one is likely to find. The choice between pathways is seen primarily as between the cell-wall–endodermal pathway and the symplastic pathway. Unfortunately, there still has been no accumulation of the types of data that will allow an unequivocal choice between these alternate radial pathways, nor is it clear whether an "either/or" situation exists. Also, there is no reliable information indicating whether or how much the various solvent and solute fluxes might be proportioned between alternate pathways, and how this proportioning might be affected by changes in the total volume flux through the root system. Fortunately, for many purposes and analyses, it is not necessary to identify specific structures. We need only know how the systems act compared with some theoretical system with ideal properties. For example, we find that in many ways, root systems act like nothing more than semipermeable membranes with the ability to transport solutes at the expense of metabolic energy. We may, therefore, assign other membrane-related properties to root systems, even though we may not know what structure or combination of structures is responsible for these properties. Ultimately our goal is to identify and model these specific structures.

D. The Steady State

Throughout this discussion we have made implicit use of several steady-state concepts without defining or further elucidating this particular concept. Generally, the steady state is thought of as a situation in which the particular flux being examined does not change with time. In the literature, one finds numerous instances, indeed the majority of cases, in which criteria for the steady-state are never defined and the unchanging flux only implied.

When dealing with complex biological systems, it seems desirable to try and establish, in each case, rather rigid criteria that define the steady state. This is especially important in systems where more than one force or flux interact and especially where alternate pathways may be involved. In these situations it seems desirable to specify that all interacting forces and fluxes are in the steady state. This last statement is obviously relevant to plant systems in general and especially to those in the field where major environmental influences cannot generally be controlled. Lack of a true steady state is particularly troublesome in field experiments designed to determine physiological water-transport coefficients. For example, changes in wind speed, direction, solar radiation, humidity, air or soil temperatures, stomatal resistance, or even the age of the plant part itself could all, or any individually, upset the delicate steady state.

In fact, it seems unlikely that true steady-state conditions ever can exist in nature owing to the diurnicity of the system and the fact that it is continually changing size. It is not at all clear that any sort of transient analysis could be any more useful at this time. This is primarily because in a growing biological system, we would expect each response to be influenced by the special history of the individual system. That rather rigid steady-state criteria are necessary, even in the laboratory, may be illustrated by the recent work of Bunce (1978) on soybeans and cotton. He found that in both species leaf water potential did not reach steady values until 3 hours after the transpiration rate had ceased to change. It is clear that if the transpiration rate had been used as the sole steady-state criterion, interpretation of the observed relationships could easily become confused.

E. Induction and Hysteresis

Another area related to the steady-state problem is the so-called induction and hysteresis involved in water flow through root systems. The name "induction" was bestowed by Kuiper (1974) and derives from the fact that when root systems of some species are subjected to stepwise changes in hydrostatic pressure, the flow rate does not immediately reach a new steady value. Rather, there is a period of time, frequently several hours, during which flow increases gradually to its new steady state. Conversely, when the pressure is decreased suddenly, the new steady flow rate may not be achieved immediately (induction) and may sometimes be observed to be somewhat higher than the original rate at that same pressure (hysteresis). Both of these phenomena have been attributed to some effect of flow rate on the hydraulic conductance of the system. Our observations and those of Markhart et al. (1979) indicate that when sufficient care is taken to achieve steady-state rates of solute as well as solvent flux, we do not observe hysteresis. In addition, we feel that the gradual rather than stepwise changes in flow rate when the pressure is changed may be due to the time required for the solute fluxes, and thus the osmotic forces, to come to the steady state, rather than to changes in the hydraulic conductance.

F. Soil- vs. Solution-Grown Roots

Most of our discussion of root systems has dealt with those grown in growth chambers or greenhouses in nutrient solution. We must finally address the question of how these systems act compared with soil-grown systems. Unfortunately, the comparisons cannot be direct, since extant data were generally not gathered for the same purpose. However, we may make some rough comparisons based on observed flow rates, measured potential differences, and assumed root-area/leaf-area ratios. We will compare other data with those of Fiscus and Markhart (1979) for beans and soybeans. The hydraulic conductances measured according to Fiscus (1977) ranged from 0.8×10^{-6} to 6.1×10^{-6} cm s^{-1} bar^{-1} for beans, and from 2.2×10^{-6} to 5.6×10^{-6} cm s^{-1} bar^{-1} for soybeans. These determinations varied considerably because of differences associated with root system size and age (Fiscus, 1979).

For comparative purposes, we may draw on several pieces of data in the literature. First, we can taken the single soil-grown sunflower root system used by Boyer (1974) in the pressure chamber and estimate an L_p based on his Fig. 4 and a 1:1 root-area/projected-leaf-area ratio. The result, based on an estimate of the limiting differential resistance, is $L_p = 1.8 \times 10^{-6}$ cm s^{-1} bar^{-1}, which is well within the range of our measurements.

Another opportunity for comparison is presented by the work of Neumann et al. (1973) on corn, sunflowers, and soybeans grown in a silica rooting medium. Here again we will assume a 1:1 root/leaf area ratio and that the root transport characteristics control the overall flow pattern within the plant. Based on these assumptions, the inverse of the limiting differential resistance (L_p) may be calculated from their figures. For soybeans the L_p are found to range from 0.8×10^{-6} to 2.3×10^{-6} cm s^{-1} bar^{-1}. For sunflowers, this range is 0.8×10^{-6} to 2.2×10^{-6} cm s^{-1} bar^{-1}, and the normalized composite (Neumann's Fig. 3) yields a value of 1.2×10^{-6} cm s^{-1} bar^{-1} for corn.

In addition to these data, a further comparison made by Fiscus and Markhart (1979) also shows that, based on the flow rates encountered in our experiments, root system coefficients may be calculated by the method of Taylor and Klepper (1975), which are similar to the value of 1×10^{-6} cm d^{-1} they calculated for their cotton plants grown outdoors in soil.

The above comparisons indicate, at least within the large degree of uncertainty inherent in such maneuvers, that soil- and solution-grown root systems may act in a very similar manner with respect to their water-transport properties.

IV. SUMMARY

In most plant systems the stomata control the loss of water from the plant. However, it is the rate of water transport to the leaves that controls, to a large degree, the stomata. Of course it is the root-water-transport characteristics that control the flow to the leaves, so there is adequate rationale for expecting to be able to genetically manipulate the water transport in the plant by selecting for root characteristics. However, it is not root resistance alone that must be selected but eventually combinations of root resistance and stomatal response. Thus far there is vary little firm data concerning the interactions of root resistance and stomatal response characteristics, and for this and other reasons it is not always clear what constitutes a desirable water-transport trait. It will probably develop that a wide range of combinations, from plants that offer practically no impedance to water flow to those that severely restrict it, will be useful in various production situations.

Production strategy will also be important in that different transport characteristics will be desirable, depending upon whether we wish to maximize production or to stabilize it.

Finally, we cannot deal with water transport exclusive of other plant functions. We may be successful in selecting for L_p, for example, and find that the system is dominated by growth rates and patterns. A high value of L_p is useless if the plant is unable to explore an adequate volume of soil or if the cost in terms of growth-related resources is too high.

LITERATURE CITED

Aston, J. J., and D. W. Lawlor. 1979. The relationship between transpiration, root water uptake, and leaf water potential. J. Exp. Bot. 30:169–181.

Boyer, J. S. 1974. Water transport in plants: mechanism of apparent changes in resistance during absorption. Planta 117:187–207.

Bunce, J. A. 1978. Effects of shoot environment on apparent root resistance to water flow in whole soybean and cotton plants. J. Exp. Bot. 29:595–601.

Fiscus, E. L. 1975. The interaction between osmotic- and pressure-induced water flow in plant roots. Plant Physiol. 55:917–922.

Fiscus, E. L. 1977. Determination of hydraulic and osmotic properties of soybean root systems. Plant Physiol. 59:1013–1020.

Fiscus, E. L., and P. J. Kramer. 1975. Liquid phase resistance to water flow in plants. What's New in Plant Physiol. 7(3):1–4.

Fiscus, E. L., and A. H. Markhart. 1979. Relationships between root system water transport properties and plant size in *Phaseolus*. Plant Physiol. 64:770–773.

Gradmann, H. 1928. Untersuchungen uber die wasserverhaltnisse des Bodens als grundlage des pflanzenwachstums. Jahrb. Wiss. Bot. 69:1–100.

Huber, B. 1924. Die beuteilung des wasserhaushaltes der pflanze. Ein beitrag zur vergleichenden physiologie. Jahrb. Wiss. Bot. 64:1–120.

Katchalsky, A., and P. F. Curran. 1965. Nonequilibrium termodynamics in biophysics. Harvard University Press, Cambridge.

Kramer, P. J. 1942. Species differences with respect to water absorption at low soil temperatures. Am. J. Bot. 29:828–832.

Kramer, P. J. 1948. Root resistance as a cause of decreased water absorption by plants at low temperatures. Plant Physiol. 15:63–79.

Kuiper, P. J. C. 1964. Water uptake of higher plants as affected by root temperature. Meded. Landbouwhogesch. Wageningen 64:1–11.

Kuiper, F., and P. J. C. Kuiper. 1974. Permeability and self-induction as factors in water transport through bean roots. Physiol. Plant. 31:159–162.

Lopushinsky, W. 1964. Effects of water movement on ion movement into the xylem of tomato roots. Plant Physiol. 39:494–501.

Markhart, A. H., E. L. Fiscus. A. W. Naylor, and P. J. Kramer. 1979. Effect of temperature on water and ion transport in soybean and broccoli systems. Plant Physiol. 64:83–87.

Mees, G. C., and P. E. Weatherley. 1957. The mechanism of water absorption by roots: I. Preliminary studies on the effects of hydrostatic pressure gradients. Proc. R. Soc. London Ser. B. 147:367–388.

Michel, B. E. 1977. A model relating root permeability to flux and potentials: application to existing data from soybean and other plants. Plant Physiol. 69:259–264.

Neumann, H. H., G. W. Thurtell, and K. R. Stevenson. 1973. In situ measurements of leaf water potential and resistance to water flow in corn, soybean, and sunflower at several transpiration rates. Can. J. Plant Sci. 54:175–184.

Newman, E. I. 1974. E. W. Carson (ed.) Root and soil water relations in the plant root and its environment. University of Virginia Press, Charlottesville. p. 363–440.

Newman, E. I. 1976a. Water movement through root systems. Philos. Trans. R. Soc. London Ser. B. 273:463–478.

Newman, E. I. 1976b. Interaction between osmotic and pressure induced water flow in plant roots. Plant Physiol. 57:738–739.

Richter, H. 1973. Frictional potential losses and total water potential in plants: a re-evaluation. J. Exp. Bot. 24:983–994.

Taylor, H. M., and B. Klepper. 1975. Water uptake by cotton root systems: an examination of assumptions in the single root model. Soil Sci. 120:57–67.

van den Honert, T. H. 1948. Water transport in plants as a catenary process. Discuss. Faraday Soc. 3:146–153.

CHAPTER 5A

Canopy Gas Exchange: An Overview

EDGAR LEMON

Agricultural Research Service, USDA
Ithaca, New York (now retired)

I. INTRODUCTION

A little over ten years ago I suggested that micrometeorologists and agronomists were at the threshold of understanding many aspects of the physical processes that define the environment within and immediately above crop canopies (Lemon, 1969). This threshold has indeed been crossed in the intervening years in a number of areas. In fact, the experimental results have led to sufficient understanding that a number of complex, plant-canopy models have been developed and tested.

The success of the micrometeorological research efforts in the past decade has been so great that from the agronomist's viewpoint interest in both understanding and measuring natural aerodynamic gas exchange in field crops has waned. Two reasons come to mind. For one, the development of the canopy models have allowed their makers to find suitable answers for many agronomic questions by fiddling with the aerodynamic parameters. There appears to be little need for greater, detailed understanding. Secondly, the exchange of the two main gases of interest, CO_2 and H_2O, are wed so rigidly to solar-energy-dependent processes that aerodynamic processes become secondary.

Nevertheless, the aerodynamic properties of a crop canopy can have a substantial effect on the efficient use of water by that crop. Even though aerodynamic properties appear to have a secondary effect on evapotranspiration, there are still large ambiguities about water vapor transport at the soil surface and in the canopy itself. Each of these topics is considered in depth by the discussants of this chapter. Also, the aerodynamic processes may have profound influences on the efficient use of water through their influence on other aspects of crop production. For example, NH_3 and N_2O are two gases that are not rigidly coupled to solar energy exchange, yet they are very much a part of the nitrogen cycle and their transport is potentially linked to the crop's well-being. Also, the introduction of a number of gaseous and particulate matters into the atmosphere has

Limitations to Efficient Water Use in Crop Production.

reopened the questions of gas exchange at vegetative surfaces. There is a growing awareness that plants and soils play a role in soaking up and deactivating some of the gaseous garbage that man spews into the air. The aerodynamic properties of the crop canopy are also important in the dispersion of insect pheromones and insecticides either at application or revolitalization. Finally, benign control of crop pests and diseases requires more aerodynamic knowledge to predict crop environmental factors such as humidity, temperature, and wind speed.

All of the above processes can have significant effects on the performance of crops, and thereby on the efficient use of water by that crop. Here I intend to point out, briefly, a few aerodynamically related phenomena that are significant but not too widely recognized for their importance to the agronomist.

II. TURBULENCE AND PRESSURE FLUCTUATIONS

In the field, gases move chiefly by turbulence once they leave the near-surface domain of molecular-diffusion control. Parcels of air, called eddies, move relatively quickly in all directions with an organized circular motion, carrying gases and heat to and from plant and soil surfaces. Usually there is a predominant mean horizontal motion of these eddies giving the mean wind. Nonetheless, the vertical components of flow in these eddies provide rapid vertical gas exchange right down through the crop canopy to the soil surface.

Figure 1 intends to show this. If one carefully measures small changes in air pressure at, say, the soil surface, there are pronounced fluctuations around the mean. In the idealized case a sine wave is made by passing eddies that come and go through the crop. Downward motion on one side of an eddy adds pressure, upward motion the reverse. The mean horizontal flow of all eddies over a longer period of time, say 30 min or more, gives a distributed smooth wind speed profile shown at the right in the figure.

Eddies are made by two processes near the ground. So-called "forced convections" create eddies from friction between the flowing mass of air and the rough elements on the surface. "Free convection" is initiated by differential heating of the air by the surface, causing density gradients where warm air rises and cool air settles. Figure 2 shows this. The proportionate importance of these two processes has great significance to the understanding and aerodynamic measurement of gas exchange, which we shall return to later.

Kimball and Lemon (1970) conducted studies of the eddy pressure-fluctuation enhancement of gas exchange in soils. This was prompted by observation that increased water vapor flow through the top layers of soil coincided with increased wind speed. Their investigations covered spectra of air pressure and wind speed at or near the soil surface in the frequency range of 10^{-4}–10^{2} Hz. Figure 3 shows four spectra at a bare soil surface when the wind speed was 290 cm s^{-1} at 50 cm above the ground. The square of pres-

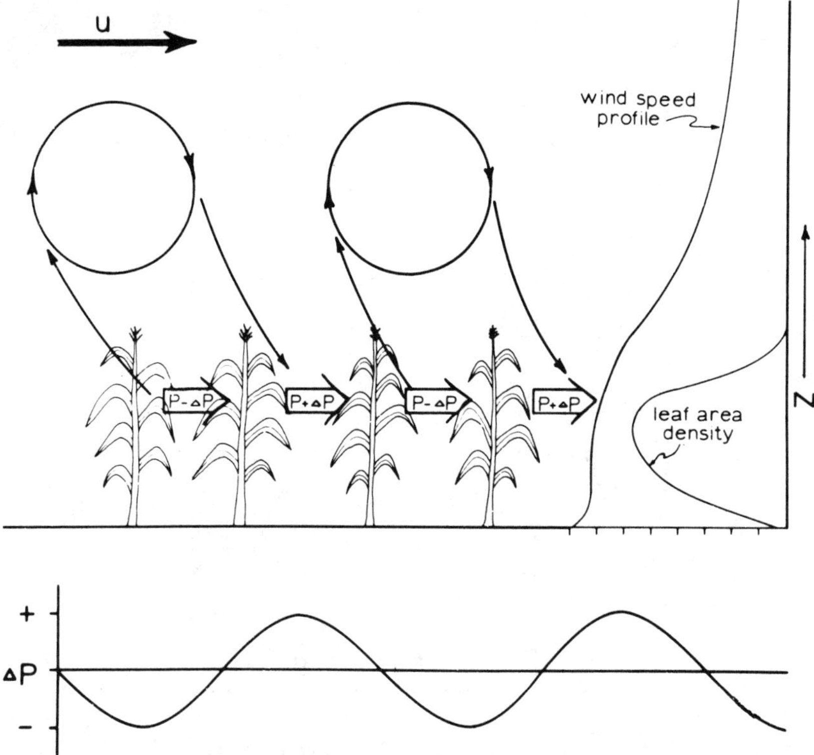

Fig. 1—Idealized concept of how atmospheric eddies create pressure changes down into a vegetation canopy. A mean wind profile and leaf area density profile are also presented. A sine wave of pressure change indicates the effects of the above eddies.

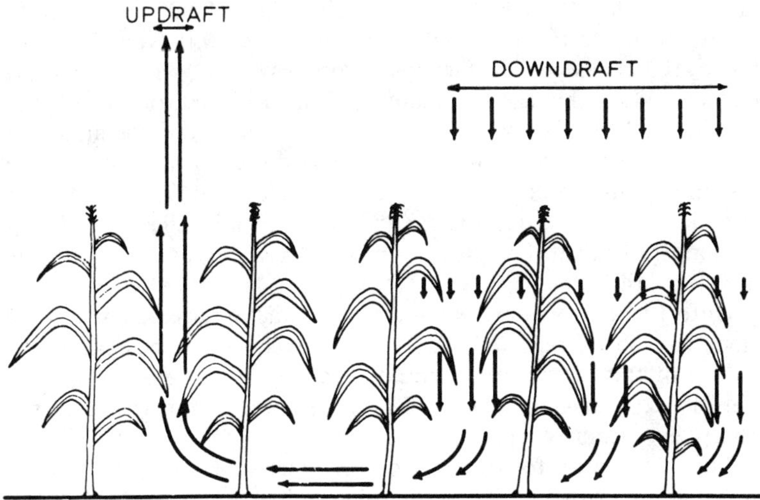

Fig. 2—Idealized concept of how eddies are formed in "free convection" due to differential heating and cooling in a vegetation canopy.

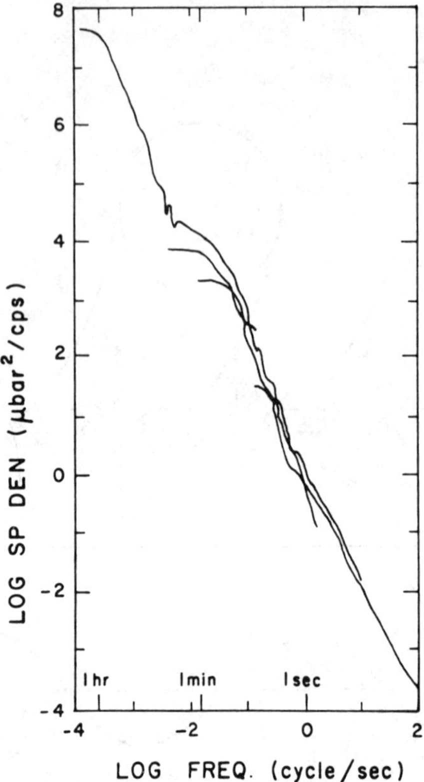

Fig. 3—Spectra of air-pressure fluctuations above a bare soil (Kimball & Lemon, 1970).

sure per unit frequency (spectral density) decreases with the $-6/3$ power of frequency (cycles s^{-1}, or Hz). Evidently this "law" holds for all conditions from about 10^{12} to 10^{-2} μbar Hz^{-1} over the frequency range from < 1 cycle d^{-1} to > 100 cycles s^{-1}. The law holds with or without vegetation above the soil. Those eddies that are created by near-surface wind friction and heating are in the high-frequency range, $> 10^{-2.5}$ Hz. Here the magnitude of the spectral density is very sensitive to wind speed (i.e., there is about a 10^3 increase in pressure spectral density with an increase of windspeed from 100 to 500 cm s^{-1}). The plotted spectra are almost identical because windspeeds were almost identical. Increasing wind displaces the curve upward, but the $-6/3$ slope still holds.

Unfortunately pressure-enhancement studies of gas exchange at the soil surface have been reported only for relatively low winds, thus the importance of this phenomenon could be greater than realized in, say, the Great Plains of the United States. Kimball elaborates further on the physics of this phenomenon in Chapter 5C.

My chief reason for dwelling on this phenomena is to caution the researcher about methods of measuring gas exchange at the soil surface. Since eddy-induced pressure fluctuations extend well into the soil and into sam-

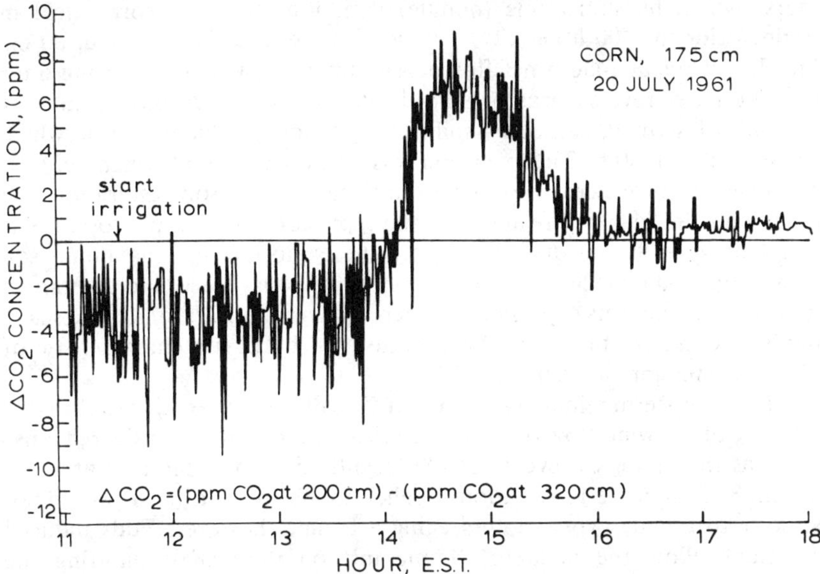

Fig. 4—Differential CO_2 concentration between 200 and 320 cm above a corn crop grown in Ellis Hollow, N.Y.

pling boxes, one needs to be aware of errors created by mass flow. Too little is known about such errors, especially in high-wind regimes.

A common method of measuring soil-surface gas exchange entails placing a small box, open on the bottom, against the soil surface, and slowly drawing air through the box to some analytical trap. Workers who measured flow rate through the box have discovered a correlation between flow fluctuation and air turbulence outside the box. This, of course, proves there is a close coupling between turbulent pressure changes outside the box, through the soil, to the box interior. The actual surface area sampled by the box, and needed for correct flux calculations, is thus in doubt.

Because of the intermittent nature of the pressure fluctuations and their natural influence on gas exchange, natural variability in time and space become serious problems in sampling the desired natural and real phenomenon as well. This physically caused variability is compounded in many cases by biologically caused variability.

I call to your attention Fig. 4 to demonstrate the intermittent nature of both meteorological and biological events. Here is plotted the difference in CO_2 concentration between two sampling levels above an active corn crop. Sampling was made by two perforated suction hoses strung horizontally for several meters at two levels above the crop (Wright & Lemon, 1966). The air from the two hoses led into the two cells of an infrared gas analyzer, thus the output signal registers the difference in gas concentration between the two sampling levels. Negative values indicate CO_2 is less in the lower hose. This is indicated by a downward gradient when photosynthesis is occurring. A positive gradient indicates an upward flow of CO_2 or an excess of respiration. In this case two main biological events occurred on the time scale of hours, while the short time (minute) variations were meteorological in origin. Prior to 1400 h net CO_2 flux density downwards was about 80 mg dm^{-2} h^{-1}, while at 1430 h net flux density upwards was about 26 mg dm^{-2} h^{-1}. The latter reversal was presumed due to a soil microbiological "explosion" of activity, caused by soil wetting from sprinkle irrigation which was started at 1130 h. The explosion was all over by 1600 h when net CO_2 exchange was near zero. I estimate that "base level" soil respiration was about 4 mg dm^{-2} h^{-1} before and after the explosion, thus net photosynthesis would be 4 + 80 mg dm^{-2} h^{-1} before the explosion and maximum soil respiration would be 26 + 84 mg dm^{-2} h^{-1} at 1430 h. It needs to be pointed out however, that this explosion is a mere "burp" on a seasonal time scale, highly intermittent in nature. Such occasional burps are characteristic of other soil emissions as well, i.e., N_2O.

The short-term minute variations of CO_2 difference are due to the eddy structure of the wind flow over the corn field and would be considered noise as far as the biological events of photosynthesis and respiration are concerned. Such short-term variations in pheromone concentrations would not be noise to a male gypsy moth seeking a female, however. Eddy-induced variations allow the male to "home in" on the female emitting the pheromone. So, on a scale of time and space, signal and detection become difficult research sampling problems in the field.

III. THE AERODYNAMIC DISCREPANCY

Wright and Brown (1967), in comparing 1962 above-canopy flux measurements in the Ellis Hollow, N.Y., corn field, found a discrepancy between the diffusivity coefficients determined by momentum and energy balances. At the top of the crop at wind speeds above 200 cm s^{-1}, the energy balance diffusivity, K, values were twice as large as momentum balance values. Other studies of tall vegetation, to date, have consistently shown that the K values for water vapor and heat are about twice that of the momentum K value. This poses serious problems to the researcher who uses flux-profile methods of gas exchange where equality of K is assumed for momentum and the other properties. Since this phenomenon seems restricted to just above high vegetation, and no suitable explanation has been put forward, our understanding of canopy gas exchange is evidently in doubt. Much more is known about the mean flow characteristics in the canopy and will be discussed by Cionco in Chapter 5B.

The cause for the discrepancy in K has to originate in the canopy because it disappears if one makes measurements well above the canopy. Several possible explanations for the discrepancy have been put forward:

a. Assumption of a Common Zero-Plane Displacement Parameter, *d*, for all Entries in the Canopy—This constant is usually found through analysis of neutral wind speed profiles and indicates the mean level of momentum absorption within the vegetation canopy. Other entity sources and sinks are assumed to have the same mean level in the canopy as that of momentum absorption. No effects of thermal stability are allowed either. The former condition is unrealistic, the latter, questionable.

b. The Application to Tall Vegetation of Free-Convection Stability Correction Found over Short Vegetation—It is more than likely that stability correction is specific for site, for crop, and for each entity. Since, as a practical matter, thermal effects on the zero-plane displacement are really lumped into the stability parameter, the complexity is compounded. The stability parameter thus becomes a surface parameter as well. Unfortunately, not enough information is available to sort all of these out.

c. The Possibility of Horizontal Inhomogeneous Temperature Structure in Tall Canopies—Here discrete hot-air plumes or chimneys that are fairly large and fairly stationary enhance free convection even in near-neutral conditions. This can be visualized in Fig. 2. While this idea is fairly attractive, there are two difficulties arising out of our own experience in the Ellis Hollow cornfield. First, the aerodynamic discrepancy was observed (Lemon et al., 1971) despite horizontal long-path spacial sampling of temperature and water vapor profiles. Secondly, wind direction is never constant, so that considerable spacial integration is always taking place out-of-doors. Even wind profiles taken from a single mast over a time frame of 20–30 min is the integration of a wide area around the mast. Figure 5 demonstrates this. Here wind direction is given at 4 m over an active corn crop when mean wind speed is somewhat <1 m s^{-1}. Obviously over a fairly "steady period" between, say, 1430 and 1500 h, the wind swept an arc of 90° between north and west. With a wind fetch over the corn field of about 400 m, such an arc represents a sizable area of corn being naturally integrated over the 30-min sampling period.

d. Spacial Inhomogeneity of Wind Speed Near the Surface Roughness Elements at the Canopy Top—Wind tunnel studies have indicated that individual element wakes extend downwind much more than formerly realized and give rise to wind speed variability near the rough canopy surface. Our experience with mean wind profiles out-of doors doesn't support this phenomenon as a cause of the "aerodynamic discrepance." We have

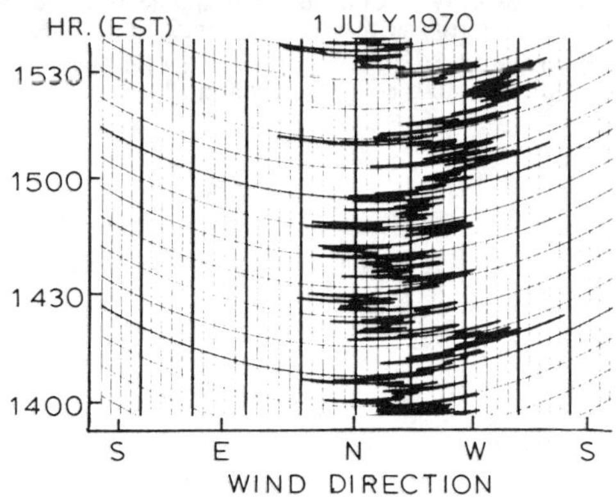

Fig. 5—Wind direction at 3 m above a corn crop grown in Ellis Hollow, N.Y.

found the semilog wind-profile plot of experimental points to obey linearity well down into the top of the roughness elements. Evidently the spacial integration performed by the ever-changing direction of the wind wipes out, over time, the individual wake effects. This natural feature of the out-of-doors probably invalidates extrapolations about wakes and the like from wind tunnel studies.

IV. SUMMARY

As an agronomist I make the plea that we do not lessen our interest in aerodynamic processes of gas exchange in the field. There are reasons why the interest has waned in the 1970's. Nonetheless there will be future problems in the field production of crops, as well as environmental problems, that will come to the door of the agronomist to be solved by micrometeorological measurement or other measurement, demanding an understanding of the natural gas exchange processes. Here I have outlined briefly a caution in making soil surface gas exchange measurements because of the intermittent nature of natural air pressure fluctuation. Secondly, I point out our woeful lack of understanding of canopy gas-exchange processes and their consequences to flux-gradient measurement. An unexplained discrepancy between aerodynamic momentum balance and energy balance creates a factor of two difference in gas-flux calculations. Several reasons have been proposed in the past but none are very satisfactory.

LITERATURE CITED

Kimball, B. A., and E. R. Lemon. 1970. Spectra of air pressure fluctuations at the soil surface. J. Geophys. Res. 75(33):6771–6777.

Lemon, E. R. 1969. Gaseous exchange in crop stands. p. 117–137. *In* J. D. Eastin and F. A. Haskins (ed.) Physiological aspects of crop yield. Am. Soc. of Agron., Madison, Wis.

Lemon, E. R., D. W. Stewart, and R. W. Shawcroft. 1971. The sun's work in a cornfield. Science 174:371–378.

Wright, J. L., and K. W. Brown. 1967. Comparison of momentum and energy balance methods of computing vertical transfer with a crop. Agron. J. 59:427–432.

Wright, J. L., and E. R. Lemon. 1966. Photosynthesis under field conditions: IX. Vertical distribution of photosynthesis within a corn crop. Agron. J. 58:265–268.

CHAPTER 5B

Canopy Gas Exchange: Canopy Flow Modeling and Its Applications

RONALD M. CIONCO

U.S. Army Atmospheric Sciences Laboratory
White Sands Missile Range, New Mexico

I. INTRODUCTION

Once water vapor moves out of the immediate domain of the soil surface or leaf surface, but is not yet in that of the canopy boundary layer, the transfer of water vapor is controlled by complex aerodynamic processes. The aerodynamic properties of the vegetative canopy themselves, therefore, have a direct bearing on evapotranspiration. Even though there is still considerable ambiguity about the aerodynamics of a vegetative canopy, as previously discussed by Lemon (Chapter 5A), recent progress has led to models that estimate mean air flow in a canopy. These models now appear to be at a sufficient stage of development to give considerable insight into the mean movement of water vapor in the vegetative canopy.

In this brief paper, a canopy air flow model is presented describing the vertical variation of the mean horizontal wind speed at any height in the canopy. The coupling of wind speeds in the canopy with ambient flow above the canopy is also discussed. Finally, variations in turbulent intensity in vegetative canopies are considered. Knowledge of these aerodynamic properties can be applied directly to problems of water vapor transfer around crop and soil surfaces.

II. THE MODEL AND ITS PROPERTIES

In modeling air flow in the surface boundary layer, the canopy domain is defined as the volume spanning the region from the top of the vegetation down to the ground surface. The equations of motion for a viscous fluid can be derived from Newton's second law, where the product of the mass and acceleration is equal to the source of the external forces:

$$\varrho(\mathrm{D}V/\mathrm{D}t) = F_g + F_p + F_f + F_c + \ldots \text{etc.,} \qquad [1]$$

Limitations to Efficient Water Use in Crop Production.

where

F_g = a body force due to gravity, g;
F_p = a surface force due to pressure, p;
F_f = a surface force due to friction, f;
F_c = a body force due to the influence of the rotation of the earth;
ϱ = the density of air; and
DV/Dt = the local time (t) rate of change of velocity, V, plus the change due to translation.

A statement describing the behavior of air flow in vegetative canopies can be developed from the basic equations of motion and the continuity equation, providing several simplifying assumptions are made (see Cionco, 1971, for full development of equations). Initially it is assumed that air is an incompressible viscous fluid of uniform density. The flow is turbulent, not laminar. The steady-flow condition prevails and the thermal-stability condition is neutral. For these assumptions, the most general equation of motion for unaccelerated canopy flow in three dimensions becomes:

$$(1/\varrho)\,\nabla P = (1/\varrho)\,\tau' + (1/\varrho)\,S(\overline{V}), \qquad [2]$$

where velocity is permitted to change direction with height (Z), τ' is the divergence of the shearing stress for turbulent flow, and $S(\overline{V})$ is the source or sink function expressing the loss of momentum due to the drag effect (C_D') of the vegetation parts upon the air flow ($S = [\varrho/2]\, C_D' A\,|\overline{V}|\,\overline{V}$). Equation [2] is applicable to both simple and complex canopies if the leaf area per unit volume term (A) is permitted to vary with height (Cionco, 1971).

A special case of this equation of motion is the simple canopy flow condition where the pressure gradient, ∇P, is negligible for small areas and small canopy heights compared with the shear and drag terms. For simple canopies, only vertical variations are important, so that the formulation (Cionco et al., 1963; Cionco, 1965) is applicable in terms of mixing length (l), canopy drag, and wind speed (U):

$$(\partial/\partial Z)[l(\partial U/\partial Z)]^2 = (1/2)\,C_D' A U^2 \qquad \text{for } 0 \le Z \le H, \qquad [3]$$

where H is the height of the canopy elements and Z is the height above the reference plane.

Cionco et al. (1963) formulated a canopy flow model of roughness elements that were ideally uniform in geometry, distribution, and behavior. The basic assumptions of this idealized canopy model are that the scale length of turbulence and the drag coefficient are constant with height. The idealized canopy is, therefore, defined as having:

1) a uniform vertical leaf area distribution,
2) a uniform vertical distribution of the drag coefficient of the canopy elements, and
3) a drag coefficient that is independent of the local Reynolds number.

The resulting flow relationships are that, within the canopy, wind speed is an exponential function of height according to

$$U = U_H \exp\left[a\left(\frac{Z}{H} - 1\right)\right], \quad [4]$$

and above the vegetative surface, wind speed is a logarithmic function of height. The attenuation coefficient, α, is an index of the air flow's response to the presence of the vegetation in terms of mixing properties, total and local drag properties, and element density and flexibility (Cionco, 1971, 1972). Inoue (1963) also suggested an exponential wind profile relationship in the canopy.

The properties of this model are:

1) the wind speed is an exponential function of height;
2) the mixing length is constant with height except very near the ground, $l \neq F(Z)$;
3) the turbulence intensity is also essentially constant with height, $(\overline{U^2})^{1/2}/\overline{U} \neq F(Z)$; and
4) there are two factors relating the index to mixing, drag, density, and flexibility properties,

$$a = (HAC_D')/(2C_D), \text{ and } a^3 = (H^3AC_D')/(4l^2).$$

For simple canopies, these relationships and properties are valid from the top of the vegetation down to $0.1H$.

Figure 1 shows the effects of vegetative canopies upon the flow at the surface. These profiles graphically emphasize the loss of momentum due to the drag effect of a wheat canopy upon the air flow as compared with unobstructed flow over a smooth, flat ice surface (Cionco, 1971). Examples of canopy wind profiles (Cionco, 1971, 1972) are presented in Fig. 2 for several simple and complex canopies. Note that the profiles are normalized as unit profiles (U/U_H) for a unit canopy (Z/H).

III. INDEX OF CANOPY FLOW

A least squares fit of these wind speed profiles (Fig. 2) will provide values of the canopy flow index (attenuation coefficient, a) where

$$a = \ln(U/U_H)/[(Z/H) - 1]. \quad [5]$$

Table 1 presents a list of average index values for simple natural and wind-tunnel roughness elements. In general, the index increases with canopy density and element flexibility. Certainly, for the simple canopies of concern, the index increases as both density and flexibility increase. However, as elements streamline with the flow, the index can also vary with wind speed.

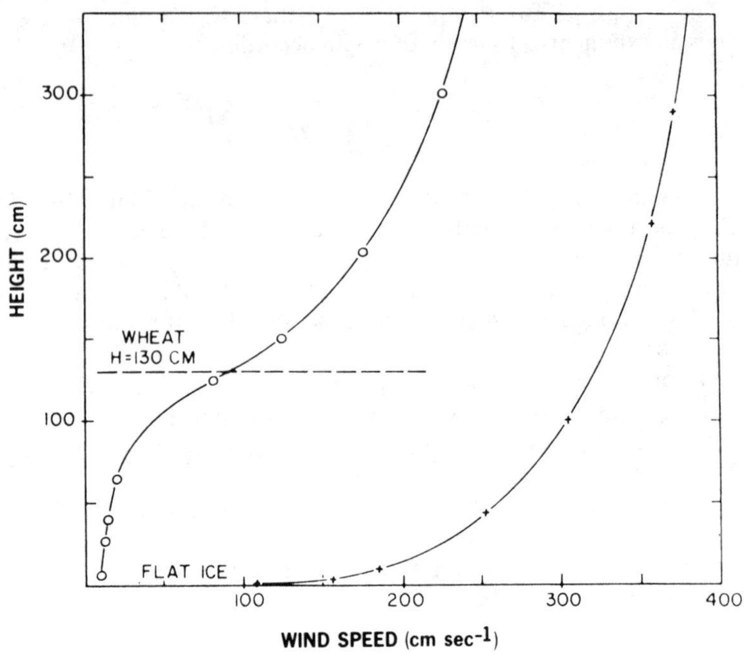

Fig. 1—A comparison of wind speed profiles for flow through a wheat canopy to that for a flat ice surface.

A micrometeorological field experiment was conducted (Lemon et al., 1968) to specifically address the density effect and hold the flexibility aspect constant. A wind-tunnel experiment was also conducted for the same reasons using wooden pegs (Kawatani & Meroney, 1968). An analysis of the wind profiles and indices for each density array of these experiments produced some unexpected results. The results of these two quite different canopy-density experiments are shown in Fig. 3 (Cionco, 1978). The top curve shows how the canopy index varies for different densities of corn plants, while the other curve presents the index variation for similar densi-

Table 1—A compilation of averaged canopy flow index values ($\bar{a}$) for simple canopies differing in density and element flexibility.

Canopies	Source	$\bar{a}$ value
Immature corn	Uchijima & Wright (1963)	2.82
Oats	Allen (1965)	2.80
Wheat	Inoue et al. (1964, unpublished data)	2.45
Corn	Tan & Ling (1961)	1.97
Plastic strips	Plate & Cermak (1963)	1.67
Rice	Nakagawa (1956)	1.62
Sunflower	Allen (1965)	1.32
Christmas trees	Stearns (1964)	1.06
Larch trees	Allen (1968)	1.00
Wooden pegs	Plate & Quraishi (1965)	0.79
Citrus orchard	Brooks & Schultz (1958)	0.44
Bushel baskets	Kutzbach (1961)	0.36

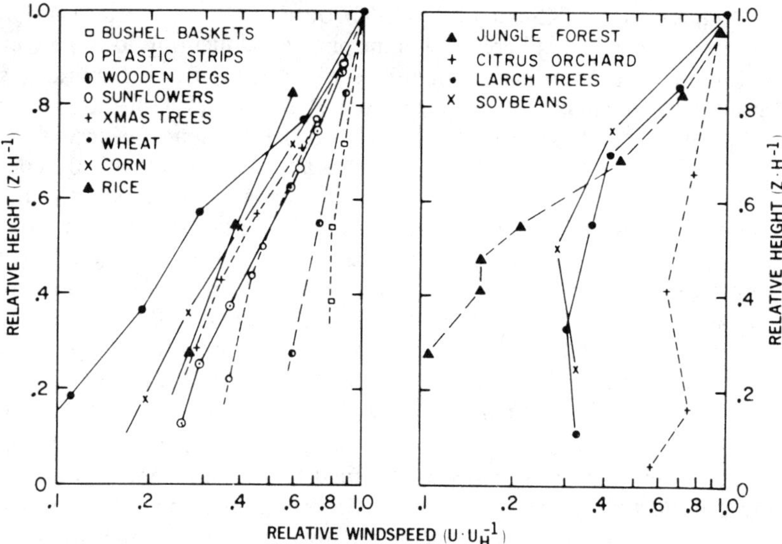

Fig. 2—Normalized plots of wind profiles within simple (left) and complex (right) canopies.

ties of wooden pegs in a large wind tunnel. The similar behavior of the two canopies is quite striking.

The maxima that occur at the 36% density arrays of each canopy are significant and can be satisfactorily explained in a qualitative manner. In the most dense arrays (≥80%), the roughness elements are so closely planted that the ambient flow senses a new zero plane displaced upward to the top of the dense canopy; therefore, eddy penetrations and interactions are usually limited to the upper portions of the canopy and only part of the

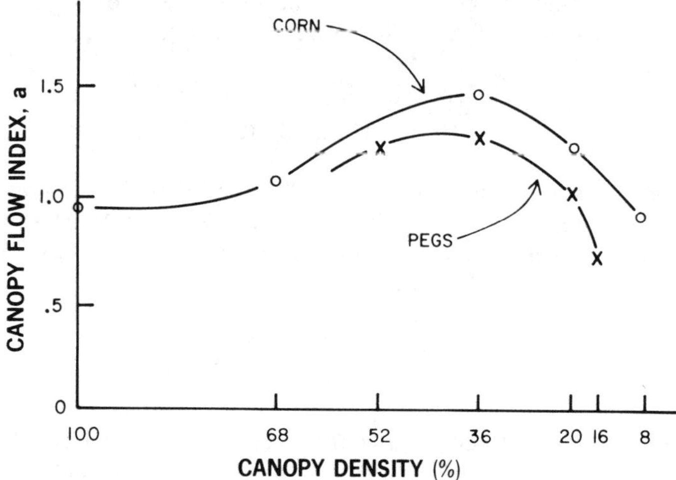

Fig. 3—The variation of canopy flow index (a) with canopy density for natural (corn) and artificial (wooden pegs) roughness elements.

flow is "canopy flow." Schlichting (1960) observed similar effects upon flow for dense arrays of spheres and hemispheres. Schlichting found a corresponding decrease in effective roughness when he validated Nikuradse's sand grain experiments. The converse gross effect occurs with the sparser arrays ($\leq 20\%$). For these cases, the horizontal element spacing exceeds the vertical dimensions of the roughness elements and the individual wakes deteriorate before they can interact. The ground surface now exerts a much greater influence on flow than the very few existing vertical elements do. Virtually none of the flow can be considered canopy flow. The moderately spaced arrays are relatively optimum compared with the extreme cases. A good balance between resistive elements and airspace exists. Based upon the data available, the 36% density arrays probably represent the optimum meteorological canopy flow regime. The canopy flow index is most representative and useful for these moderate densities.

The 0.2 offset of the index at the 36% density is mainly due to different turbulence levels occurring between the field and wind tunnel. The corn canopy is about 18% more turbulent than the wooden peg canopy. This correlates well with the index for the corn canopy that is 16% higher than the wooden peg index at the maximum.

IV. COUPLING OF CANOPY FLOW TO AMBIENT FLOW

Another important aspect of canopy flow modeling is the degree of coupling of air flow in the canopy to the ambient flow of the surface boundary layer. A canopy flow coupling ratio parameter, R_c, has been defined (Cionco, 1979) to describe the degree of coupling of canopy flow (U at $Z = 0.25H$) to the ambient flow (U at $Z = 1.4H$) for a unit canopy. This rationale employs the canopy's exponential wind-height relationship and the logarithmic wind-height relationship for the ambient air close to the vegetative surface, as shown in Fig. 4. The parameter is stated as:

$$R_c = \frac{U(0.25H)}{U(1.40H)} \equiv \frac{U_H \exp[a(Z/H - 1)]}{[U_*/k] \ln[(Z - D)/Z_o]}. \quad [6]$$

From previous studies, approximations of Stanhill's (1969) zero-plane displacement height, D, can be taken as $0.7H$; Kung's (1961) roughness height, Z_o, as $0.14H$; and Von Karman's constant as $k = 0.4$, so that the coupling ratio parameter simplifies to:

$$R_c = \frac{U_H/U_*}{[2.5 \ln 5.0][\exp 0.75a]}. \quad [7]$$

Cionco (1979) addressed how the degree of coupling is related to canopy density and canopy types, as well as to the flow upwind and downwind of well-defined canopies and through the zone of established canopy flow. Figure 5 shows that canopy flow coupling is relatively independent of

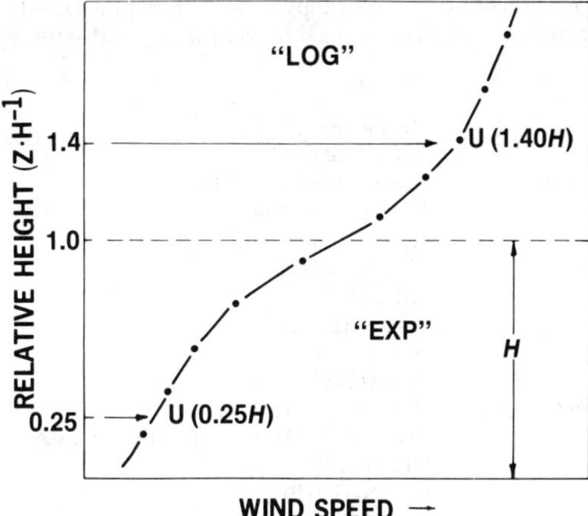

Fig. 4—Example profile defining the exponential and logarithmic wind reference levels essential to the canopy flow coupling ratio parameter formulation. The symbols *U, Z,* and *H* are defined in the text.

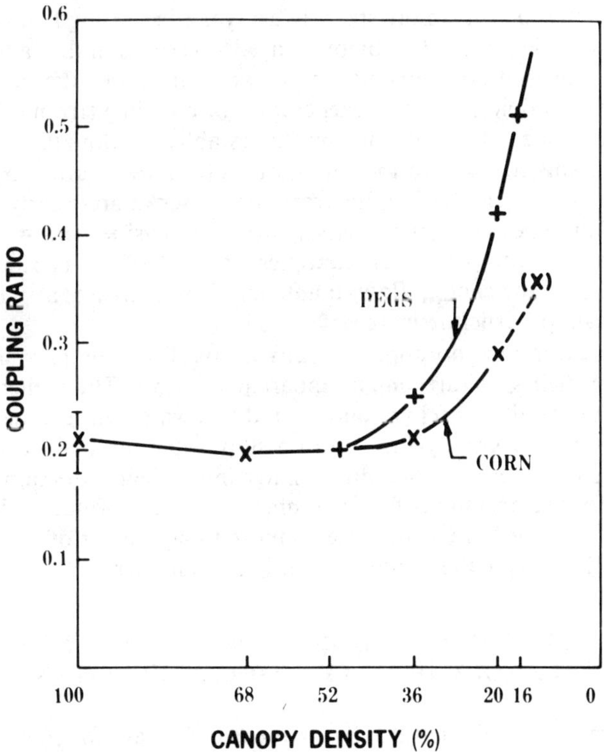

Fig. 5—The variation of the canopy flow coupling ratio with canopy densities for natural (corn) and artificial (wooden pegs) roughness elements.

Table 2—A compilation of averaged canopy flow coupling index values for a variety of natural canopies of increasing complexity and structure.

Canopy	Source	Upwind-canopy†
Citrus orchard	Brooks & Schultz (1958)	0.49
Sunflower	Allen (1965)	0.29
Deciduous forest—winter	Tourin & Shen (1969)	0.28
Rice	Nakagawa (1965)	0.24
Soybean	Allen (1965)	0.23
Red-white pine forest	Martin (1971)	0.23
Oats	Allen (1965)	0.22
Corn	Cionco (1978)	0.21
Oak-gum forest	Shinn (1971)	0.19
Larch plantation	Allen (1968)	0.16
Red-white pine forest	Raynor (1971)	0.68 -0.15
Wheat	Inoue et al. (1964, unpublished data)	0.14
Scot pine forest	Oliver (1971)	0.14
Black spruce forest	Cionco (1979)	0.13
Deciduous forest—summer	Tourin & Shen (1969)	0.12
Jungle forest	Cionco (1979)	0.09
Gum-maple forest	Melpar Inc. (1968)	0.07
Tropical forest	Bendix Corp. (1963)	0.03 (est.)

† Upwind = upwind of canopy domain; canopy = within the canopy domain.

canopy density for the more dense arrays, but increases rapidly (from 0.2 to 0.5) as densities become more sparsely arrayed (decreasing from 45 to 16% of full canopy density). The obvious was also verified in that a high degree of coupling existed both upwind and downwind of well-defined field and wind tunnel canopies, while a degeneration of coupling (from 0.35 to 0.12) was clearly evident through the zone of established flow (Cionco, 1979). Table 2 presents coupling ratio computations for a wide variety of canopies. As one might expect, the coupling ratio decreases in an orderly manner as these canopy types vary from sparsely arrayed densities ($R_c \cong 0.5$) to the more dense and complex forest canopies ($R_c \leq 0.19$). Very simply stated, the coupling of the canopy flow to ambient flow decreases as the complexity of the canopy structure increases.

Some exceptions do appear in this array. Two canopies are coupled better than their structures might apparently permit. The higher degree of coupling for the citrus orchard and the red and white pine can be attributed to the presence of gravity flow and possibly some upslope flow. Another anamoly to consider is the seasonal change in the degree of coupling that is apparent for the deciduous forest summer and winter values. The amount of coupling for the bare winter trees is more than a factor of 2 greater than that occurring during the full-leaf condition of summer.

V. VARIATIONS OF TURBULENCE INTENSITY WITH CANOPY DENSITY AND CANOPY TYPES

Figure 6 presents the turbulence intensity (i) data for these same corn and wooden peg canopies (Cionco, 1978). These results indicate that the two sets of turbulence values behave in a similar manner with maxima oc-

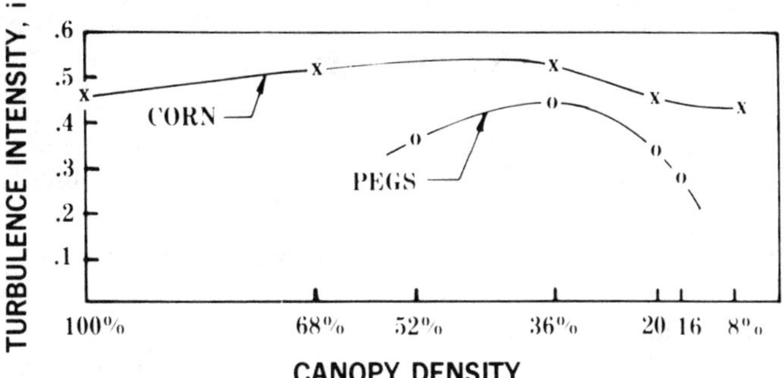

Fig. 6—The variation of turbulence intensity (i) with canopy densities for natural (corn) and artificial (wooden pegs) roughness elements.

curring again for the optimum 36% densities. The corn canopy exhibits a smaller variation of turbulence than the peg canopy as the density changes, but it is readily apparent that the density of canopy elements has a discernible effect upon the production of turbulence in canopies.

In a gross way, one can also detect the effect of canopy density upon turbulence intensity levels as canopy types evolve from simple agronomic crops to the more complex forest canopies. Turbulence intensity values for a variety of 14 canopy types were computed and averaged for an analysis of this effect. Table 3 exhibits that there is a definite gross effect of canopy complexity and structure upon the level of turbulence within these natural canopies. This list shows an orderly increase of averaged turbulence intensity values as both the density and structural complexity of the vegetation increase. Also, note the seasonal effect of producing turbulence in a deciduous forest: in summer $\bar{i}$ = 0.80, and in winter $\bar{i}$ = 0.53. The letter V behind some of the $\bar{i}$ values in Table 3 indicate that $\bar{i}$ may have varied moderately with height. These variations correspond to changes of leaf areas with height.

Table 3—A compilation of averaged turbulence intensity values ($\bar{i}$) for natural canopies of increasing complexity and structure.†

Canopy	Source	$\bar{i}$
Rice	Nakagawa (1956)	0.33
Vineyard	Weiss & Allen (1976)	0.45
Wheat	Inoue et al. (1964, unpublished data)	0.50V
Conifer, black spruce	Cionco (1972)	0.51V
Corn	Cionco (1978)	0.52
Soybean	Allen (1965)	0.52V
Immature corn	Uchijima & Wright (1963)	0.52V
Deciduous—winter	Tourin & Shen (1969)	0.53V
Pine plantation	Raynor (1971)	0.53V
Japanese larch	Allen (1968)	0.54
Sunflower	Allen (1965)	0.59V
Tropical forest	Cionco (1972)	0.76V
Deciduous—summer	Tourin & Shen (1969)	0.80V
Jungle forest	Melpar Inc. (1968)	1.10V

† Values followed by the letter V indicate that variations with height occurred.

VI. SUMMARY

What the idealized canopy flow model provides and how other features and factors of canopy flow can relate back to the model are summarized as follows:

1) The model can predict wind profiles within simple canopies.
2) A least-squares fit to canopy wind profiles can provide canopy flow index values.
3) One can then apply the profile data and index values to relate mean air flow to canopy-density arrays and canopy types in terms of structure and complexity of vegetation components.
4) The profile data and index values can also define the degree of coupling of canopy flow to ambient flow through the use of canopy density and types of relationships.
5) A level of turbulence in canopies can be inferred from the canopy density and types of relationships to the turbulence intensity index.

The realization of a coupling ratio parameter makes it quite possible to formulate a complete boundary-layer wind model whereby the vertical canopy wind profile can be coupled to the surface layer of a mesoscale wind-flow model to produce a continuous vertical wind profile down to the bottom of the canopy at each grid point of the mesoscale model.

LITERATURE CITED

Allen, L. H., Jr. 1965. A comparison of airflow pattern within and above five different vegetative canopies. 1965 Annu. Rep. Agric. Res. Serv., Ithaca, N.Y.

Allen, L. H., Jr. 1968. Turbulence and wind speed spectra within a Japanese Larch plantation. J. Appl. Meteorol. 7:73–78.

Bendix Corporation. 1963. Jungle canopy penetration. Vegetation and Meteorological Studies II. Ann Arbor, Mich.

Brooks, F. A., and H. B. Schultz. 1958. Observations and investigations of nocturnal density currents. p. 272–279. *In* Climatology and micrometeorology. Proc. Canberra Symp., Arid Zone, Canberra, Australia. UNESCO, Paris.

Cionco, R. M. 1965. A mathematical model for air flow in a vegetative canopy. J. Appl. Meteorol. 4:517.

Cionco, R. M. 1971. Application of the ideal canopy flow concept to natural and artificial roughness elements. Res. and Dev. Tech. Rep. ECOM-5372. Atmos. Sci. Lab., USAECOM, White Sands Missile Range, N.M. (AD730638[1]).

Cionco, R. M. 1972. A wind-profile index for canopy flow. Boundary-Layer Meteorol. 3:255.

Cionco, R. M. 1978. Canopy index values for various canopy densities. Boundary-Layer Meteorol. 15:81–93.

Cionco, R. M. 1979. A summary of an analysis of canopy flow coupling for a variety of canopy types. p. 105–106. *In* Extended abstracts: 14th Conf. on Agric. and Forest Meteorol. and 4th Conf. on Biometeorol. Minneapolis, Minn. 2–6 Apr. 1979. Am. Meteorol. Soc., Boston, Mass.

Cionco, R. M., W. D. Ohmstede, and J. F. Appleby. 1963. A model for air flow in an idealized vegetative canopy. Meteorol. Res. Notes no. 5. USAERDAA, Fort Huachuca, Ariz. (AD422705[1]).

Inoue, E. 1963. On the turbulent structure of airflow within crop canopies. J. Meteorol. Soc. Jpn. Sec. II. 41:317.

Kawatani, T., and R. N. Meroney. 1968. Structure of canopy flow field. Fluid Dynamics and Diffusion Lab. Rep. CER 67-68-TK-66. Colorado State University, Fort Collins (AD678591[1]).

Kung, E. C. 1961. Derivation of roughness parameters from wind-profile data over tall vegetation. *In* Report on studies of the three-dimensional nature of the planetary boundary layer. Dep. Meteorol., University of Wisconsin, Madison. (AD273053[1]).

Kutzbach, J. E. 1961. Investigation of the modification of wind profiles by artificially controlled surface roughness. Annu. Rep. 1961. Sect. 7. Department of Meteorology, University of Wisconsin, Madison. (AD273053[1]).

Lemon, E. R., et al. 1968. Microclimate investigations. p. 86–102. *In* 1968 Annu. Progress Rep. USDA-ARS, Ithaca, N.Y.

Martin, H. C. 1971. Average winds above and within a forest. J. Appl. Meteorol. 10:1132.

Melpar, Inc. 1968. Diffusion under a jungle canopy. Vol. I and II, and Final Report. Meteorol. Res. Lab., Melpar, Inc., Falls Church, Va. (see also Allison et al., 1968. Wind and turbulent diffusion in and above a jungle canopy. Bull. Am. Meteorol. Soc. 49:765).

Nakagawa, Y. 1965. Studies on the airflow amongst the stalks in a paddy field. J. Agric. Meteorol. (Tokyo, Japan) 12:61.

Oliver, H. R. 1971. Wind profiles in and above a forest canopy. Q. J. R. Meteorol. Soc. 97: 548.

Plate, E. J., and J. E. Cermak. 1963. Investigations to develop wind tunnel techniques for measuring atmospheric gaseous diffusion in model vegetative surface. Gluid Dynamics and Diffusion Lab., Colorado State University, Fort Collins.

Plate, E. J., and A. A. Quraishi. 1965. Modeling of velocity distributions inside and above tall crops. J. Appl. Meteorol. 4:400–408.

Raynor, G. S. 1971. Wind and temperature structure in a coniferous forest and a contiguous field. For. Sci. 17:351–363.

Schlichting, H. 1960. Boundary layer theory. 4th ed. McGraw-Hill Book Co., Inc., New York.

Shinn, J. H. 1971. Steady-state two-dimensional flow in forests and the disturbance of surface layer flow by a forest wall. Res. and Dev. Tech. Rep. ECOM-5583. Atmos. Sci. Lab., White Sands Missile Range, N. Mex. (AD730681[1]).

Shinn, J. H., and R. M. Cionco. 1972. A note on observations of turbulence and mean flow in vegetative canopies. Proc. Symp. on Air Pollution, Turbulence and Diffusion. 7–10 Dec. 1971. Atmos. Sci. Lab., USAECOM, White Sands Missile Range, N. Mex.

Stanhill, G. 1969. A simple instrument for the field measurement of turbulent diffusion flux. J. Appl. Meteorol. 8:509.

Stearns, C. 1964. Report on wind profile modification experiments using fields of Christmas trees on the ice of Lake Mendota. Annual Report 1964. *In* H. H. Lettau (ed.) Studies of the effects of variations in boundary conditions on the atmospheric boundary layer. University of Wisconsin, Madison.

Tan, H. S., and S. C. Ling. 1961. A study of atmospheric turbulence and canopy flow. TAR-TR 611. Cooperative Research Program (Therm Advanced Research, Inc.; USDA; and Cornell University), Ithaca, N.Y. Published by ThermAdv. Res., Inc. Also available in Lemon et al., 1963. (AD605527[1])

Tourin, M. H., and W. C. Shen. 1969. Deciduous forest diffusion study. Vol. 3. Detailed technical data supplement. Final Report no. 3004. Applied Science Division, Litton Systems, Inc., Minneapolis, Minn. (AD856705L[1])

Uchijima, A., and J. L. Wright. 1963. An experimental study of air flow in a corn plant-air layer. Bull. Natl. Inst. Agric. Sci. (Japan). Ser. A11. p. 19–66.

Weiss, A., and L. H. Allen, Jr. 1976. Air flow patterns in vineyard rows. Agric. Meteorol. 16: 329–342.

[1]Paper available with this reference number from Defense Technical Information Center (DTIC), Defense Logistic Agency, Cameron Station, Alexandria, VA 22314.

CHAPTER 5C

Canopy Gas Exchange: Gas Exchange with Soil[1]

B. A. KIMBALL

U.S. Water Conservation Laboratory, USDA
Phoenix, Arizona

I. INTRODUCTION

A large portion of the water used in crop production is lost by evaporation at or beneath the soil surface. Any progress in reducing this gaseous loss of water will be aided by understanding the mechanisms that cause movement of soil gases. Several other soil gases are essential for crop production, such as O_2 for root respiration, so understanding these mechanisms is doubly important. In this chapter I discuss several such mechanisms and the significance of their contribution to soil gas exchange.

Turbulent mixing of air is the primary mechanism for gas exchange above and within plant canopies, where the transport coefficient typically ranges from 10,000 to 100 $cm^2\ s^{-1}$. At the soil surface, however, there is a drastic reduction in soil air movement, and the transport coefficient for soil gas exchange typically ranges from 0.1 to 0.01 $cm^2\ s^{-1}$. The lower limit for the transport coefficient is, of course, the molecular diffusion coefficient. Any movement of the soil air, however, will increase the effective value of the transport coefficient above this lower limit. Therefore, I will discuss diffusion and several other mechanisms that augment diffusion by causing soil air movement.

To further complicate matters, water moves in soil as both a liquid and a gas. Evaporation from a saturated soil takes place at the soil surface with all soil water movement occurring in the liquid phase. As the soil dries, the liquid water is replaced by gases, and evaporation can occur at progressively greater depths in the soil; progressively more water then moves as vapor (Jackson, 1964). Dry soil strongly adsorbs water, however, and a molecule of water that evaporates from deep in the soil profile will probably condense and re-evaporate many times before it reaches the soil surface. The evaporation of water is a physical process, and consequently, the equilibrium concentration of water vapor in the gaseous phase of the soil depends

[1] Contribution from the Agricultural Research Service, USDA, Phoenix, Ariz.

Limitations to Efficient Water Use in Crop Production.

upon soil temperature and water potential in a way that can be described mathematically. The equation is (Philip & deVries, 1957):

$$\varrho = \varrho_o \exp(\Psi M_w)/(\varrho_l RT), \quad [1]$$

where ϱ_o is the saturated water vapor density (kg m^{-3}), Ψ is the water potential of the soil (bars), M_w is the molecular weight of water (0.018016 kg mol^{-1}), ϱ_l is the density of liquid water (kg m^{-3}), R is the universal gas constant (8.3144×10^{-5} bars m^3 mol^{-1} °K^{-1}), and T is the absolute temperature (°K).

II. MECHANISMS FOR SOIL GAS EXCHANGE

Early investigators (Buckingham, 1904; King, 1907, 1913; Romell, 1922; Keen, 1931) considered several mechanisms by which soil gases could be exchanged with the atmosphere. They reasoned that molecular diffusion was most important and that mass flow of soil gases caused by various meteorological conditions was unimportant. Since then much more data have been obtained and less presumptive theories have been developed, so let us consider each of several postulated mechanisms in the light of more recent works.

To first set the order of magnitude on the required soil-atmosphere exchange rate, estimates can be obtained by considering the exchange of CO_2. Romell (1922), as described by Keen (1931), used 7 liters CO_2 m^{-2} day^{-1} (80 mm^3 CO_2 m^{-2} s^{-1}; 150 μg m^{-2} s^{-1}) as a standard respiration rate for consideration of various mechanisms. This is a typical rate for cropland, although values from 0.10 to 10 times this have been reported (Wesseling, 1962; Currie, 1970). At this rate, the air in the top 20 cm of a soil with 25% air porosity is exchanged every hour, if the average CO_2 concentration in the top 20 cm is a typical 0.6%. This corresponds to an air exchange rate of 7/0.006 = 1,200 liter air m^{-2} day^{-1}.

A. Diffusion

The basic equation describing the diffusion of one gas into another with no soil involved is generally written for one dimension as

$$q = -D_o(d\varrho/dZ), \quad [2]$$

Where q is the rate of movement (kg m^{-2} s^{-1}), ϱ is the gas concentration (kg m^{-3} of gaseous volume), Z is distance (m), and D_o is the diffusion coefficient (cm^2 s^{-1}). Equation [2] applies strictly to isothermal conditions only, so the partial-pressure form (Buckingham, 1904; Penman, 1940a) has often been used instead:

$$q = -\frac{D_o M_w}{RT}\frac{dp}{dZ} \div -D_o'\frac{dp}{dZ}, \quad [3]$$

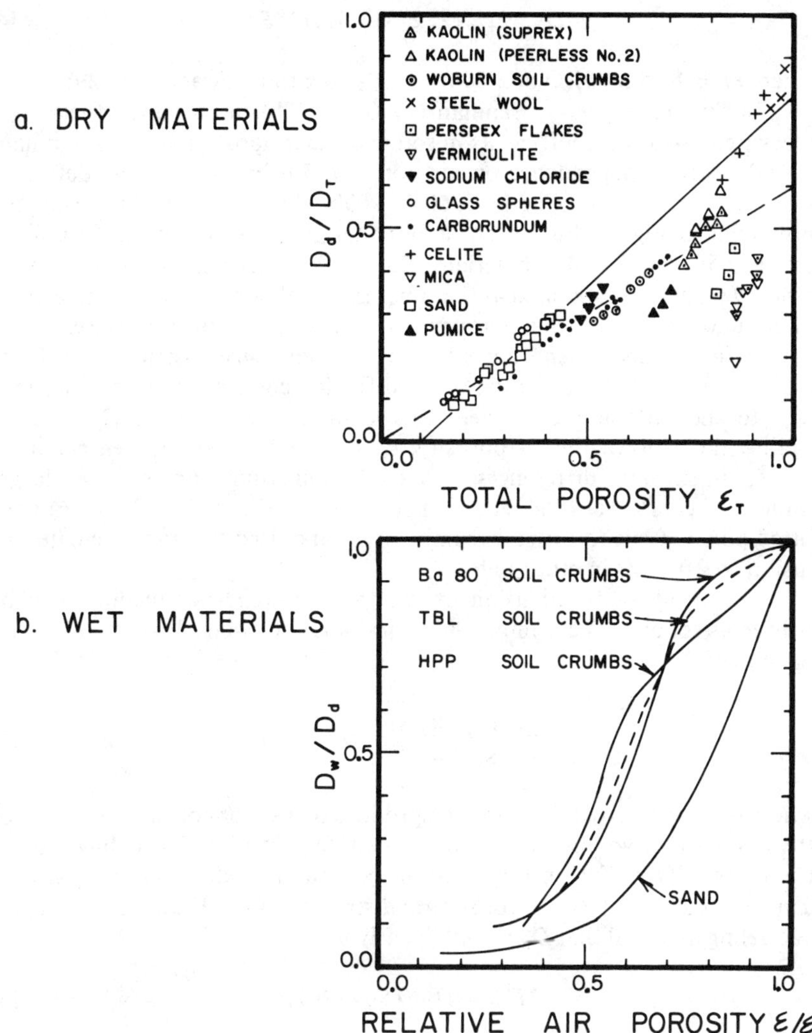

Fig. 1—(*A*) the ratio of the diffusion coefficient in several dry porous materials, D_d, to the diffusion coefficient in air, D_T, vs. the total porosity, ϵ_T. (*B*) The ratio of the diffusion coefficient in wet porous materials, D_w, to the diffusion coefficient in the same material when dry, D_d, vs. the ratio of the air porosity, ϵ, to the total porosity, ϵ_T. From Currie (1960b, 1961).

where p (bars) is the partial pressure. For a reference temperature of 273°K and pressure of 1.013 bars, Boynton and Brattain (1929) list 0.220 and 0.138 $cm^2\ s^{-1}$ as values of D_o for the diffusion into air of H_2O and CO_2, respectively. The D_o for another gas diffusing into air can be estimated by multiplying one of these D_o values times the square root of the ratio of molecular weights of H_2O or CO_2 (depending on which of the D_o values is being used) to the new gas. The diffusion coefficient, D_T, at another pressure, P, and temperature, T, can be computed from

$$D_T = D_o(T/273)^m (1.013/P), \quad [4]$$

where $m = 1.75$ for H_2O or 2.00 for CO_2 (Boynton & Brattain, 1929).

Buckingham (1904), Penman (1940a, 1940b), and many others have measured diffusion coefficients in soil and other porous materials and have found a strong dependence on air porosity. Probably the most definitive work has been that of Currie (1960a, 1960b, 1961); some of his results are presented in Fig. 1. Figure 1A shows that the ratio of the diffusion coefficient in dry soil-like materials, D_d, to the diffusion coefficient in air alone, D_T, can be estimated as 0.6 times the total porosity, ϵ_T, up to a value of about 0.7 of ϵ_T. When a porous material is wetted, however, the effect on the diffusion coefficient depends on the pore size distribution of the material. Figure 1B shows ratio of the diffusion coefficient in wet materials, D_w, to the diffusion coefficient in the same dry material, D_d, plotted against the ratio of the air porosity, ϵ, to ϵ_T. Obviously no general curve exists, but since the differences between the soil crumb curves are not large, some workers (Rose, 1968; Jackson et al., 1974; Kimball et al., 1976) have fitted polynomials to the soil crumb curves and used them to calculate diffusion coefficients in other soils.

The steady-state diffusion of gas through a porous medium can be written as follows (Wesseling, 1962; Kirkham & Powers, 1972, p. 442; Van Bavel, 1951):

$$\frac{d}{dZ}\left(D_w \frac{d\varrho}{dZ}\right) + \alpha = 0, \quad [5]$$

where α is the "activity" or rate of production (+) or consumption (−) of the gas per unit volume (kg s^{-1} m^{-3}). Wesseling and Van Bavel have solved Eq. [5] for CO_2 diffusion subject to the boundary conditions $\varrho = \varrho_o$ at $Z = 0$ and $d\varrho/dZ = 0$ at $Z = L$ for several distributions of D_w and α with depth. Wesseling assumed that D_w varied linearly with ϵ:

$$D_w = D_T(0.9\epsilon - 0.1); \quad [6]$$

and that ϵ varied as a power function with depth:

$$\epsilon = \epsilon_o(1 - a_k Z^k); \quad [7]$$

where ϵ_o is the porosity at the soil surface, and k (superscript) and a_k are empirical constants. He also assumed that the CO_2 production, α, has a distribution given by

$$\alpha_z = \alpha_o[1 - (Z/L)^{1/4}], \quad [8]$$

and that the depth L, (where activity stops and the concentration no longer increases with depth) is the depth where the air porosity equals 0.1, so that the diffusion coefficient becomes 0 as per Eq. [6]. Following Kirkham and Powers (1972), Wesseling's solution is

$$\varrho = \varrho_o + \frac{BI_k}{0.9\, D_T \epsilon_o a_k L^{(k-1)}}, \qquad [9]$$

where ϱ_o is the atmospheric concentration, B is the total rate of CO_2 emanation from the soil surface (kg m^{-2} s^{-1}), and I_k is an integral that is a function of Z/L. Curves of I_1 and I_2 are plotted in Fig. 2.

Figure 2 also shows how the CO_2 concentration varies with soil depth for the steady-state diffusion conditions assumed by Wesseling (1962). It decreases almost linearly near the surface and then tapers to zero gradient at $Z = L$. The highest concentrations are associated with the least porous (low ϵ_o) and deepest (small a_2) soils. The value of B (42.8 μg m^{-2} s^{-1}) is lower than most field measurements (Wesseling, 1962), and generally CO_2 production does not become zero at the depth where air porosity happens to equal 0.1. Actual field CO_2 profiles are not as smooth as the curves in Fig. 2 and in the spring they have maxima near the soil surface (deJong & Schappert, 1972). Nevertheless, the curves in Fig. 2 illustrate the principles of steady-state production and diffusion of CO_2 in soils.

These curves (Fig. 2) and equations (Eq. [2], [5], and [9]) also illustrate that diffusion alone is capable of producing the 7 liters of CO_2 m^{-2} day^{-1} of gas exchange required for aeration (Romell, 1922). If the diffusion coefficient decreases or CO_2 production increases, the gradient automatically increases to compensate (Eq. [2]).

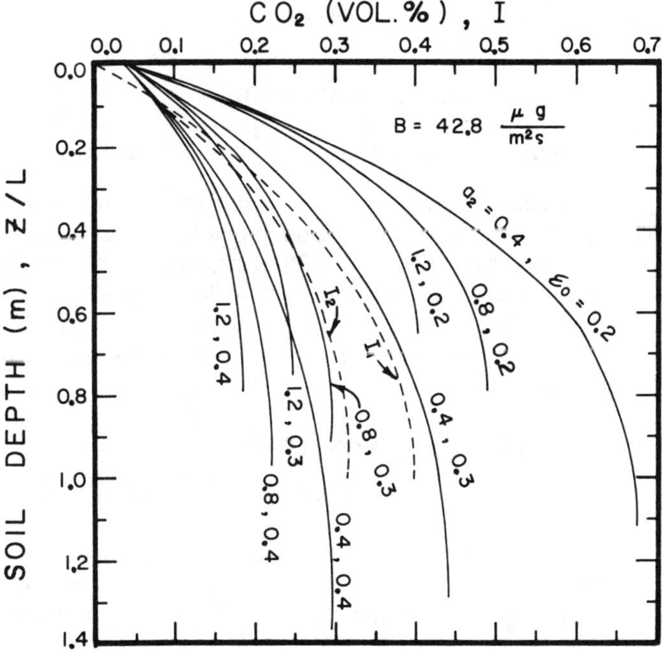

Fig. 2—Carbon dioxide concentration vs. depth from Eq. [9] for various values of the air porosity at the soil surface, ϵ_o, and change of porosity with depth parameter, a_2. Also shown are values of the integrals I_1 and I_2 vs. relative depth, Z/L. From Wesseling (1962).

B. Unequal Production and Consumption of Gases

If, at any point in the soil, gas production occurs at a different volumetric rate than consumption, a difference in total pressure must be created. The analysis by Evans[2] showed that usually, but not generally, the production/consumption rates are low enough that diffusion dominates the mass flow caused by these processes. However, in the case of a wet soil with a water table at 1 m, Kimball[3] has shown that significant mass flow occurs when the O_2 concentration decreases to zero while CO_2 is still being produced. The resultant CO_2 concentration profiles are sigmoidal rather than concave, as predicted by diffusion (Fig. 2).

C. Liquid Water Content Changes

Obviously rain, irrigation, evapotranspiration, drainage, and deep percolation are processes involving soil liquid and gas exchanges. To exchange the standard aeration of 7 liters of CO_2 m^{-2} day^{-1} (Romell, 1922) at an average 0.6% CO_2, however, the water content of the soil profile would have to change by 1.2 m day^{-1}. Thus any significant gas exchange caused by a water content change would be a very transitory phenomenon, as Keen (1931) concluded. Such a case can occur, however. Lance et al. (1973) found that mass flow accounted for 30–40% of the O_2 that entered an initially saturated, 2-m-long soil column during drainage.

D. Temperature Effects

1. EXPANSION AND CONTRACTION

King (1913) discusses an "incessant temperature breathing" of the upper layers of soil from contraction and expansion of soil air due to the diurnal temperature wave. However, a 20°K change in temperature of the upper 0.2 m of a soil with 25% air porosity and 300°K average temperature in one-half day would cause an air content change of about 7 liters m^{-2} day^{-1}, which is much below the previously mentioned standard of 1,200 liters of air m^{-2} day^{-1}.

2. BUOYANCY AND NATURAL CONVECTION

Romell (1922), as reviewed by Keen (1931), tried to assess the amount of natural convection in soils due to temperature gradients by imagining a cylinder of warm soil air suspended in colder surrounding soil air. Both authors concluded that natural convection could account for at most one-tenth of normal aeration. Since then Elder (1967) has shown that as long as

[2] D. D. Evans, 1952. Effect of combined pressure and concentration gradient on gaseous flow through soils. Ph.D. Thesis. Iowa State University, Ames.

[3] B. A. Kimball, 1965. Soil atmosphere composition as related to microbial activity and diffusion coefficient. Unpublished M.S. Thesis. Iowa State University, Ames.

a dimensionless parameter (called the Rayleigh number) does not exceed a value of 40, natural convection does not occur in porous materials. For a soil 1 m thick, with a 10°C temperature difference from bottom to top, a low thermal conductivity of 0.2 W m^{-1} $°C^{-1}$, and a high permeability of 3,300 μm^2, the Rayleigh number is 0.4. Thus this newer, more rigorous analysis supports the old conclusions.

3. WATER VAPOR DISTILLATION

A gradient of temperature in the soil will cause water vapor pressure gradients to exist concurrently, so water vapor will move from warmer to cooler soil (Nielsen et al., 1972, p. 98–119). This enhances evaporation at night and retards it during the day. Much research has been devoted to predicting the magnitude of the effects on evaporation, and a review is beyond the scope of this chapter. Calculations with field data (Jackson et al., 1974) indicated that daytime evaporation in the falling-rate stage of soil drying would be 2–10 times larger if water vapor movement due to thermal gradients did not oppose water movement due to water content gradients.

E. Pressure Fluctuations

1. THEORY

Both barometric pressure fluctuations and wind- or air-turbulence-caused pressure fluctuations have long been suggested as a possible mechanism for soil gas exchange (Buckingham, 1904; King, 1907, 1913), but theoretical calculations of their effects on soil gas exchange have not been satisfyingly definitive. The characteristics of natural pressure fluctuations at the soil surface are illustrated in Fig. 3, which shows pressure variance spectra from Gossard (1960) and Kimball and Lemon (1970), along with a pressure amplitude curve that was obtained from an average of Gossard's spectra using $a = (S\Delta f/2)^{1/2}$, where a is the amplitude of a pressure wave (μbars), S is the spectral density ($\mu bars^2$ $cycle^{-1}$ s), and Δf is the frequency increment chosen to be 0.1 on a log scale. The curves show that the low-frequency barometric pressure fluctuations on the left side are several orders of magnitude larger than the high-frequency wind-caused pressure fluctuations on the right side, but there is a smooth transition at about the -2 power of frequency all the way from weekly to millisecond phenomena.

Buckingham (1904) derived an equation describing the penetration of sinusoidal pressure waves into the soil and calculated that barometric pressure fluctuations penetrate readily into soil. He assumed that the depth of penetration of atmospheric air would be $\Delta Pd/P$ where ΔP is the pressure change at the surface and d is the depth to an impermeable barrier, such as a water table. From barometric pressure data he concluded that only for very deep (>15 m), low-porosity soils would mass flow caused by barometric pressure fluctuations be significant compared with diffusion, and that generally these fluctuations would have no influence. Recently, Clements and Wilkening (1974) and Duwe (1976) used numerical methods to predict

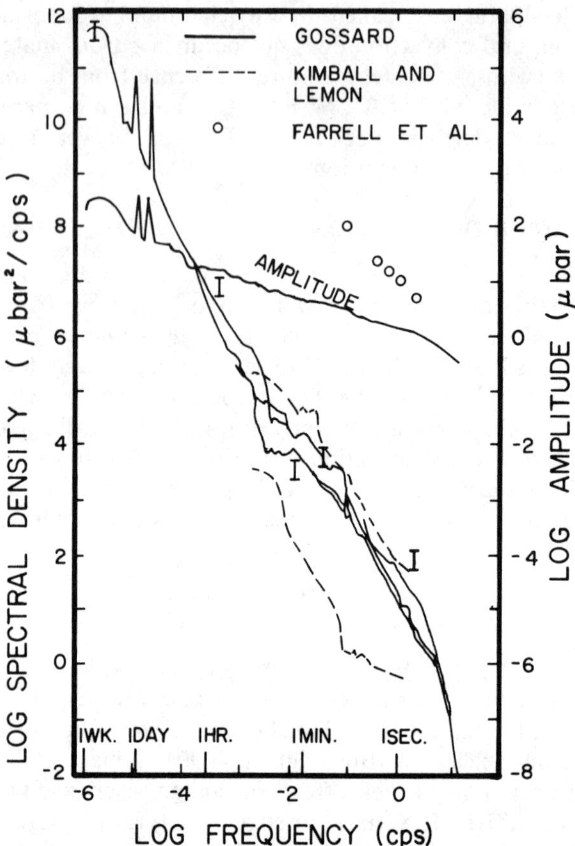

Fig. 3—Spectra of air pressure fluctuations at the soil surface from Gossard (1960) and Kimball and Lemon (1970). The prominent pair of peaks are from the diurnal and semidiurnal pressure waves. The upper Kimball and Lemon spectrum was obtained during daytime when the wind speed was 552 cm s^{-1} at a height of 50 cm, and the lower was obtained at night when the wind speed was 102 cm at a 150-cm height. Also shown is an amplitude curve derived from the average of the Gossard spectra as well as the amplitude of five pressure waves measured by Farrell et al. (1966).

successfully the changes in radon (Rn) flux from linear changes in barometric pressure and from diurnal and half-diurnal barometric waves. The Rn flux perturbations were as much as 60%. However, the soil they studied apparently was several hundred meters deep and dry to bedrock, so their result showing that barometric pressure is important does not conflict with Buckingham's conclusion. Workers should be aware of this depth dependence.

The effects of air turbulence have been harder to assess. Buckingham (1904), Fukuda (1955), Farrell et al. (1966), and Kimball and Lemon (1972) have derived progressively more sophisticated analytical equations for calculating the velocity of soil air movement from surface pressure fluctuations, and Duwe (1976) has used a numerical procedure for the same purpose. The equations of Kimball and Lemon predict that the effects of the high-frequency pressure fluctuations will attenuate rapidly with depth, but there will be an oscillatory flow near the soil surface.

Laboratory studies with pure sinusoidal flows have shown how an oscillatory flow enhances the dispersion of gases in porous media (Scotter et al., 1967; Scotter & Raats, 1968), particularly with crusted soils (Currie, 1970). Scotter and Raats (1969) used these laboratory data to calculate the effects of soil air velocities predicted by the equation of Farrell et al. (1966) with the pure sinusoidal waves shown in Fig. 3. They predicted vapor movement through a 2-cm mulch of 5-mm-diam glass beads can be increased 2–4 times over molecular diffusion, but the wave amplitudes they used were rather large (Fig. 3). Kimball (1970) attempted to use these same laboratory dispersion data along with soil air velocities predicted from more realistic spectra of pressure fluctuations (Fig. 3) by the equations of Kimball and Lemon (1972). His results predicted little increase over diffusion even for media with particles 10 mm in diam.

Duwe (1976) also used the whole spectrum (Fig. 3), but she assumed that the effective diffusion coefficient was equal to the molecular diffusion coefficient plus a summed velocity × displacement term. Her analysis predicted an even smaller increase than Kimball's. These theoretical predictions that surface air turbulence will cause little increase over diffusion, even with particles 10 mm in diam, conflict with some field observations.

2. FIELD OBSERVATIONS

Field observations of the effects of barometric pressure fluctuations on soil gas exchange have been somewhat more definitive than the theories. The ^{222}Rn data of Kraner et al. (1964) and Clements and Wilkening (1974) have shown conclusively that barometric pressure fluctuations are important for very deep soils. Delwiche et al. (1978) found that N_2O concentrations at the surface of a deep, N-fertilized soil were slightly correlated with the rate of decreasing pressure. Duwe (1976) examined the Rn data from several previous workers and concluded that Rn flux follows a diurnal pattern of a broad nighttime minimum, an increase in the morning to an average value of about 2.5 times the minimum, a decrease during the early afternoon, and a second increase during the late afternoon to an average value of about 1.5 times the minimum. Using a numerical model, she was able to attribute the flux variability observed by Clements over deep soil to diurnal plus semidiurnal barometric pressure waves (Fig. 3). On the other hand, some of the earlier data of Pearson et al. (1965), which she used to infer the diurnal flux pattern, were from a shallow wet soil, and the flux pattern could be attributed to neither barometric nor wind-caused pressure fluctuations. However, these data were obtained with a collector of questionable design (Kanemasu et al., 1974). Kraner et al. (1964) obtained data in a moist orchard soil with a water table at 5 m in addition to the deep soil data mentioned above, and they did not report any effects of barometric pres-

[4] W. E. Clements, 1974. The effect of atmospheric pressure variation on the transport of radon-222 from the soil to the atmosphere. Ph.D. thesis. New Mexico Inst. of Mining and Technol., Socorro.

sure fluctuations with the shallow soil. Apparently, Buckingham's (1904) conclusion, that barometric pressure changes are unimportant for exchanging gases in soils < 15 m deep, is correct. For deep soils such as those found in arid regions, barometric pressure fluctuations can be expected to alter the rate of gas exchange by 20–60% for pressure changes of 1–2% day^{-1} (Clements & Wilkening, 1974).

Some workers have attempted to measure the influence of wind or higher-frequency pressure fluctuations $> 10^{-2.5}$ cycles s^{-1} (Fig. 3) on soil gas exchange under field conditions. These results have not been as definitive as desired, but have indicated more influence than predicted by the theories discussed previously. Kimball and Lemon (1971) used a specially designed "vapor exchange meter" and found that the rate of gas exchange through 2-cm layers of straw, gravel, and sand was strongly influenced by pressure fluctuations and wind. When a 2-cm soil layer was tried, there was little correlation with wind, but the flux was about double the rate for diffusion. Kimball (1973) measured the flux of water vapor through 0.5-, 1.0-, and 2.0-mm glass bead mulches and found little correlation with wind speed, but the rate was about 26% faster than the rate for diffusion. Kimball et al. (1976), Jackson et al. (1974), Wierenga et al. (1969), and Allmaras et al. (1977) have tested the diffusion-based heat and water transport theories of Philip and de Vries (1957) and de Vries (1958, 1963) with field soils and measured fluxes that were nearly the same or smaller than the theoretical fluxes. However, when the soil porosity was >0.61, Allmaras et al. (1977) measured thermal conductivities larger than those predicted by theory, and they attributed the increase to air turbulence. Kraner et al. (1964) observed changes in the dispersion of ^{222}Rn by about 25%, even to a depth of 1 m in a dry desert soil, depending on whether the atmosphere was stable or unstable. Thus, the higher-frequency pressure fluctuations associated with wind and air turbulence can increase the gas movement through coarse mulches several times over diffusion. With sand-size particles and dry soils an increase of at least 25% is still possible, but it appears that in soils moist enough to grow plants, the wind-caused pressure fluctuations are not very important.

Pressure effects must always be considered in studies of soil gas exchange, however. Any device placed on the soil surface to measure the flux of a gas will alter the very thing it is supposed to measure. Kanemasu et al. (1974) have shown that blowing air through a sampling box gives results very different than sucking the air through does. A much better design to minimize pressure effects is the vented enclosure with no air circulation of Hutchinson and Mosier (1981).

III. SUMMARY

Several suggested mechanisms for the exchange of soil gases were discussed. Molecular diffusion can account for most exchanges of soil gases because the concentration gradient adjusts to accommodate variable production/consumption rates and variale diffusion coefficients. Unequal production and consumption of gases, liquid water content changes, and temperature effects are shown to be usually unimportant except that temperature gradients affect water vapor movement. Barometric pressure fluctuations can cause perturbations of up to 60% from the diffusion rate in deep (>100 m) soils, but probably are unimportant in soils with <15 m to the water table or other barrier. Wind or air turbulence can increase gas movement several times over diffusion through coarse, shallow mulches. In sand-size mulch or dry granulated soils, an increase of at least 25% is still possible, but it appears that in soils moist enough to grow plants, winds normally have little effect on soil gas movement.

LITERATURE CITED

Allmaras, R. R., E. A. Hallauer, W. W. Nelson, and S. D. Evans. 1977. Surface energy balance and soil thermal property modifications by tillage-induced soil structure. University of Minnesota, Agric. Exp. Stn. Tech. Bull. 306. St. Paul.

Boynton, W. P., and W. H. Brattain. 1929. Interdiffusion of gases and vapors. *In* National Research Council (ed.) International critical tables of numerical data, physics, chemistry, and technology. Vol. 5. McGraw-Hill Book Co., Inc., New York.

Buckingham, E. 1904. Contributions to our knowledge of the aeration of soils. USDA, Bureau of Soils Bull. no. 25. U.S. Government Printing Office, Washington, D.C.

Clements, W. E., and M. H. Wilkening. 1974. Atmospheric pressure effects on Rn^{222} transport across the earth-air interface. J. Geophys. Res. 79:5025–5029.

Currie, J. A. 1960a. Gaseous diffusion in porous media: Part 1. A non-steady state method. Br. J. Appl. Phys. 11:314–317.

Currie, J. A. 1960b. Gaseous diffusion in porous media: Part 2. Dry granular materials. Br. J. Appl. Phys. 11:318–324.

Currie, J. A. 1961. Gaseous diffusion in porous media: Part 3. Wet granular materials. Br. J. Appl. Phys. 12:275–281.

Currie, J. A. 1970. Movement of gases in soil respiration. Sorption and transport processes in soils. Monogr. Soc. Chem. Industry 37:152–171.

DeJong, E., and H. J. V. Schappert. 1972. Calculation of soil respiration and activity from CO_2 profiles in the soil. Soil Sci. 113:328–333.

Delwiche, C. C., S. Bissell, and R. Virginia. 1978. Soil and other sources of nitrous oxide. p. 459–476. *In* D. R. Nielsen and J. G. MacDonald (ed.) Nitrogen in the environment. Vol. 1. Nitrogen Behavior in the Field Soil. Academic Press, New York.

de Vries, D. A. 1958. Simultaneous transfer of heat and moisture in porous media. Trans. Am. Geophys. Union 39:909–916.

de Vries, D. A. 1963. Thermal properties of soils. p. 210–235. *In* W. R. Van Wijk (ed.) Physics of the plant environment. John Wiley & Sons, Inc., New York.

Duwe, M. P. 1976. The diurnal variation in radon flux from the soil due to atmospheric pressure change and turbulence. Ph.D. Thesis. University of Wisconsin (Diss. Abstr. 37b(8): 4012).

Elder, J. W. 1967. Steady free convection in a porous medium heated from below. J. Fluid Mech. 27:29–48.

Farrell, D. A., E. L. Greacen, and C. G. Gurr. 1966. Vapor transfer in soil due to air turbulence. Soil Sci. 102:305–313.

Fukuda, H. 1955. Air and vapor movement in soil due to wind gustiness. Soil Sci. 79:249–256.

Gossard, E. E. 1960. Spectra of atmospheric scalars. J. Geophys. Res. 65:3339–3351.

Hutchinson, G. D., and A. R. Mosier. 1981. Improved soil cover method for field measurement of nitrous oxide fluxes. Soil Sci. Soc. Am. J. 45:311–316.

Jackson, R. D. 1964. Water vapor diffusion in relatively dry soil: I. Theoretical considerations and sorption experiments. Soil Sci. Soc. Am. Proc. 28:172–176.

Jackson, R. D., R. J. Reginato, B. A. Kimball, and F. S. Nakayama. 1974. Diurnal soil-water evaporation: comparison of measured and calculated soil-water fluxes. Soil Sci. Soc. Am. Proc. 38:861–866.

Kanemasu, E. T., W. L. Powers, and J. W. Sij. 1974. Field chamber measurements of CO_2 flux from soil surface. Soil Sci. 118:233–237.

Keen, B. A. 1931. The physical properties of soil. Longmans, Green, and Co., New York.

Kimball, B. A. 1970. Effects of air turbulence upon gas exchange from soil. Ph.D. Thesis. Cornell Univeristy, Ithaca, N.Y. (Diss. Abstr. 31b(3):1008).

Kimball, B. A. 1973. Water vapor movement through mulches under field conditions. Soil Sci. Soc. Am. Proc. 37:813–818.

Kimball, B. A., R. D. Jackson, R. J. Reginato, F. S. Nakayama, and S. B. Idso. 1976. Comparison of field-measured and calculated soil-heat fluxes. Soil Sci. Soc. Am. J. 40:18–25.

Kimball, B. A., and E. R. Lemon. 1970. Spectra of air pressure fluctuations at the soil surface. J. Geophys. Res. 75:6771–6777.

Kimball, B. A., and E. R. Lemon. 1971. Air turbulence effects upon soil gas exchange. Soil Sci. Soc. Am. Proc. 35:16–21.

Kimball, B. A., and E. R. Lemon. 1972. Theory of soil air movement due to pressure fluctuations. Agric. Meteorol. 9:163–181.

King, F. H. 1907. A text book of the physics of agriculture. F. H. King, Madison, Wis.

King, F. H. 1913. The soil: its nature, relations, and fundamental principles of management. The Macmillan Co., London.

Kirkham, D., and W. L. Powers. 1972. Advanced soil physics. Wiley-Interscience, N.Y.

Kraner, H. W., G. L. Schroeder, and R. D. Evans. 1964. Measurements of the effects of atmospheric variables on Radon-222 flux and soil-gas concentrations. p. 191–215. *In* J. A. S. Adams and W. M. Lowder (ed.) The natural radiation environment. William Marsh Rice University by the University of Chicago Press.

Lance, J. C., F. D. Whisler, and H. Bouwer. 1973. Oxygen utilization in soils flooded with sewage water. J. Environ. Qual. 2:345–350.

Nielsen, D. R., R. D. Jackson, J. W. Cary, and D. D. Evans (ed.). 1972. Soil water. Am. Soc. of Agron. and Soil Sci. Soc. of Am., Madison, Wis.

Pearson, J. E., D. H. Rimbey, and G. E. Jones. 1965. A soil-gas emanation measurement system used for radon-222. J. Appl. Meteorol. 4:349–356.

Penman, H. L. 1940a. Gas and vapor movements in the soil: I. The diffusion of vapors through porous solids. J. Agric. Sci. 30:438–462.

Penman, H. L. 1940b. Gas and vapor movements in the soil. II: The diffusion of carbon dioxide through porous solids. J. Agric. Sci. 30:570–581.

Philip, J. R., and D. A. de Vries. 1957. Moisture movement in porous materials under temperature gradients. Trans. Am. Geophys. Union 38:222–232.

Romell, L. G. 1922. Luftvaxlungen i marken s. ekol. faktor. Medd. Stat. Skogsforsoks 19: 125–359.

Rose, C. W. 1968. Water transport in soil with a daily temperature wave: I. Theory and experiment. Aust. J. Soil Res. 6:31–44.

Scotter, D. R., and P. A. C. Raats. 1968. Dispersion in porous mediums due to oscillating flow. Water Resour. Res. 4:1201–1206.

Scotter, D. R., and P. A. C. Raats. 1969. Dispersion of water vapor in soil due to air turbulence. Soil Sci. 108:170–176.

Scotter, D. R., G. W. Thurtell, and P. A. C. Raats. 1967. Dispersion resulting from sinusoidal gas flow in porous materials. Soil Sci. 104:306–308.

Van Bavel, C. H. M. 1951. A soil aeration theory based on diffusion. Soil Sci. 72:33–46.

Wesseling, J. 1962. Some solutions of the steady state diffusion of carbon dioxide through soils. Neth. J. Agric. Sci. 10:109–117.

Wierenga, P. J., D. R. Nielsen, and R. M. Hagan. 1969. Thermal properties of a soil based on field and laboratory measurements. Soil Sci. Soc. Am. Proc. 33:354–360.

CHAPTER 6A

Physiological Consequences of Cellular Water Deficits[1]

THEODORE C. HSIAO AND KENT J. BRADFORD

University of California
Davis, California

I. INTRODUCTION

Numerous changes in plant cells and tissue can be elicited by water deficit or stress. In a previous review (Hsiao, 1973a), it was pointed out that almost any parameter one cares to look at in a plant is altered by water stress, provided that the stress is sufficiently severe and sustained. The wide spread and interlocking nature of the responses to water stress was attributed to two basic causes: the myriad roles played by water in cellular processes, and the integrative nature of living organisms. Since life evolved in the medium of water, the number of functions of water in the whole molecular machinery is huge; any enumeration of these roles inevitably would understate the case. Thus, when there is sufficient water stress, many cellular processes must be directly affected. On the other hand, the plant, like all other organisms, is highly integrated. Diverse processes in the cell must be linked, either tightly or loosely, by feedback and feedforward loops. Primary perturbation of one process effected by stress must lead, in time, to modulation of many others. It follows that, for the purpose of this book, emphasis will be placed on judiciously selected physiological changes caused by water stress; an attempt to list all the known changes would serve no useful purpose. Further, it is paramount to consider stress severity and time course when examining causal relations and the consequences of water stress.

To permit the analysis that follows, the first section of this chapter elaborates on water deficit levels and some important fundamentals of cell- and tissue-water relations. The subsequent sections deal with physiological consequences of water stress. Effects of water deficits on crop productivity are best analyzed in terms of CO_2 assimilation and the partition of assimilates into harvestable yield, integrated over the growing season (Hsiao et al., 1976b). Thus, the underlying physiological changes elicited by water stress

[1] This chapter was written in late 1980. A number of relevant papers have since been published, including a review by A. D. Hanson and W. D. Hitz (Metabolic Responses of Mesophytes to Plant Water Deficits. 1982. Ann. Rev. Plant Physiol. 33:163–203), and a symposium volume edited by L. G. Paleg and D. Aspinall (*Physiology and Biochemistry of Drought Resistance in Plants.* 1982. Academic Press, New York).

Limitations to Efficient Water Use in Crop Production.

are discussed within the reference frame of crop assimilation whenever possible. Expansive growth is discussed first because it is the basis of the development of the assimilation surface. This section includes two key processes underlying expansive growth: cell wall and protein synthesis. Next, the more direct stress effects on assimilation rates in leaves are discussed in relation to stomatal behavior. The biochemical and molecular aspects of stress effects on assimilation are covered by Pearcy in Chapter 6C. The next section is devoted to hormonal changes, in which we consider their possible links to senescence and other aspects of plant development. Osmotic adjustment is another consequence of water stress and is elaborated on by Radin in Chapter 6B.

The intent of this chapter is to cover highlights pertinent to crop productivity as affected by water supply. No pretense is made to present a comprehensive review or to deal with the latest developments in physiology and metabolic biology in depth. For such purposes, general (Begg & Turner, 1976; Boyer, 1976; Boyer & McPherson, 1975; Bradford & Hsiao, 1982; Hsiao, 1973a; Mussell & Staples, 1979; Turner & Kramer, 1980) and specialized (Bewley, 1979; Davies et al., 1980; Jarvis & Mansfield, 1982; Zimmerman, 1978) references should be consulted.

II. LEVELS OF WATER DEFICITS AND FUNDAMENTALS OF CELL- AND TISSUE-WATER RELATIONS

A. Levels of Water Deficits

What are the cellular and metabolic consequences of water deficiency in plants? Before answering, we need to know what is meant by the seemingly simple term "water deficiency." As is made clear by the preceding chapters, water uptake by plants is a passive process, down the gradient of water potential, Ψ, which in turn develops as the result of transpirational loss. In other words, a reduction in plant water content and Ψ is an intrinsic part of the diurnal cycle the plant undergoes in the natural environment. This being the case, how far must water content and Ψ fall in order for the plant to be called water-deficient? The answer to this question largely depends on what physiological processes are evaluated, whether one is dealing with short- or long-term responses, and the preconditioning or growth history of the plant. Some of these points are amplified below. For now, it suffices to point out that in certain instances, remarkably small reductions may cause clear deficiency responses. Not being able to define the upper threshold of water deficiency in absolute terms, we often resort to the use of relative levels, with well-irrigated plants growing under the same conditions as the standard.

The next problem is what physical parameters or indicators to use to quantify plant water status. There is no question that water potential is well-based on thermodynamics, and its gradient is all important for water transport. From the view of physiology and mechanisms, however, a direct relation between Ψ and plant functions is not at all clear (see Section IIC). Nonetheless, we will use tissue Ψ as a convenient indicator because it is the most commonly measured, can be related to other parameters (see Section IIB) more closely linked to function, and is correlatable to productivity and yield (Hiler et al., 1972). Another indicator of plant water status is *relative water content* (RWC), which is the amount of water in the tissue relative to the amount the tissue holds at full turgor. Relative water content is readily measurable and in some ways is more easily linked to plant form and functions than water potential. As pointed out before (Hsiao, 1973a), however, it suffers from being an insensitive indicator, with a low signal-to-noise ratio. With plants of high water status, marked effects on physiological processes can be caused by reductions in RWC of a few percent, yet this is about the size of the random error in RWC measurements (Barrs, 1968).

Using Ψ and RWC, it is helpful to make some generalizations regarding the levels of water deficits characteristic of physiological changes. The remarks here are applicable to most herbaceous crop species grown outdoors and having little or no growth history of osmotic adaptation. On sunny days without strong advection, the exposed leaves of many crops growing on well-irrigated soils exhibit Ψ values at midday in the range of -9 to -13 bars, with RWC probably in the range of 85–95%. Between midnight and dawn, leaf Ψ should be -1 or -2 bars and RWC close to 100%. When there is obvious wilting or pronounced stomatal closure induced by leaf water deficits, leaf Ψ generally falls in the range of -13 to -17 bars. This range of leaf water status corresponds approximately to the point where cell turgor drops to nearly zero. Lowering Ψ to -30 to -35 bars, corresponding to RWC in the 45% range, soon results in the death of cells.

B. Relationship Among Water Content, Water Potential, and Component Potentials

To examine plant behavior as influenced by its water status, it is necessary to have an understanding of some of the physical aspects of water in the cell and tissue. How do water potential, its various components, and related physical parameters change as cells or tissues lose water? An example of the relation between leaf water potential and RWC is shown in Fig. 1A. As a fully hydrated leaf (RWC = 100%) loses water, the corresponding drop in Ψ is very steep at first, and then becomes more and more gentle as RWC continues to be reduced, to the point of zero turgor. As Ψ declines, the chemical activity of water (a_w) decreases also. Quantitatively, a_w is related to Ψ by the fundamental equation.

$$\Psi = (RT/\bar{v}_w) \ln a_w, \quad [1]$$

where R is the universal gas constant, T is absolute temperature, and $\bar{v}_w$ is the partial molal volume of water. The equation indicates that a_w changes little as Ψ changes because their relationship is logarithmic.

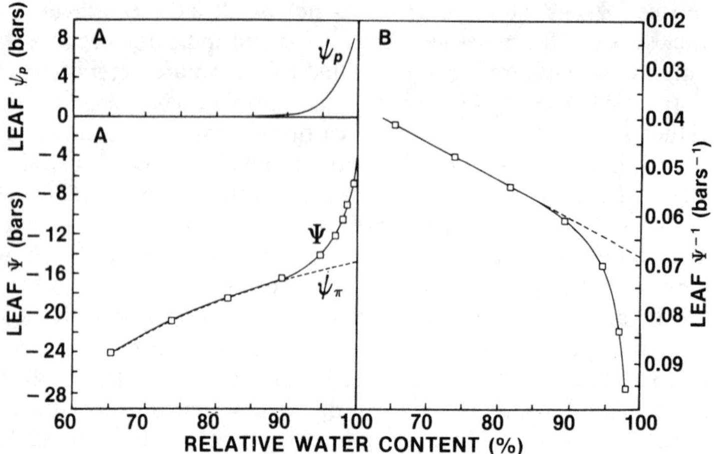

Fig. 1—Relation of water potential and components to relative water content for a field-grown cotton leaf. (*A*) Ψ, ψ_π, and ψ_p vs. RWC. (*B*) Inverse of Ψ vs. RWC. Data were obtained from a single leaf using the pressure chamber technique (Scholander et al., 1965; Boyer, 1969). Dotted line represents ψ_π calculated according to Eq. [3] in (*A*), or $1/\psi_\pi$ in (*B*). Deviation of ψ_π from Ψ is due to the pressure component (ψ_p) of water potential (Hsiao, unpublished data.)

In the protoplasm enclosed by the plasmalemma and cell wall, the two major components of water potential are pressure potential (ψ_p) and osmotic or solute potential (ψ_π). Pressure potential represents the effect of pressure on the energy status of water. The pressure results from protoplasmal water uptake and volume increase against the constraining effect of the cell wall. The pressure inside the protoplasm is usually above the reference atmospheric pressure outside the cell wall ($\psi_p > 0$), and there is a pressure differential across the plasmalemma–cell-wall complex. On the other hand, ψ_π represents the depressive effect of dissolved solutes on water potential. Being a colligative property, ψ_π changes in accordance with well-known laws when the protoplasm loses water. For the concentration range of concern in plant-water relations, the equation of Morse is more accurate than that of Van't Hoff (Crafts et al., 1949) and can be expressed as

$$\psi_\pi = -(RT\, n_s/V_w), \tag{2}$$

where n_s is the effective number of moles of solutes and V_w is the volume of solvent water. Figure 1A shows an example of estimated ψ_π values as water status varies in a leaf. The water volume, V_w, has been normalized and expressed as RWC. Note that ψ_π changes gradually as RWC changes. Figure 1A also shows Ψ and ψ_p as functions of RWC and is really a modern version of the classical Höfler diagram. At or near full turgor (100% RWC), Ψ declines steeply as the tissue loses water. This is because ψ_p declines steeply (Fig. 1A) as the cell volume decreases and the cell deflates. In constructing the curve for ψ_p in Fig. 1A, it is assumed $\psi_p = \Psi - \psi_\pi$; i.e., the matrix component of water potential (ψ_m) is zero or negligible (Boyer, 1967; Wiebe, 1966).

The accurate estimation of ψ_π and ψ_p curves for leaves such as those in Fig. 1A makes use of a version of the inverse of Morse's equation:

$$\frac{1}{\psi_\pi} = -\frac{V_w^\circ}{RTn_s}\,\text{RWC} + \frac{V_w^\circ}{RTn_s}\,U, \qquad [3]$$

where V_w° is the total volume of tissue water at full turgor and U is the fraction of V_w° located or held in such a way as to make it unavailable or ineffective as solvent. Water held in the cell wall and xylem of the tissue is ineffective as solvent, since most solutes are confined to the protoplasm. The last term in Eq. [3] corrects for this apoplastic water. Relative water content (as a fraction) is used for normalization and equals V_w/V_w°. If n_s and U remain nearly constant as RWC changes, Eq. [3] is a linear function and is useful for the estimation of ψ_π and ψ_p at any RWC. Instead of the inverse of ψ_π, the inverse of Ψ is plotted against RWC (Scholander et al., 1965), as in Fig. 1B. Where the line is linear, Eq. [3] applies and ψ_π is presumably the only significant component of Ψ. Deviation from linearity when the tissue approaches saturation is attributed to ψ_p. Thus, both ψ_π and ψ_p at any given RWC can be estimated from the plot. Also, U can be estimated by extrapolating the line to $1/\Psi = 0$ (Boyer, 1969). When $1/\Psi$ (hence, $1/\psi_\pi$) equals zero, RWC $= U$, in accordance with Eq. [3]. This extrapolation, however, involves much uncertainty because the linear portion of data points usually spans between RWC values of 0.8–0.6 and has to be extended for a long distance to reach the extrapolated U value, which falls around 0.1 or 0.2 RWC.

Pressure potential is a function of the protoplasm volume and the elasticity of the cell wall. By extending Hooke's law to a three-dimensional mechanical stress-strain relationship, a volumetric modulus of elasticity (ϵ) can be defined for the wall of a cell (Dainty, 1976):

$$d\psi_p = \epsilon(dV_{prot}/V_{prot}),$$

where V_{prot} is the protoplasm volume. For the small ranges of V_{prot} near full turgor, and where V_{prot} is due mostly to V_w, dV_{prot} is approximated by dV_w and V_{prot} by V_w°. Recognizing that $d(V_w/V_w^\circ) = d(\text{RWC})$,

$$d\psi_p \cong \epsilon\, d(\text{RWC}). \qquad [4]$$

Equation 4 shows that the slope of a plot of ψ_p vs. RWC yields values for ϵ. From the changing slope of such a plot (Fig. 1A), it is clear that ϵ increases with RWC and with ψ_p. That is, the wall becomes stiffer or less elastic as it stretches. It is important to point out that by definition, elastic behavior refers to the reversible deformations caused by forces such as pressure. The irreversible deformation characterizing expansive growth (Section III) is not directly relatable to the modulus of elasticity.

Strictly speaking, the foregoing discusses water-related parameters in a single cell. The available data, such as those shown in Fig. 1, however, have been taken almost exclusively on tissue or organs consisting of many thousands of heterogeneous cells. Heterogeneity among cells prevails even in rather uniform tissues such as storage roots, as illustrated by the frequency distribution of ψ_π values in cells of beet slices (Currier, 1944). So, by necessity, our real data are composite or integrated values. Interpretation of the data in terms of function in individual cells entails uncertainties (Cheung et al., 1976), which become more serious as attempts are made to accurately pinpoint the levels of water stress needed to trigger certain metabolic changes (e.g., Pierce & Raschke, 1980).

Two parameters that should be relatively homogeneous in a tissue or organ are total water potential (Ψ) and the related chemical activity of water (a_w). In tissue at water equilibrium, Ψ and a_w must be constant throughout. In tissue where there is water transport, a Ψ gradient exists and Ψ varies with location, but usually gently, monotonically, and in a predictable fashion (Molz & Boyer, 1978; Silk & Wagner, 1980).

C. Water-Related Parameters Important to Plant Functions

Considering the various parameters altered by changes in water status, it is pertinent to ask: Which ones are responsible for the observed changes in physiology and metabolism effected by water deficits? Further, how are changes in such physical parameters transduced (Hsiao et al., 1976a) into metabolic alteration? Our knowledge is extremely limited in this regard, especially when water deficit levels are not too severe and are in ranges often encountered in crops in the field. Nonetheless, a conceptual analysis is useful. Recasting the physical parameters of water in terms of considerations relevant to metabolism and physiology, we can list six possible direct effects of water loss from cells and tissue:

1) Reduction in the chemical potential or activity of water, i.e., a reduction in Ψ.
2) Concentration of small molecules and macromolecules in the cell that results from a reduction in RWC and cell volume and is reflected in a lowering of ψ_π.
3) Changes in spatial relations in membranes, organelles, or cell-wall matrix through the reduction in volume.
4) Changes in the water of hydration or in the structure of water adjacent to macromolecules such as proteins and nucleic acids.
5) Reduction in hydrostatic pressure (ψ_p).
6) Decrease in the pressure differential across the plasmalemma–cell-wall complex following a reduction in ψ_p in the protoplasm.

It is unlikely that a reduction in the chemical activity of water is responsible for changes in metabolism caused by mild to moderate water stress. According to Eq. [1], a reduction in Ψ of 10 bars reduces the chemical activity of water only 0.7%, and a reduction of 30 bars, only 2.2% (Hsiao, 1973a). Even for reactions involving water as reactants or products, such small changes in a_w cannot easily account for any marked effects on metabolism.

Effect (2), the concentration of cell contents, may play a role, but only if the affected metabolic reactions are catalyzed by allosteric enzymes extremely sensitive to small changes in the concentration of specific effectors (Hsiao, 1973a). Effect (3), changes in spatial relations, appears as a hypothetical mechanism, but direct data are lacking. Effect (4), changes in water around macromolecules, is highly speculative and seems improbable unless water stress is very severe and the cell loses the major portion of its water. Effect (5), a reduction in ψ_p, if isotropic (uniformly through space and lacking direction), should have little or no influence since pressure changes of tens to hundreds of bars are necessary to affect metabolic processes investigated so far (Vidaver, 1972; Zimmerman, 1970). Effect (6), a change in pressure differential across the membrane-wall complex, can be accepted as a proven linkage between water stress and expansive growth, possibly by modulating membrane transport processes. This will be discussed in a following section. Overall then, effects (1) and (5) can be virtually ruled out and effect (6) is widely accepted. Definitive data are badly needed to further evaluate effects (2), (3), and (4). A more thorough discussion of these possible effects, with reference to available data, was presented earlier (Hsiao, 1973a).

In terms of the physical parameters commonly estimated for work in water relations, the foregoing discussion points out the importance of ψ_p for maintaining certain plant functions, but only as a pressure differential between the cell interior and its surroundings. Also implied is the need to maintain high RWC for normal volume and shape. Commonly, high ψ_p, RWC, and Ψ go hand in hand in cells (Fig. 1A). If the plant has undergone osmotic adjustment in water-limiting conditions (see Radin, Chapter 6B), however, ψ_p and RWC may be high in spite of low Ψ values.

III. EXPANSIVE GROWTH

Expansion of cells is the basis for the development of assimilation area. The limitation of plant growth by water stress has been amply documented on both a short-term and a seasonal basis (see reviews by Hsiao et al., 1976b; Fischer & Turner, 1978; Turner, 1979). The growth of the plant as a whole is ultimately determined by the rates and extent of cell division and expansion within its tissues. Cell division is often reduced in water-stressed plants, especially if the stress is prolonged or severe (Kirkham et al., 1972; Meyer & Boyer, 1972; reviewed by Hsiao, 1973a). However, the primary effect of water stress is to reduce expansive growth (Hsiao, 1973a), which can be defined as irreversible cell enlargement. An example is the data of Cutler et al. (1977), where water stress caused an 18% reduction in leaf area, but cell density was increased by a similar amount. The total number of cells per leaf, therefore, remained relatively constant. A decrease in cell size is a general characteristic of leaves subjected to water stress during development (Ordin, 1966). Cell expansion and division are undoubtedly coordinated in the plant, and reduction of expansion may feed back and inhibit division (Hsiao et al., 1976b). Understanding how water stress inhibits expansion requires knowledge of how growth occurs, and of the factors that control the growth rate. Accordingly, these are discussed along with effects of water stress.

A. Physics of Expansive Growth

Recent research has emphasized the key roles of the cell wall and ψ_p in the growth process (Albersheim et al., 1977; Green & Cummins, 1974; Green et al., 1977). As described in Section IIB, ψ_p results from the resistance of the cell wall to volume expansion due to water uptake. In a nonexpanding cell in equilibrium with its surroundings, the difference between ψ_π inside the cell and outside ($\psi_\pi^{in} - \psi_\pi^{out}$) is exactly balanced by ψ_p. Hence, $\Psi^{in} = \Psi^{out}$ and there is no net water flux. Growth is initiated by a relaxation or "loosening" of the cell wall, as illustrated in Fig. 2. This immediately reduces ψ_p slightly. Consequently Ψ^{in} becomes < Ψ^{out} and water flows into the cell. The cell wall expands due to water uptake and the force of turgor pressure until the resistance of the wall again causes ψ_p to rise to the point where $\Psi^{in} = \Psi^{out}$. The next step, wall relaxation, is then repeated; the stepwise sequence is only presented for the ease of comprehension. In reality, during sustained growth, wall relaxation, water flux, expansion, and cell wall synthesis are occurring continuously and simultaneously, and Ψ^{in} remains below Ψ^{out} (Boyer, 1968; Lockhart, 1965; Ray et al., 1972). This assumes that ψ_π is constant as the volume of the cell increases. It follows that solutes must be accumulated or generated internally to compensate for the dilution of cellular contents and to maintain the Ψ gradient for water uptake. Otherwise, water uptake and expansion would eventually cease as the gradient diminished, regardless of the degree of wall relaxation (Grenetz & List, 1973). Thus, ψ_p, wall relaxation, and solute accumulation are mutually dependent components of growth.

The various factors that might control expansion will be first considered in an operational sense, without reference to the underlying metabolism. This approach has been utilized by Lockhart (1965) to define a relation between growth rate and ψ_p:

$$dV/Vdt = E_g(\psi_p - \psi_{p,th}); \qquad [5]$$

where V is the cell volume, dV/Vdt is the rate of growth, and $\psi_{p,th}$ is a threshold turgor pressure, below which no extension occurs; E_g is gross extensibility, a coefficient that is the slope of the curve of dV/Vdt vs. ($\psi_p - \psi_{p,th}$), and should not be confused with ϵ, the elastic modulus defined in Section IIB. Gross extensibility refers only to the irreversible or "plastic" extensibility of the cell wall during growth, while ϵ is a measure of the wall's reversible or "elastic" expansive properties. Equation [5] illustrates several points in the relationship between growth and ψ_p (Fig. 3). First, a minimum turgor pressure ($\psi_{p,th}$) must be exceeded before growth occurs. This has been observed in a variety of plant tissues (Boyer, 1968; Bunce, 1977; Green et al., 1971, 1977; Hsiao et al., 1976a; Meyer & Boyer, 1972) with values for

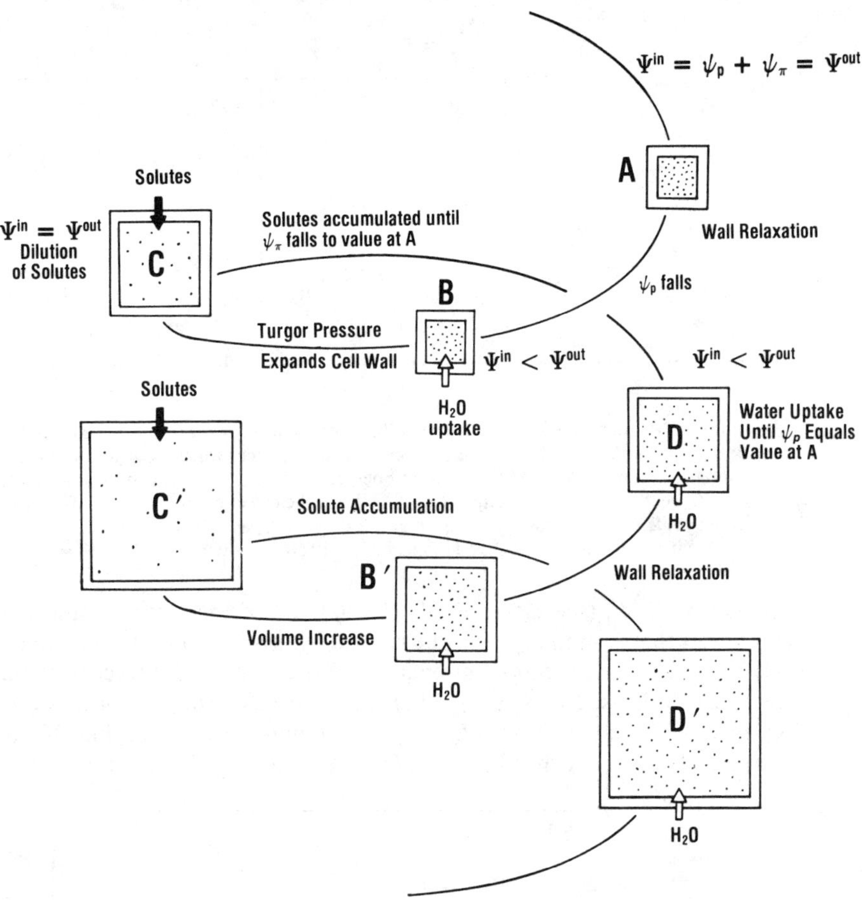

Fig. 2—Conceptual diagram of the interrelations of wall relaxation, water uptake, turgor pressure, and solute accumulation in expansive growth. Beginning at the upper right, growth is initiated by wall loosening (*A*) which lowers ψ_p and creates a Ψ gradient for water uptake (*B*). Turgor expands the wall as water enters the cell, but cellular contents are diluted (*C*). Solute accumulation restores the initial ψ_π, and water enters until ψ_p reaches the original value (*D*). The steps repeat in the next cycle (*B′*, *C′*, *D′*). The sequential presentation is artificial and is depicted for ease of understanding. In steady state growth, wall loosening, water uptake, expansion, and solute accumulation are occurring simultaneously and continuously.

$\psi_{p,th}$ ranging from 1 to 6.5 bars, depending upon the conditions employed (see below). Second, the instantaneous growth rate is seen to be sensitive only to that increment of ψ_p that exceeds $\psi_{p,th}$ (Fig. 3A). Thus, small decreases in ψ_p can cause reduction or cessation of growth, even though ψ_p may still be considerable. This is evident from the sharp decline in growth rates of leaves subjected to small (1- to 4-bar) reductions in Ψ (Acevedo et al., 1971; Sharp et al., 1979; Wenkert et al., 1978), as illustrated in Fig. 4. When Ψ of quite turgid tissue is rapidly reduced, ψ_p is the component of Ψ^{in} that declines most dramatically (Fig. 1A). The short-term changes in instantaneous growth rate following water stress can therefore be attributed primarily to

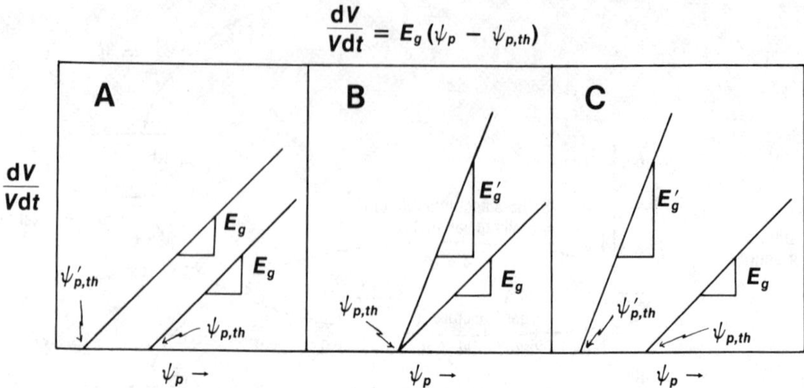

Fig. 3—Empirical growth rate relations and possible modifications. The lower curve of Graph *A* illustrates that no growth occurs until ψ_p exceeds $\psi_{p,th}$, then proceeds at a rate directly proportional to ψ_p with slope of E_g. In the upper curve of (*A*), $\psi_{p,th}$ has been lowered to $\psi'_{p,th}$ while E_g remains constant. Graph *B* demonstrates the effect on growth rate of increasing E_g to E'_g while $\psi_{p,th}$ remains the same. In graph *C*, both E_g and $\psi_{p,th}$ are altered. It is apparent that shifts in E_g and/or $\psi_{p,th}$ lead to markedly different growth rates for a given ψ_p.

the reduction in ψ_p (see Hsiao et al. [1976a] for further discussion). The third point is that E_g and $\psi_{p,th}$ may be altered by growth conditions or water stress. These conditions represent changes in the slope and intercept of the curve of d*V*/*V*d*t* vs. $(\psi_p - \psi_{p,th})$ (Fig. 3A and 3B). Adaptation to water stress may involve alterations in E_g, $\psi_{p,th}$, or in both parameters (Fig. 3C) to permit growth to occur at a lower ψ_p (Acevedo et al., 1971). Limited data

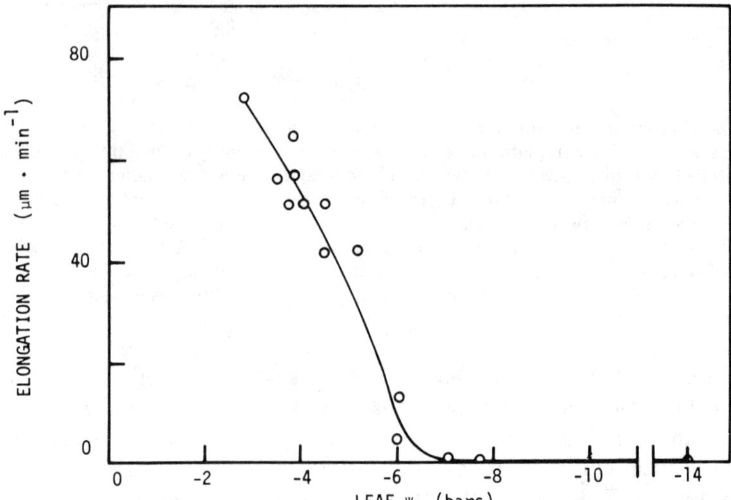

Fig. 4—Rate of leaf elongation in relation to leaf Ψ in 10-day-old maize seedlings. Points are single measurements taken in a growth chamber. Elongation was monitored with a linear variable differential transformer (LVDT) and was the youngest (third leaf, whereas Ψ was of the next oldest (second) leaf. In separate experiments, Ψ of the youngest leaf was found to be within 1.5 bar or less of the Ψ of the next oldest leaf at several stress levels. Derived from Acevedo and coworkers (1971).

(Bunce, 1977) suggest that under the very different environments of growth chamber, greenhouse, and field, $\psi_{p,th}$ and E_g change in opposite directions such as to allow overall expansion rates to remain similar despite differences in Ψ and ψ_p (Fig. 3C). More data are needed, however, to establish this as a generality.

An additional option for maintaining growth rate as Ψ declines is also inherent in the basic rate equation. Since $\psi_p = \Psi - \psi_\pi$, we can write $dV/Vdt = E_g(\Psi - \psi_\pi - \psi_{p,th})$. If the cell can raise its solute content and consequently lower ψ_π to keep pace at least partly with the Ψ decline, it can maintain $\Psi - \psi_\pi - \psi_{p,th}$ and continue to enlarge. Recent work has verified that crop plants can osmotically adjust by taking up or generating osmotica in response to water stress to maintain a positive turgor as Ψ declines (see Hsiao et al., 1976a; Turner, 1979). The adjustment can be on a diurnal or seasonal basis, varies with crops, and contributes to the ability of some crops to grow near midday in spite of the depression in leaf Ψ at that time (Acevedo et al., 1979). The effects of osmotic adjustment on growth can be accounted for by deriving more complicated forms of Eq. [5] (e.g., Grenetz & List, 1973). It should be pointed out that if ψ_p falls to zero, changes in E_g and $\psi_{p,th}$ cannot compensate sufficiently for growth to occur; only osmotic adjustment can achieve that end. Thus, evidence exists for modifications of each component of the rate equation, singly or in concert, to permit growth to continue as Ψ falls.

Having presented a straightforward method in the form of Eq. [5] for the analysis of growth as affected by osmotic adjustment, we hasten to add that there may be complications. Data are accumulating (Meyer & Boyer, 1972; Michelena & Boyer, 1980; Hsiao, unpublished; Oliveira, 1982) which indicate that in osmotically adjusted plants, expansive growth can be minimal or even nil in spite of a ψ_p nearly equal to or exceeding that in well-watered plants. Thus, modifications of E_g and $\psi_{p,th}$ can act to either increase or decrease the growth rate for a given value of ψ_p, depending upon the circumstances. Further discussion of this point is found elsewhere (Bradford and Hsiao, 1982).

Thus far, we have assumed that the permeability of the cell to water is not a factor in determining cell water uptake to sustain growth. Where single cells or slowly growing tissues are concerned, this assumption is probably valid (Ray et al., 1972). However, in tissues where some cells are at macroscopic distances from the water source (i.e., the xylem vessels), the hydraulic conductance of the tissue must also be considered, as discussed by Lockhart (1965). An equation for growth (Hsiao et al., 1976a), incorporating the conductance (C) and based on that of Lockhart, is as follows:

$$dV/Vdt = E_g C \frac{\Delta\Psi + \psi_p - \psi_{p,th}}{E_g + C}, \qquad [6]$$

where $\Delta\Psi$ is the difference in Ψ between the source of water and the sink for water in the growing cells. Boyer (1968) estimated $\Delta\Psi$ to be 1.5–2.5 bars in growing leaves. Recent analyses (Molz & Boyer, 1978; Silk & Wagner, 1980) estimate that individual cells of a growing tissue are in near Ψ equilibrium with their surroundings, but overall Ψ gradients may range from about -1 to -2 bars for the growing region. A serious uncertainty is the magnitude of C. Boyer and Wu (1978) reported that supplying indoleacetic acid (IAA) to auxin-responsive tissues increased their hydraulic conductances, presumably via effects on permeability of the cell membranes. These authors calculate that increases in E_g and C contribute about equally to the auxin-induced growth. How C may change along with ψ_p in affecting growth during water stress is unknown.

B. Mechanism of Expansive Growth

Thus far, growth and water-stress effects have been discussed in physical or operational terms without reference to the underlying cellular biology and metabolism. In this section we will briefly consider current working hypotheses for regulation of growth and turgor maintenance and how they relate to growth under water stress. This is presently a very active and controversial area of research, due both to new theories and to new experimental techniques for testing them.

1. ACID-GROWTH HYPOTHESIS AND CELL WALL METABOLISM

As described above, cell growth requires an increase in extensibility, or "loosening" of the cell wall. Agents that accelerate the rate of growth, such as indoleacedic acid (IAA) or gibberellins (GA), also increase the extensibility of the cell wall (Adams et al., 1975; Cleland, 1967; Stuart & Jones, 1977). The discovery that acid solutions would also stimulate wall loosening and expansion led to the proposal that the effects of IAA on growth were due to an acidification of the cell wall (Hager et al., 1971; Rayle & Cleland, 1972). This proposal was strengthened by the discovery that auxin could induce H^+ excretion into the medium surrounding the tissue (Cleland, 1973; Rayle, 1973). The basic tenet of the acid-growth hypothesis is that IAA stimulates the extrusion of H^+ from the plasma membrane causing the acidification of the cell wall, which in turn activates the enzymes responsible for wall loosening. The extensive evidence in support of this hypothesis, much of it on *Avena* coleoptile and pea epicotyl, has recently been reviewed (Cleland & Rayle, 1978; Rayle & Cleland, 1977). Other workers, however, have either been unable to observe acidification in response to IAA, or found acidification to be unrelated to the growth response (reviewed by Penny & Penny, 1978). One discrepancy between acid- and IAA-induced growth is that the former is of much shorter duration, lasting only 2–4 hours (see Rayle & Cleland, 1977). The possibility exists that auxin- and acid-induced growth operate via related but independent mechanisms (Pope, 1978; Vanderhoef et al., 1977), and further research is needed to resolve the conflicting data.

Extrusion of H^+ in response to IAA is presumed to be due to the activation of a plasma-membrane-associated ATPase (Hager et al., 1971; Beffagna et al., 1979). Recent experiments with the ATPase inhibitor vanadate support this conclusion (Taiz & Jacobs, 1980). A rapid osmotic shock, sufficient to slightly plasmolyze the cells, is known to affect the plasma membrane, causing an increase in passive permeability to solutes and an inhibition of auxin-induced acidification (Rubinstein, 1977; Rubinstein et al., 1977). On the other hand, hypertonic media without IAA can actually bring about a slight increase in H^+ extrusion (Cleland, 1975; Göring et al., 1978; Marrè et al., 1973). Whether these are direct effects of the reduction in ψ_p is unknown.

Changes in cell wall synthesis and turnover underlie the wall loosening associated with growth. Plant cell walls are composed primarily of cellulose microfibrils linked together by hemicelluloses and pectic polysaccharides (Albersheim, 1976). As cells grow, new wall material must be synthesized and incorporated into the expanding wall matrix. Current models of expansive growth require breakage and resynthesis of the hemicellulose cross-linkages as the cellulose microfibrils rearrange during expansion (Albersheim et al., 1977). In addition, new polysaccharides are inserted into the cellulose network to maintain wall thickness and density (Maclachlan, 1977). As would be expected, there is generally a close parallel between cell wall synthesis and elongation rate (Ray, 1962). Expansive growth is somewhat more sensitive to reduction in Ψ than is wall synthesis (Cleland, 1967), suggesting that inhibition of wall synthesis by water stress is a consequence of reduced growth. Wall loosening can occur in the absence of wall synthesis, and the loosening process apparently involves turnover of the hemicellulose components of the wall (Labavitch, 1981). Reduction in hemicellulose content of the cell wall and solubilization of its xyloglucan components are known to accompany IAA-induced growth (Labavitch & Ray, 1974a, 1974b; Loescher & Nevins, 1972). Sakurai and Masuda (1977) have correlated these changes with increased extensibility of the cell wall, and have found an interesting relationship with the strength of external osmotica. When IAA is supplied to *Avena* coleoptiles in solutions with varying concentrations of mannitol, the maximum increase in extensibility and decrease in wall noncellulosic glucose content (an estimate of hemicellulose) occur at mannitol concentrations between 0.1 and 0.2*M* (−2.5 to −5 bars), which prevent expansive growth. Higher mannitol concentrations block the IAA effects (see also Cleland, 1967; Loescher & Nevins, 1973). Thus, small reductions in ψ_p increase the activity of IAA on wall metabolism, while more severe reductions prevent the effect. Interestingly, mannitol alone (without IAA) decreased the extensibility of both *Avena* (Sakurai & Masuda, 1977) and *Regnellidium* cells (Cookson & Osborne, 1979). These results suggest that ψ_p may be a factor regulating the cellular response to IAA and subsequent cell wall metabolism. Such interactions may be related to the reduction in cell size observed under drought conditions (Cutler et al., 1977; Ordin, 1966).

2. TURGOR AND OSMOTIC REGULATION IN GROWING CELLS

As pointed out in Section IIIA, sustained growth requires the continuous importation of solutes or their precursors into the enlarging cell. In fact, despite the abundance of research on the role of cell wall extensibility in regulating growth rates, it is not certain whether this, or the ability of the cells to osmoregulate to maintain turgor, is the factor controlling elongation in intact higher plants (Cleland, 1977; Hsiao et al., 1976a). Exogenous sucrose and K^+ salts are well-known for their promotive effects on expansive growth of excised tissue (Ordin et al., 1956; Ray, 1962). The exchange of extruded H^+ for K^+ from the cell exterior, accompanied by synthesis of organic counterions such as malate, has been suggested as a mechanism for maintaining ψ_π, ψ_p, and cytoplasmic pH during growth. Fixation of CO_2 into malate was specifically stimulated by IAA in *Avena* coleoptiles, and the amount of malate accumulated was correlated with the growth response (Haschke & Lüttge, 1975a, 1975b, 1977). While malate synthesis may be a response to H^+ extrusion in order to stabilize cytoplasmic pH (Davies, 1973), in some cases the accumulated K^+-malate may be important osmotically. Dhindsa et al. (1975) reported that in elongating cotton fibers, up to 55% of the total osmotic solutes could be accounted for by K^+-malate. This component of ψ_π could fully maintain cell turgor as growth occurred.

The possible role of IAA in stimulating malate synthesis may be the basis for the discrepancy between acid- and IAA-induced growth mentioned earlier. The short duration of the acid enhancement of growth may be due to an inability to induce the osmoregulatory mechanisms within the cell, leading to a decline in ψ_p as growth proceeds (Cleland & Rayle, 1978; Taiz & Metraux, 1978; Vesper & Evans, 1979). Thus, the ability of IAA to promote malate synthesis or other osmoregulatory mechanisms may be as important to continued growth as its ability to cause H^+ extrusion and wall loosening. This again emphasizes the interrelated nature of expansive growth, requiring simultaneous wall loosening, turgor pressure, and solute importation for steady-state expansion (Fig. 2).

While K^+ and malate are important in turgor maintenance in elongating cotton fiber and in turgor changes in guard cells necessary for stomatal movement (see Section IV), they seem to be less significant in osmotic adjustment in nongrowing or slowly growing tissue affected by water stress. In cotton, although malate concentration increased dramatically as leaf Ψ declined, all organic acids, including malate, still only accounted for a small percentage of the total osmotic solutes (Cutler & Rains, 1978). In another study of diurnal osmotic adjustment in sorghum, organic acids represented <0.1 bar of a total 3.9-bar change in ψ_π, while sugars represented about 3 bars (Acevedo et al., 1979). However, it is possible that H^+-K^+ exchange and malate synthesis represent only short-term osmoregulation as growth occurs, to be largely replaced by other solutes such as sugars or amino acids for seasonal or stress-induced osmotic maintenance (Dhindsa et al., 1975; Mott & Steward, 1972; Munns et al., 1979; see also Radin, Chapter 6B, this volume).

Since osmotic level is so important in cells undergoing expansion, it must be closely regulated (see Cram, 1976), but direct data are lacking for higher plants. Here, some extremely interesting osmoregulatory mechanisms in giant algal cells are briefly outlined. It is unlikely, however, that they bear much resemblance to those in higher plants.

Many marine algae, especially those exposed to widely fluctuating water and salinity conditions, have the ability to regulate their osmotic contents to maintain relatively constant ψ_p regardless of changes in external osmolarity (for review, see Gutknecht et al., 1978). The giant marine algae, *Valonia,* have a turgor-maintenance mechanism linked to K^+ transport. The hydrostatic pressure difference across the plasma membrane controls the rate of K^+ uptake. When the pressure difference is reduced, either by hypertonic solutions or by inserting a capillary manometer tube into the vacuole to vary the pressure, the membrane potential becomes more negative and K^+ uptake into the vacuole is accelerated. Other species synthesize specific organic compounds as osmotica for turgor regulation (Kauss, 1977). Thus, the physical event (pressure change) is transduced into a biochemical event (ion transport or osmotica synthesis). In higher plants, although numerous examples of environmental and hormonal effects on ion transport and accumulation are known (Pitman & Cram, 1977), the role of ψ_p in mediating these events, with the possible exception of stomatal ion transport (Section V), remains uncertain.

A major limitation to the study of ψ_p in higher plants has been the lack of a direct method of measurement. Pressure potential is usually calculated as the difference between Ψ and ψ_π and is subject to a variety of errors. Pressure probes for direct measurement of ψ_p in single giant algal cells have been used for some time (Green, 1968). Recently, the probes have been modified to measure ψ_p in higher plant cells (Edwards & Meidner, 1977; Hüsken et al., 1978; Zimmermann et al., 1980), as well as to estimate parameters such as cell hydraulic conductance and elastic modulus. However, the new techniques have yet to be utilize to study expansive growth. Cells large enough to be penetrated by the probe have generally ceased elongation, and the viability and functionality of the cells after penetration by the probe remains to be established.

3. PROTEIN SYNTHESIS

As pointed out above, sustained growth must involve the coordination of numerous processes, including the manufacturing of cellular constituents. Many synthetic processes in the plant are known to be altered by water stress, including that of cell walls and proteins. The role of wall synthesis in expansive growth has been touched on above. The continued synthesis of protein must also be a prerequisite for maintaining growth. In addition, protein synthesis is crucial for all metabolic functions, particularly for obtaining high rates of photosynthesis, since the carboxylating enzymes constitute a major fraction of the protein in green leaves.

In general, protein synthesis in growing tissue is quite sensitive to mild water stress (see Hsiao, 1973a). Hsiao (1970) found a rapid shift of polyribosomes to monoribosomes in rapidly expanding tissue of *Zea mays* following a brief desiccation treatment that reduced Ψ only 2 or 3 bars. The shift was interpreted to reflect a drop in overall protein synthesis. Similar results have since been recorded for several other crop species (e.g., Barlow et al., 1977; Morilla et al., 1973). While continued protein synthesis is certainly required for growth (Bates & Cleland, 1979; Lado et al., 1977), a major question remains of whether inhibition of protein synthesis by water stress limits growth, or whether protein synthesis declines due to feedback effects of slowed growth rates. Rhodes and Matsuda (1976) found a good correlation between growth rate and percent polyribosomes as a function of Ψ in pea and pumpkin seedlings. Similar correlations on a daily basis were also observed in squash fruits (Cocucci et al., 1976). The last authors suggested that midday reductions in growth are due to the effects of water stress on protein synthesis. Dhindsa and Cleland (1975a), on the other hand, found protein synthesis to be slightly less sensitive than expansive growth to water stress (see also Hsiao, 1973a). Recently, Scott and coworkers (1979) compared polyribosome levels of mature and elongating wheat leaves at various levels of water stress (−8 to −30 bars). They found that the stress caused a rapid decline in polyribosome percentage in the elongating leaves but had no significant effect on polyribosomes in mature leaves. The shift from polyribosomes to monoribosomes occurred concomitant with the cessation of expansive growth in the young tissue. These data, confirmed by unpublished work of Hsiao on maize, suggest that the inhibition of protein synthesis is the result, not the cause, of the reduction in expansive growth, as hypothesized earlier (Hsiao, 1970).

There is evidence that protein synthesis in drought-hardy species is less sensitive to water stress than in drought-susceptible species (Brandle et al., 1973). Although reduced by water stress, considerable protein synthesis activity is retained in black locust seedlings even at a Ψ value of −25 bars (Brandle et al., 1977). The entire shoot used in these studies may have contained both elongating and mature tissues, thus making changes in total polyribosome levels less dramatic. It would seem worthwhile to reinvestigate drought-hardy species with careful attention to the stage of growth of the tissue. For information on protein synthesis in dessication-tolerant plants, see Bewley (1979).

Dhindsa and Cleland (1975b) tested the effect of external hydrostatic pressure on protein synthesis in osmotically stressed (−7 bars) *Avena* coleoptiles. A pressure of 8 bars caused alterations in the patterns of proteins synthesized, but was unable to restore the pattern found in the absence of osmotic stress. External hydrostatic pressure is not analogous to ψ_p (Section IIC). The ψ_p-sensing mechanism in *Valonia* is sensitive only to pressure differences across the membrane, not to total hydrostatic pressure (Hastings & Gutknecht, 1974). In addition to reducing the rate of protein synthesis, water stress can qualitatively alter the proteins that are synthesized (Dhindsa and Cleland, 1975b). Specification of the proteins involved in these alterations could shed light on the processes involved in adaptation to water stress. Relating these changes to the elevation and reduction in levels of specific enzymes (reviewed by Hsiao, 1973a) would also be useful.

For a discussion of other aspects of nitrogen metabolism as related to water stress, see Hansen and Hitz (Chapter 7C, this volume).

IV. STOMATAL BEHAVIOR

A. General Relationship Among Stomatal Behavior, Carbon Dioxide Assimilation, and Transpiration

Transpiration is an unavoidable consequence of the maximization of CO_2 assimilation by plants (Raschke, 1975). When sufficient water stress reduces transpiration, CO_2 assimilation is also suppressed. This hand-in-hand relationship between transpiration and assimilation is attributable to the fact that CO_2 and water vapor transport share much of their pathways (including those through stomata) in common, though the movements are in opposite directions. Further, both processes usually depend on radiation capture and hence are similarly sensitive to the influence of stress on the effective foliage area for light interception. The common features shared by assimilation and transpiration, as well as their differences are summarized in Fig. 5. Since transpiration has been discussed extensively in the preceding

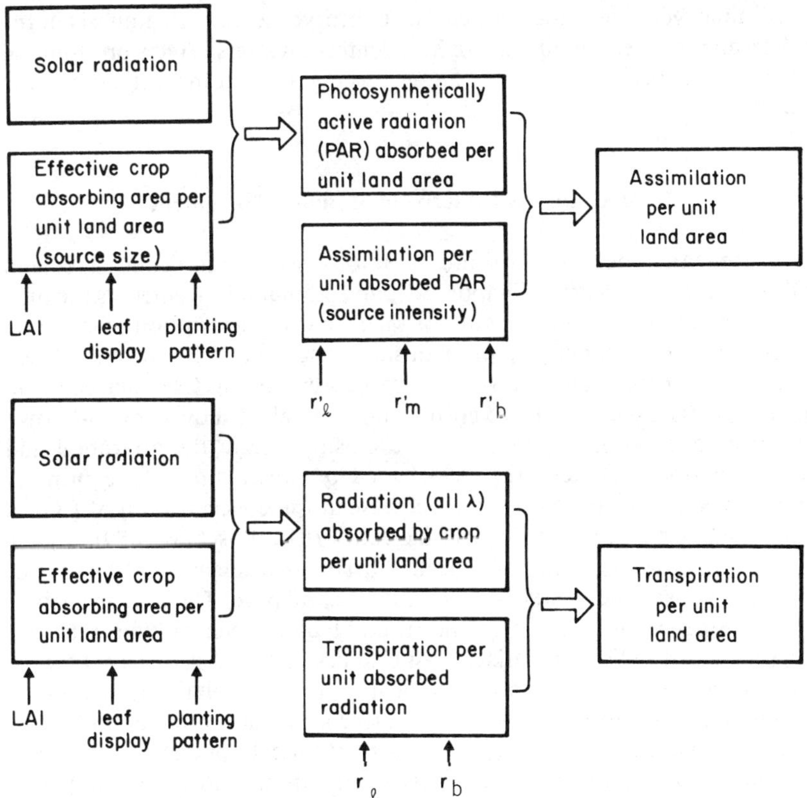

Fig. 5—Factors affecting CO_2 assimilation and transpiration per unit land area. LAI = leaf area index; r_l and r_l' = epidermal diffusion resistance to water vapor and to CO_2, respectively; r_m' = "mesophyll" (or residual) resistance to CO_2; and r_b and r_b' = boundary layer resistance to water vapor and to CO_2, respectively. Note the many common features shared between assimilation and transpiration.

chapter and elsewhere in this book, attention will be directed to stomatal behavior and CO_2 assimilation. It is valuable to keep in mind, however, that whenever water stress alters stomatal resistance or the light-intercepting area, both assimilation and transpiration will be affected similarly.

For assimilation to take place, the necessary machinery, the assimilative surface, must exist. Sustained water stress, even when mild, can reduce expansive growth and foliage area markedly (Hsiao et al., 1976b) and restrict the size of the assimilative machinery (see Fig. 5). The significance of this long-term effect on crop productivity is discussed in Chapter 7A. On the other hand, stress can affect the assimilative processes more directly. Assimilation requires the transport of CO_2 from the air through the surface boundary layer, leaf epidermis, and intercellular space; the dissolution of the gas in the cell wall water of the leaf mesophyll; the movement of the dissolved CO_2 to the site of the carboxylative enzyme; the carboxylation process itself; and finally the conversion to the intermediate and end products of photosynthesis. A priori, there is no obvious reason to rule out possible effects of water stress on any of the steps in the pathway. Although water stress can affect biochemical aspects of photosynthesis, the emphasis in the past decade of research, and again here centers on stress effects on stomatal aperture, the major control point governing CO_2 transport across the epidermis.

B. Mechanisms Underlying Stomatal Movement

Since leaf epidermis constitutes a major resistance to water vapor and CO_2 transport (see Pearcy, Chapter 6C), and stomatal movement is the only dynamic means to alter epidermal resistance, a brief description of stomatal movement and the underlying mechanism is pertinent. With the notable exception of pineapple and a few other species exhibiting Crassulacean acid metabolism (CAM), stomata of crop plants generally follow the well-known pattern of opening in the morning and closing late in the afternoon and evening, as long as they are not affected by water stress. The primary stimuli for opening are light and a decrease in the concentration of CO_2 in the intercellular space of the leaf (Hsiao et al., 1973; Raschke, 1979).

Opening results from the mechanical displacement and parting of guard cell pairs in the middle to form the stomatal pore. The displacement is the consequence of changes in the turgor balance and volume relations between guard cells and adjacent subsidiary cells. With dicot type of stomata, increases in turgor and volume in guard cells relative to that in the surrounding cells causes the guard cells, attached to each other at both ends, to bend outward in the middle, forming an elliptical pore. With grass type of stomata, the dumbbell-shaped guard cells inflate mostly at the paired bulbous ends, pushing their middles (handle of dumbbell) apart to form a slit opening. An in-depth review of the mechanics and anatomical aspects of stomatal movement was presented by Raschke (1979) (see also Cooke, Chapter 4B, this volume).

Due to their adjacent locations and absences of impervious barriers to water, Ψ in guard cells usually must be very similar to Ψ in the surrounding cells. It follows that the sum of ψ_p and ψ_π (the only important components of Ψ) in guard cells must be nearly equal to the sum of ψ_p and ψ_π in the adjacent cells.[2] For stomata to open, large differences in ψ_p between guard cells and their immediate neighbors are necessary (Edwards & Meidner, 1977; Raschke, 1979). This means there must be a corresponding difference in their ψ_π. An exciting development in plant physiology in the last score of years has been the identification of ion transport, especially that of potassium (K^+), and the associated changes in anions as the dominant mechanisms for changing ψ_π in guard cells (reviewed by Hsiao, 1976). Additional recent data can be found in the article by Raschke (1979) and in a recent symposium volume (Jarvis & Mansfield, 1981). Briefly, it appears that all the sustained changes in stomatal aperture in response to environmental factors arise from shifts in ψ_π (and the consequent shifts in ψ_p) in guard cells relative to ψ_π of adjoining cells. In some dicots, most of the ψ_π change takes place in guard cells, with only minor alterations in adjoining cells (Raschke, 1979). In grasses such as maize, however, changes in ψ_π in the adjoining subsidiary cells, opposite in direction to those in guard cells and therefore additive to them, appear to be equally important in contributing to the ψ_π shift (Raschke & Fellows, 1971). The change in ψ_π underlying major alterations in stomatal aperture is very large, ranging from 10 or 20 bars (Hsiao, 1976) to >50 bars (Raschke, 1979). Generally, K^+ transport into or out of guard cells accounts for more than half of the change in guard cell ψ_π (Hsiao, 1976). This means that guard cell K^+ concentration can be close to 1 eq liter^{-1} in the case of wide-open stomata! Changes in anions presumably account for much of the remainder. With the exception of onion guard cells (Raschke, 1979), there is no tight specificity for anion species as there is for the cation. Stomata in isolated leaf epidermis open equally well on a solution of K^+ whether Cl^-, Br^-, or NO_3^- is supplied as the anion (Humble & Hsiao, 1969). In the leaf, Cl^- often is transported by guard cells in association with K^+ and can account for as much as one-half of the negative charge needed to balance the positive charge of the transported K^+ (Raschke & Fellows, 1971). Still, it appears that organic anions, such as malate produced by the leaf, have a prevailing role since Cl^- is not commonly found in large amounts in crops grown in soils low in that ion (Hsiao, 1976). Apparently, malate and Cl^- can be used almost interchangeably, depending on the availability of each (Raschke, 1979).

What mechanism is responsible for this massive transport of ions? Proton extrusion by guard cells appears to be an intimate part of the stomatal opening process (Raschke, 1979). It is not established, however, whether the transport of K^+ is motivated by proton extrusion (Zeiger & Bloom, 1978) or the inward transport of anions (Raschke, 1979). In fact, it seems even too early to rule out the active transport or direct pumping of K^+ since good guard cell electrical potential data are only now being obtained.

---

[2] The near equilibrium in Ψ between guard cells and adjoining cells apparently does not hold true when transpiration or water uptake changes suddenly and sharply. In that case, water status in the adjoining epidermal cells changes ahead of that in guard cells, resulting in turgor inbalance between the two cell types and leading to transitory stomatal movements.

C. Stomatal Response to Water Stress and Humidity

Current views consider stomata to be regulatory valves operating under two often conflicting demands: maximizing assimilation and minimizing the reduction in plant water status to levels that may cause substantial damage to the tissue. The system has evolved to achieve reasonable balance between the two opposing demands. When leaf water status is high and evaporative demand is low, the opening responses to low internal CO_2 and to light ensure that stomata are wide open and represent a minimum resistance to CO_2 assimilation. When low water status develops in leaves, stomata respond by closing, with the consequent reduction in transpiration. Also, stomata close partly in many cases when humidity in the air is low and the evaporative demand is excessive, even though the water status of the leaf as a whole is favorable for stomatal opening. The closing responses to low leaf water status and to high evaporative demand are both underlain by changes in turgor and hence solute balance between guard cells and surrounding epidermal cells. They differ in specifics, however, as will be made clear below.

1. THE PHENOMENA

The response to low leaf water status is described first. Two kinds of effects of low leaf water status on stomatal behavior may be distinguished: short-term and long-term. The former concerns the familiar behavior of closing and opening within minutes or a fraction of an hour after changes in leaf water status. The latter persists for some time after the water stress is released and includes any preconditioning effect or aftereffect of stress.

When leaf water content or Ψ is reduced sufficiently in a short time, stomata begin to close in a matter of minutes. In many cases, a threshold level of water stress in the bulk leaf must be exceeded before closure will occur (Beadle et al., 1973; Davies et al., 1982; Hsiao, 1973a; Turner, 1974). If leaf water status is reduced but remains above the threshold, stomatal opening is either not affected or reduced only slightly. When leaf water status drops below this threshold, stomata close rather abruptly, and leaf diffusion resistance rises many-fold upon a further reduction in leaf water status of a few bars (Fig. 6). The threshold behavior is consistent with the concept of stomata acting as an overriding protective mechanism against excessive leaf water deficit that is not triggered into operation until the deficit reaches a critical level. Midday stomatal closure in the field is probably an example of such a response. For recently matured leaves of many herbaceous crops growing under high radiation in the field, the threshold Ψ for stomatal closure lies in the range of -12 to -16 bars, if the crop has had little or no recent history of water stress. When other environmental factors are less than optimal, the threshold Ψ for closure apparently may be raised to more positive values. The threshold differs for leaves of different age and prior exposure to radiation (Jordan et al., 1975).

A plot such as Fig. 6 characterizes the short-term stomatal response for a wide range of leaf Ψ. Other than the location of the threshold, important

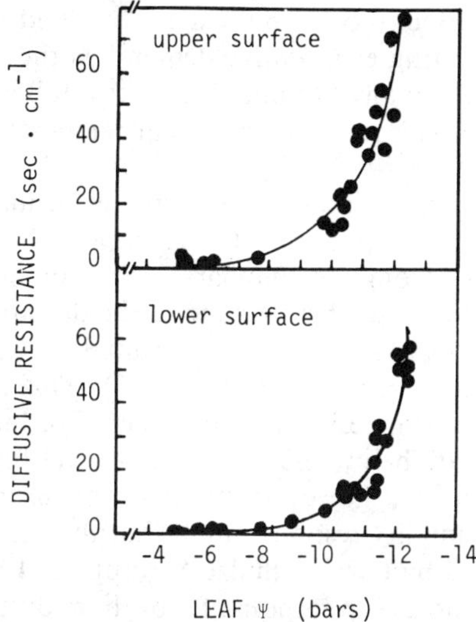

Fig. 6—Diffusion resistances of the upper and lower leaf surfaces of tomato as functions of bulk leaf Ψ. Derived from Duniway (1971).

features are the maximum attainable opening and the Ψ value at which stomata are more or less fully closed. A leaf resistance of 40–50 sec cm^{-1} may be taken as indicative of full closure. Assimilation at this point should be nil or only a few percent of the rate attained at high leaf water status. A history of water stress, either severe or mild and prolonged, often alters the shape of the short-term stomatal response curve and displaces the key features. When leaves are preconditioned by mild and prolonged water stress and undergo substantial osmotic adjustment, the threshold Ψ for stomatal closure is lower (more negative) than that in leaves without the stress history and the Ψ value for full closure is also lower (Ackerson, 1980). On the other hand, the maximum opening when Ψ is above the threshold is likely to be less in the preconditioned leaves (Ludlow, 1980). After a severe stress, stomata may not regain their maximum opening potential for as long as 5–7 days (Fischer et al., 1970). It is not known whether this aftereffect of stress is related to osmotic adjustment. Although the evidence for a threshold type of closing response to leaf water status is extensive, there are a number of other studies showing an apparent lack of the threshold behavior (Boyer, 1970). In many cases, this apparent lack is not unexpected, since the factors affecting threshold value and the maximum opening attainable were not controlled, yet data were plotted with leaf resistance as a function only of leaf Ψ. This situation is commonly encountered in field studies. Particularly confounding are data composited over a span of time during water stress when threshold Ψ and maximum attainable opening are probably shifting due to stress preconditioning.

Stomata can also close partially in response to increases in the humidity difference between the leaf interior and the bulk air outside, even though water status of the bulk leaf remains above the threshold level for closure (reviewed by Hall et al., 1976). Work by Maier-Maercker (1979) showed that increasing air velocity over the leaf accentuated the response to low humidity, thus suggesting evaporative demand as the eliciting factor. The response is observed in many but not all studies (Hall et al., 1976), and can be eliminated or minimized by a change in temperature in some species (Hall & Kaufmann, 1975).

Generally, the closure induced by increases in humidity difference is not very pronounced in crop species. For example, when the vapor pressure difference is increased from 8 to 30 mbars, corresponding to changing from a humid atmosphere of a temperate zone to the dry atmosphere of a semi-arid zone with an ambient temperature of around 30°C, leaf resistance may increase from 1 to 3 s cm^{-1} (Hall et al., 1976). This is in contrast to the many-fold increase of leaf resistance, to 30 s cm^{-1} or higher, when leaf Ψ is reduced to several bars below the threshold Ψ level for closure. For some trees and native species, however, behavior of stomata appears to be dominated by the humidity response and is hardly influenced by leaf water status, as exemplified by hazel (Schulze & Küppers, 1979). In many cases, the partial stomatal closure in response to low humidity can lead to a reduction in transpiration in spite of the increase in the driving force for transpiration (humidity difference) (Farquhar, 1978). Inevitably, there is a reduction in CO_2 assimilation corresponding to the increase in leaf epidermal resistance (Hall & Kaufmann, 1975; Schulze & Küppers, 1979).

2. UNDERLYING MECHANISMS

For response to bulk leaf water status, the notion is still held by some that stress leads directly to stomatal closure because leaf turgor is reduced and opening is a turgor-mediated process. In reality, the mechanism is much more complex and is, in the terms of Stålfelt, "hydroactive" (involves movement of solutes and water) rather than just "hydropassive" (involves only passive movement of water). Many years ago, Stålfelt (1955) reported that as water stress developed in leaves, stomatal closure and solute loss from guard cells occurred simultaneously. In most cases, the loss of guard cell solutes (measured plasmolytically) went hand in hand with the loss of water from the leaf.

What solutes are involved? As expected, K^+ apparently is the major solute transported in water-stress-induced stomatal closure (Ehret & Boyer, 1979; Hsiao, 1973b). As the leaf loses water, stomata close concurrently with decreases in stainable K^+ in guard cells. The changes in stomatal aperture and guard cell K^+ have been completely correlated, both as stomata close during water stress and as they reopen upon the recovery from stress (Hsiao, 1973b). When stress is effected or relieved quickly, the changes in stomatal aperture lag much behind the changes in tissue water content but are synchronous with the changes in K^+ content of guard cells. In fact, aperture and guard cell K^+ continue to change for many minutes after the leaf has stablized at the new water status. This behavior indicates that the hydropassive component is insignificant in the sustained stomatal movement induced by changes in leaf water status. Water stress apparently triggers the efflux of K^+ and anions from guard cells with the consequent loss of guard cell turgor and stomatal closure.

How do changes in leaf water status modulate or even reverse the direction of net ion transport underlying stomatal movement? Since exogenous abscisic acid (ABA) causes rapid stomatal closure, and endogenous ABA in leaves increases dramatically during water stress, the key role of modulating guard cell ion transport has been attributed to ABA (reviewed by Hsiao, 1976). This hypothesis is widely accepted, and Raschke (1979) recently presented its latest refinement along with a review of the relevant experimental data. Some observations, however, appear to contradict the hypothesis, at least on the surface. When stress develops suddenly, the accumulation of ABA in the leaves often lags behind stomatal closure (Beardsell & Cohen, 1975; Walton et al., 1977). Upon the release of water stress, ABA level may decline to that in the control leaf 1 or 2 days prior to full recovery of stomatal opening (Beardsell and Cohen, 1975; Loveys and Kriedemann, 1973). These apparent discrepancies may be explained by the use of whole leaf tissue for ABA measurements, while the site of ABA action must be at the guard cells, which only constitute about 1% of the leaf volume (Raschke, 1979). Nonetheless, the discrepancies are disturbing and their resolution awaits the development of techniques for the analysis of picogram amounts of ABA in guard cells. Also lacking is a reasonable understanding of how ABA acts to alter the rate and direction of ion transport in the stomatal complex. For more details, the readers are referred to the reviews by Hsiao (1976) and Raschke (1979).

One set of observations that may be taken as additional evidence for the hypothesis of ABA mediation is related to osmotic adjustment. As mentioned earlier, stomatal closure begins when ψ_p of the bulk leaf falls to near zero (Turner, 1974). For ABA to accumulate substantially in stressed tissue, bulk Ψ must also drop below a threshold value that is near the point of zero turgor (Pierce & Raschke, 1980). When osmotic adjustment due to water-stress preconditioning shifted the zero turgor point to a lower Ψ value, the Ψ value at which ABA increased sharply also shifted accordingly (Pierce & Raschke, 1980). It appears that this may account for the lowering of threshold Ψ for stomatal closure observed in osmotically adjusted leaves (Brown et al., 1976; Jones & Turner, 1978).

Very little is known about the mechanisms underlying stomatal responses to changes in evaporative demand at high leaf water status. In the only study that came to our attention, Lösch and Schenk (1978) found that upon a sudden change in air humidity, the response in stomatal aperture of *Valerianella locusta* preceded changes in guard cell K^+ by about 20 min. Notable change in K^+ took place only after the aperture change was nearly complete. They discussed two possibilities: (i) that an initial hydropassive aperture change was later stabilized and adjusted by K^+ transport, and (ii) that solutes other than K^+ may have been responsible for the initial response. Unfortunately, guard cell ψ_π was not measured concurrently to distinguish between the two possibilities. It is not known whether ABA is involved in stomatal response to humidity.

V. GROWTH REGULATORS

Plant growth regulators are known to be involved in virtually every phase of growth and development. It is not surprising, therefore, that water stress has been found to markedly alter hormone levels within the plant. The discussion that follows will of necessity be brief and can only serve to point out current research and hypotheses. For more specific treatment with respect to water stress, the reader is referred to recent reviews of the subject (Aspinall, 1980; Itai & Benzioni, 1976; Vaadia, 1976; Wright, 1978).

A. Indoleacetic Acid

Indoleacetic acid (IAA) has already been referred to extensively in connection with growth. Data are limited, however, concerning the effects of water stress on endogenous levels of IAA. Hartung and Witt (1968) reported that the diffusable auxin content of *Helianthus annuus* L. decreased as soil water content declined. Darbyshire (1971) measured a linear increase in IAA oxidase activity with decreasing Ψ between -2 and -15 bars in tomato and suggested that leaf IAA content would be reduced as a result. These and other data taken together indicate that water stress may cause a moderate reduction in IAA. There is evidence that basipetal auxin transport is inhibited by water stress ($\Psi < -10$ bars) (Kaldewey et al., 1974; Davenport et al., 1977). Reduced IAA transport to the abcission zone is involved in the promotion of leaf abscission by water stress (McMichael et al., 1972; Davenport et al., 1977). In the intact plant, as Cleland (1977) has pointed out, ψ_p is more likely to be limiting growth than is the supply of IAA. Water-stress-induced reductions in IAA content or transport, therefore, probably affect crop productivity through abscission and reduced leaf area rather than through growth inhibition.

B. Abscisic Acid

As was discussed in Section IV C, research on ABA has focused on its involvement in stomatal control. The dramatic induction of de novo ABA synthesis in water-stressed tissues (e.g., Hiron & Wright, 1973; Walton, 1980) raises the question of possible effects on the physiology of cells other than stomata. This is also suggested by the ability of both roots (Walton et al., 1976) and tissue cultured cells (Loveys et al., 1975) to synthesize ABA in response to osmotic stress. In fact, the concept is emerging of ABA as a factor in physiological responses to a variety of environmental stresses. This section will briefly survey a few of the areas where ABA seems implicated in nonstomatal adaptations to stress conditions (see also Davies et al., 1980).

A well-known effect of water stress is to increase the root/shoot ratio. Continued root growth to explore a larger soil volume, and reduced shoot growth to limit transpirational area, are both important adaptations to drought (Hsiao & Acevedo, 1974). ABA may well be involved in both of these processes. Quarrie and Jones (1977) compared the effects of water stress with those of ABA applications on the growth and development of wheat. Artificially elevating ABA levels in well-watered plants resulted in reduced leaf growth, smaller epidermal cell size, fewer stomata per leaf, increased differentiation of trichomes, and fewer spikelets per ear, all symptomatic of water stress. These morphological changes, in addition to stomatal closure, would all tend to improve survival when water is limiting.

On the other hand, root initiation (Chin et al., 1969) and root elongation (Aspinall et al., 1967) can both be stimulated by ABA. Good evidence exists for the synthesis of ABA in mature leaves and its export to sink regions (Setter et al., 1980; Zeevaart, 1977). In addition, Walton et al. (1976) have demonstrated that roots of water-stressed (−4-bar osmotic solution) bean plants can increase in ABA content even when the stem is girdled to prevent import of ABA from the shoot. Solute accumulation in roots is enhanced by water stress, allowing turgor maintenance and continued growth (Greacon & Oh, 1972; Sharp & Davies, 1979). One is tempted to speculate on the involvement of ABA in this process, based on studies of the effects of water stress (from 98 to 88% RWC) and ABA ($10^{-6}M$) on ion uptake and transport in roots (Cram & Pitman, 1972; Pitman et al., 1974b). Both treatments had no effect on ion uptake, but transport from roots to shoots was reduced, resulting in a net accumulation of ions in the roots. However, conflicting effects of ABA and/or water stress on ion uptake and transport have been reported (e.g., Karmoker & van Steveninck, 1978). Apparently the results of these experiments are highly dependent upon growth conditions, degree of stress, methods of measurement, ions employed, species, and other factors (Pitman et al., 1974a). An important piece of information missing from these studies is knowledge of the endogenous levels of ABA accompanying low root Ψ. Since many hormonal effects can be either promotive or inhibitory, depending upon the concentrations employed (e.g., Aspinall et al., 1967), the use of realistic concentrations of ABA would seem to be essential. More information on endogenous ABA levels and movement within the plant, and on growth responses to realistic concentrations of ABA in nonstressed plants, is needed to establish the role of ABA in altering the root/shoot ratio.

Water stress can alter cell permeability to water (Lee-Stadelmann & Stadelmann, 1976). Though the functional significance is not clear, this does raise the question of ABA effects on root hydraulic conductance (C). Measurements of C are often confounded by changes in ion transport during the experiment, and are subject to the same sources of variability cited above. Thus, despite considerable research on the topic, there is still no consensus of opinion on either the direction or magnitude of changes in C in response to ABA (Glinka, 1977; Markhart et al., 1979; Pitman & Wellfare, 1978). It seems apparent that ABA can influence the permeability of the root to water and ions. The specific changes, on the other hand, are highly variable and dependent upon both physiological and environmental parameters. Until these parameters have been identified and controlled, it is difficult to assign ABA a role in altering root conductance during water stress.

Since both ion transport and *C* are membrane phenomena, ABA may have a general effect on membrane properties. It has been proposed that a major role of ABA might be to protect cell membranes (in some unknown way) from damage under stress conditions (Markhart et al., 1979). It is known that a variety of conditions in addition to water stress will cause increases in ABA, including nutrient, salt, chilling, and waterlogging (anaerobic) stresses (Boussiba & Richmond, 1976; Hiron & Wright, 1973; Rikin et al., 1976). Furthermore, applications of ABA can increase the tolerance of plants to subsequent stress (Mizrahi & Richmond, 1972; Rikin et al., 1976). In some cases, the increased tolerance can be attributed to stomatal closure and no additional membrane effects need to be postulated. However, the ability of ABA to protect seedlings from chilling injury certainly implicates membrane changes (Rikin & Richmond, 1976; Rikin et al., 1976), although water loss may be involved here also (Wright, 1974). Thus, there is indirect evidence that ABA may have a more general role in alleviating cellular distress in addition to its effect on water balance.

C. Cytokinins

Interest in the involvement of cytokinins (CK) in water-stress symptoms was sparked by their ability to prevent leaf senescence. Cytokinins are thought to be primarily synthesized in the roots and transported to the shoots (reviewed by Van Staden & Davey, 1979). Itai and coworkers (1968) suggested that leaf senescence during water stress could be due to a reduction in the supply of CK from the roots and demonstrated lower CK levels in xylem exudate from stressed roots. Subsequent work has, in general, supported the idea that CK levels fall rapidly in water-stressed tissues (for review, see Itai & Benzioni, 1976; Vaadia, 1976). Applications of CK, on the other hand, have often resulted in a small stimulation of transpiration and stomatal opening (Livne & Vaadia, 1965; Luke & Freeman, 1968). Since the changes in CK content and subsequent effects on water balance during stress are generally the opposite of those of ABA, it has been hypothesized that shifts in the ABA/CK ratio are the dominant hormonal responses to water stress (Tal & Imber, 1971; Itai & Benzioni, 1976). High ratios result in stomatal closure and increased root permeability, while low ratios promote stomatal opening and reduce root permeability. This hypothesis offers an attractive mechanism for intraplant regulation of water balance, but several points should be raised. Little recent work has been done to confirm and extend the study of Itai and Vaadia (1971) on CK content of water-stressed shoots. The role of CK in regulation of stomatal behavior is not clear-cut because the effect is not as dramatic as that of ABA. The effects of CK on root permeability and ion transport are fraught with the same problems mentioned previously for ABA. Finally, the hypothesis tends to minimize the role of other growth regulators, although

IAA and ethylene have recently been incorporated into the overall scheme (Tal et al., 1979). As a working hypothesis, it does provide a useful framework for stimulating new research.

D. Ethylene

Increased ethylene production is a general response of plant tissues to environmental stresses, including water stress (reviewed by Yang and Pratt, 1978). Wright (1977) compared the sensitivity of Ethylene and ABA synthesis with water stress in wheat leaves. As Ψ declined, ethylene production began to increase between -7 and -8 bars, while ABA started to rise between -8 and -9 bars (Fig. 7A). Ethylene synthesis was therefore more sensitive to a slowly developing water stress. Conversely, if leaves were quickly stressed to -12 bars and maintained in the wilted condition. ABA accumulated more rapidly than did ethylene. Aharoni (1978) also found ethylene production to be an earlier response to water stress than ABA accumulation and demonstrated that the high ethylene levels did not induce ABA synthesis. However, both CK and ABA can modify the rate of ethylene synthesis in water-stressed wheat leaves, emphasizing the importance of considering hormonal interactions (Wright, 1980). McMichael, Jordan, and coworkers (reviewed by Hsiao, 1973a) showed that an elevation in ethylene production plays a role in water-stress-induced abscission in cotton. Aharoni (1978) found that leaves with abscission zones maintained high rates of ethylene production during water stress, while those lacking abscission zones showed a transient increase followed by a decline to the control level. This is consistent with the involvement of ethylene in leaf abscission (Jackson & Osborne, 1970) and senescence (Aharoni & Lieberman, 1979). On the other hand, ethylene has little direct effect on stomata (Pallaghy & Raschke, 1972). The major effects of water-stress-induced ethylene synthesis on plant productivity are, therefore, probably the results of leaf senescence, chlorophyll degradation, and abscission.

E. Gibberellins

Changes in endogenous gibberellins (GA) as a function of water stress have only recently been measured (Aharoni et al., 1977). The results indicate that GA activity rapidly declines in stressed leaves, and rises again during a subsequent recovery period. The loss of GA precedes the rise in ABA content of the same tissue (Fig. 7B). These data are somewhat complicated by the use of detached leaves, which are simultaneously undergoing senescence. Nonetheless, water stress (10% reduction in RWC) accelerated the rate of decline in GA (Aharoni & Richmond, 1978). Treatment with GA did not enhance stomatal opening, but did retard closure during a stress treatment (Aharoni et al.,1977). Based on the limited data, GA seem to be affected by water stress in a manner similar to that discussed above for CK.

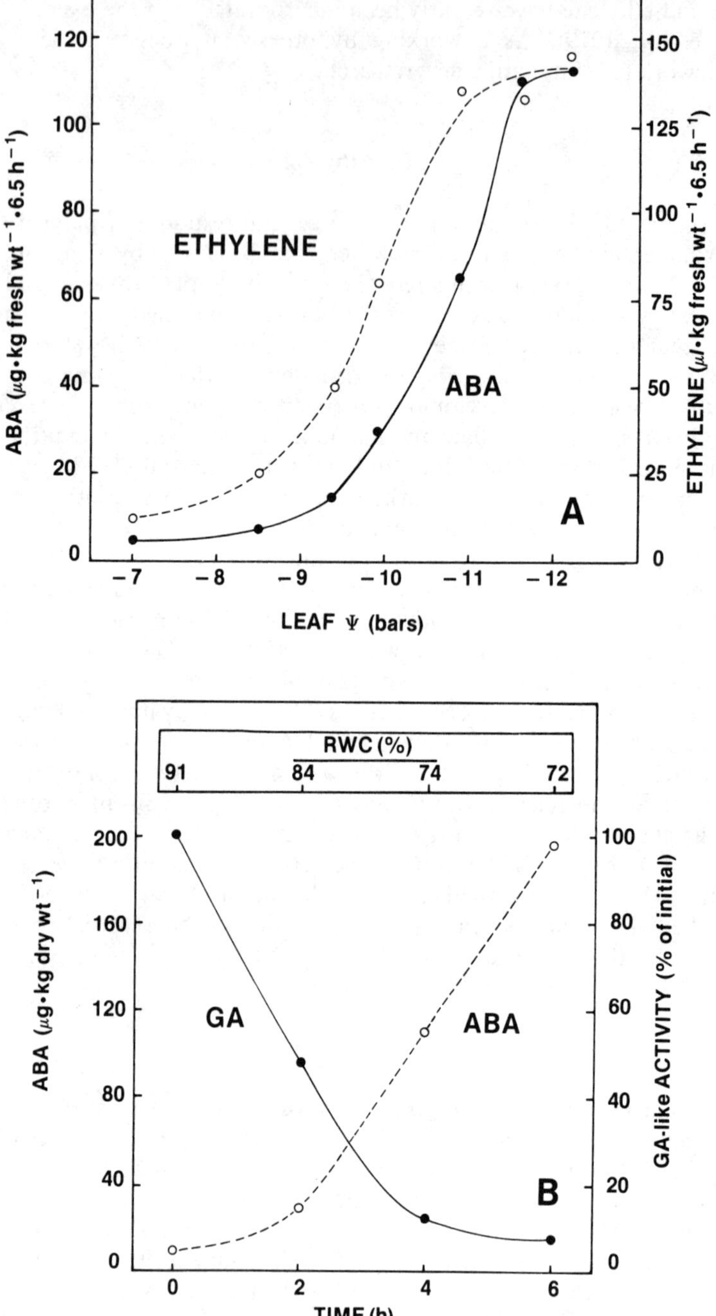

Fig. 7—Changes in endogenous growth regulators during water stress. (*A*) Relationship of ethylene and ABA synthesis to Ψ in excised wheat leaves; redrawn from Wright (1977). (*B*) Time course of changes in RWC, ABA, and GA in excised lettuce leaves; redrawn from Aharoni and Richmond (1978).

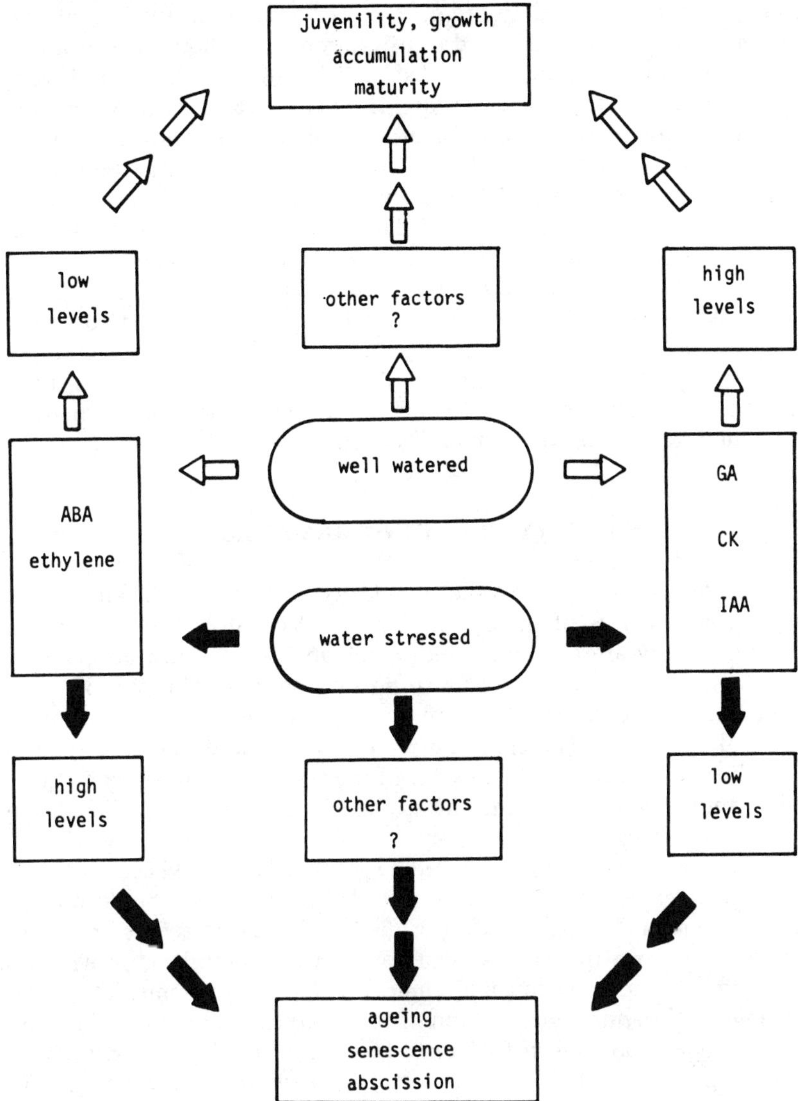

Fig. 8—Generalized responses of plant growth regulators to water stress and their probable significances. The scheme is based on prevailing literature and exceptions exist. These changes may be related to the regulation of growth and assimilation in the absence of stress, and to senescence and abscission in the presence of stress.

F. Hormonal Interactions During Water Stress

Although each of the plant growth regulators has been discussed separately, it is important to remember that the effects of a given hormone are largely dependent upon its interactions with the others. Thus, the approach of measuring simultaneous changes in several growth regulators is certainly the most informative (e.g., Fig. 7). In addition, further information on the sites of synthesis and translocation of the hormones as a function of water stress is definitely needed. As a broad generalization, there are two groups of growth regulators that differ in their response to water stress. One group, comprised of ABA and ethylene, increases as a result of stress. The other group, IAA, CK and GA, tends to decrease under stress. This is consistent with the reciprocal or antagonistic effects of these two groups on many aspects of plant growth and development. ABA and ethylene are associated with inhibition of growth and promotion of senescence, while IAA, CK, and GA usually accelerate growth and retard senescence. Overall effects of water stress on the state of the plant as related to growth regulators are hypothesized and generalized in Fig. 8. Our task for the future is to determine the extent to which this generalization holds, refine the model for specific crop species, and utilize this information to manipulate plant development for agronomic benefit.

VI. CONCLUDING REMARKS

Our discussion highlights the complexity of the metabolic and physiological consequences of plant water deficit. Virtually all aspects of the plant's economy are altered, including metabolism of substrates, growth regulators, and macromolecules; transport of ions and solutes; growth and development of cells, leaves, and roots; and responses to the environment (stomatal behavior). The complexity could easily lead one to dispair of ever achieving an accurate knowledge of the system. Fortunately, complete understanding is not a prerequisite to the effective practical utilization of the basic concepts.

While the emphasis in the preceding sections has often been on gaps in our knowledge in order to point out needed research, we should not lose sight of the remarkable progress over the last two decades. The broad outlines of plant growth and assimilation are well-established, providing a physiological framework to guide studies relating to environmental stress or agronomic utilization. Each section of this chapter has been based on conceptual models, now taken for granted, which often did not exist a score of years ago. Section II presents cellular and tissue water relations in terms of Ψ, chemical activity, and elasticity with a sound thermodynamic basis, a considerable improvement over previous concepts such as "diffusion pressure deficit" or "suction force." Recognition of these physical parameters has permitted the development of new techniques for measuring plant water status that are largely responsible for the rapid expansion of research on water stress. Accurate measurement of ψ_p in intact higher plant tissues remains elusive, however, and technological advance in this area is sorely needed. Basic research on expansive growth is experiencing an extremely active and productive phase at present, and although this work involves model plant material far removed from crop production, there is hope of characterizing many of the key processes on a biochemical level (Labavitch,

1981). The importance of expansive growth for development of assimilatory surface and harvestable yield, as well as its extreme sensitivity to water deficit, make this topic crucial to the crop scientists hoping to optimize plant growth in the field. Metabolic adaptations such as osmotic adjustment represent the dynamic aspect of plant responses to stress. The replacement of static concepts such as "permanent wilt point" by an appreciation that plants actively modify their internal status to adjust to changing environmental conditions represents a key conceptual advance of recent years.

As pointed out in Section IV, our understanding of the mechanism of stomatal movement has increased dramatically, to the point where models of the role of stomata in the optimization of carbon uptake vs. water loss can be based on physiological control points (Cowan & Farquhar, 1977). While stomatal behavior often still appears to be eccentric and temperamental (Hsiao, 1976), we do have a fairly good idea of the important environmental stimuli and at least a hint of how they might be transduced into movements of guard cells (Raschke, 1979). In the field of growth regulators, there is virtually a knowledge explosion, although understanding has often lagged behind the accumulation of observations. However, there is great potential for agronomic utilization of this information, as the widespread use of herbicides, thinning agents, growth retardants, defoliants, and ripening agents will attest. Thus, while serious gaps still exist in our basic knowledge of plant physiology, the achievements of the last 20 years forecast an optimistic outlook for future research and understanding.

Optimism is also warranted from the viewpoint of production agriculture. At the first glance, the myriad changes elicited by water stress appears too numerous and diverse to be of use in the analysis of crop productivity and water use. We suspect, however, that there is more order underlying this apparent chaos. No change in a plant can be totally isolated, regardless of whether it is at the molecular, cellular, or organ level. Coordination and integration distinguish processes and functions carried out by organisms from pure chemical reactions no matter how complex the latter may be. It is highly probable that the array of water-stress responses in a plant are tightly orchestrated through metabolic controls as well as the long-distance communication network of plant hormone, water, and assimilate transport. Thus, in the context of crop water use and productivity, it is not necessary to keep track of all the changes in a crop undergoing stress; a few well-chosen signposts will do. For the signpost parameters, recent field experience pinpoints at least the following: ground cover or effective foliage area (source size and the result of expansive growth), assimilation rate or at least stomatal opening at midday, degree of leaf senescence as the crop matures, and reproductive behavior of grain or fruit crops. The significance of these and related parameters in productivity has been reviewed (Fischer & Turner, 1978; Hsiao et al., 1976b; Turner, 1979) and is further discussed by Jordan in Chapter 7A.

While field studies of yield and water relations may not require a comprehensive knowledge of the underlying metabolic events, such studies do lead to the identification of embarrassing gaps in the physiology of water-stress responses. For example, the crop and irrigation literature has alluded to flowering time as a critical stage for water-stress effect on yield for many years (Salter & Goode, 1967). Yet, we know very little about how pollination and fertilization are affected by stress and how much stress is required to cause sterility. Another example is stress-induced leaf senescence, a key factor in determining yield, especially near the time of maturation (Fischer, 1973; Hsiao et al., 1976; Jordan, Chapter 7A). The only information we have is that on stress-induced ethylene evolution and abscission in cotton (Section V D). Details on other facets of senescence under water stress are lacking. Thus, physiologists need the broad outline of agronomy to maintain perspective. On the other hand, agronomists need a physiological underpinning to sharpen strategy and refine concepts arising out of specific crops and locations. Interactions between agronomists and physiologists are highly synergistic and should ensure rapid progress in the decade to come.

LITERATURE CITED

Acevedo, E., E. Fereres, T. C. Hsiao, and D. W. Henderson. 1979. Diurnal growth trends, water potential, and osmotic adjustment of maize and sorghum leaves in the field. Plant Physiol. 64:476–480.

Acevedo, E., T. C. Hsiao, and D. W. Henderson. 1971. Immediate and subsequent growth responses of maize leaves to changes in water status. Plant Physiol. 48:631–636.

Ackerson, R. C. 1980. Stomatal response of cotton to water stress and abscisic acid as affected by water stress history. Plant Physiol. 65:455–459.

Adams, P. A., M. J. Montague, M. Tepfer, D. L. Rayle, H. I. Kuma, and P. B. Kaufman. 1975. Effect of gibberellic acid on the plasticity and elasticity of *Avena* stem segments. Plant Physiol. 56:757–760.

Aharoni, N. 1978. Relationship between leaf water status and endogenous ethylene in detached leaves. Plant Physiol. 61:658–662.

Aharoni, N., A. Blumenfeld, and A. E. Richmond. 1977. Hormonal activity in detached lettuce leaves as affected by leaf water content. Plant Physiol. 59:1169–1173.

Aharoni, N., and M. Lieberman. 1979. Ethylene as a regulator of senescence in tobacco leaf discs. Plant Physiol. 64:801–804.

Aharoni, N., and A. E. Richmond. 1978. Endogenous gibberellin and abscisic acid content as related to senescence of detached lettuce leaves. Plant Physiol. 62:224–228.

Albersheim, P. 1976. The primary cell wall. p. 225–274. *In* J. Bonner and J. E. Varner (ed.) Plant biochemistry. 3rd ed. Academic Press, New York.

Albersheim, P., M. McNeil, and J. N. Labavitch. 1977. The molecular structure of the primary cell wall and elongation growth. p. 1–12. *In* P. E. Pilet (ed.) Plant growth regulation. Springer-Verlag, New York.

Aspinall, D. 1980. Role of abscisic acid and other hormones in adaptation to water stress. p. 155–172. *In* N. C. Turner and P. J. Kramer (ed.) Adaptation of plants to water and high temperature stress. Wiley-Interscience, New York.

Aspinall, D., L. G. Paleg, and F. T. Addicott. 1967. Abscisin II and some hormone-regulated plant responses. Aust. J. Biol. Sci. 20:869–882.

Barlow, E. W. R., R. Munns, N. S. Scott, and A. H. Reisner. 1977. Water potential, growth and polyribosome content of the stressed wheat apex. J. Exp. Bot. 28:909–916.

Barrs, H. D. 1968. Determination of water deficits in plant tissues. p. 235–368. *In* T. T. Kozlowski (ed.) Water deficits and plant growth. Vol. 1. Development, control and measurement. Academic Press, New York.

Bates, G. W., and R. E. Cleland. 1979. Protein synthesis and auxin-induced growth: inhibition studies. Planta 145:437–442.

Beadle, C. L., K. R. Stevenson, H. H. Neumann, G. W. Thurtell, and K. M. King. 1973. Diffusive resistance, transpiration, and photosynthesis in single leaves of corn and sorghum in relation to leaf water potential. Can. J. Plant Sci. 53:537–544.

Beardsell, M. F., and D. Cohen. 1975. Relationships between leaf water status, abscisic acid levels, and stomatal resistance in maize and sorghum. Plant Physiol. 56:207–212.

Beffagna, N., E. Marrè, and S. N. Cocucci. 1979. Cation-activated ATPase activity of plasmalemma-enriched membrane preparations from maize coleoptiles. Planta 146:387–391.

Begg, J. E., and N. C. Turner. 1976. Crop water deficits. Adv. Agron. 28:161–217.

Bewley, J. D. 1979. Physiological aspects of dessication tolerance. Annu. Rev. Plant Physiol. 30:195–238.

Boussiba, S., and A. E. Richmond. 1976. Abscisic acid and the after effect of stress in tobacco plants. Planta 129:217–219.

Boyer, J. S. 1967. Matric potentials of leaves. Plant Physiol. 42:213–217.

Boyer, J. S. 1968. Relationships of water potential to growth of leaves. Plant Physiol. 43: 1056–1062.

Boyer, J. S. 1969. Measurement of the water status of plants. Annu. Rev. Plant Physiol. 20: 351–364.

Boyer, J. S. 1970. Differing sensitivity of photosynthesis to low leaf water potentials in corn and soybean. Plant Physiol. 46:236–239.

Boyer, J. S. 1976. Photosynthesis at low water potentials. Philos. Trans. R. Soc. London Ser. B. 273:501–512.

Boyer, J. S., and H. G. McPherson. 1975. Physiology of water deficits in cereal crops. Adv. Agron. 27:1–23.

Boyer, J. S., and G. Wu. 1978. Auxin increases the hydraulic conductivity of auxin-sensitive hypocotyl tissue. Planta 139:227–237.

Bradford, K. J., and T. C. Hsiao. 1982. Physiological responses to moderate water stress. *In* O. L. Lange et al. (ed.) Encyclopedia of plant physiology, N.S. Physiological plant ecology II. Water relations and carbon assimilation. Springer-Verlag, Berlin/New York (in press).

Brandle, J. R., T. M. Hinckley, and G. N. Brown. 1977. The effects of dehydration-rehydration cycles on protein synthesis in black locust seedlings. Physiol. Plant. 40:1–5.

Brandle, J. R., P. D. Schnare, T. M. Hinckley, and G. N. Brown. 1973. Changes in polysomes of black locust seedlings during dehydration-rehydration cycles. Physiol. Plant. 29:406–409.

Brown, K. W., W. R. Jordan, and J. C. Thomas. 1976. Water stress induced alterations of the stomatal response to decreases in leaf water potential. Physiol. Plant. 37:1–5.

Bunce, J. A. 1977. Leaf elongation in relation to leaf water potential in soybean. J. Exp. Bot. 28:156–161.

Cheung, Y. N. S., M. T. Tyree, and J. Dainty. 1976. Some possible sources of error in determining bulk elastic moduli and other parameters from pressure-volume curves of shoots and leaves. Can. J. Bot. 54:758–765.

Chin, T., M. M. Meyer, Jr., and L. Beever. 1969. Abscisic-acid-stimulated rooting of stem cuttings. Planta 88:192–196.

Cleland, R. E. 1967. A dual role of turgor pressure in auxin-induced cell elongation in *Avena* coleoptiles. Planta 77:182–191.

Cleland, R. E. 1973. Auxin-induced hydrogen ion excretion from *Avena* coleoptiles. Proc. Nat. Acad. Sci. USA 70:3092–3093.

Cleland, R. E. 1975. Auxin-induced hydrogen ion excretion: correlation with growth, and control by external pH and water stress. Planta 127:233–242.

Cleland, R. E. 1977. The control of cell enlargement. p. 101–116. *In* D. H. Jennings (ed.) Integration of activity in the higher plant. Cambridge University Press, Cambridge.

Cleland, R. E., and D. L. Rayle. 1978. Auxin, H^+-excretion and cell elongation. Bot. Mag. Tokyo, Special Issue. 1:125–139.

Cocucci, S., N. Cocucci, and C. Poma Treccani. 1976. Effect of water deficit on the growth of squash fruit. Physiol. Plant. 36:379–382.

Cookson, C., and D. J. Osborne. 1979. The effect of ethylene and auxin on cell wall extensibility of the semi-aquatic fern, *Regnellidium diphyllum*. Planta 146:303–307.

Cowan, I. R., and G. D. Farquhar. 1977. Stomatal function in relation to leaf metabolism and environment. Symp. Soc. Exp. Biol. 31:471–505.

Crafts, A. H., H. B. Currier, and C. R. Stocking. 1949. Water in the physiology of plants. The Ronald Press Company, New York.

Cram, W. J. 1976. Negative feedback regulation of transport in cells. The maintenance of turgor, volume and nutrient supply. p. 284–316. *In* U. Lüttge and M. G. Pitman (ed.) Encyclopedia of plant physiology. New Series, Vol. 2. Transport in plants II. Part A. Cells. Springer-Verlag, Berlin.

Cram, J. W., and M. G. Pitman. 1972. The action of abscisic acid on ion uptake and water flow in plant roots. Aust. J. Biol. Sci. 25:1125–1132.

Currier, H. B. 1944. Water relations of root cells of *Beta vulgaris*. Am. J. Bot. 31:378–387.

Cutler, J. N., and D. W. Rains. 1978. Effects of water stress and hardening on the internal water relations and osmotic constituents of cotton leaves. Physiol. Plant. 42:261–268.

Cutler, J. N., D. W. Rains, and R. S. Loomis. 1977. The importance of cell size in the water relations of plants. Physiol. Plant. 40:255–260.

Dainty, J. 1976. Water relations of plant cells. p. 12–35. *In* U. Lüttge and M. G. Pitman (ed.) Encyclopedia of plant physiology. New series, Vol. 2. Transport in Plants II. Part A. Cells. Springer-Verlag, New York.

Darbyshire, B. 1971. Changes in indoleacetic acid oxidase activity associated with plant water potential. Physiol. Plant. 25:80–84.

Davenport, T. L., P.W. Morgan, and W. R. Jordan. 1977. Auxin transport as related to leaf abscission during water stress in cotton. Plant Physiol. 59:554–557.

Davies, D. D. 1973. Control of and by pH. Symp. Soc. Exp. Biol. 27:513–529.

Davies, W.J., T. A. Mansfield, and A. R. Wellburn. 1980. A role for abscisic acid in drought endurance and drought avoidance. p. 242–253. *In* F. Skoog (ed.) Plant growth substances 1979. Springer-Verlag, Berlin.

Davies, W. J., J. A. Wilson, R. E. Sharp, and O. Osonubi. 1982. Control of stomatal behaviour in water-stressed plants. *In* P. G. Jarvis and T. A. Mansfield (ed.) Stomatal physiology. Cambridge University Press, Cambridge (in press).

Dhindsa, R. S., C. A. Beasley, and I. P. Ting. 1975. Osmoregulation in cotton fiber. accumulation of potassium and malate during growth. Plant Physiol. 56:394–398.

Dhindsa, R. S., and R. E. Cleland. 1975a. Water stress and protein synthesis. I. Differential inhibition of protein synthesis. Plant Physiol. 55:778–781.

Dhindsa, R. S., and R. E. Cleland. 1975b. Water stress and protein synthesis. II. Interaction between water stress, hydrostatic pressure, and abscisic acid on the pattern of protein synthesis in *Avena* coleoptiles. Plant Physiol. 55:782–785.

Duniway, J. M. 1971. Water relations of *Fusarium* wilt in tomato. Physiol. Plant Pathol. 1: 537–546.

Edwards, M., and H. Meidner. 1977. Direct measurements of turgor pressure potentials III. A. Injections of solutions into stomatal cells; B. Comparisons of pressure potentials on stomatal cells of different species. J. Exp. Bot. 28:669–677.

Ehret, D. L., and J. S. Boyer. 1979. Potassium loss from stomatal guard cells at low water potentials. J. Exp. Bot. 30:225–234.

Farquhar, G. D. 1978. Feedforward responses of stomata to humidity. Aust. J. Plant Physiol. 5:787–800.

Fischer, R. A. 1973. The effect of water stress at various stages of development on yield processes in wheat. p. 233–241. *In* Plant response to climatic factors. Proc. Symp., Uppsala, Sweden. 1970. UNESCO, Paris.

Fischer, R. A., T. C. Hsiao, and R. M. Hagan. 1970. After-effect of water stress on stomatal opening potential. I. Techniques and magnitudes. J. Exp. Bot. 21:371–385.

Fischer, R. A., and N. C. Turner. 1978. Plant productivity in the arid and semiarid zones. Annu. Rev. Plant Physiol. 29:277–317.

Glinka, Z. 1977. Effects of abscisic acid and of hydrostatic pressure gradient on water movement through excised sunflower roots. Plant Physiol. 59:933–935.

Göring, H., W. Bleiss, and H. Kretschmer. 1978. Stimulated elongation growth of coleoptile segments as a consequence of activated H^+ secretion after temporary turgor reduction. Biochem. Physiol. Pflanz. 173:373–376.

Greacon, E. L., and J. S. Oh. 1972. Physics of root growth. Nature New Biol. 235:24–25.

Green, P. B. 1968. Growth physics in *Nitella*: a method for continuous *in vivo* analysis of extensibility based on a micro-manometer technique for turgor pressure. Plant Physiol. 43: 1169–1184.

Green, P. B., K. Bauer, and W. R. Cummins. 1977. Biophysical model for plant cell growth: auxin effects. p. 30–45. *In* A. M. Jungreis et al. (ed.) Water relations in membrane transport in plants and animals. Academic Press, New York.

Green, P. B., and W. R. Cummins. 1974. Growth rate and turgor pressure: auxin effect studied with an automated apparatus for single coleoptiles. Plant Physiol. 54:863–869.

Green, P. B., R. O. Erickson, and J. Buggy. 1971. Metabolic and physical control of cell elongation rate. *In vivo* studies in *Nitella.* Plant Physiol. 47:423–430.

Grenetz, P. S., and A. List, Jr. 1973. A model for predicting growth response of plants to changes in external water potential: *Zea mays* primary root. J. Theor. Biol. 39:29–45.

Gutknecht, S., D. F. Hastings, and M. A. Bisson. 1978. Ion transport and turgor pressure regulation in giant algal cells. p. 125–174. *In* G. Giebisch et al. (ed.) Membrane transport in biology. Vol. 3. Springer-Verlag, New York.

Hager, A., H. Menzel, and A. Krauss. 1971. Versuche und hypothesen zur primarwirking des auxins beim streckungswachstum. Planta 100:47–75.

Hall, A. E., and M. R. Kaufmann. 1975. Stomatal response to environment with *Sesamum indicum* L. Plant Physiol. 55:455–459.

Hall, A. E., E.-D. Schulze, and O. L. Lange. 1976. Current perspectives of steady-state stomatal responses to environment. p. 169–188. *In* O. L. Lange et al. (ed.) Ecological studies. Analysis and synthesis. Vol. 19. Water and plant life. Springer-Verlag, New York.

Hartung, W., and J. Witt. 1968. The influence of soil moisture on the auxin content of *Anastatica hierochuntica* and *Helianthus annuus.* Flora Jena Abt. B. 157:603–614.

Haschke, H.-P., and U. Lüttge. 1975a. Interaction between IAA, potassium and malate accumulation, and growth in *Avena* coleoptile segments. Z. Pflanzen. 76:450–455.

Haschke, H.-P., and U. Lüttge. 1975b. Stochiometric correlation of malate accumulation with auxin-dependent K^+-H^+ exchange and growth in *Avena* coleoptile segments. Plant Physiol. 56:696.

Haschke, H.-P., and U. Lüttge. 1977. Action of auxin on CO_2 dark fixation in *Avena* coleoptile segments as related to elongation growth. Plant Sci. Lett. 8:53–58.

Hastings, D. F., and J. Gutknecht. 1974. Turgor pressure regulation: modulation of active potassium transport by hydrostatic pressure gradients. p. 79–83. *In* U. Zimmermann and J. Dainty (ed.) Membrane transport in plants. Springer-Verlag, Berlin.

Hiler, E. A., C. H. M. van Bavel, M. M. Hossain, and W. R. Jordan. 1972. Sensitivity of southern peas to plant water deficit at three growth stages. Agron. J. 64:60–64.

Hiron, R. W., and S. T. C. Wright. 1973. The role of endogenous abscisic acid in the response of plants to stress. J. Exp. Bot. 24:769–781.

Hsiao, T. C. 1970. Rapid changes in levels of polyribosomes in *Zea mays* in response to water stress. Plant Physiol. 46:281–285.

Hsiao, T. C. 1973a. Plant responses to water stress. Annu. Rev. Plant Physiol. 24:519–570.

Hsiao, T. C. 1973b. Effects of water deficit on guard cell potassium and stomatal movement. Plant Physiol. Suppl. 51:9.

Hsiao, T. C. 1976. Stomatal ion transport. p. 195–221. *In* U. Lüttge and M. G. Pitman (ed.) Encyclopedia of plant physiology. New Series, Vol. 2. Transport in plants II. Part B. Tissues and organs. Springer-Verlag, New York.

Hsiao, T. C., and E. Acevedo. 1974. Plant responses to water deficits, water use efficiency and drought resistance. Agric. Meteorol. 14:59–84.

Hsiao, T. C., E. Acevedo, E. Fereres, and D. W. Henderson. 1976a. Water stress, growth and osmotic adjustment. Phil. Trans. R. Soc. London Ser. B. 273:479–500.

Hsiao, T. C., W. G. Allaway, and L. T. Evans. 1973. Action spectra for guard cell Rb^+ uptake and stomatal opening in *Vicia faba.* Plant Physiol. 51:82–88.

Hsiao, T. C., E. Fereres, E. Acevedo, and D. W. Henderson. 1976b. Water stress and dynamics of growth and yield of crop plants. p. 281–305. *In* O. L. Lange et al. (ed.) Ecological studies. Analysis and synthesis. Vol. 19. Water and plant life. Springer-Verlag, New York.

Humble, G. D., T. C. Hsiao. 1969. Specific requirement of potassium for light-activated opening of stomata in epidermal strips. Plant Physiol. 44:230–234.

Hüsken, D., E. Steudle, and U. Zimmermann. 1978. Pressure probe technique for measuring water relations of cells in higher plants. Plant Physiol. 61:158–163.

Itai, C., and A. Benzioni. 1976. Water stress and hormonal response. p. 176–188. *In* O. L. Lange et al. (ed.) Ecological studies. Analysis and synthesis. Vol. 19. Water and plant life. Springer-Verlag, New York.

Itai, C., A. Richmond, and Y. Vaadia. 1968. The role of root cytokinins during water and salinity stress. Isr. J. Bot. 17:187–195.

Itai, C., and Y. Vaadia. 1971. Cytokinin activity in water-stressed shoots. Plant Physiol. 47: 87–90.

Jackson, M. B., and D. J. Osborne. 1970. Ethylene, the natural regulator of leaf abscission. Nature 225:1019–1022.

Jarvis, P. G., and T. A. Mansfield (ed.). 1982. Stomatal physiology. Cambridge University Press, Cambridge.

Jones, M. M., and N. C. Turner. 1978. Osmotic adjustment in leaves of sorghum in response to water deficits. Plant Physiol. 61:122–126.

Jordan, W. R., K. W. Brown, and J. C. Thomas. 1975. Leaf age as a determinant in stomatal control of water loss from cotton during water stress. Plant Physiol. 56:595–599.

Kaldewey, H., U. Ginkel, and G. Wawczyniak. 1974. Auxin transport and water stress in pea (*Pisum sativum* L.). Ber. Dtsch. Bot. Ges. 7:563–576.

Karmoker, J. L., and R. F. M. van Steveninck. 1978. Stimulation of volume flow and ion flux by abscisic acid in excised root systems of *Phaseolus vulgaris* w. Redland Pioneer. Planta 141:37–43.

Kauss, H. 1977. Biochemistry of osmotic regulation. p. 119–140. *In* D. H. Northcote (ed.) International review of biochemistry. Plant Biochemistry II. Vol. 13. University Park Press, Baltimore.

Kirkham, M. B., W. R. Gardner, and G. C. Gerloff. 1972. Regulation of cell division and cell enlargement by turgor pressure. Plant Physiol. 49:961–962.

Labavitch, J. M. 1982. Cell wall turnover in plant development. Annu. Rev. Plant Physiol. Vol. 32 (in press).

Labavitch, J. M., and P. M. Ray. 1974a. Turnover of cell wall polysaccharides in elongating pea stem segments. Plant Physiol. 53:669–673.

Labavitch, J. M., and P. M. Ray. 1974b. Relationship between promotion of xyloglucan metabolism and inducation of elongation by indoleacetic acid. Plant Physiol. 54:499–502.

Lado, P., R. Rasi-Caldogno, and R. Colombo. 1977. Effect of cycloheximide on IAA- or FC-induced cell enlargement in pea internode segments. Plant Sci. Lett. 9:93–101.

Lee-Stadelmann, O. Y., and E. J. Stadelmann. 1976. Cell permeability and water stress. p. 268–280. *In* O. L. Lange et al. (ed.) Ecological studies. Analysis and synthesis. Vol. 19. Water and plant life. Springer-Verlag, New York.

Livne, A., and Y. Vaadia. 1965. Stimulation of transpiration rate in barley leaves by kinetin and gibberellic acid. Physiol. Plant. 18:658–664.

Lockhart, J. A. 1965. An analysis of irreversible plant cell elongation. J. Theor. Biol. 8:264–276.

Loescher, W. H., and D. J. Nevins. 1972. Auxin-induced changes in *Avena* coleoptile cell wall composition. Plant Physiol. 50:556–563.

Loescher, W. H., and D. J. Nevins. 1973. Turgor-dependent changes in *Avena* coleoptile cell wall composition. Plant Physiol. 52:248–251.

Lösch, R., and B. Schenk. 1978. Humidity responses of stomata and the potassium content of guard cells. J. Exp. Bot. 29:781–787.

Loveys, B. R., C. J. Brien, and P. E. Kriedemann. 1975. Biosynthesis of abscisic acid under osmotic stress: studies based on a dual labelling technique. Physiol. Plant. 33:166–170.

Loveys, B. R., and P. E. Kriedemann. 1973. Rapid changes in abscisic acid-like inhibitors following alterations in vine leaf water potential. Physiol. Plant. 28:476–479.

Ludlow, M. M. 1980. Adaptive significance of stomatal responses to water stress. p. 123–238. *In* N. C. Turner and P. J. Kramer (ed.) Adaptation of plants to water and high temperature stress. John Wiley & Sons, New York.

Luke, H. H., and T. E. Freeman. 1968. Stimulation of transpiration by cytokinins. Nature 217:873–874.

Maclachlan, G. A. 1977. Cellulose metabolism and cell growth. p. 13–20. *In* P. E. Pilet (ed.) Plant growth regulation. Springer-Verlag, Berlin.

Maier-Maercker, U. 1979. Peristomatal transpiration and stomatal movement: a controversial view. II. Observation of stomatal movements under different conditions of water supply and demand. Z. Pflanzenphysiol. 91:157–172.

Markhart, A. H., III, E. L. Fiscus, A. W. Naylor, and P. J. Kramer. 1979. Effect of abscisic acid on root hydraulic conductivity. Plant Physiol. 64:611–614.

Marrè, E., P. Lado, F. Rasi-Caldogno, and R. Colombo. 1973. Correlation between cell enlargement in pea internode segments and decrease in the pH of the medium of incubation. I. Effects of fusicoccin, natural and synthetic auxins and mannitol. Plant. Sci. Lett. 1: 179–184.

McMichael, B. L., W. R. Jordan, and R. D. Powell. 1972. An effect of water stress on ethylene production by intact cotton petioles. Plant Physiol. 49:658–660.

Meyer, R. F., and J. S. Boyer. 1972. Sensitivity of cell division and cell elongation to low water potentials in soybean hypocotyls. Planta 108:77–87.

Michelena, V. A., and J. S. Boyer. 1980. Growth and osmotic adjustment in maize leaves having low water potentials. Plant Physiol. Suppl. 65:8.

Mizrahi, Y., and A. E. Richmond. 1972. Hormonal modification of plant response to water stress. Aust. J. Biol. Sci. 25:437–442.

Molz, F. S., and J. S. Boyer. 1978. Growth-induced water potentials in plant cells and tissues. Plant Physiol. 62:423–429.

Morilla, C. A., J. S. Boyer, and R. H. Hageman. 1973. Nitrate reductase activity and polyribosomal content of corn (*Zea mays* L.) having low leaf water potentials. Plant Physiol. 51:817–824.

Mott, R. L., and F. L. Steward. 1972. Solute accumulation in plant cells. I. Reciprocal relations between electrolytes and non-electrolytes. Ann. Bot. 36:621–639.

Munns, R., C. J. Brady, and E. W. R. Barlow. 1979. Solute accumulation in the apex and leaves of wheat during water stress. Aust. J. Plant Physiol. 6:379–389.

Mussel, H., and R. C. Staples (ed.). 1979. Stress physiology in crop plants. Wiley-Interscience, New York.

Oliveira, E. C., Jr. 1982. Growth and adaptation of cotton in the field under water deficient conditions. Ph.D. Thesis. University of California, Davis. Univ. Microfilm no. 00000. Ann Arbor, Mich.

Ordin, L. 1966. Cell division and enlargement related to water status: angiosperms. p. 493–499. *In* P. L. Altman and D. S. Dittmer (ed.) Environmental biology. Fed. Am. Soc. Biol., Bethesda, Md.

Ordin, L., T. H. Applewhite, and J. Bonner. 1956. Auxin-induced water uptake by *Avena* coleoptile sections. Plant Physiol. 31:44–53.

Pallaghy, C. K., and K. Raschke. 1972. No stomatal response to ethylene. Plant Physiol. 49: 275–276.

Penny, P., and D. Penny. 1978. Rapid responses to phytohormones. p. 537–597. *In* D. S. Letham et al. (ed.) Phytohormones and related compounds—a comprehensive treatise. vol. II. Phytohormones and the development of higher plants. Elsevier, Amsterdam.

Pierce, M., and K. Raschke. 1980. Correlation between loss of turgor and accumulation of abscisic acid in detached leaves. Planta 148:174–182.

Pitman, M. G., and W. S. Cram. 1977. Regulation of ion content in whole plants. p. 391–424. *In* D. H. Jennings (ed.) Integration of activity in the higher plant. Cambridge University Press, Cambridge.

Pitman, M. G., U. Lüttge, L. Lauchli, and E. Ball. 1974a. Action of abscisic acid on ion transport as affected by root temperature and nutrient status. J. Exp. Bot. 25:147–155.

Pitman, M. G., N. Schaefer, and R. A. Wilder. 1974b. Effect of abscisic acid on fluxes of ions in barley roots. p. 391–396. *In* U. Zimmermann and J. Dainty (ed.) Membrane transport in plants. Springer-Verlag, Berlin.

Pitman, M. G., and O. Wellfare. 1978. Inhibition of ion transport in excised barley roots by abscisic acid; relation to water permeability of the roots. J. Exp. Bot. 29:1125–1138.

Pope, D. G. 1978. Does indoleacetic acid promote growth via cell wall acidification? Planta 140:137–142.

Quarrie, S. A., and H. G. Jones. 1977. Effects of abscisic acid and water stress on development and morphology of wheat. J. Exp. Bot. 29:192–203.

Raschke, K. 1975. Stomatal action. Annu. Rev. Plant Physiol. 26:309–340.

Raschke, K. 1979. Movements of stomata. p. 383–441. *In* W. Haupt and M. E. Feinleib (ed.) Encylclopedia of plant physiology. New Series. Vol. 7. Physiology of movements. Springer-Verlag, New York.

Raschke, K., and M. P. Fellows. 1971. Stomatal movement in *Zea mays*: shuttle of potassium and chloride between guard cells and subsidiary cells. Planta 101:296–316.

Ray, P. M. 1962. Cell wall synthesis and cell elongation in oat coleoptile tissue. Am. J. Bot. 49: 928–939.

Ray, P. M., P. B. Green, and R. E. Cleland. 1972. Role of turgor in plant cell growth. Nature 239:163–164.

Rayle, D. L. 1973. Auxin-induced hydrogen-ion secretion in *Avena* coleoptiles and its implication. Planta 114:63–73.

Rayle, D. L., and R. E. Cleland. 1972. The *in vitro* acid-growth response: relation to *in vivo* growth responses and auxin action. Planta 104:282–296.

Rayle, D. L., and R. E. Cleland. 1977. Control of plant cell enlargement by hydrogen ions. p. 187–214. *In* A. A. Moscona and A. Monroy (ed.) Current topics in development biology. Vol. 11. Academic Press, New York.

Rhodes, P. R., and K. Matsuda. 1976. Water stress, rapid polyribosome reductions, and growth. Plant Physiol. 58:631–635.

Rikin, A., A. Blumenfeld, and A. E. Richmond. 1976. Chilling resistance as affected by stressing environment and ABA. Bot. Gaz. 137:307–312.

Rikin, A., and A. E. Richmond. 1976. Amelioration of chilling injuries in cucumber seedlings by abscisic acid. Physiol. Plant. 38:95–97.

Rubinstein, B. 1977. Osmotic shock inhibits auxin-stimulated acidification and growth. Plant Physiol. 59:369–371.

Rubinstein, B., P. Mahar, and T. A. Tattar. 1977. Effects of osmotic shocks on some membrane-regulated events of oat coleoptile cells. Plant Physiol. 59:365–368.

Sakurai, N., and Y. Masuda. 1977. Effect of indole-3-acetic acid on cell wall loosening: changes in mechanical properties and noncellulosic glucose content of *Avena* coleoptile cell wall. Plant Cell Physiol. 18:587–594.

Salter, P. J., and J. E. Goode. 1967. Crop responses to water at different stages of growth. commonw. Bur. Hortic. and Plantation Crops, Farnham Royal, Bucks, England.

Scholander, P. F., H. T. Hammel, E. D. Bardstreet, and E. A. Hemmingsen. 1965. Sap pressure in vascular plants. Science 148:339–346.

Schulze, E.-D., and M. Küppers. 1979. short-term and long-term effects of plant water deficits on stomatal response to humidity in *Corylus avellana* L. Planta 146:319–326.

Scott, N. S., R. Munns, and E. W. R. Barlow. 1979. Polyribosome content in young and aged wheat leaves subjected to drought. J. Exp. Bot. 30:905–911.

Setter, T. L., W. A. Brun, and M. L. Brenner. 1980. Effect of obstructed translocation on leaf abscisic acid, and associated stomatal closure and photosynthesis decline. Plant Physiol. 65:1111–1115.

Sharp, R. E., and W. J. Davies. 1979. Solute regulation and growth by roots and shoots of water-stressed maize plants. Planta 147:43–49.

Sharp, R. E., O. Osonubi, W. A. Wood, and W. J. Davies. 1979. A simple instrument for measuring leaf extension in grasses, and its application in the study of the effects of water stress on maize and sorghum. Ann. Bot. 44:35–45.

Silk, W. K., and K. Wagner. 1980. Growth sustaining water potentials in the primary corn root: a noncompartmented continuum model. Plant Physiol. 66:859–863.

Stålfelt, M. G. 1955. The stomata as a hydrophotic regulator of the water deficit of the plant. Physiol. Plant. 8:572–593.

Stuart, D. A., and R. L. Jones. 1977. Roles of extensibility and turgor in gibberellin- and dark-stimulated growth. Plant Physiol. 59:61–68.

Taiz, L., and M. Jacobs. 1980. Vanadate inhibition of auxin-enhanced H^+ secretion and elongation in pea epicotyls and oat coleoptiles. Plant Physiol. Suppl. 65:719.

Taiz, L., and J.-P. Metraux. 1978. The bidirectional growth kinetics of etiolated pea stem sections in response to acid, auxin and fusicoccin. Plant Physiol. Suppl. 61:239.

Tal, M., and D. Imber. 1971. Abnormal stomatal behavior and hormonal imbalance in *flacca*, a wilty mutant of tomato. III. Hormonal effects on the water status in the plant. Plant Physiol. 47:849–850.

Tal, M., D. Imber, A. Erez, and E. Epstein. 1979. Abnormal stomatal behavior and hormonal imbalance in *flacca*, a wilty mutant of tomato. V. Effect of abscisic acid on indoleacetic acid metabolism and ethylene evolution. Plant Physiol. 63:1044–1048.

Turner, N. C. 1974. Stomatal behavior and water status of maize, sorghum and tobacco under field conditions. Plant Physiol. 53:360–365.

Turner, N. C. 1979. Drought resistance and adaptation to water deficits in crop plants. p. 343–372. *In* H. Mussel and R. C. Staples (ed.) Stress physiology in crop plants. Wiley-Interscience, New York.

Turner, N. C., and P. J. Kramer (ed.). 1980. Adaptation of plants to water and high temperature stress. Wiley-Interscience, New York.

Vaadia, Y. 1976. Plant hormones and water stress. Phil. Trans. R. Soc. London Ser. B. 273: 513–522.

Vanderhoef, L. N., T.-Y. S. Lu, and C. A. Williams. 1977. Comparison of auxin-induced and acid-induced elongation in soybean hypocotyl. Plant Physiol. 59:1004–1007.

Van Staden, J., and J. E. Davey. 1979. The synthesis, transport and metabolism of endogenous cytokinins. Plant Cell and Environ. 2:93–106.

Vesper, M. J., and M. L. Evans. 1979. Nonhormonal induction of H^+ efflux from plant tissues and its correlation with growth. Proc. Natl. Acad. Sci. USA 76:6366–6370.

Vidaver, W. 1972. Effects of pressure on the metabolic processes of plants. p. 159–174. *In* The effects of pressure on organisms. Symp. Soc. Exp. Biol. no. 26. Academic Press, New York.

Walton, D. C. 1980. Biochemistry and physiology of abscisic acid. Annu. Rev. Plant Physiol. 31:453–489.

Walton, D. C., E. Galson, and M. A. Harrison. 1977. The relationship between stomatal resistance and abscisic acid levels in leaves of water-stressed bean plants. Planta 133:145–148.

Walton, D. C., M. A. Harrison, and P. Coté. 1976. The effects of water stress on abscisic acid levels and metabolism in roots of *Phaseolus vulgaris* L. and other plants. Planta 131: 141–144.

Wenkert, W., E. R. Lemon, and T. R. Sinclair. 1978. Leaf elongation and turgor pressure in field-grown soybean. Agron. J. 70:761–764.

Wiebe, H. H. 1966. Matric potential of several plant tissues and biocolloids. Plant Physiol. 41: 1439–1442.

Wright, M. 1974. The effect of chilling on ethylene production, membrane permeability and water loss of leaves of *Phaseolus vulgaris.* Planta 120:63–69.

Wright, S. T. C. 1977. The relationship between leaf water potential (ψ_{leaf}) and the levels of abscisic acid and ethylene in excised wheat leaves. Planta 134:183–189.

Wright, S. T. C. 1978. Phytohormones and stress phenomena. p. 497–536. *In* D. S. Letham et al. (ed.) Phytohormones and related compounds: a comprehensive treatise. Vol. II. Phytohormones and the development of higher plants. Elsevier, Amsterdam.

Wright, S. T. C. 1977. The relationship between leaf water potential (Ψ_{leaf}) and the levels of abscisic acid and ethylene in excised wheat leaves. Planta 134:183–189.

Yang, S. F., and H. K. Pratt. 1978. The physiology of ethylene in wounded plant tissues. p. 595–622. *In* G. Kahl (ed.) Biochemistry of wounded plant tissues. Walter de Gruyter and Co., Berlin.

Zeevaart, J. A. D. 1977. Sites of abscisic acid synthesis and metabolism in *Ricinus communis* L. Plant Physiol. 59:788–791.

Zeiger, E., and A. J. Bloom. 1978. Ion transport in stomatal guard cells: a chemiosmotic hypothesis. What's New in Plant Physiol. 9:29–32.

Zimmerman, A. M. (ed.). 1970. High pressure effects on cellular processes. Academic Press, New York.

Zimmermann, U. 1978. Physics of turgor- and osmoregulation. Annu. Rev. Plant Physiol. 29: 121–148.

Zimmermann, U., D. Hüshken, and E.-D. Schultze. 1980. Direct turgor pressure measurements in individual leaf cells of *Tradescantia viginiana.* Planta 149:445–453.

CHAPTER 6B

Physiological Consequences of Cellular Water Deficits: Osmotic Adjustment

JOHN W. RADIN

Western Cotton Research Laboratory, USDA
Phoenix, Arizona

I. INTRODUCTION

Higher plants have evolved numerous strategies for survival and propagation in the face of suboptimal or irregular and unpredictable, water supply. These drought-resistance strategies have been classified as drought-escape, -avoidance, or -tolerance mechanisms (Levitt, 1972). Although such a division has been criticized as imprecise (Turner, 1979) and artificial (Steponkus et al., 1980), the terms retain some value, and the concepts of tolerance and avoidance will be used in this chapter in discussing osmotic adjustment. Specifically, these terms will refer to the ability of a plant to maintain favorable water status by increasing water uptake or decreasing its loss in an environment of high evaporative demand (avoidance), and to continue to function despite low tissue Ψ (tolerance).

Osmotic adjustment can be defined as an accumulation of solutes that decreases ψ_π and helps to maintain turgor at low Ψ. As the word "adjustment" indicates, it is an adaptive process that occurs in response to a stress. Because its effect is to maintain turgor and (presumably) turgor-dependent processes, it can be classified as a tolerance mechanism in the scheme above. Begg and Turner (1976) considered it to be the most important tolerance mechanism, although there are some crops that apparently do not display it (Turner et al., 1978a; Turk & Hall, 1980).

Osmotic adjustment has long been recognized as a tolerance mechanism for salinity (Eaton, 1927; Bernstein, 1961; Slatyer, 1961). Only recently has it also been accepted as an adaptive response to water stress (Begg & Turner, 1976; Hsiao et al., 1976a). It should be recognized that there are some important differences between the restraints to water uptake imposed by excess salinity and those imposed by water deficits. As a consequence, there is considerable divergence in the implementation of osmotic adjustment under these two stresses. First, a plant growing under saline conditions (or any conditions in which the soil solution has a high solute content) has available to it a ready supply of solutes for osmotic adjustment, i.e., the soil

Limitations to Efficient Water Use in Crop Production.

solution itself. A water-stressed plant may have lesser supplies available, especially if the nutrients are concentrated in the upper part of the soil profile, which tends to dry first (Viets, 1972). Thus, osmotic adjustment in the latter case results largely from internally generated osmotica. This subject is discussed further in Section IV.

A second important difference between salinity and drought is that under saline conditions, a decreased internal ψ_π may be an absolute necessity for a plant to absorb any water; once the requisite lowering of Ψ is achieved, the water supply may be nonlimiting (especially in certain ecosystems such as salt marshes). For a plant that gradually experiences drought, though, the situation is different. During drought, soil water is freely available at first, becoming depleted in the root zone as the drought progresses. Thus the extension of the root zone by continued root growth is a possible means of resupplying water to the plant. This is considered a drought-avoidance strategy. In addition, there exist many other types of drought-avoidance responses in which water losses are minimized by leaf senescence, changes in leaf shape or orientation, increased stomatal and cuticular resistances to transpiration, and pubescence that minimizes the radiation load.

Considering the many possible ways to achieve water balance, the question "Which way is best?" can have no general answer. The complexities of plant growth are such that multiple resistance mechanisms are usually brought into play at the same time. Indeed, enhanced root growth in a drying soil (drought avoidance) depends upon turgor maintenance, which in turn may depend upon osmotic adjustment (drought tolerance). The interrelationship between osmotic adjustment and other responses to stress is considered, in a simplified fashion, in Section IV.

II. CRITERIA TO ESTABLISH OSMOTIC ADJUSTMENT

Simply defined, osmotic adjustment is the accumulation of solutes. When considering this process in the context of water stress, it is not enough to show that ψ_π of a tissue decreases in response to stress. Such a change can arise solely by the passive concentration of solutes during dehydration and consequent cellular shrinkage. This phenomenon is shown schematically in Fig. 1A. As the relative water content (RWC) is decreased, ψ_π is also decreased. A true osmotic adjustment would be evidenced by a downward shift (toward more negative values) in ψ_π at all RWC (Fig. 1B).

Although osmotic adjustment results in turgor maintenance at low Ψ, it must be emphasized that turgor maintenance also arises for other reasons. For example, tissues with high elasticity exhibit large changes in volume per unit change in turgor (Fig. 1C). This allows the RWC to decrease to a relatively low value before turgor is completely lost, and the cell solutes can thus be passively concentrated to a greater degree. However, elasticity changes of this type do not affect ψ_π at any fixed RWC (compare Fig. 1C with Fig. 1A).

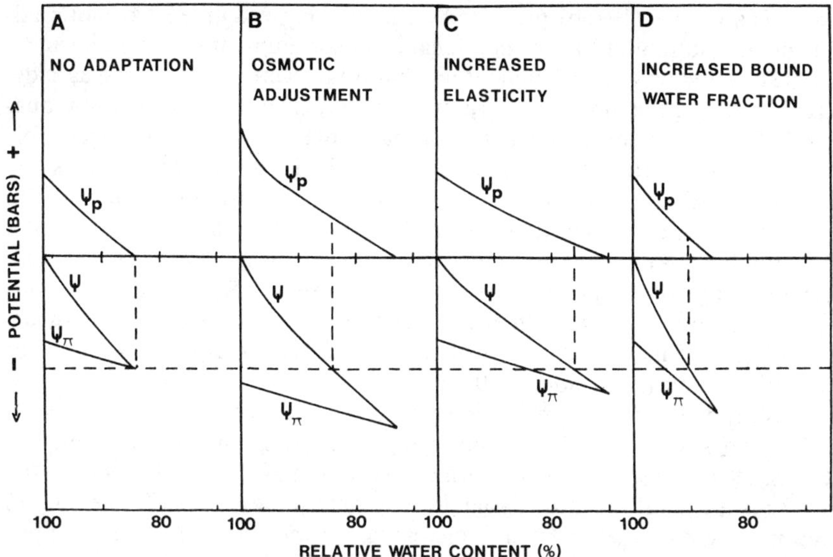

Fig. 1—Representations of the relationships among relative water content and the components of water potential in a tissue. *A*: no adaptation to drought. *B, C,* and *D*: turgor maintenance by osmotic adjustment, increase in elasticity, and increase in bound-water fraction, respectively. The horizontal dashed line indicates the Ψ that in *A* is associated with zero turgor. In *B, C,* and *D* the Ψ for zero turgor is considerably lower. The vertical dashed lines within each diagram indicate the relative water content and turgor associated with the constant Ψ. Note that for constant Ψ the relative water content is unchanged in *B*, decreased in *C,* and increased in *D.*

Another method for apparent turgor maintenance is a change in the partitioning of water into bound and free fractions. The former fraction is considered to be uninvolved in the dissolution of cellular solutes, and is probably located mostly in the cell wall (although other sites with matric forces may also be involved). This water is held very tightly, and is the last to be lost during drying. Thus, as the RWC decreases from 100%, initially all of the water lost is from the osmotically active (free) fraction. If two tissues have equal water contents, but it is partitioned so that one has a much greater bound-water fraction, then the one with the greater bound water will show more rapid concentration of solutes as the RWC is decreased (Fig. 1D). In this case the slope of the ψ_π curve is greater with respect to RWC than in any of the other cases (compare Fig. 1D with Fig. 1A). Because of these differences in slope, there must be a "crossover" point at which the ψ_π is the same for both tissues. As Fig. 1A and 1D are drawn, this crossover occurs at 100% RWC. Thus, this mechanism for turgor maintenance fails the test for osmotic adjustment in that ψ_π is not lowered at all RWC. It must also be pointed out that measuring ψ_π on frozen and thawed tissue introduces a systematic error, because the two fractions of water are mixed (Fereres et al., 1978). A pressure-volume technique such as that described by Jones and Turner (1980) appears to circumvent these mixing problems.

Hsiao et al. (1976b) presented an interesting example of osmotic adjustment resulting in turgor maintenance in sorghum. It occurred during a single day and was of such magnitude that turgor actually increased as Ψ decreased during the morning. Sorghum also adjusted osmotically on a long-term basis to provide complete turgor maintenance, with no significant role for changes in water partitioning (Fereres et al., 1978). Similarly, Jones and Turner (1978) presented a well-documented case of osmotic adjustment in sorghum during drought adaptation. In cotton, though, changes in water partitioning may be the most important mechanism of adaptation, with osmotic adjustment playing a lesser role (Cutler & Rains, 1978). In this species, adaptive changes in water partitioning were associated with accumulation of massive amounts of starch and the consequent shrinkage of osmotic volume (Ackerson, 1980).

Both types of turgor maintenance—osmotic adjustment (Fereres et al., 1978; Jones & Turner, 1978) and changes in water partitioning (Cutler & Rains, 1978)—were associated with decreases in tissue elasticity. Such a decrease is a typical result of tissue development under stress (Turner, 1979) and may be intimately related to increased cell wall thickness (Cutler et al., 1977) and dry matter content (Fereres et al., 1978). In contrast, Elston et al. (1976) found that water stress increased elasticity in beans. Also, Davies and Lakso (1979) reported turgor maintenance in apple trees by the unusual combination of osmotic adjustment and an increase in elasticity.

These examples show that all three types of turgor maintenance described in Fig. 1 have been reported as adaptive responses to drought, and may occur in combinations as well. They emphasize the point that turgor maintenance at low Ψ (or maintenance of stomatal opening, a commonly measured indicator of adaptation) is not sufficient evidence to suggest osmotic adjustment.

III. OSMOTIC ADJUSTMENT IN RESPONSE TO SALINITY

This subject has been thoroughly discussed in recent reviews (Hellebust, 1976; Flowers et al., 1977), so there is little need to treat it comprehensively here. In addition, a volume resulting from a recent symposium (Rains et al., 1980) offers a complete discussion of problems related to salt tolerance in both higher and lower plants. Generally speaking, osmotic adjustment in response to salinity involves the accumulation of ions from he rooting medium. Halophytes (plants that tolerate large amounts of NaCl) frequently accumulate Na^+ and Cl^- (Flowers et al., 1977). Glycophytes (plants that cannot tolerate large amounts of NaCl) seldom accumulate Na^+ in the tops, although there may be some active uptake into the roots (Shepherd & Bowling, 1973). The most prominent accumulated inorganic ion in salt-stressed glycophytes is frequently K^+ (Bernstein, 1963). If accumulation of cations such as K^+ results in a charge imbalance, then malate is often synthesized as a counterion (Hellebust, 1976). Rush and Epstein (1976) demonstrated all of these principles in an elegant comparison of two species of tomato differing in salt tolerance.

For the most part, ion accumulation is localized in the cell vacuole. The best evidence for this compartmentation is indirect and inductive, involving the effects of salts at high concentrations on enzyme activities in vitro. In extracts of *Plantago maritima,* a halophyte, 500 m*M* NaCl inhibited activities of several enzymes known to be compartmented in the cytosol, mitochondria, and chloroplasts (Ahmad et al., 1979). Only acid phosphatase, a vacuolar enzyme (Butcher et al., 1977; Nishimura & Beevers, 1978), was unaffected. High concentrations of Na^+ inhibited enzymes of both halophytes and glycophytes (Flowers, 1972; Greenway & Osmond, 1972). Even K^+, which is normally a more innocuous ion, is harmful at the bulk concentrations to which it can accumulate (Greenway & Sims, 1974). Other approaches to ion localization, such as compartmental efflux analysis and electron probe analysis, have not yet been especially productive but tend to confirm the vacuole as the site of highest concentration (Flowers et al., 1977).

All compartments of the cell are subjected to equal Ψ and equal turgor. If ions accumulate in the vacuole, thereby lowering that organelle's ψ_π, it follows that the cytosol must also have lowered ψ_π. The typical pattern is for a "compatible solute" to accumulate. This term arises from the ability of certain kinds of solutes to build up to high concentrations without injuring the enzymes or metabolism of the cytosol. This class of compounds includes proline, sorbitol, and various quaternary ammonium compounds such as glycine betaine (Stewart & Lee, 1974; Storey & Wyn Jones, 1975; Ahmad et al., 1979). Compatible solutes can accumulate to concentrations of 0.5–1 osmol kg^{-1}, assuming that they are in the (estimated) 5–10% of the cell volume that comprises the cytosol (Stewart & Lee, 1974; Ahmad et al., 1979). For comparison, sea water is about 1 osmol kg^{-1}.

IV. OSMOTIC ADJUSTMENT IN RESPONSE TO DROUGHT

The beneficial principle of osmotic adjustment in response to drought is the same as for adjustment in response to salinity, i.e., turgor maintenance through lowered ψ_π. However, not all water deficits will elicit the adjustment response. First, osmotic adjustment apparently does not occur in some species such as soybean (Turner et al., 1978a) and cowpea (Turk & Hall, 1980). Second, osmotic adjustment may be absent or very weak in plants grown in pots for laboratory experimentation (as opposed to field plants). One possible reason for this behavior is that adaptation results from repeated slow stress cycles, as occur in the field (Fereres et al., 1978; Jones & Rawson, 1979; Jones & Turner, 1978). Pot-grown plants that have severely restricted root volumes may become stressed too rapidly, thus precluding the adaptive response. Turner et al. (1978b) estimated the rate of osmotic adjustment in field-grown sorghum and sunflower at 1 bar d^{-1}. At this rate, stress must obviously progress over at least 7 days to produce 7 bars of osmotic adjustment.

The slow development of a net decrease in ψ_π from day to day must be contrasted with the rapid osmotic adjustment seen during a single diurnal

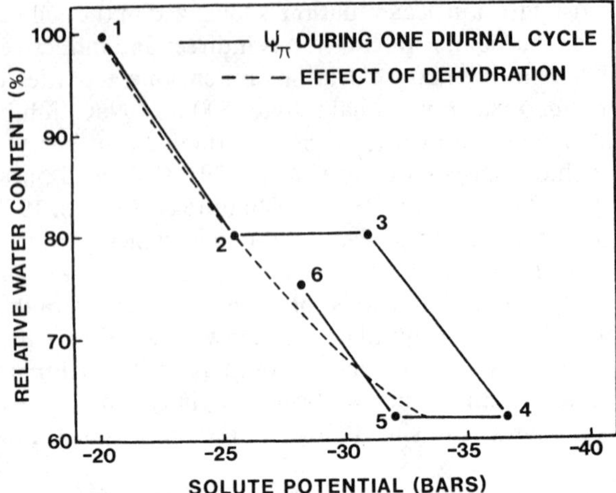

Fig. 2—Relationship between ψ_π and relative water content for apple leaves (data from Table 1 of Davies & Lakso, 1979). Points were obtained during a single day at the following times: (*1*) predawn, (*2*) 0645h, (*3*) 0845h, (*4*) 1155h, (*5*) 1445h, and (*6*) 1645 h. The effect of dehydration was estimated by the Van't Hoff approximation.

cycle of water stress. In sorghum, solute accumulation was rapid enough to increase turgor during the morning when Ψ was declining (Hsiao et al., 1976b). This daily accumulation may perhaps reflect photosynthetic and translocation patterns of the leaves, with the peak of photosynthetic activity preceding that of translocation by a few hours. Data of Davies and Lakso (1979) show a similar trend in apple leaves. Because Davies and Lakso also measured RWC during a diurnal cycle, they were able to distinguish between change in ψ_π resulting from simple dehydration and that resulting from osmotic adjustment. A plot of ψ_π vs. RWC shows that early in the day there was a pronounced shift away from the curve for dehydration alone (Fig. 2). This shift, representing solute accumulation, disappeared in the afternoon. The hysteresis in this relationship was also seen in plots of Ψ vs. RWC, giving rise to more or less constant turgor (Davies & Lakso, 1979). Where such hysteresis occurs in other crops (e.g., Jordan & Ritchie, 1971, for cotton), it may well suggest transient solute accumulation during the morning.

Osmotic adjustment in water-stressed plants is primarily from accumulation of organic solutes (Hsiao et al., 1976a; Begg & Turner, 1976; Hellebust, 1976), although inorganic solutes can also play a major role (Jones et al., 1980). Glucose and organic acid anions, especially malate, are frequently found as major osmotica. This use of organic solutes presents a problem: because of the high molecular weights of the typical organic solute compared with the solutes accumulated under salinity, large amounts (on a weight basis) must be accumulated. A ψ_π of -10 bars from glucose alone would imply (on the basis of some reasonable assumptions) that glucose constituted 25% of the tissue's dry weight. Such accumulations are rarely found. For example, in osmotically adjusting sorghum and sunflower, soluble sugars did not exceed 12–14% of dry weight, and only about half of that amount was associated with osmotic adjustment (Turner et al., 1978b). Calculations for organic acids lead to similar conclusions, i.e., that the amounts required for extensive adjustment are much greater than the accumulations actually observed. Thus, it seems reasonable to believe that osmotic adjustment of >2 or 3 bars must involve simultaneous accumulation of a variety of solutes. This idea is borne out by the diversity of solutes accumulated within most tissues (Munns et al., 1979; Cutler & Rains, 1978; Jones et al., 1980).

What are the limits to osmotic adjustment? Most reports of adjustment indicate no more than 5–7 bars decrease in ψ_π, and some plants such as rice (Cutler et al., 1980a) and sunflower (Jones & Turner, 1980) may be limited to less than that. Morgan (1977) found that osmotically adjusting wheat could maintain turgor down to a Ψ of about −30 bars, but no lower. The question can perhaps be best approached by considering osmotic adjustment as an aspect of whole plant development. The primary source of accumulated solutes most likely is current photosynthesis (Begg & Turner, 1976; Munns & Pearson, 1974). Accumulation of assimilates is possible because of the decline in leaf expansion and growth of other important sinks during drought, which frees some photosynthate for other uses (Barlow & Boersma, 1976). The magnitude of these changes is illustrated in Table 1 (representative data of Krieg & Sung, 1979). In field-grown cotton, water stress inhibited photosynthesis of individual leaves by about 35% (on a leaf area basis) but substantially increased the percentage of photosynthate retained in the blade. Overall, leaves of stressed plants retained about 15–20 mg more photosynthate g^{-1} dry wt d^{-1} (13 mg in the example shown) than did leaves of unstressed plants. If all the extra retained photosynthate were glucose, this could produce osmotic adjustment of about 0.6 bars d^{-1}. Such a rate of adjustment could account for more than the observed changes in ψ_π of cotton as soil moisture availability decreased (Ackerson et al., 1977).

It must be noted that no rigorous proof exists establishing that osmotic adjustment is a passive result of changes in sink demands. However, the only requirement for such a hypothesis is that leaf growth (the major sink in a vegetative plant) be more sensitive to stress than is photosynthesis. This is indeed the case (Boyer, 1968; Acevedo et al., 1971). Furthermore, it has long been recognized that accumulations in roots of stressed plants result from declines in assimilate utilization for shoot growth (cf. Curtis & Clark, 1950, p. 444).

Table 1—Characterization of photosynthesis by individual leaves of field-grown cotton. Data are from Krieg and Sung (1979).

Treatment	Ψ	Leaf area	Photosynthetic rate	Photosynthate produced	Photosynthate retained
	bars	dm^2	mg dm^{-2} $hour^{-1}$	mg $leaf^{-1}$ d^{-1}	mg g dry wt^{-1}
Unstressed	−15 to −17	0.83	37	139	61
Stressed	−22 to −24	0.76	24	83	74

If osmotic adjustment is simply a result of changes in assimilate demands, then the limits to adjustment must be set by the balance between photosynthetic rate and assimilate demand at any particular Ψ. On the basis of this analysis, Turner (1979) predicted that plants that do not adjust osmotically will be those in which leaf expansion and photosynthesis are equally sensitive to stress. To go further, if one could characterize photosynthesis, growth, and other important sink (or source) activities as functions of water status, one could presumably predict the degree of adjustment elicited by any particular stress regime. Unfortunately, these processes are not related to Ψ in a simple fashion. For example, Cutler et al. (1980b) found that stress adaptation in rice was associated with a decreased growth rate under favorable conditions, partly because the turgor necessary to maintain leaf growth rate was increased. A great deal more such knowledge must be developed before the limits to osmotic adjustment can be fully understood.

No discussion of osmotic adjustment is complete without mention of proline. Excellent evidence exists for its role as a compatible solute in salinity-induced adjustment (see Section III). Proline accumulation during drought is also well-documented (Barnett & Naylor, 1966; McMichael & Elmore, 1977; Tully et al., 1979), although there is no proof that it serves as a compatible solute as described in Section III. The best current evidence suggests that proline accumulation is incidental to some other changes induced by water stress, especially senescence (Hanson, 1980). This idea is reinforced by the recent findings that N deficiency (Goring & Thien, 1979) and treatment with exogenous abscisic acid (Rajagopal & Andersen, 1978) also cause proline accumulation. It is also pertinent that compatible solutes have survival value during salinity stress because they protect the cytosol from damaging effects of concentrated inorganic ions. The organic nature of solutes accumulated during water stress may obviate the need for such protection.

LITERATURE CITED

Acevedo, E., T. C. Hsiao, and D. W. Henderson. 1971. Immediate and subsequent growth responses of maize leaves to changes in water status. Plant Physiol. 48:631–636.

Ackerson, R. C. 1980. Osmoregulation and water stress adaptation in cotton. Plant Physiol. 65(suppl.):154.

Ackerson, R. C., D. R. Krieg, T. D. Miller, and R. E. Zartman. 1977. Water relations of field grown cotton and sorghum: temporal and diurnal changes in leaf water, osmotic, and turgor potentials. Crop Sci. 17:76–80.

Ahmad, I., F. Larher, and G. R. Stewart. 1979. Sorbitol, a compatible osmotic solute in *Plantago maritima.* New Phytol. 82:671–678.

Barlow, E. W. R., and L. Boersma. 1976. Interaction between leaf elongation, photosynthesis, and carbohydrate levels of water-stressed corn seedlings. Agron. J. 68:923–926.

Barnett, N. M., and A. W. Naylor. 1966. Amino acid and protein metabolism in Bermuda grass during water stress. Plant Physiol. 41:1222–1230.

Begg, J. E., and N. C. Turner. 1976. Crop water deficits. Adv. Agron. 28:161–217.

Bernstein, L. 1961. Osmotic adjustment of plants to saline media. I. Steady state. Am. J. Bot. 48:909–918.

Bernstein, L. 1963. Osmotic adjustment of plants to saline media. II. Dynamic phase. Am. J. Bot. 50:360–370.

Boyer, J. S. 1968. Relationships of water potential to growth of leaves. Plant Physiol. 43: 1056–1062.

Butcher, H. C., G. J. Wagner, and H. W. Siegelman. 1977. Localization of acid hydrolases in protoplasts. Examination of the proposed lysosomal function of the mature vacuole. Plant Physiol. 59:1098–1103.

Curtis, O. F., and D. G. Clark. 1950. Introduction to plant physiology. McGraw-Hill, New York.

Cutler, J. M., and D. W. Rains. 1978. Effects of water stress and hardening on internal water relations and osmotic constituents of cotton leaves. Physiol. Plant. 42:261–268.

Cutler, J. M., D. W. Rains, and R. S. Loomis. 1977. The importance of cell size in the water relations of plants. Physiol. Plant. 40:256–259.

Cutler, J. M., K. W. Shahan, and P. L. Steponkus. 1980a. Dynamics of osmotic adjustment in rice. Crop Sci. 20:310–314.

Cutler, J. M., K. W. Shahan, and P. L. Steponkus. 1980b. Influence of water deficits and osmotic adjustment on leaf elongation in rice. Crop Sci. 20:314–318.

Davies, F. S., and A. N. Lakso. 1979. Diurnal and seasonal changes in leaf water potential components and elastic properties in response to water stress in apple trees. Physiol. Plant. 46:109–114.

Eaton, F. M. 1927. The water-requirement and cell-sap concentration of Australian saltbush and wheat as related to the salinity of the soil. Am. J. Bot. 14:212–226.

Elston, J., A. J. Karamanos, A. H. Kassam, and R. M. Wadsworth. 1976. The water relations of the field bean crop. Phil. Trans. R. Soc. London Ser. B. 273:581–591.

Fereres, E., E. Acevedo, D. W. Henderson, and T. C. Hsiao. 1978. Seasonal changes in water potential and turgor maintenance in sorghum and maize under water stress. Physiol. Plant. 44:261–267.

Flowers, T. J. 1972. Salt tolerance in *Suaeda maritima* (L.) Dum. The effect of sodium chloride on growth, respiration, and soluble enzymes in a comparative study with *Pisum sativum* L. J. Exp. Bot. 23:310–321.

Flowers, T. J., P. F. Troke, and A. R. Yeo. 1977. The mechanism of salt tolerance in halophytes. Annu. Rev. Plant Physiol. 28:89–121.

Goring, H., and B. H. Thien. 1979. Influence of nutrient deficiency on proline accumulation in the cytoplasm of *Zea mays* seedlings. Biochem. Physiol. Pflanz. 174:9–16.

Greenway, H., and C. B. Osmond. 1972. Salt responses of enzymes from species differing in salt tolerance. Plant Physiol. 49:256–259.

Greenway, H., and A. P. Sims. 1974. Effects of high concentrations of KCl and NaCl on responses of malate dehydrogenase (decarboxylating) to malate and various inhibitors. Aust. J. Plant Physiol. 1:15–29.

Hanson, A. D. 1980. Interpreting the metabolic responses of plants to water stress. HortScience 15:623–629.

Hellebust, J. A. 1976. Osmoregulation. Annu. Rev. Plant Physiol. 27:485–505.

Hsiao, T. C., E. Acevedo, E. Fereres, and D. W. Henderson. 1976a. Water stress, growth, and osmotic adjustment. Phil. Trans. R. Soc. London Ser. B. 273:479–500.

Hsiao, T. C., E. Fereres, E. Acevedo, and D. W. Henderson. 1976b. Water stress and dynamics of growth and yield of crop plants. p. 281–305. *In* O. L. Lange et al. (ed.) Water and plant life. Problems and modern approaches. Springer-Verlag, Berlin.

Jones, M. M., C. B. Osmond, and N. C. Turner. 1980. Accumulation of solutes in leaves of sorghum and sunflower in response to water deficits. Aust. J. Plant Physiol. 7:193–205.

Jones, M. M., and H. M. Rawson. 1979. Influence of rate of development of leaf water deficits upon photosynthesis, leaf conductance, water use efficiency, and osmotic potential in sorhgum. Physiol. Plant. 45:103–111.

Jones, M. M., and N. C. Turner. 1978. Osmotic adjustment in leaves of sorghum in response to water deficits. Plant Physiol. 61:122–126.

Jones, M. M., and N. C. Turner. 1980. Osmotic adjustment in expanding and fully-expanded leaves of sunflower in response to water deficits. Aust. J. Plant Physiol. 7:181–192.

Jordan, W. R., and J. T. Ritchie. 1971. Influence of soil water stress on evaporation, root absorption, and internal water status of cotton. Plant Physiol. 48:783–788.

Krieg, D. R., and F. J. M. Sung. 1979. Source-sink relations of cotton as affected by water stress during boll development. p. 302–304. *In* J. M. Brown (ed.) Proc. 1979 Beltwide Cotton Prod. Res. Conf., Phoenix, Ariz. Natl. Cotton Council, Memphis, Tenn.

Levitt, J. 1972. Responses of plants to environmental stresses. Academic Press, New York.

McMichael, B. L., and C. D. Elmore. 1977. Proline accumulation in water stressed cotton leaves. Crop Sci. 17:905–908.

Morgan, J. M. 1977. Differences in osmoregulation between wheat genotypes. Nature 270: 234–235.

Munns, R., C. J. Brady, and E. W. R. Barlow. 1979. Solute accumulation in the apex and leaves of wheat during water stress. Aust. J. Plant Physiol. 6:379–389.

Munns, R., and C. J. Pearson. 1974. Effect of water deficit on translocation of carbohydrate in *Solanum tuberosum*. Aust. J. Plant Physiol. 1:529–537.

Nishimura, M., and H. Beevers. 1978. Hydrolases in vacuoles from castor bean endosperm. Plant Physiol. 62:44–48.

Rains, D. W., R. C. Valentine, and A. Hollaender (ed.). 1980. Genetic engineering of osmoregulation: impact on plant productivity for food, chemicals, and energy. Plenum Press, New York.

Rajagopal, V., and A. S. Andersen. 1978. Does abscisic acid influence proline accumulation in stressed leaves? Planta 143:85–88.

Rush, D. W., and E. Epstein. 1976. Genotypic responses to salinity. Differences between salt-sensitive and salt-tolerant genotypes of the tomato. Plant Physiol. 57:162–166.

Shepherd, U. H., and D. J. F. Bowling. 1973. Active accumulation of sodium by roots of five aquatic species. New Phytol. 72:1075–1080.

Slatyer, R. O. 1961. Effects of several osmotic substrates on the water relations of tomato. Aust. J. Biol. Sci. 14:519–540.

Steponkus, P. L., J. M. Cutler, and J. C. O'Toole. 1980. Adaptation to water stress in rice. p. 401–418. *In* N. C. Turner and P. J. Kramer (ed.) Adaptation of plants to water and high temperature stress. John Wiley & Sons, New York.

Stewart, G. R., and J. A. Lee. 1974. The role of proline accumulation in halophytes. Planta 120:279–289.

Storey, R., and R. G. Wyn Jones. 1975. Betaine and choline levels in plants and their relationship to sodium chloride stress. Plant Sci. Lett. 4:161–168.

Tully, R. E., A. D. Hanson, and C. E. Nelson. 1979. Proline accumulation in water-stressed barley leaves in relation to translocation and the nitrogen budget. Plant Physiol. 63:518–523.

Turk, K. J., and A. E. Hall. 1980. Drought adaptation of cowpea. II. Influence of drought on plant water status and relations with seed yield. Agron. J. 72:421–427.

Turner, N. C. 1979. Drought resistance and adaptation to water deficits in crop plants. p. 333–372. *In* H. Mussell and R. C. Staples (ed.) Stress physiology in crop plants. John Wiley & sons, New York.

Turner, N. C., J. E. Begg, H. M. Rawson, S. D. English, and A. B. Hearn. 1978a. Agronomic and physiological responses of soybean and sorghum crops to water deficits. III. Components of leaf water potential, leaf conductance, $^{14}CO_2$ photosynthesis, and adaptation to water deficits. Aust. J. Plant Physiol. 5:179–194.

Turner, N. C., J. E. Begg, and M. L. Tonnet. 1978b. Osmotic adjustment of sorghum and sunflower crops in response to water deficits and its influence on the water potential at which stomata close. Aust. J. Plant Physiol. 5:597–608.

Viets, F. G. 1972. Water deficits and nutrient availability. p. 217–239. *In* T. T. Kozlowski (ed.) Water deficits and plant growth. Vol. III. Academic Press, New York.

CHAPTER 6C

Physiological Consequences of Cellular Water Deficits: Nonstomatal Inhibition of Photosynthesis by Water Stress

ROBERT W. PEARCY

University of California
Davis, California

I. INTRODUCTION

It has long been recognized that water stress has profound effects on the photosynthetic performance of crop and other plant species. As stress is imposed, net photosynthesis is often observed to be at first relatively independent of leaf water potential (ψ_L) but then to decrease markedly, often reaching 0 at greater stress levels. While the strong inverse correlation between the decline in photosynthesis and leaf resistance to water vapor loss as water stress increases implies an important role for stomatal closure, there is considerable evidence that direct effects of water stress on the photosynthetic apparatus may also be of importance in limiting net CO_2 uptake. The objective of this chapter will be to review the available evidence concerning the nature of these nonstomatal sources of inhibition of net CO_2 uptake.

II. RESISTANCE MODEL FOR CARBON DIOXIDE UPTAKE

For the purposes of considering the inherent limitations to CO_2 uptake by a leaf and the influences of environmental stresses on these limitations, the photosynthetic apparatus is most easily thought of in terms of a resistance model (Fig. 1). This type of model, which was first utilized extensively by Gaastra (1959), partitions the photosynthetic apparatus into a series of steps, with each having an associated resistance representing a limitation to CO_2 uptake and the CO_2 concentration drop occurring across this resistance. Considering the entire system, the net photosynthetic CO_2 uptake rate (N_L) can be expressed in terms of the total resistance to CO_2 uptake (r_t') and the concentration gradient across the apparatus as:

$$N_L = \{[CO_2]_a - [CO_2]_c\}/r_t', \qquad [1]$$

Limitations to Efficient Water Use in Crop Production.

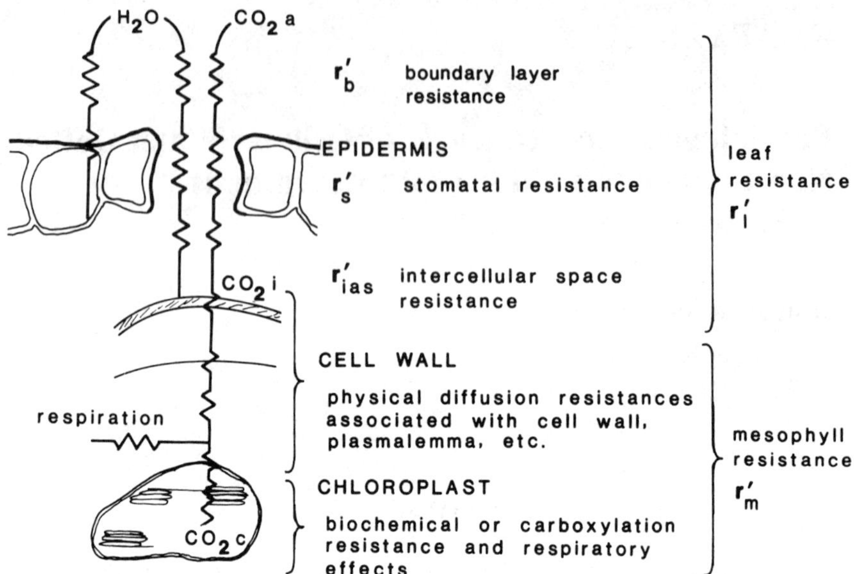

Fig. 1—Pathway and resistances involved in CO_2 uptake by a leaf. For comparative purposes, the pathway for water vapor is also shown.

where $[CO_2]_a$ and $[CO_2]_c$ are the atmospheric CO_2 concentration and the CO_2 compensation point (the CO_2 concentration at which $N_L = 0$), respectively. Note that in Eq. [1], r_t' includes not only the physical diffusion resistances in the vapor and liquid phases, but also a resistance expressing the biochemical limitation to CO_2 uptake. Since resistances in series are additive, Eq. [1] can be rewritten in terms of the component boundary layer (r_b'), stomatal (r_s'), intercellular (r_{ias}'), and mesophyll (r_m') resistances as:

$$N_L = \frac{[CO_2]_a - [CO_2]_c}{r_b' + r_s' + r_{ias}' + r_m'} = \frac{[CO_2]_a - [CO_2]_c}{r_l' + r_m'}, \quad [2]$$

where the sum of $r_b' + r_s' + r_{ias}'$ equals r_l', the leaf resistance to CO_2 uptake. Since the net fluxes across each part of the series are equal, Eq. [2] can be rewritten in terms of the component paths and associated resistances as:

$$N_L = \frac{[CO_2]_a - [CO_2]_i}{r_l'} = \frac{[CO_2]_i - [CO_2]_c}{r_m'}, \quad [3]$$

where $[CO_2]_i$ is the intercellular airspace CO_2 concentration. The resistances r_l' and r_m' thus express the gas-phase limitations to CO_2 uptake, which in most leaves are primarily stomatal, and the nonstomatal limitations associated with the liquid phase, including the physical diffusion of CO_2, the biochemical limitations, and the effects of dark respiration and photorespiration (Fig. 1). A determination of the individual contributions of the various liquid-phase resistances to r_m' is not possible from gas-exchange measurements alone. However, there is good reason to believe that the biochemical components are a major part of r_m' (Collatz, 1977).

Determination of r_l' and r_m' is facilitated by simultaneous measurements of transpiration and photosynthesis, since, with the exception of cuticular transpiration, the path for water vapor and CO_2 from the interior cell walls to the external atmosphere is the same (Fig. 1). Thus, if leaf resistance to water vapor loss, r_l, is determined, r_l' can be calculated as:

$$r_l' = (D_{H_2O}/D_{CO_2})r_l, \quad [4]$$

where (D_{H_2O}/D_{CO_2}) is the ratio of the diffusion coefficients for water vapor and CO_2 and has a numerical value of 1.6 for the combined stomatal and boundary-layer path. Cuticular transpiration can generally be ignored because it is very much smaller than the stomatal fluxes. However, the validity of this assumption under water-stress conditions when stomatal resistance is quite high has been questioned (Moldau, 1973). Subtraction of r_l' from the total resistance determined from Eq. [1] gives:

$$r_m' = r_t' - r_l'. \quad [5]$$

Alternatively, r_m' can be determined as the reciprocal of the initial slope of the curve relating N_L to $[CO_2]_i$, with the latter calculated from Eq. [2]. Both procedures are analogous, but the latter does not make assumptions about the linearity of the photosynthetic response between the $[CO_2]_c$ and $[CO_2]_i$ levels. Boyer (1976) has criticized assumptions concerning the relative importance of r_l' and r_m' calculated as a residual as in Eq. [5], since in this procedure there is no indication that CO_2 and, therefore, diffusion are actually limiting CO_2 uptake. For sunflower, he found a decrease in photosynthesis and an increase in r_l' during a drying cycle, suggesting a causal relationship. However, after induction of water stress, CO_2 uptake was relatively independent of concentration, suggesting a greater importance of a nonstomatal inhibition. In other species, however, $(CO_2)_i$ dependence curves are linear up to concentrations above normal intercellular levels both before and after substantial inhibition by water stress (Mooney et al., 1977; Jones, 1973b; Collatz et al., 1976), and thus, either method in these species should give comparable results.

A major limitation of the use of a resistance model is that r_m' is calculated as a residual. Clearly, any error in r_l' leads to an error in r_m'. This also holds for r_m' calculated from the CO_2 dependence curves, since r_l' is used to calculate $[CO_2]_i$. These sources of error and their significance have been discussed in detail by others (Jarvis, 1971; Jones, 1973a; Koller, 1970). It suffices to say here that the major sources of error are in leaf temperature measurements required to calculate water vapor concentrations in the leaf and in the use of average transpiration rates from both leaf surfaces for amphistomatous leaves. For the latter, the two surfaces should be treated as parallel paths, and the resistances calculated accordingly (Jarvis, 1971). Fortunately, these and other errors involved in the measurements of leaf gas exchange tend to compensate for one another so that, overall, the error associated with the estimation of r'_m may not be too serious (Jones, 1973a).

III. WATER STRESS AND MESOPHYLL RESISTANCE

Responses of r'_m to water stress observed in a variety of species appear to fall into several categories. First, in some species, little or no change in r'_m is noted in response to a relatively rapid increase in water stress, although r'_l increases markedly. These results have led to the suggestion that changes in r'_m are unimportant in determining response to water stress and that stomatal changes are the primary cause for decreased photosynthesis. In many other studies, however, r'_m has been observed to increase, sometimes substantially, as water stress increases, particularly as higher levels of stress are reached. Finally, there is evidence that pretreatment conditions can have a large effect on the response.

Studies with cotton illustrate the variety of responses that can be obtained depending on the nature of the experiment. Troughton (1969) and Slatyer (1973), using gas-exchange measurements with r'_m calculated as the residual, showed that r'_m was constant with a relative water content (RWC) change from 92 to 75%, but increased markedly with a further decrease in RWC to 56%. In other experiments, Troughton and Slatyer (1969) attempted to eliminate possible errors in r'_m determinations associated with r'_l measurements by establishing a pressure gradient across the leaf and determining N_L from the CO_2 concentration drop and the flow rate through the leaf. The $[CO_2]_i$ was also determined from the concentration drop. In these flow-through experiments, no increase in r'_m occurred at low RWC levels. Unfortunately, ψ_L was specified only at the lowest RWC (56% = ψ_L of -25 bars). Jones (1973b) stressed cotton over longer periods (1 or 2 weeks) using osmotic soil cells. In these plants, r'_m, measured by forcing air through the leaf, increased substantially, particularly after 2 weeks of stress. Measurements were made after rewatering, allowing for some decrease in r'_l so as to facilitate the measurements; rewatering apparently did not change r'_m. Both N_L at normal ambient CO_2 levels and $N_L$max measured at saturating CO_2 were affected by stress, with the latter clearly indicating an inhibition of the nonstomatal components since gas-phase diffusion would not be limiting. Interestingly, this inhibition apparently occurred in the stressed plants at a ψ_L of -8 to -16 bars, which is far above the values where Troughton and Slatyer's measurements showed no short-term effects. Increases in r'_l were a major limitation to CO_2 uptake in all of the experiments, but where stress was imposed for relatively long periods, increases in r'_m apparently became a significant factor. This conclusion is also supported by the strong decline in N_L, measured by $^{14}CO_2$ uptake, with no concomitant increase in r_l as field-grown cotton underwent a gradual (>1 month) drying from -8 to -30 bars (Ackerson et al., 1977). With these field plants, a nonstomatal inhibition appeared to be the major factor accounting for the decline in N_L.

Radin and Ackerson (1981) also found a marked increase in r'_m at ψ_L below -12 bars in cotton subjected to a relatively slow development of stress. These experiments also had an interesting control of exogenously applied abscisic acid (ABA), which closed the stomata but had no effect on the calculated r'_m values, suggesting that errors in r'_l measurements were not of significance in the increased r'_m found after the development of water stress. As with other slowly stressed plants, r'_m appeared to be a major factor in limiting N_L since $[CO_2]_i$ actually increased slightly with decreasing ψ_L, even though r'_l increased markedly.

The responses of most other species to short-term water stress are generally similar to those observed by Troughton (1969) and Slatyer (1973) in cotton. That is, as ψ_L or RWC decreases, r'_m remains constant until some threshold level is reached and then increases, sometimes considerably. In nearly all species, however, r'_l increases before there is any effect on r'_m, and usually limits photosynthesis to a greater extent. Boyer (1970), using maize and soybeans, found increases in r'_m increased about twofold at -40 bars while r'_l increased by ninefold. For maize, the increases in r'_m and r'_l at -16 bars were about two- and fivefold, respectively. Similarly, Slatyer (1973) reported that r'_m increased below about 80% RWC in maize; at 62% RWC (-26 bars) r'_m had increased by 14-fold over the initial no-stress level. However, he found no such increase in r'_m as RWC decreased to about 60% in wheat and millet or to 55% in two *Atriplex* species.

Thus, there are considerable species differences in the water-stress levels required to bring about an increase in r'_m. The most sensitive species investigated so far appears to be tobacco, which exhibited a marked linear increase in r'_m as RWC decreased from 95 to 83% (Redshaw & Meidner, 1972). Moreover, this increase in r'_m accounted for about 50% of the total inhibition of CO_2 uptake. On the other hand, Mederski et al. (1975) reported no stress-induced inhibition of r'_m in soybeans, maize, and mung beans. However, measurements were made only at RWC >80% in the latter two, while for soyeans, the leaf was detached and dried very rapidly (approximately 70 min from 90 to 40% RWC). Also, light intensities were quite low so that the results may not be strictly comparable to others. Moldau (1973) also found no increase in r'_m with increasing water stress in beans, but the minimum ψ_L reached was only -13 bars.

As with cotton, pretreatment history and the rate of stress development have a considerable influence on the nonstomatal photosynthetic limitations in several other species. Sorghum grown in large containers, where ψ_L was decreased at a rate of 1.5 bars d^{-1} during a drought cycle, maintained higher photosynthetic rates at any given ψ_L than did more rapidly stressed plants (7 or 12 bars d^{-1}) grown in smaller containers (Jones & Rawson, 1979). Stomata remained more open at any ψ_L in the slowly stressed as compared with the rapidly stressed plants. Additionally, there was indirect evidence suggesting that nonstomatal inhibition of CO_2 uptake was less in the slowly as compared with the rapidly stressed plants. Similar but more direct evidence of a prehistory effect has been demonstrated in *Larrea divaricata*, an evergreen desert shrub that is capable of maintaining positive net photosynthesis in the field at a branch Ψ as low as −78 bars (Strain, 1970). However, in laboratory-grown, rapidly stressed plants, photosynthesis was strongly inhibited at branch Ψ below −20 bars with 75–80% of the inhibition accounted for by nonstomatal components (Mooney et al., 1977). In contrast, only a relatively small inhibition of photosynthesis occurred over the year in field plants at Death Valley, even though branch Ψ changed from −26 to −49 bars. Furthermore, all of this photosynthetic inhibition could be accounted for by changes in r_l' with no change in the nonstomatal components. These results demonstrate a rather remarkable capacity for acclimation to low water potential in *Larrea* when stresses develop slowly over a season. Mooney et al. (1977) suggest that osmotic adjustment may play an important role in this acclimation.

Recovery of photosynthesis from the effects of water stress has received much less attention than the stress induction phase, but there is some evidence that it may be an important consideration in the overall nonstomatal limitations on photosynthesis. Doley and Trivett (1974) found an increase in both r_m' and r_l' during drying cycles in mitchell grass, a drought-tolerant Australian forage species. Rewatering resulted in a rapid and complete recovery of r_l', while r_m' remained unchanged. Similarly, stomatal recovery was more rapid than recovery of r_m' following long-term water stress in cotton (Jones, 1973b). Thus, while stomatal closure usually seems to be the most significant limitation to CO_2 uptake during drying, slow recovery of r_m' following rewatering may represent a more significant limitation to CO_2 uptake during this phase. This, however, may not be true for all species, since in sorghum, recovery of photosynthesis and r_l' were virtually identical following rewatering, despite evidence for an inhibition of r_m' during the imposition of stress (Jones & Rawson, 1979).

Recent work with other arid-zone shrubs provides interesting contrasts with results from crop species and from *Larrea.* Collatz et al. (1976) subjected *Eucalyptus socialis* to a series of drought cycles and determined the responses of N_L, r_l', and r_m'. Both r_m' and r_l' remained relatively constant between ψ_L of −3 and −10 or −15 bars, but at lower ψ_L values, both increased markedly. During the first stress cycle, r_l' increased much more than r_m', but in the second and subsequent cycles, r_l' and r_m' exhibited similar increases. Surprisingly, the long-term stress cycle treatments, which resulted in substantial leaf morphological changes, did not further alter the responses. While N_L was somewhat higher at ψ_L values between −15 and −25 bars in the second cycle as compared with the first, this was almost entirely due to the altered stomatal response; no differences were apparent in the effect of ψ_L on r_m'. After rewatering a rapid recovery to essentially prestress r_m' and r_l' values occurred.

Eucalyptus differs from the crop species discussed above in that increased r_m' represents a much larger fraction of the total limitation to CO_2 uptake at decreased ψ_L, and in the lack of an apparent aftereffect of water stress. Possibly, a greater intrinsic capacity for recovery of the photosynthetic apparatus after stress in this species may mean that stomatal closure to prevent excessively low ψ_L is less important. Consequently, stomata can remain more open, allowing for higher rates of CO_2 uptake. This adaptive mechanism may also be operative in *Simmondsia chinensis,* an evergreen desert shrub of the southwestern United States, in which r_m' and r_i' show parallel and quantitatively similar increases with decreased ψ_L (Collatz, 1977).

IV. WATER STRESS AND PHOTOSYNTHETIC PARTIAL PROCESSES

A parallel approach to determination of mesophyll resistances involves investigation of water-stress effects on the components of the photosynthetic apparatus. Most generally, chloroplasts or photosynthetic enzymes are isolated from desiccated and nondesiccated tissue, or chloroplasts, enzymes, etc., are treated after isolation with osmotica to give different water potentials. This type of approach has been extensively reviewed by Boyer (1976), so only the salient features will be discussed here (see also Chapter 7B).

One point that must be made early in this discussion is that only certain types of lesions can be detected in this manner. Because the in vitro activities of enzymes isolated from stressed and unstressed tissue are the same, it does not follow that the in vivo activities were unaffected by the water deficit. Any readily reversible inhibition would not be detected. Similarly, any differences detected in vitro might result from the fact that the isolation was from very different sources—stressed and unstressed tissue—rather than from an in vivo difference. Addition of an osmotica lowers water potential but does not duplicate other changes in the cellular environment during desiccation. These constraints pose real limitations on the interpretations of stress effects on subcellular processes and their relationships to whole leaf photosynthetic responses.

Despite these limitations, there is good evidence accumulating that, depending on species and conditions, water stress can have a considerable impact on photosynthetic partial processes. Photosystem II (PSII) and photophosphorylation appear to be particularly sensitive steps in some species. Boyer and his colleagues (Boyer & Bowen, 1970; Keck & Boyer, 1974) found that chloroplasts isolated from stressed sunflower leaves had reduced activities for both of these reactions, while Photosystem I (PSI) was somewhat less inhibited. Both the onset and magnitudes of inhibition of PSII correlated closely with inhibition of CO_2 uptake by water stress at both high- and low-light levels. Later, a good correlation was shown between the inhibition by water stress of the quantum yields for CO_2 uptake and the inhibition of PSII (Mohanty & Boyer, 1976). Reduced quantum yields at low ψ_L have also been found in *Larrea divaricata* grown under growth-chamber conditions. However, quantum yields were nearly constant over a wide range of ψ_L under field conditions (Mooney et al., 1977), further illustrating the rather remarkable capacity of this species to acclimate to water stress.

There may, however, be considerable species differences in inhibition of electron transport and/or photophosphorylation and in their relationship to inhibition of r'_m. Beadle and Jarvis (1977) could find no inhibition of PSI or PSII electron transport in chloroplasts isolated from sitka spruce needles exhibiting a wide range of water potentials, although there was an apparent inhibition in r'_m.

Studies of water-stress effects on enzymes associated with dark CO_2 fixation pathways have almost all been concerned primarily with ribulose-1,5-bisphosphate carboxylase (RuBPCase). Extractable activities of RuBPCase are frequently reduced in water-stressed leaves (Jones, 1973b; O'Toole et al., 1976; Johnson et al., 1974). Attempts to correlate RuBPCase activities to increases in mesophyll resistance, however, have led to somewhat ambiguous results. For beans, the decline in RuBPCase seems to be greater than the increase in r'_m (O'Toole et al., 1976). In sitka spruce, on the other hand, RuBPCase activities remained constant over a wide range of ψ_L, while r'_m apparently decreased (Beadle & Jarvis, 1977).

While RuBPCase has a central role in the photosynthetic C fixation pathways and may be an important determinant of photosynthetic capacity in some plants, there is no a priori reason to believe that it is necessarily the most sensitive step under water-stress conditions. Other enzymes, notably carbonic anhydrase (Jones, 1973b), ribulose-5-phosphate kinase, and phospho(enol) pyruvate carboxylase, also decrease with increasing water stress (Huffaker et al., 1970). To date, however, effects of water deficit on other photosynthetic enzymes have not been studied. Particularly interesting to examine would be fructose-1,6-diphosphatase, since it has been suggested as a possible limiting step in CO_2 fixation in C_3 species (Björkman & Badger, 1977).

V. MESOPHYLL RESISTANCE AND WATER-USE EFFICIENCY

All other conditions being equal, increases in r'_m will always increase the transpiration ratio (TR) of a leaf (g H_2O mg^{-1} CO_2) because net photosynthesis decreases while transpiration remains unchanged. Thus, in this context, an r'_m that is insensitive to water stress will clearly be advantageous. In contrast, increases in r_l have much more complex effects and can lead to either increases or decreases in the transpiration ratio because of induced changes in the leaf energy balance, and hence leaf temperature, as well as on transpiration and photosynthesis. The effects of changes in r_l on TR are discussed in detail in Chapter 1A of this volume and elsewhere (Cowan, 1977; Fischer & Turner, 1978). From these treatments, it is apparent that TR can either increase or decrease depending on the relative effects of changes in r_l and r'_m.

Measurements indeed show a variety of changes in TR with increasing stress. Data for barley and wheat show decreased TR as ψ_L declines (Johnson et al., 1974). Any advantage accrued is questionable, however, since N_L was greatly restricted at the low ψ_L levels. In contrast, TR increased in soybeans during a drought cycle because of increased leaf temperatures (Rawson et al., 1978). Increased TR during drying cycles seen in corn under field conditions by Sinclair et al. (1975) and in sorghum by Jones and Rawson (1979) may also be due, at least in part, to increased r'_m. In the latter, TR remained constant or increased only slightly if stress developed slowly, while the increase was much more dramatic under rapid stress development. This pattern is essentially consistent with the apparent changes in r'_m in these plants.

VI. SUMMARY AND FUTURE RESEARCH NEEDS

I have presented considerable evidence to support an influence of water stress on the photosynthetic apparatus of higher plants. While at least in rapidly stressed plants lower water potentials are required to inhibit the nonstomatal components of CO_2 uptake than to cause stomatal closure, the former can be, under some circumstances, important in limiting CO_2 uptake and water-use efficiency. Nonstomatal inhibition may be an important factor, especially when stress develops slowly, as would usually occur under field conditions.

Despite the work to date, our understanding of the nature and significance of nonstomatal inhibition is very limited. Conclusions concerning increases in r'_m during drought stress must be tempered by the fact that nearly all studies, especially those with crop species, have dealt with short-term rapidly induced stress. The rather marked differences in the responses of field- and laboratory-grown *Larrea* plants to water stress found by Mooney et al. (1977) dramatically illustrate the effects of preconditioning and rates of stress development on nonstomatal inhibition. A role for osmotic adjustment has been suggested in these differences, but despite the considerable interest in its function in stomatal behavior (Brown et al., 1976; Fereres et al., 1978), there have been no intensive studies of its role in altering the nonstomatal inhibition of photosynthesis. Similarly, little attention has been given to field photosynthetic responses of crops under drought conditions. Rawson et al. (1978) have reported a marked diurnal increase in r'_m up to values that considerably exceeded r'_l in water-stressed soybeans under field conditions, but indicate that midday cloudiness may be partly responsible. Their studies illustrate not only the difficulties associated with determining nonstomatal inhibition under field conditions, but also the need for further studies to clarify its role in limiting crop productivity and water-use efficiency.

At the chloroplast and biochemical levels, water stress has been shown to influence several component reactions, notably, electron transport, photophosphorylation and RuBPCase. The evidence for a role for the inhibition of the light-harvesting reactions in limiting CO_2 uptake at least at low light is convincing because of the parallel decrease in quantum yields. Moreover, there is apparently considerable variation in the sensitivity of chloroplast electron transport and quantum yield to water stress among species. As with the nonstomatal effects on CO_2 exchange, an important area for future research lies in determining the nature of these preconditioning effects on the sensitivity of the chloroplast partial reactions to water stress.

LITERATURE CITED

Ackerson, R. C., D. R. Krieg, C. L. Haring, and N. Chang. 1977. Effects of plant water status on stomatal activity, photosynthesis, and nitrate reductase activity of field grown cotton. Crop Sci. 17:81–84.

Beadle, C. L., and P. G. Jarvis. 1977. The effects of shoot water status on some photosynthetic partial processes in sitka spruce. Physiol. Plant. 41:7–13.

Björkman, O., and M. Badger. 1977. Thermal stability of photosynthetic enzymes in heat- and cool-adapted C_4 species. Carnegie Inst. Washington Yearb. 76:346–354.

Boyer, J. S. 1970. Differing sensitivity of photosynthesis to low leaf water potentials in corn and soybean. Plant Physiol. 46:236–239.

Boyer, J. S. 1976. Water deficits and photosynthesis. p. 153–190. *In* T. T. Kozlowski (ed.) Water deficits and plant growth IV. Academic Press, New York.

Boyer, J. S., and B. L. Bowen. 1970. Inhibition of oxygen evolution in chloroplasts isolated from leaves with low water potentials. Plant Physiol. 45:612–615.

Brown, K. W., W. R. Jordan, and J. C. Thomas. 1976. Water stress induced alterations of the stomatal response to decreases in leaf water potential. Physiol. Plant. 37:1–5.

Collatz, G. J. 1977. Influence of certain environmental factors on photosynthesis and photorespiration in *Simmondsia chinensis.* Planta 134:127–132.

Collatz, J., P. J. Ferrar, and R. O. Slatyer. 1976. Effects of water stress and differential hardening treatments on photosynthetic characteritics of a xeromorphic shrub *Eucalyptus socialis*, F. Muell. Oecologia 23:95–105.

Cowan, I. R. 1977. Stomatal behavior and environment. Adv. Bot. Res. 4:117–228.

Doley, D., and N. B. A. Trivett. 1974. Effects of low water potentials on transpiration and photosynthesis in Mitchell Grass (*Astrebla lappacea*). Aust. J. Plant Physiol. 1:539–550.

Fereres, E., E. Acevedo, D. W. Henderson, and T. C. Hsiao. 1978. Seasonal changes in water potential and turgor maintenance in sorghum and maize under water stress. Physiol. Plant. 44:261–267.

Fischer, R. A., and N. C. Turner. 1978. Plant productivity in the arid and semiarid zones. Annu. Rev. Plant Physiol. 29:277–317.

Gaastra, P. 1959. Photosynthesis of crop plants as influenced by light, carbon dioxide, temprature, and stomatal diffusion resistance. Meded. Landbouwhogesch. Wageningen 59:1–68.

Huffaker, R. C., T. Radin, G. E. Kleinkopf, and E. L. Cox. 1970. Effects of mild water stress on enzymes of nitrate assimilation and enzymes of the carboxylative phase of photosynthesis in barley. Crop Sci. 10:471–474.

Jarvis, P. G. 1971. The estimation of resistances to carbon dioxide transfer. p. 566–632. *In* Z. Sestak et al. (ed.) Plant photosynthetic production, manual of methods. W. Junk N.V., The Hague, Netherlands.

Johnson, R. R., N. M. Frey, and D. N. Moss. 1974. Effect of water stress on photosynthesis and transpiration of flag leaves and spikes of barley and wheat. Crop Sci. 14:728–731.

Jones, H. G. 1973a. Gas exchange in plant leaves having different transfer resistances through their two surfaces. Aust. J. Biol. Sci. 26:1045–1055.

Jones, H. G. 1973b. Moderate term water stresses and associated changes in some photosynthetic parameters in cotton. New Phytol. 72:1095–1105.

Jones, H. G. 1976. Crop characteristics and the ratio between assimilation and transpiration. J. Appl. Ecol. 13:605–622.

Jones, M. M., and H. M. Rawson. 1979. Influence of rate of development of leaf water deficit upon photosynthesis, leaf conductance, water use efficiency, and osmotic potentil in sorghum. Physiol. Plant. 45:103–111.

Keck, R. W., and J. S. Boyer. 1974. Chloroplast response to low leaf water potentials. III. Differing inhibition of electron transport and photophosphorylation. Plant Physiol. 53: 474–479.

Koller, D. 1970. The partitioning of resistances to photosynthetic CO_2 uptake in the leaf. New Phytol. 69:971-981.

Mederski, H. J., L. H. Chen, and R. B. Curry. 1975. Effect of leaf water deficit on stomatal and nonstomatal regulation of net carbon dioxide assimilation. Plant Physiol. 55:589-593.

Mohanty, P., and J. S. Boyer. 1976. Chloroplast response to low leaf water potentials. IV. Quantum yield is reduced. Plant Physiol. 57:704-709.

Moldau, H. 1973. Effects of various water regimes on stomatal and mesophyll conductances of bean leaves. Photosynthetica 7:1-7.

Mooney, H. A., O. Björkman, and G. J. Collatz. 1977. Photosynthetic acclimation to temperature and water stress in the desert shrub, *Larrea divaricata*. Carnegie Inst. Washington Yearb. 76:328-335.

O'Toole, J. C., R. K. Crookston, K. J. Treharne, and J. L. Ozbun. 1976. Mesophyll resistance and carboxylase activity. A comparison under water stress conditions. Plant Physiol. 57: 465-468.

Radin, J. W., and R. C. Ackerson. 1981. Water relations of cotton plants under nitrogen deficiency: III. Stomatal conductance, photosynthesis, and abscisic acid accumulation during drought. Plant Physiol. 67:115-119.

Rawson, H. M., N. C. Turner, and J. E. Begg. 1978. Agronomic and physiological responses of soybean crops to water deficits: IV. Photosynthesis, transpiration and water use efficiency of leaves. Aust. J. Plant Physiol. 5:195-209.

Redshaw, A. J., and H. Meidner. 1972. Effects of water stress on the resistance to uptake of carbon dioxide in tobacco. J. Exp. Bot. 23:229-240.

Sinclair, T. R., G. E. Bingham, E. R. Lemon, and L. H. Allen, Jr. 1975. Water use efficiency of field grown maize during moisture stress. Plant Physiol. 56:245-249.

Slatyer, R. O. 1973. Effects of short periods of water stress on leaf photosynthesis. p. 271-275. *In* R. O. Slatyer (ed.) Plant response to climatic factors. Proc. Symp., Uppsala, Sweden. 15-20 Sept. 1970. UNESCO, Paris.

Strain, B. R. 1970. Field measurements of tissue water potential and carbon dioxide exchange in the desert shrubs *Prosopus juliflora* and *Larrea divaricata*. Photosynthetica 4:118-122.

Troughton, J. H. 1969. Plant water status and carbon dioxide exchange of cotton leaves. Aust. J. Biol. Sci. 22:289-302.

Troughton, J. H., and R. O. Slatyer. 1969. Plant water status, leaf temperature and the calculated mesophyll resistance to carbon dioxide of cotton leaves. Aust. J. Biol. Sci. 22:815-827.

CHAPTER 7A

Whole Plant Response to Water Deficits: An Overview

W. R. JORDAN

Texas Agricultural Experiment Station, Blackland Research Center
Temple, Texas

I. INTRODUCTION

During the past decade we have seen a resurgence in research devoted to the enrichment of our understanding of plant responses to soil water deficits and to the management of limited water resources to achieve high yields. For convenience sake, the topics of plant response have been separated from overall crop management in this treatise (see Chapters 8–11), but it is acknowledged that plant responses to water deficits depend greatly upon management decisions made prior to and during each cropping season. These decisions may either supplement or detract from the genetic flexibility in growth and reproductive capacity of the crop, particularly when grown under rainfed conditions.

How limited soil water supply is transduced by the plant to alter development, activity, and duration of various "sources" and "sinks" is central to an understanding of productivity under dryland conditions. A great deal of literature on the topic is available from controlled environment studies, but there is a growing concern that these results do not transfer quantitatively to field conditions (Begg & Turner, 1976). This failure to transfer is believed to lie in major differences in radiation and/or severity, duration, and rate of development of plant water deficits. The gradual development of water deficits by plants growing in most field situations allows slow and continual adjustments in physiological processes, eventually manifested as alterations in growth and development.

This chapter presents views on whole-plant response to water deficits resulting from limited soil water availability. Perspectives gained from the recent contributions of many others, especially those of J. S. Boyer (Boyer, 1969; Boyer, 1976; Boyer & McPherson, 1975), T. C. Hsiao (Hsiao, 1973; Hsiao et al., 1976a, 1976b; Hsiao & Acevedo, 1974), N. C. Turner (Turner, 1979; Begg & Turner, 1976; Fischer & Turner, 1978), and R. A. Fischer (Fischer, 1979; Fischer & Turner, 1978), are acknowledged. Effects of water deficit on C and N metabolism are discussed more fully in Chapters 7B and 7C.

Limitations to Efficient Water Use in Crop Production.

II. TOTAL DRY MATTER PRODUCTION

Results from field experiments with several very different crops suggest a common linear association between total dry matter production and total solar radiation intercepted during the growing season so long as environmental and cultural constraints are minimal (Monteith, 1977). Since total solar radiation during the growing season is a feature of each particular climate and differs little from the long-term average, light interception by the crop depends primarily on the seasonal distribution of the leaf area index (LAI). Early in the growing season energy capture in photosynthesis is limited by low LAI's; light interception increases rapidly to about 95% as LAI increases to about 4 for cereal crops (Evans et al., 1975), but for row crops in general, row spacing, plant density, and plant geometry interact to determine the precise relation of LAI to light interception (Arkin et al., 1978). After effective canopy closure, photosynthesis is no longer dependent upon further increases in LAI, but is most closely correlated with total photosynthetically active radiation (PAR).

Total dry matter production in rainfed areas is usually below the maximum for irrigated crops in the same climate. There are many possible explanations for these yield depressions, but the example cited by Monteith (1977) provides a novel basis for exploring these possibilities in a general manner (Fig. 1). In this example, total dry matter (DM) is given as a linear function of total seasonal light interception (R) such that the efficiency of light utilization, l_e, which is simply DM/R in energy units, has a constant value. The minimum and maximum yields in this specific climate are given

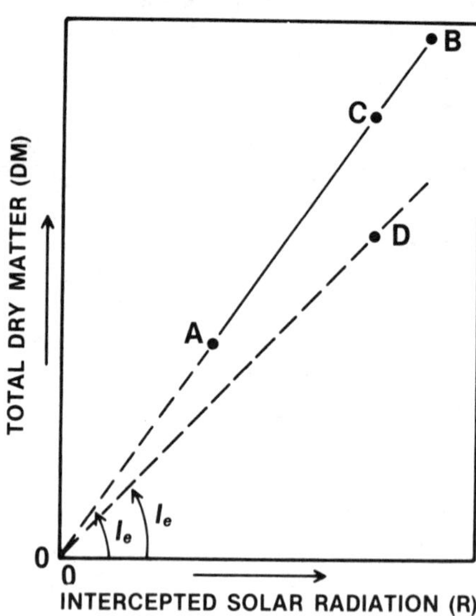

Fig. 1—Diagrammatic representation of the relation between total seasonal dry matter production (DM) and total seasonal intercepted solar radiation (R). Points A, B, C, and D represent specific cases discussed in the text; l_e and l_e' are photosynthetic efficiencies. Modified from Monteith (1977).

by A and B, respectively. If, for example, DM and R for a specific crop were represented by point C, the crop used light with the maximum efficiency, but yield was below maximum because of a failure to intercept sufficient solar radiation. If, on the other hand, crop performance yielded point D, reduced yield would be attributed to both reduced light interception (B–C) and reduced photosynthetic efficiency, $l_{e'}$ (Legg et al., 1979).

One could anticipate that this simple analysis in terms of light interception and efficiency factors could well be valid for crops where total dry matter constitutes the economic yield, but this is not the case with most crops of agronomic importance. While yield of harvestable products may or may not be strongly correlated with total dry matter production, the importance of leaf area development to yield cannot be denied. Reduced efficiency as defined in Fig. 1 may result from several factors ranging from stomatal closure to failures in reproduction. The remainder of this chapter attempts to place these and other water-related limitations to productivity into perspective in terms of whole-plant performance in natural environments.

III. STAND ESTABLISHMENT

The introduction of seed into soil is the first crucial step in establishing the life-long union of a crop with its aqueous environment. The brief period during which germination and emergence occurs is especially critical in arid and semiarid regions where water availability within the seedbed may change dramatically within a single day (Reginato, 1975; Chapter 2B). Transient decreases in soil water availability may reduce both rate and ultimate degree of germination and emergence, leading to uneven or spotty stands. Several aspects of the overall influence of soil water availability on stand establishment are examined in the following sections.

A. Soil Water Potential and Seed Hydration

The dynamic nature of water availability within the seedbed system is illustrated in Fig. 2. Water absorbed during hydration (imbibition) is supplied by a relatively small sphere of soil (of radius, r) surrounding the seed (Dasberg, 1971). Available theory suggests that water transmission through the soil should not limit the rate of seed hydration so long as ψ_{soil} exceeds about -15 bars (Hillel, 1972), since the soil water diffusivity is one to two orders of magnitude greater than that of seed (Doering, 1965; Philip, 1957; Phillips, 1968). Under actual conditions, however, seeds hydrate more slowly than predicted as a result of reduced seed-soil contact, low seedcoat permeability, and internal resistance of the seed to water uptake (Dasberg & Mendel, 1971; Hadas, 1969; Hadas, 1976; Hadas & Russo, 1974a, 1974b; Phillips, 1968). Actual ψ_{soil} at seed depth, d, depends on evaporation rate and on water transmission characteristics of the soil. Up-

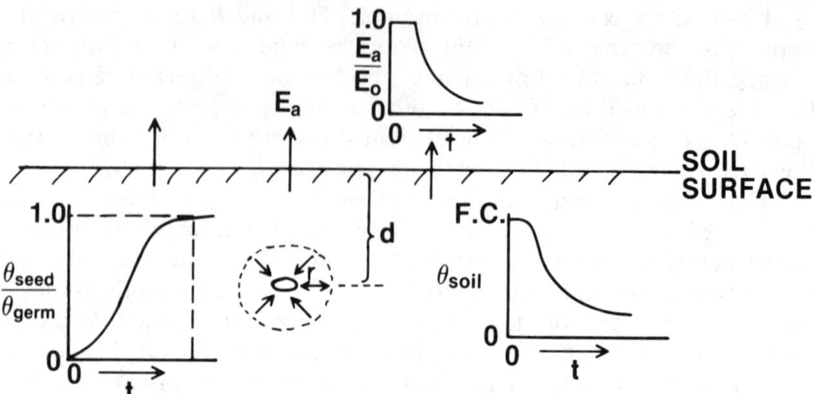

Fig. 2—Diagrammatic representation of important aspects of the seed-seedbed system with regard to water supply during germination. Depicted are general trends for water loss from the surface relative to potential evaporation (E_a/E_o), seed hydration relative to that necessary for germination ($\theta_{seed}/\theta_{germ}$), and soil water content (θ_{soil}) at planting depth (d), all as a function of time. Water absorbed during germination is supplied by a relatively small sphere of soil with radius, r.

ward flux from the surface, E_a, depends on potential evaporation, E_o, and on volumetric water content, θ_{soil}, of the surface layer. Typical trends in E_a/E_o are shown as a function of time, as are θ_{soil} at seed depth and seed water content, θ_{seed}, relative to that which allows germination, θ_{germ}.

B. Water Requirements in Germination

Germination appears to require the uptake of a certain minimum amount of water (θ_{germ} in Fig. 2), suggesting that some minimum seed water potential (ψ_{seed}) must be exceeded for a period of time. This minimum ψ_{seed} is usually inferred from equilibrium systems where the water potential of the germination medium is lowered by matric forces or by addition of an osmoticum. The use of soil in these equilibrium systems is usually desirable from an agronomic view, but presents special problems. Equilibrium soil systems are frequently nonsterile and seeds may deteriorate before germination, especially at low ψ_{soil} where rates of water transmission are very low. Values derived from these studies must, of necessity, have a finite time component, which, if not sufficiently long, leads to an overestimation of the minimum ψ_{seed}. Table 1 shows typical results derived from soil systems with several species. For the most part, these represent practical rather than absolute values for the minimum ψ_{seed} for germination.

Minimum ψ_{seed} determined from osmotic systems may differ greatly from those determined in matric systems (Collis-George & Sands, 1962; Hillel, 1972). Osmotica that penetrate seeds may reduce the effective Ψ difference between seed and medium leading to a serious underestimate of the minimum ψ_{seed} necessary for germination. Also, specific ion toxicity may lead to gross overestimates of the minimum ψ_{seed}. Germination tests

Table 1—Minimum seed water potentials (ψ_{seed}) for germination of seeds of several species in soil.

Species	ψ_{seed}†	Comments and reference
	bars	
Corn	−12.7	
Rice	− 8.0	25 °C, four soil types
Soybean	− 6.7	Hunter & Erickson (1952)
Sugar beet	− 3.5	
Wheat	< −20	Lindstrom et al. (1976); Owen (1952)
Grain sorghum	−10	Lyles & Fanning (1964)
Cotton	< −10	Wanjura & Buxton (1972a)
Crested wheatgrass	−3.5 to −5.5	Johnson & Asay (1978)
Barley	< −30	Sepaskhah & Ardekani (1978)
Perennial ryegrass	< − 9	
Abruzzi rye	< − 9	
Annual ryegrass	−3 to −6	21 °C, 8 days
Kentucky 31 fescue	−3 to −6	Wright et al. (1978)
Kentucky bluegrass	−3 to −6	
Crownvetch	−1/3 to −3	

† First seven values are taken as equivalent to soil matric potential at which 50% germination occurred; remaining values are assumed to be equal to the soil matrix potential at which 50% emergence occurred.

with high-molecular-weight, nonpenetrating osmotica appear to produce results more similar to those with soil than do a variety of salts and lower-molecular-weight osmotica. For the present, osmotic evaluations of the minimum ψ_{seed} for germination appear best suited for situations where rapid evaluation of genetic stocks is required.

Failure of the seedbed to maintain ψ_{soil} above the minimum ψ_{seed} for germination often results in serious stand failures. Some seeds are capable of undergoing several hydration-desiccation cycles without appreciable loss of viability, provided pathogens do not pose a problem. Seeds pass from an insensitive to a sensitive stage with respect to desiccation about the time cell division and vacuolation begins in the embryo (Bewley, 1979). Desiccation during the insensitive stage results in a reversible loss of polysomes, which re-form on subsequent hydration. Specific sites of damage from desiccation during the sensitive period are not well-understood, but since each species seems to have a unique, minimum ψ_{seed} for germination, genetic variability may exist for desiccation tolerance at these sites.

Many claims have been made for beneficial effects of presowing "hardening" treatments wherein seeds are taken through one or more wetting-drying cycles under controlled conditions (Bewley, 1979; Hegarty, 1978). Seeds treated in this manner germinate more rapidly and uniformly than untreated seeds, suggesting some initial germination processes, once completed, need not be repeated. Although there seems to be little evidence for other substantial long-term growth advantages resulting from these treatments, there may be potentially important benefits of rapid, synchronous germination for species that must be planted at shallow depths, in cold soils, or in poorly prepared seedbeds (Akalehiywot & Bewley, 1977; Hegarty, 1978).

C. Emergence

Even though seeds may be capable of germination, emergence is slow and final emergence percentage is usually reduced at ψ_{soil} below a few bars (Table 1; Doneen & McGillivray, 1943). In addition to the dynamic nature of the ψ_{soil} at planting depth during germination (Fig. 2), other direct and indirect effects of soil water availability to the young seedling greatly affect emergence.

1. GROWTH OF THE PRE-EMERGED SEEDLING

Seedling growth before emergence occurs in such a manner so as to provide access to the more freely available water at greater soil depths. The water potential of the seedling before emergence should reflect that of the wettest region of soil in contact with the radicle, not the ψ_{soil} at planting depth. Furthermore, regulatory systems within the young seedling are capable of altering the flow of seed reserves among expanding organs such that roots grow at higher relative rates than shoots as ψ_{soil} falls (Hadas & Stibbe, 1977; Sharpe & Davies, 1979; Wanjura & Buxton, 1972b).

While results from equilibrium soil systems may provide information on the lower limit of ψ_{soil} for emergence (Table 1), consideration of changes in θ_{soil} (or ψ_{soil}) with depth and time are more appropriate for actual field situations. Examples of this approach are provided by Wiegand (1962) and by H. Gardner (Chapter 2B).

2. SURFACE CRUSTS AND HIGH-STRENGTH SUBSURFACE LAYERS

Upon drying, many soils form surface crusts of high strength that may impede emergence. The strength of soil crusts is determined by a complex interaction of many physical and chemical factors including soil texture, soil structure, organic matter content, and water content (Taylor, 1974). Emergence through crusts requires that the seedling "find" a fissure or be capable of generating enough "growth pressure" (Gill & Bolt, 1955) in the shoot to fracture the crust.

Soil moisture not only influences the strength of the crust (Hadas & Stibbe, 1977), but also the ability of the seedling to generate growth pressure. By isolating the crust from underlying soil in which seeds were planted, Hanks and Thorpe (1957) found that the crust strength required to reduce emergence of wheat, sorghum, and soybeans by 50% fell three- to eightfold as available soil water was reduced to 0.25 of field capacity.

The mechanism through which an external impedance is translated into a growth response, at least for epicotyls or hypocotyls of dicotyledonous species, appears to be mediated by phytohormones, especially ethylene. Goeschl et al. (1966) found that a sharp increase in ethylene evolution and radial expansion occurred when growth of pea epicotyls was mechanically impeded. The increase in endogenous ethylene (C_2H_4) production and epicotyl diameter were proportional to the degree of impedance, thereby suggesting a sensitive mechanism to modify the emergence force (growth pressure) to prevailing soil conditions. The extent to which this association between ethylene production and mechanical stress holds for other species is not known, but exogenous ethylene causes swelling of epicotyl and hypocotyl tissues of a number of species (Goeschl & Kays, 1975). Similar swelling of soybean (Knittle & Burris, 1979) and cotton (Garner & Bowen, 1966) hypocotyls occurred during emergence through compacted soil. No comparable mechanism is known to exist for monocotyledonous species. Maximum growth pressures generated by members of the *Poaceae* appear quite similar. Calculations by Hadas and Stibbe (1977) placed the maximum growth pressure of wheat coleoptiles in the range of 9–14 bars. Earlier work (Parker & Taylor, 1965; Taylor et al., 1966) provided evidence that emergence of several *gramineous* species was reduced by 50% by soil strengths between 6 and 12 bars, with no emergence at 18 bars, as measured with a soil penetrometer.

The influence of soil strength on crop performance and crop-water relations is not restricted to limitations to emergence imposed by crusts. Root growth (Eavis et al., 1969; Taylor & Gardner, 1963; Taylor & Ratliff, 1969) and radial stem growth (Mathers & Welch, 1964; Taylor et al., 1964) may be restricted by high-strength layers either above or below planting depth. Inhibition of either axial or radial root growth or stem-diameter increase could produce local regions of high resistance to water flow and thereby contribute to development of low plant water potentials when evaporative demand is high.

IV. MODULATION OF LEAF AREA DEVELOPMENT BY PLANT WATER DEFICIT

In the simplest analysis, growth, expressed as dry matter accumulation rates, reduces to the difference between rates of production and utilization of photosynthate. Effects of water deficits on photosynthate production are well-known, but relatively little is known about effects of water stress on utilization of photosynthate in growth and maintenance of existing biomass.

A. Photosynthate Production

The initial reduction in CO_2-uptake rates during water stress usually parallels a reduction in stomatal conductance resulting from changes in turgor of guard and/or subsidiary cells. Current evidence suggests there is no unique ψ_L at which stomatal closure occurs in any species or variety; the critical ψ_L for stomatal closure varies with leaf position (Jordan et al., 1975; Kanemasu et al., 1974; Millar & Denmead, 1976; Turner, 1974a, 1974b), plant or leaf age (Frank et al., 1973; Thomas et al., 1976), and growth conditions, especially the history of water stress during growth (Begg & Turner, 1976; Brown et al., 1976; Cutler & Rains, 1977; McCree, 1974). However, in each case studied closely, stomatal closure occurred within a range of turgors between 0 and 3 bars (Turner, 1974b), suggesting the importance of the osmotic potential of leaves in fixing threshold or critical values of ψ_L for stomatal closure.

Nonstomatal inhibition of CO_2 uptake may occur concomitant with stomatal closure (Boyer, 1976), but more often nonstomatal inhibition becomes significant only after severe or prolonged stress (Begg & Turner, 1976). The precise nature of nonstomatal inhibition of photosynthesis is not known, but may involve basic biochemical steps at the chloroplast level (Boyer, 1976). A more complete discussion of nonstomatal inhibition of photosynthesis is presented in Chapter 6C.

B. Photosynthate Utilization

Water deficits may affect photosynthate utilization by altering either the efficiency with which photosynthate is converted to new growth, or the rate of photosynthate use in maintenance of existing dry matter. Effects of water deficits on respiration rates of various plants species have been summarized recently by Boyer (1976), with the conclusion that the ultimate effect of water stress was to decrease both light and dark components of respiration. A recent study by Wilson et al. (1980) further separated substrate utilization into both growth and maintenance components based on whole-plant carbon balance measurements of young sorghum plants (Fig. 3). As water stress became more severe, rates of carbon utilization for both growth and maintenance fell, but the growth component was more severely affected. The decline in growth component resulted from reduced photosynthate production during stress, while the growth conversion efficiency (units of new dry matter carbon formed per unit substrate carbon consumed) remained constant at 0.70. Factors contributing to the decline in maintenance respiration are not clear, but possibly result from alterations in the general metabolic activity of cells as the plant is forced to adapt to a stressful environment (Penning de Vries, 1975). Organic solutes do accumulate during adaptation to stress (Acevedo et al., 1979; Cutler & Rains, 1977; Jones & Turner, 1978), but the consequences in terms of maintenance respiration are unknown. In this specific example with sorghum, the loss in active biomass due to leaf senescence may have been the most influencial factor in reduced maintenance requirements.

C. Development, Display, and Maintenance of Green Leaf Area

The production of photosynthetically active leaf area by field crops is one of the most important factors affecting crop productivity (Legg et al., 1979; Monteith, 1977; Yoshida, 1972; Fig. 1), but it is also the component of growth that is perhaps the most sensitive to water stress (Boyer, 1970;

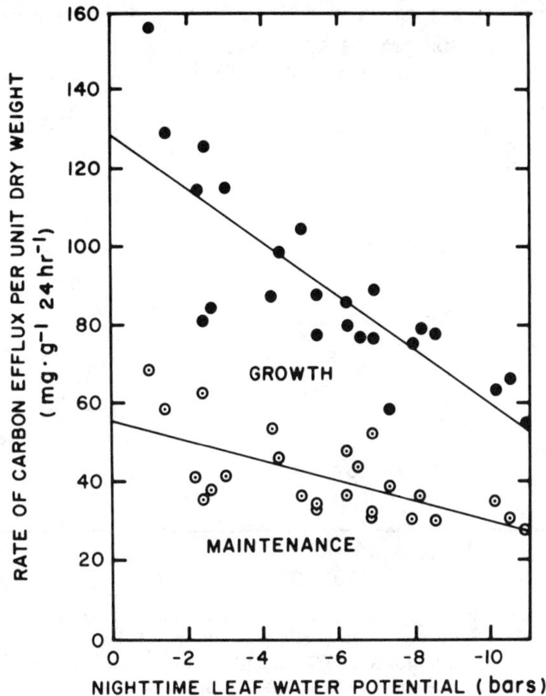

Fig. 3—Daily rates of carbon efflux per unit dry weight as a function of mean nighttime leaf water potential for young sorghum plants. The upper curve is the total efflux, the lower curve is the maintenance component of the efflux, and the difference between the two curves represents the growth component. Redrawn from Wilson et al. (1980).

Hsiao, 1973). The reduced rate of leaf area accumulation usually associated with growth in dryland environments may be associated with a smaller size of individual leaves or with production of fewer leaves. Additionally, plant water deficits may alter light interception through effects on display and duration of green leaf area. Each of these topics will be considered separately in the following sections, but the overall effects of one form of limitation on the others must be recognized. To simplify discussion, only growth alterations where water deficits develop slowly over a period of many days will be considered, unless otherwise indicated.

1. LEAF EXPANSION AND LEAF AREA INDEX DEVELOPMENT

Once a leaf primordium is initiated, the final size attained is determined by the rates and duration of cell division and expansion. The extreme sensitivity of leaf expansion to decreases in ψ_L is well-known (Acevedo et al., 1971; Boyer, 1968, 1970; Hsiao, 1973). At least some reports place the sensitivity of cell division equal to that for cell enlargement (McCree & Davis, 1974; Meyer & Boyer, 1972), while others suggest cell division may continue at water potentials that reduce or prevent enlargement (Hsiao, 1973; Slatyer, 1974).

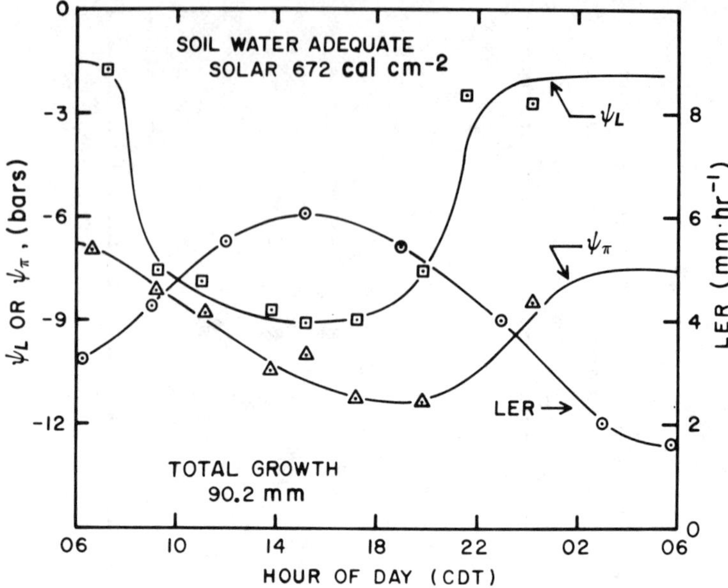

Fig. 4—Diurnal changes in ψ_L, ψ_π, and leaf elongation rate (LER) of sorghum growing on Houston Black Clay soil (fine, montmorillonitic, thermic Udic Pellusterts) with adequate soil water. Total solar radiation was 672 cal cm^{-2} and total leaf extension for the 24-hour period was 90.2 mm. Data points for ψ_L and ψ_π are means of three-six observations on similar plants. (Unpublished data of W. R. Jordan and G. F. Arkin, Texas Agric. Exp. Stn., Temple, Tex.)

In controlled environments, rates of leaf enlargement may be reduced by 75% or more as ψ_L falls to -4 or -6 bars (Boyer, 1968, 1970; Hsiao et al., 1976a, 1976b). However, ψ_L of well-watered crops growing in a field environment may fall to -10 to -18 bars, yet these crops maintain high growth rates throughout the day. This apparent discrepancy between controlled environment and field studies can best be explained through a closer examination of the daily growth pattern for leaves of field-grown plants. Growth rate and total water and component potentials of the 12th leaf of 'Tophand' hybrid grain sorghum are illustrated in Fig. 4 for field-grown plants with adequate water. Under well-watered conditions, ψ_L was normally depressed to about -9 to -10 bars at midday for plants of this age, with slightly lower midday values attained later in the season. Parallel, but not completely offsetting, changes also occur in osmotic potential (ψ_π) such that positive turgor is maintained throughout the day although reduced to only 1 to 2 bars near solar noon. General aspects of this pattern are similar to those for sorghum and corn grown at Davis, Calif., but complete turgor maintenance reported for sorghum was not found in our studies (Acevedo et al., 1979; Fereres et al., 1978). Leaf extension rates, which were initially low at sunrise, accelerated to reach a peak of 6 mm h^{-1} or more near 1500 h when turgor was ≤ 2 bars. Similar data from other leaves under a variety of field conditions support the finding that maximum growth rates occur near midday when turgor is typically 1 or 2 bars. The findings suggest that the minimum turgor needed for expansion of sorghum leaves is very near 0 bars, and a turgor of 1 to 2 bars appears sufficient to sustain high leaf growth rates. Growth at night may be limited more by temperature than by turgor (Acevedo et al., 1979; Loomis, 1934; Watts, 1974; Wenkert et al., 1978), although possible limitations due to photosynthate supply (Christ, 1978) and phytohormone availability at night cannot be discounted.

The diurnal growth pattern illustrated in Fig. 4 is in agreement with data of others for sorghum, corn, sudax, and soybean (Acevedo et al., 1979; Chu & Kerr, 1977; Johnson, 1967; Wenkert et al., 1978). However, for potatoes, the diurnal growth pattern may be inverted so that leaf expansion occurs mostly at night (Gandar & Tanner, 1976). Apparently potatoes, or at least the variety studied, are unable to actively accumulate solutes to the extent necessary to maintain positive turgors as ψ_L is depressed during the day. Similarly, the extreme sensitivity of chamber- and greenhouse-grown plants (Acevedo et al., 1971; Boyer, 1968, 1970; Jordan, 1970) to decreases in ψ_L suggests that conditions during growth or the rate at which stress is imposed precluded solute accumulation necessary to maintain turgor. As a whole, these and other findings suggest an important role for osmoregulation in maintenance of high growth rates as ψ_L falls either diurnally or as soil water availability is reduced (see discussion by Radin, Chapter 6B).

A progressive reduction of available soil water is usually accompanied by a gradual depression in the rate of leaf area development. Leaf area accumulation for sorghum grown on stored soil water is compared with that for an irrigated crop in Fig. 5. Leaf area accumulation rates were depressed

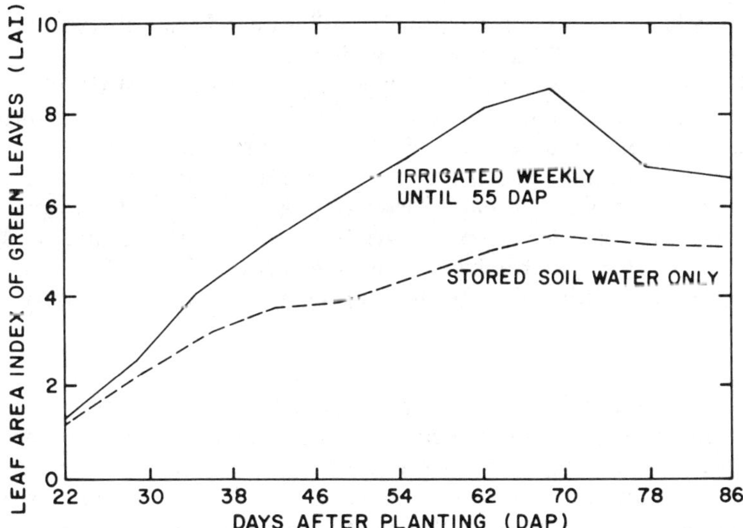

Fig. 5—Leaf area accumulation of a grain sorghum crop growing only on stored soil water compared with that of an irrigated crop. The last irrigation was applied 55 days after planting (DAP). Both treatments were covered by a rain shelter. (J. T. Ritchie, R. G. C. Smith, J. E. Begg, and W. E. Lonkerd, 1978. Blackland Research Center, Temple, Tex. Unpublished data.)

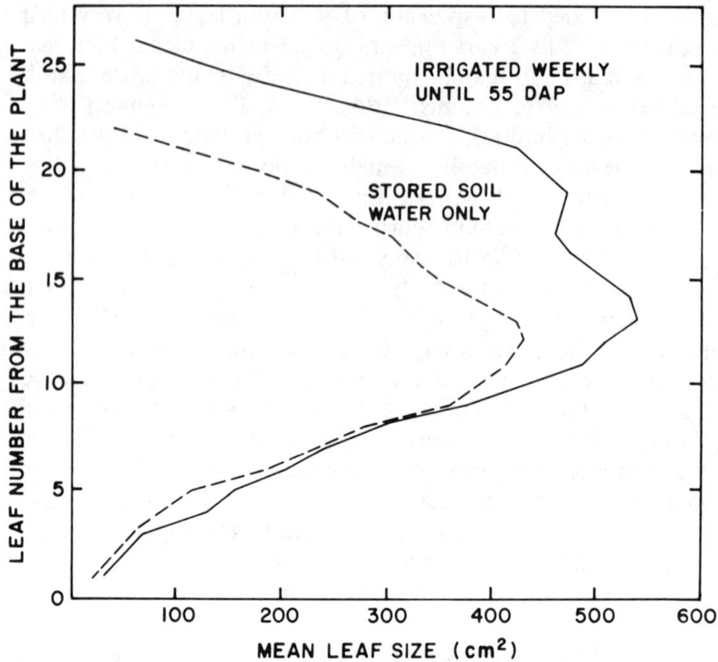

Fig. 6—Vertical distribution of leaf area for sorghum plants grown on stored soil water or irrigated weekly until 55 days after planting (DAP). Irrigated plants formed an average of four more leaves before heading. (J. T. Ritchie, R. G. C. Smith, J. E. Begg, and W. E. Lonkerd, 1978. Blackland Research Center, Temple, Tex. Unpublished data.)

very early even though total available soil water was not different between treatments until >30 days after planting (DAP). Midday ψ_L between wet and dry treatments differed by only about 0.5 bars at 38 DAP, illustrating that gradual development of stress allows close coordination of growth between evaporative surfaces and water uptake surfaces. The lower maximum green leaf area index (LAI) was due to several factors, but reduced size of individual leaves was a major factor (Fig. 6). Total root length per plant growing on stored water was greater from 38 to 60 DAP, and, combined with the smaller LAI, resulted in a much higher root length/LAI ratio commonly observed for plants growing with a limited supply of soil water.

2. LEAF DISPLAY

Changes in foliage display are common features of crops experiencing moderate to severe water deficits, however, little quantitative information exists regarding the consequent effects on light interception and water vapor and CO_2 transport. The more obvious changes can be grouped into two categories. The first is shape changes arising from an overall loss of turgor and not involving a specialized structure or mechanism. The typical wilting and drooping of many dicot leaves would fall into this category. When turgor is very low or zero, the tissue loses mechanical strength, the lamina folds, and the petiole bends downward resulting in a smaller projected area

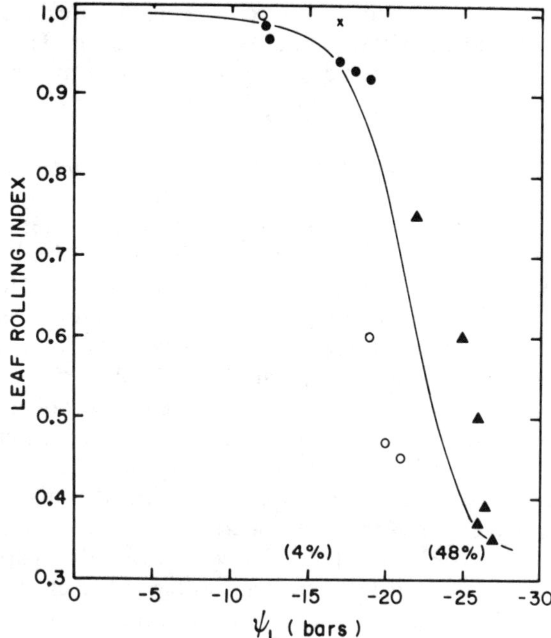

Fig. 7—Leaf-rolling response for a sorghum crop subjected to soil water stress. The leaf-rolling index is the projected planar area of leaves relative to well-watered unrolled leaves. Values in parentheses are percentage light transmission to the soil surface (● = irrigated plants; ▲ = stressed plants growing on stored soil water; ○ = stressed plants after re-irrigations). (J. T. Ritchie, R. G. C. Smith, J. E. Begg, and W. E. Lonkerd, 1978. Blackland Research Center, Temple, Tex. Unpublished data.)

on the horizontal plane for each leaf (Begg & Turner, 1976). If the LAI is not very high, a droopy canopy would intercept less radiation than a turgid one when the sun is overhead, leading to a smaller energy load and evaporative demand in the crop (Rawson, 1979). On the other hand, the amount of radiation intercepted early in the morning and late in the afternoon may not be much affected or even increased. Since the water status in a wilted leaf is most likely low enough to markedly reduce stomatal conductance and photosynthesis, it is improbable that substantial further decline in photosynthesis would result from the reduction in intercepted radiation at midday.

The second category of shape changes induced by water stress includes all of those underlain by movements apparently associated with specialized anatomic features. A well-known shape change is leaf-rolling in certain grass species. Rolling is presumed to be due to volume changes in the bulliform or motor cells (e.g., Esau, 1965; Duncan, 1975; Bull & Glasziou, 1975), commonly distributed longitudinally in rows along the adaxial side of the blade. Rolling was recognized as a means of reducing transpiration by A. Tschirch in the 19th century (Oppenheimer, 1960). More recently, rolling of leaves of thickspike wheatgrass was observed by Ripley and Redmann (1976) to correlate well with leaf water stress (as reflected in leaf ψ_π).

In the case of the sorghum study discussed in the previous section, the threshold ψ_L for leaf-rolling was about -17 bars (Fig. 7). The normally displayed canopy of the irrigated crop allowed only 4% of incident PAR to penetrate to the soil surface at midday, compared with about 48% for the dry plot.

Rice, especially the semidwarf cultivars, exhibits pronounced rolling. When under severe water stress, the leaves can be tightly rolled into needles. This also forces the blades to be held nearly vertical. A threshold ψ_L falling in the range of -9 to -12 bars must be reached before rolling is observed (O'Toole & Cruz, 1980). Rolling appears to start at about the same ψ_L as stomatal closure (O'Toole & Cruz, 1980), but contributes to reducing transpiration independently of stomatal closure. O'Toole et al. (1979) demonstrated a marked decline in transpiration when a well-watered leaf was rolled artificially by tying with a thread. Beneficial effects of rolling in terms of improving plant water balance are: (i) less radiation load on the crop due to more erect leaves and smaller exposed leaf area, and (ii) a reduction in transpiring surface because once rolled, the adaxial side of the leaf is in a protected, enclosed atmosphere that would inhibit water vapor escape. Boundary-layer conductance for water vapor (inverse of r_b) should be increased by tight rolling (estimated according to Nobel, 1974). Nonetheless, the product of boundary-layer conductance $\times$ effective leaf area should be reduced, accounting for a part of the reduction in transpiration.

Another shape change is known to be induced by water stress and apparently depends on specialized anatomic features. When stressed, leaflets of some legumes tend to orient parallel to the sun's rays (Dubetz, 1969; Begg, 1980; Shackel & Hall, 1979). Leaflets of well-watered plants show a tendency to be diaheliotropic (sun-tracking), whereas those of stressed plants are paraheliotropic (sun-avoiding) throughout the day. Only a very small reduction in ψ_L (about 1 bar) appears sufficient to initiate the response (T. C. Hsiao, personal communication). Judging by appearance, the reorientation of the lamina seems to be based on movement of pulvini at the base of the leaflets. Presumably, this change in shape reduces the energy load and transpiration. At the same time, CO_2 assimilation falls because of a decrement in radiation interception.

3. LEAF NUMBERS

If ψ_L is slightly depressed for an extended time, leaf initiation rates may be reduced to the extent that fewer leaves are formed even if additional water becomes available through rainfall or irrigation. For determinate crops such as the cereal grains, initiation of floral primordia signals an end to the leaf-formation period for each culm. In the sorghum experiment cited above, plants in the dry plot formed an average of four fewer leaves per plant (Fig. 6). Although a precise evaluation was not possible, the formation of up to four fewer leaves probably had much less influence on total LAI than did the effect of water deficit on final size of those leaves present.

For nondeterminate crops, the effect of water deficit on total numbers of leaves presents a more complex problem. In field studies with four cotton cultivars, mean daily leaf water potentials of −11 bars reduced rates of leaf extension and vegetative node production by as much as 50% (McMichael & Hesketh, 1978). Whether or not the ultimate leaf number is reduced depends upon the length of time following relief from stress during which leaves remain the primary sinks for photosynthate. Complete recovery appears possible only if fruiting is also delayed, since competition among reproductive and vegetative sinks for photosynthate normally limits leaf growth under the best of conditions (see discussion of vegetative "cutout" by Mauney, 1979). Additional growth restrictions as a result of water deficits appear to accelerate and accentuate the cutout phenomenon (Jordan, 1979).

4. LEAF SENESCENCE

"Water stress accelerates senescence" is a statement often heard in general discussions concerning plant-water relations, yet there is surprisingly little evidence that supports a direct role of other-than-extreme water deficits in the senescence process. Much of the present confusion regarding the role of water deficits in senescence appears to stem from a failure to recognize basic differences in plant response before and after anthesis. Leaf senescence for the irrigated and dry sorghum crops before anthesis illustrated in Fig. 5 is given in Table 2. Even though soil water availability in the dry treatment greatly reduced LAI, the fractional LAI lost to senescence was approximately the same for both treatments until 62 DAP. Irrigation of the wet treatment was stopped 55 DAP and soil water was rapidly depleted as a result of the high LAI. This rapid fall in available soil water caused greater LAI loss in the previously irrigated plot so that by 86 DAP the green LAI's of both treatments were converging. These results suggest that the leaf area duration of individual leaves is not seriously altered by water deficits that develop gradually, but rapid development of water deficits may accelerate senescence of lower leaves. Similar results were obtained from irrigation frequency studies with corn (Wilson & Allison, 1978).

Compared with the example discussed above, water deficits that develop after anthesis appear to elicit a more dramatic response in terms of

Table 2—Green leaf area index (LAI) and fraction of total LAI lost to senescence for sorghum growing on stored soil water from 17 days after planting (dry) or irrigated weekly until 55 days after planting (wet).†

	Days after planting						
	38	46	54	62	70	78	86
	Green LAI						
Wet	4.5	5.8	6.9	8.1	8.4	6.9	6.7
Dry	3.3	3.8	4.2	4.9	5.3	5.1	5.1
	Percent LAI lost						
Wet	8.2	13.4	17.9	19.0	25.0	40.0	42.2
Dry	7.0	13.6	16.0	16.9	20.9	28.2	28.2

† Unpublished data of R. G. C. Smith, J. T. Ritchie, J. E. Begg, and W. E. Lonkerd, collected at the Blackland Research Center at Temple, Tex., in 1978.

senescence (Fischer & Kohn, 1966). In the presence of a strong reproductive sink the demand for photosynthate and reduced N may exceed that available through current metabolic channels, especially since water deficits may limit nutrient uptake and photosynthate production (Chapter 7C). In this case, progressive senescence upward from oldest leaves may be accelerated and labile preanthesis assimilates may form a greater proportion of the total stored in grain (Fischer, 1979).

A more direct role for water deficits is suggested in leaf abscission of cotton. Leaf shedding is a common feature of growth of dicotyledonous species in arid or semiarid regions (Kozlowski, 1976). In container-grown cotton plants, leaf abscission was related to ψ_L in a linear manner with an apparent threshold of about -10 to -15 bars (McMichael et al., 1972). The mechanism through which turgor loss or dehydration triggers the abscission process is not fully understood, but appears to involve changes in production and/or transport rates of several phytohormones (see review by Jordan, 1979). Where water deficits develop rapidly, little assimilate recovery from senescing leaves is likely before changes in the abscission zone block phloem transport. Whether or not slowly developed water deficits induce leaf shed directly through a hormonal mechanism or through an assimilate demand similar to that for cereals remains an open question.

V. REGULATION OF REPRODUCTIVE CAPACITY

Almost without exception the actual number of floral primordia formed greatly exceeds the final numbers of fruit or grains that are brought to maturity, even under seemingly optimal conditions. Reasons for this normal reproductive failure are not known, but the phenomenon appears to provide each species some flexibility to match reproductive load to environmental conditions prior to and during anthesis. It is difficult to generalize about effects of water deficit on reproductive potential because reproductive patterns differ markedly among crops. Again, cotton and cereal grain crops will be used as examples of species with nondeterminate and determinate growth habits.

A. Nondeterminate Crops

The reproductive period for relatively nondeterminate crops such as cotton and soybeans may extend for several weeks. Because of the rainfall pattern in most areas where these two crops are grown, there is a high probability that water stress will occur during some portion of the fruiting period. Also, since the production of potential fruiting sites depends on the production of additional nodes on the main stem or fruiting branches, a close coordination between vegetative (plant height) and reproductive growth (squares per plant) is observed (Fig. 8, based on data by Bruce and Romkens [1965] as cited by Baker et al. [1979]). As the soil dried slowly over a period of approximately 25 days, plant height and numbers of po-

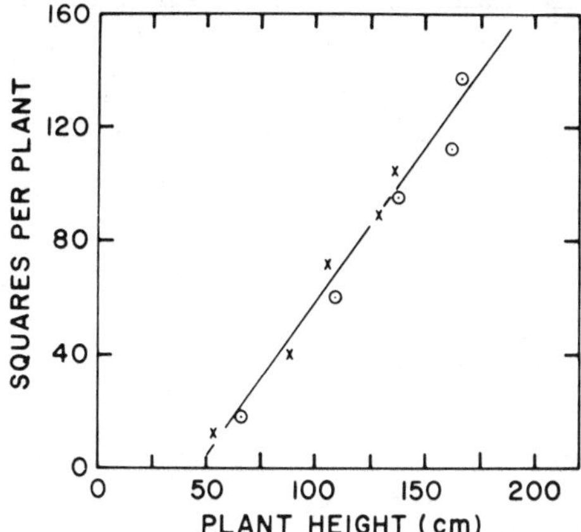

Fig. 8—The relation between squares formed per plant and plant height during the early to peak reproductive periods. Symbols X and ⊙ relate to watering treatments that resulted in differences in plant height. Data of Bruce and Romkens (1965), as cited by Baker et al. (1979).

tential fruiting sites, squares, and bolls were all reduced by approximately the same extent. In addition, water stress shortened the period during which high growth and fruiting rates occurred, and eventually limited both potential and final numbers of bolls.

The capacity of cotton to resume vegetative and reproductive growth following relief from water stress appears to be one characteristic that contributes to the species' drought resistance. However, if water deficits develop during peak fruiting, growing-season limitations and higher insect pressures may render ineffective many late-season attempts at yield recovery. The importance of early fruit set to final yield cannot be overemphasized for those production areas where midseason limitations in available soil water are common.

Data of Ritchie and Burnett (1971) further suggest that if plant water deficits develop slowly, the fraction of total dry matter partitioned to lint remains relatively stable from year to year. These findings all suggest that a dynamic balance exists between assimilate production and utilization throughout the entire fruiting period under a variety of soil water conditions (see reviews by Baker et al., 1979; Guinn, 1979; Mauney, 1979).

B. Determinate Crops

Initiation of floral primordia by the previously vegetative apex signals the first step in inflorescence development of determinate crops such as the cereal grains. Tiller and spikelet development from the time of floral initiation to anthesis are illustrated in Fig. 9 for barley (Gallagher et al., 1976). The total number of floral primordia formed within a period of about 30 days fixes the reproductive potential of the crop (N_m × number of ears), although the actual florets which produce grain (N_f) are typically only 50–60% of the potential. Because reproductive development among tillers is

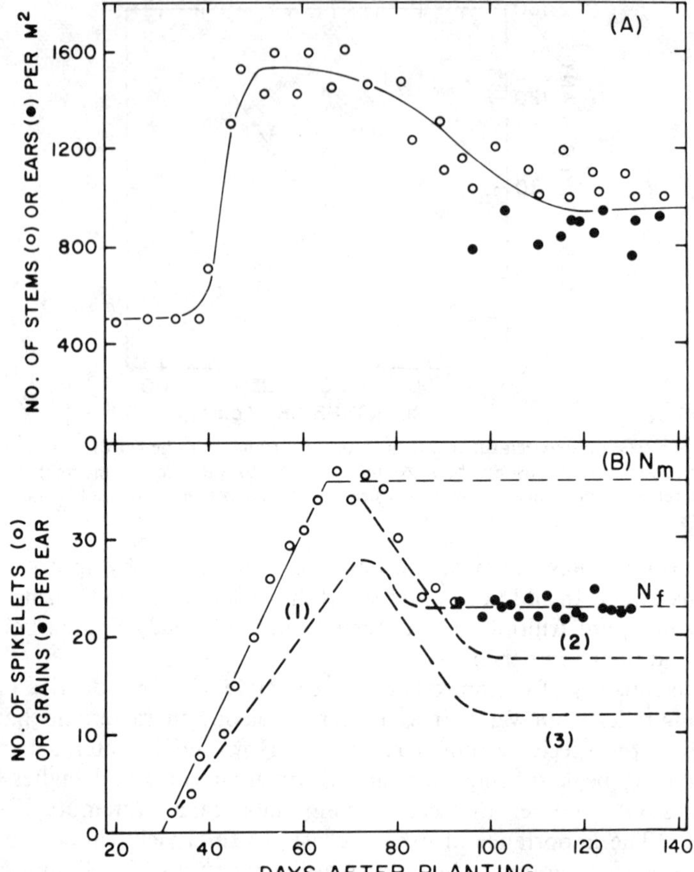

Fig. 9—Potential and actual density of reproductive axes (*A*) and reproductive structures per axis (*B*) for a barley crop grown with adequate water. Dashed lines in (*B*) show possible effects of water stress where (*1*) the rate of spikelet formation and potential spikelet numbers (N_m) are reduced, but numbers of grains per ear (N_f) are unchanged; (*2*) N_m are unaffected, but N_f are reduced, and (*3*) both N_m and N_f are reduced. Modified from Gallagher et al. (1976).

highly synchronized, reproductive flexibility to offset losses due to severe stress during this period is low.

The total reproductive potential per ear (N_m in Fig. 9B) is a product of the rate of formation and duration of the formation period. Present evidence indicates both rate and duration may be altered by water deficits as suggested by curve 1.

Just as in the case with leaves, even mild water stress may reduce the rate of initiation of floral primordia (Gates, 1968; Husain & Aspinall, 1970; Nicholls & May, 1963), but, at least in some crops, relief from mild stress allows full recovery in terms of numbers of primordia. The key to species differences may lie in the specific pattern of development of primordia once initiated. In barley, initiation of primordia ceases with the appearance of stamen initials on the first-formed spikelets (Nichols & May, 1963; Williams, 1966). Therefore, the total number of primordia produced depends on the relative rates of initiation and development. Since development rate appears less sensitive to stress than initiation rate (Husain & Aspinall, 1970), either prolonged or severe water stress would reduce the total number of primordia formed.

The influence of water stress on the length of the period between floral initiation and anthesis has received some attention, but results are varied (Salter & Goode, 1967). Data of Angus and Moncur (1977) suggest moderate water stress (ψ_L of -20 to -30 bars) after floral initiation may accelerate, while more severe stress ($\psi_L < -30$ bars) delays, anthesis. While the delay in anthesis may result simply from reduced rates of extension and development, the acceleration is more difficult to rationalize. Possibly the effect is indirect in the sense that mild or moderate water stress may cause leaf (apex?) temperatures to increase as a result of partial stomatal closure, thus accelerating growth.

The period shortly before head or ear emergence and anthesis appears to be especially sensitive to water deficits in wheat, barley, sorghum, and corn (Aspinall et al., 1964; Fischer, 1973, 1979; Inuyama et al., 1976; Moss & Downey, 1971; Wilke & Rosenow, 1976). Observed effects in "blasted" heads vary between complete absence of floral parts to aborted, but fully formed, florets. Grain set may also be reduced by fertilization failure at anthesis from an inhibition of silk elongation in corn (Moss & Downey, 1971) or failure to completely exsert the head in sorghum (W. R. Jordan, personal observation; Inuyama et al., 1976).

Possible effects of water deficits similar to those described in the preceding paragraphs are illustrated by curves 1, 2, and 3 in Fig. 9B. The postulated curves are only qualitative in nature, since quantitative data for field-grown crops are not available.

An alternate view of the effect of water deficit on development of reproductive potential of dryland wheat was presented recently by Fischer (1979). This view holds that the key to understanding the determination of kernel number (KNO) lies not with the appearance and fate of components from which KNO is derived (plants, tillers, spikelets, florets), but rather with photosynthate production and its allocation to spikes during the preanthesis period. Environmental influences on KNO (radiation, temperature, water stress) are especially critical during the 25 days or so before anthesis, during which most of the dry matter of the spikes is accumulated. Thus, the greater the dry matter accumulation rate during the period, the higher the KNO. Over a wide range of sites, sowing dates and years, KNO was found to be directly related to total dry matter at anthesis, although the relation differed among genotypes. Not withstanding the possibility of direct effects of water deficit discussed previously, Fischer (1979) concluded that KNO reductions resulting from gradual development of water deficits are mediated through effects on reduced photosynthate production. This conclusion reinforces the importance of leaf area accumulation to potential yield stressed in earlier sections of this chapter.

VI. REALIZATION OF YIELD IN DETERMINATE CROPS

The preceeding sections have sought to illustrate the significance of water deficits during the preanthesis period to the formation of yield potential. However, consequences of water deficit during the preanthesis period are also felt postanthesis as the crop attempts to fill the actual grains or fruit set. Boyer (1976) and Fischer (1979) both emphasize that grain yield is determined more by the total photosynthesis taking place during the entire season than by photosynthesis occurring during the grain-filling period alone.

One means to approach the problem is to consider the integrity of both sources and sinks after fruit or grain set. Postanthesis photosynthate and labile preanthesis metabolic products stored in leaves, stems, and roots comprise the sources (Austin et al., 1977; Bidinger et al., 1977; Gallagher et al., 1976; Stoy, 1965). Grain or fruit storage must be considered the major sink, although respiratory losses in growth and maintenance of all metabolically active tissue reduces the total supply of assimilates from sources. The literature on relative importance of sources and sinks to yield under a variety of conditions is indeed impressive (Chapter 7B), and no attempt will be made to repeat recent interpretations available elsewhere (Begg & Turner, 1976; Boyer, 1976; Evans & Wardlaw, 1976; Hsiao et al., 1976b; Fischer, 1979). Instead, I would like to draw attention to effects of water deficit on the integration of source and sink strength throughout the entire season. Gradual development of water deficits before anthesis provides ample opportunity for feedback regulation such that the potential sink strength at anthesis closely matches total source capabilities. This is reflected in the dependence of kernel number on total dry matter at anthesis found by Fischer (1979) with wheat (see Section V, Part B).

If, due to rainfall, the available soil water should increase after anthesis, postanthesis photosynthate appears sufficient to provide all but about 10–15% of the final grain dry weight, the remainder contributed from stored, preanthesis assimilate. Under these conditions, growth rates and duration of filling of individual grains are similar to those of a continuously well-watered crop. Total grain yield will still be lower than that of a fully irrigated crop because of limited sink potential. If, on the other hand, soil water availability should limit postanthesis photosynthate production, the relative contribution of stored assimilate to final grain yield increases. There is some question as to whether the absolute amount of preanthesis assimilate appearing in grain increases as a result of drought stress (Bidinger et al., 1977; Gallagher et al., 1976; Rawson et al., 1977), but the rate of withdrawal is accelerated and leaf area duration rapidly falls. The total amount of assimilate removed from storage sites may be limited by the length of time the plant lives after anthesis (Passioura, 1977). Since preanthesis growth is also reduced by water stress, less total labile assimilate is available from storage sources.

VII. CROP IMPROVEMENT FOR DROUGHT RESISTANCE

Improvement of crops for enhanced performance during periods of drought presents a complex problem to crop physiologists, agronomists, and plant breeders alike. Research efforts of many are devoted to a more complete understanding of plant response to water deficits, but relatively few of these programs concern themselves with extending these findings to actual crop improvement. A basic reason for this lack of program extension lies in the need for a truly multidisciplinary approach wherein elements of micrometeorology, soil physics, crop physiology, agronomy, and plant breeding all contribute to a common solution. This is not to suggest that these disciplines may not now coexist at a common location, but rather that this approach requires strong administrative leadership and support of a widespread nature not apparent today (see discussion by Townley-Smith & Hurd, 1979).

Demonstrated successes of conscious efforts to enhance drought resistance of crops are few. Approaches that offer the greatest chance of success appear to be those that incorporate a firm understanding of the probable pattern of soil water availability during the growing season (Blum, 1979; Jordan & Miller, 1980; O'Toole & Chang, 1979; Turner, 1979). Also, improvement for characteristics that enhance avoidance of deleterious water deficits appear to be more straight forward than those which impart greater tolerance of water deficits that actually develop. Clean separations of these two aspects of overall drought resistance may be difficult to achieve (Turner, 1979).

Where the growing season is composed of a distinct wet-dry pattern, adjustment of crop maturity to achieve full use of water available at planting offers the most readily available demonstration of benefits from avoidance of water deficits (Blum, 1970). An alternate, though untested, approach is to alter the root resistance to water flow such that more soil water will be available following anthesis (Passioura, 1972, 1976, 1977). Higher plant (root) resistance effectively lowers ψ_L during periods of high evaporative demand, and plant growth and total crop water use between planting and anthesis is reduced. Limiting preanthesis growth reduces risk of complete crop failure due to early depletion of available soil water, although yield may be reduced if soil water supplies are greater than anticipated. It would appear that certain crop management options such as proper selection of row width and planting density may achieve these same ends.

Where early-season water limitations may restrict crop establishment, ability to germinate and emerge at low ψ_{soil} (ψ_{plant}) may be crucial as is the capacity for rapid seedling root extension. This problem is especially critical for successful establishment of range species.

Incomplete extraction of stored soil water may pose a limitation to high productivity in regions dependent on summer rainfall (Jordan & Miller, 1980). Enhancement of deep rooting may be a means to avoid deleterious water deficits since genetic variability in root growth rates and field rooting patterns are present in crops such as wheat, soybeans, and sorghum (Hurd, 1974; Jordan et al., 1979; Jordan & Miller, 1980; O'Brien, 1979; Taylor et al., 1978). The success of this approach may lie in restrictions to productivity, if any, caused by the additional investment of photosynthate necessary to form and maintain larger root systems. Based on sorghum rooting patterns and estimates of root dry matter, Jordan and Miller (1980) calculated that an increase in L_v from 1 to 2 cm cm^{-3} in a 1.25-m soil layer (between the 0.50- and 1.75-m soil depth) would require at least 800 kg ha^{-1} more dry matter. The net effect of an additional demand of this magnitude on yield depends largely on whether grain yield is sink- or source-limited.

More emphasis is needed to extend to fruition additional research efforts such as those described above. Enhanced performance in terms of yield stability offers advantages to both producer and consumer. While proper crop and soil management may likewise promise greater water-use efficiency, use of drought-resistant varieties is one means of achieving efficient water use without additional inputs by producers in terms of time and energy.

VIII. FUTURE THRUSTS

The key to successful crop management to achieve efficient use of available water resources lies in our ability to correctly integrate crop response with specific environmental circumstances. The conceptual framework presented by Hsiao et al. (1976a, 1976b) provides an excellent starting point upon which to build our understanding of plant response to internal water deficits. Deficiencies in our knowledge of effects of water deficit on specific plant processes or responses have been elucidated in preceding sections of this chapter and by others in this book.

Apart from these specific problems, one significant limitation that pervades nearly all large-scale studies on this topic relates to whether or not we have a useful measure of plant water status available to us today. Historically, soil water content or potential were used to quantify plant response to soil water availability. The last two decades have been a rapid development of technology related to the quantitative description of the plant water balance (Barrs, 1968; Boyer, 1969; Slavik, 1974). Finally, ψ_L seems to have emerged as the concensus descriptor for several reasons, including the desire for uniformity with soil scientists (Taylor, 1968) and development of the pressure bomb (Scholander et al., 1965).

However, the usefulness of ψ_L as a means to quantify plant response to soil water availability depends on the end product desired. The discussion of leaf growth presented in Section IV, Part C illustrates the dependence of growth on turgor maintenance, but turgor cannot be inferred directly from measurements of ψ_L alone. Similar conclusions may be reached regarding the relation between ψ_L and stomatal closure. When the capacity of plants to osmoregulate is considered, we are faced with the present-day dilemma that quantitative expressions of the dependence of physiological processes on ψ_L derived under one set of conditions do not translate *quantitatively* to other circumstances. The lack of quantitative generality creates untold problems for those attempting to use ψ_L relationships in crop modeling and irrigation timing. While we readily acknowledge that a plant-based descriptor is desirable when quantifying plant response, these users must, for the most part, now rely on the use of descriptions based on some measure of soil water availability.

The rediscovery of the important role solute accumulation plays in turgor maintenance (osmotic adjustment or osmoregulation) in the face of drought has again focused attention on basic tissue-water relations. An in-depth discussion of recent studies on this topic is presented by Radin in Chapter 6B, but a few additional perspectives may be in order.

The active accumulation of solutes is presumed to lead to the complete or partial maintenance of turgor-dependent processes as tissue water potentials are lowered either temporally or seasonally. Indeed, exposure of several species to conditions leading to internal water deficits resulted in maintenance of stomatal opening, photosynthetic rate, and leaf extension at reduced water potentials. While the importance of maintaining a positive carbon balance during drought seems self-evident, one must question a strategy that supports continued leaf expansion as soil water becomes limiting. Since shoot growth is usually reduced by internal water deficits to a larger extent than root growth, does this mean that those species or cultivars that maintain leaf growth relatively better also maintain net photosynthesis to a greater degree, or do they partition relatively less photosynthate to roots? If the latter were true, maintenance of leaf growth during stress at the expense of root growth may lead to drought susceptibility rather than drought resistance (avoidance).

Even though solute accumulation in response to drought appears to be a characteristic of most species studied in detail, little evidence suggests osmoregulation may play a major role in cultivar differences in drought resistance. Jones and Turner (1978) found no substantial difference in osmotic adjustment between a drought-susceptible grain sorghum ('Shallu') and a resistant one ('RS610') when subjected to similar soil water deficits. Cutler et al. (1980) reported differences in solute accumulation of only a few bars for rice cultivars subjected to water stress. While cultivar differences were observed in maintenance of stomatal opening in sorghum under stress, the differences were small (Ackerson et al., 1980). On the other hand, Morgan (1977) found substantial differences in osmoregulation among wheat lines. Additional studies are required to adequately fix the limits of osmoregulation and to determine if true genetic variability (as opposed to the apparent variability arising from genotype × environment interactions) exists within cultivars of important crop species. Only then can the relative worth of osmoregulation to drought resistance be critically assessed. If careful study demonstrated true genetic variability for this characteristic, we are then faced with the task of devising evaluation techniques suitable for application to large numbers of samples required in crop improvement programs (Chapter 8C). Transfer of research findings into the technologies that can be applied in crop improvement programs stands as one of the greatest challenges to crop physiologists seeking to assist in development of crops with enhanced drought resistance.

ACKNOWLEDGMENT

The original contribution of T. C. Hsiao and K. J. Bradford to Section IV, Part C 2 of this chapter is gratefully acknowledged.

LITERATURE CITED

Acevedo, E., E. Fereres, T. C. Hsiao, and D. W. Henderson. 1979. Diurnal growth trends, water potential, and osmotic adjustment of maize and sorghum leaves in the field. Plant Physiol. 64:476-480.

Acevedo, E., T. C. Hsiao, and D. W. Henderson. 1971. Immediate and subsequent growth responses of maize leaves to changes in water status. Plant Physiol. 48:631-636.

Ackerson, R. C., D. R. Krieg, and F. J. M. Sung. 1980. Leaf conductance and osmoregulation of field-grown sorghum genotypes. Crop Sci. 20:10-14.

Akalehiywot, T., and J. D. Bewley. 1977. Promotion and synchronization of cereal grain germination by osmotic pretreatment with polyethylene glycol. J. Agric. Sci. (Camb.) 89: 503-506.

Angus, J. F., and M. W. Moncur. 1977. Water stress and phenology in wheat. Aust. J. Agric. Res. 28:177-181.

Arkin, G. F., J. T. Ritchie, and S. J. Maas. 1978. A model for calculating light interception by a grain sorghum canopy. Trans. ASAE 21:303-308.

Aspinall, D., P. B. Nicholls, and L. H. May. 1964. The effects of soil moisture stress on the growth of barley. I. Vegetative development and grain yield. Aust. J. Agric. Res. 15: 729-745.

Austin, R. B., J. A. Edrich, B. A. Ford, and R. D. Blackwell. 1977. The fate of dry matter, carbohydrates and ^{14}C lost from the leaves and stems of wheat during grain filling. Ann. Bot. 41:1309-1321.

Baker, D. N., J. A. Landivar, and J. R. Lambert. 1979. Model simulation of fruiting. p. 261-264. *In* J. M. Brown (ed.) Proc. Beltwide Cotton Prod. Res. Conf., Phoenix, Ariz. 7-11 Jan. 1979. National Cotton Council, Memphis, Tenn.

Barrs, H. D. 1968. Determination of water deficits in plant tissues. p. 235-368. *In* T. T. Kozlowski (ed.) Water deficits and plant growth. Vol. I. Academic Press, New York.

Begg, J. E. 1980. Morphological adaptations of leaves to water stress. p. 33-42. *In* N. C. Turner and P. J. Kramer (ed.) Adaptation of plants to water and high temperature stress. Wiley-Interscience, New York.

Begg, J. E., and N. C. Turner. 1976. Crop water deficits. Adv. Agron. 28:161-217.

Bewley, J. D. 1979. Physiological aspects of desiccation tolerance. Annu. Rev. Plant Physiol. 30:195-238.

Bidinger, F., R. B. Musgrave, and R. A. Fischer. 1977. Contribution of stored preanthesis assimilate to grain yield in wheat and barley. Nature 270:431-433.

Blum, A. 1970. Effect of plant density and growth duration on grain sorghum yield under limited water supply. Agron. J. 62:333-336.

Blum, A. 1979. Genetic improvement of drought resistance in crop plants: a case for sorghum. p. 447-464. *In* H. Mussell and R. C. Staples (ed.) Stress physiology in crop plants. Wiley-Interscience, New York.

Boyer, J. S. 1968. Relationship of water potential to growth of leaves. Plant Physiol. 43: 1056-1062.

Boyer, J. S. 1969. Measurement of the water status of plants. Annu. Rev. Plant Physiol. 20: 351-364.

Boyer, J. S. 1970. Leaf enlargement and metabolic rates in corn, soybean, and sunflower at various leaf water potentials. Plant Physiol. 46:233-235.

Boyer, J. S. 1976. Water deficits and photosynthesis. p. 153–190. *In* T. T. Kozlowski (ed.) Water deficits and plant growth. Vol. IV. Academic Press, New York.

Boyer, J. S., and H. G. McPherson. 1975. Physiology of water deficits in cereal crops. Adv. Agron. 27:1–23.

Brown, K. W., W. R. Jordan, and J. C. Thomas. 1976. Water stress induced alterations of the stomatal response to decreases in leaf water potential. Physiol. Plant. 37:1–5.

Bruce, R. R., and M. J. M. Romkens. 1965. Fruiting and growth characteristics of cotton in relation to soil moisture tension. Agron. J. 57:135–140.

Bull, T. A., and K. T. Glasziou. 1975. Sugar cane. p. 51–72. *In* L. T. Evans (ed.) Crop physiology: some case histories. Cambridge University Press, Cambridge, England.

Christ, R. A. 1978. The elongation rate of wheat leaves. I. Elongation rates during day and night. J. Exp. Bot. 29:603–610.

Chu, A. C. P., and J. P. Kerr. 1977. Leaf water potential and leaf extension in a sudax crop. N.Z. J. Agric. Res. 20:467–470.

Collis-George, N., and J. C. Sands. 1962. Comparison of the effects of the physical and chemical components of soil water energy on seed germination. Aust. J. Agric. Res. 13:575–584.

Cutler, J. M., and D. W. Rains. 1977. Effects of irrigation history on responses of cotton to susequent water stress. Crop Sci. 17:329–335.

Cutler, J. M., K. W. Shahan, and P. L. Steponkus. 1980. Alteration of the internal water relations of rice in response to drought hardening. Crop Sci. 20:307–310.

Dasberg, S. 1971. Soil water movement to germinating seeds. J. Exp. Bot. 22:999–1008.

Dasberg, S., and K. Mendel. 1971. The effect of soil water and aeration on seed germination. J. Exp. Bot. 22:992–998.

Doering, E. J. 1965. Soil-water diffusivity by the one-step method. Soil Sci. 99:322–326.

Doneen, L. D., and J. H. MacGillivray. 1943. Germination (emergence) of vegetable seeds as affected by different soil moisture conditions. Plant Physiol. 18:524–529.

Dubetz, S. 1969. An unusual photonastism induced by drought in *Phaseolus vulgaris*. Can. J. Bot. 47:1640–1641.

Duncan, W. G. 1975. Maize. p. 23–50. *In* L. T. Evans (ed.) Crop physiology: some case histories. Cambridge University Press, Cambridge, England.

Eavis, B. W., L. F. Ratliff, and H. M. Taylor. 1969. Use of a dead-load technique to determine axial growth pressure. Agron. J. 61:640–643.

Esau, K. 1965. Plant anatomy. 2nd ed. John Wiley & Sons, New York.

Evans, L. T., and I. F. Wardlaw. 1976. Aspects of the comparative physiology of grain yield in cereals. Adv. Agron. 28:301–359.

Evans, L. T., I. F. Wardlaw, and R. A. Fischer. 1975. Wheat. p. 101–149. *In* L. T. Evans (ed.) Crop physiology. Cambridge University Press, Cambridge, England.

Fereres, E., E. Acevedo, D. W. Henderson, and T. C. Hsiao. 1978. Seasonal changes in water potential and turgor maintenance in sorghum and maize under water stress. Physiol. Plant 44:261–267

Fischer, R. A. 1973. The effect of water stress at various stages of development on yield processes in wheat. p. 233–241. *In* R. O. Slatyer (ed.) Plant response to climatic factors. UNESCO, Paris.

Fischer, R. A. 1979. Growth and water limitations to dryland wheat yield in Australia. A physiological framework. J. Aust. Inst. Agric. Sci. 45:83–94.

Fischer, R. A., and G. D. Kohn. 1966. The relationship of grain yield to vegetative growth and post-flowering leaf area in the wheat crop under conditions of limited soil moisture. Aust. J. Agric. Res. 17:281–295.

Fischer, R. A., and N. C. Turner. 1978. Plant productivity in the arid and semi-arid zones. Annu. Rev. Plant Physiol. 29:277–317.

Frank, A. B., J. F. Power, and W. O. Willis. 1973. Effect of temperature and plant-water stress on photosynthesis, diffusion resistance, and leaf-water potential in spring wheat. Agron. J. 65:777–780.

Gallagher, J. N., P. V. Briscoe, and R. K. Scott. 1976. Barley and its environment. VI. Growth and development in relation to yield. J. Appl. Ecol. 13:563–583.

Gandar, P. W., and C. B. Tanner. 1976. Leaf growth, tuber growth, and water potential in potatoes. Crop Sci. 16:534–538.

Garner, T. H., and H. D. Bowen. 1966. Plant mechanics in seedling emergence. Trans. ASAE 9:650–653, 660.

Gates, C. T. 1968. Water deficits and growth of herbaceous plants. p. 135–190. *In* T. T. Kozlowski (ed.) Water deficits and plant growth. Vol. II. Academic Press, New York.

Gill, W. R., and G. H. Bolt. 1955. Pfeffer's studies of the root growth pressures exerted by plants. Agron. J. 47:166–168.

Goeschl, J. D., and S. K. Kays. 1975. Concentration dependencies of some effects of ethylene on etiolated pea, peanut, bean, and cotton seedlings. Plant Physiol. 55:670–677.

Goeschl, J. D., L. Rappaport, and H. K. Pratt. 1966. Ethylene as a factor regulating the growth of pea epicotyls subjected to physical stress. Plant Physiol. 41:877–884.

Guinn, G. 1979. Hormonal relations in flowering, fruiting and cutout. p. 265–276. *In* J. M. Brown (ed.) Proc. Beltwide Cotton Prod. Res. Conf., Phoenix, Ariz. 7–11 Jan. 1979. National Cotton Council, Memphis, Tenn.

Hadas, A. 1969. Effects of soil moisture stress on seed germination. Agron. J. 61:325–327.

Hadas, A. 1976. Water uptake and germination of leguminous seeds under changing external water potential in osmotic solutions. J. Exp. Bot. 27:480–489.

Hadas, A., and D. Russo. 1974a. Water uptake by seeds as affected by water stress, capillary conductivity, and seed-soil water contact. I. Experimental study. Agron. J. 66:643–647.

Hadas, A., and D. Russo. 1974b. Water uptake by seeds as affected by water stress, capillary conductivity, and seed-soil water contact. II. Analysis of experimental data. Agron. J. 66: 647–652.

Hadas, A., and E. Stibbe. 1977. Soil crusting and emergence of wheat seedlings. Agron. J. 69:547–550.

Hanks, R. J., and F. C. Thorpe. 1957. Seedling emergence of wheat, grain sorghum and soybeans as influenced by soil crust strength and moisture content. Soil Sci. Soc. Am. Proc. 21:357–359.

Hegarty, T. W. 1978. The physiology of seed hydration and dehydration and the relation between water stress and the control of germination: a review. Plant Cell Environ. 1:101–119.

Hillel, D. 1972. Soil moisture and seed germination. p. 65–89. *In* T. T. Kozlowski (ed.) Water deficits and plant growth, Vol. III. Academic Press, New York.

Hsiao, T. C. 1973. Plant respones to water stress. Annu. Rev. Plant Physiol. 24:519–570.

Hsiao, T. C., and E. Acevedo. 1974. Plant responses to water deficits, water use efficiency and drought resistance. Agric. Meteorol. 14:59–84.

Hsiao, T. C., E. Acevedo, E. Fereres, and D. W. Henderson. 1976a. Water stress, growth, and osmotic adjustment. Philos. Trans. R. Soc. London Ser. B. 273:479–500.

Hsiao, T. C., E. Fereres, E. Acevedo, and D. W. Henderson. 1976b. Water stress and dynamics of growth and yield of crop plants. p. 282–305. *In* O. L. Lange et al. (ed.) Water and plant life. Ecol. Stud. 19. Springer-Verlag, New York.

Hunter, J. R., and A. E. Erickson. 1952. Relation of seed germination to soil moisture tension. Agron. J. 44:107–109.

Hurd, E. A. 1974. Phenotype and drought tolerance in wheat. Agric. Meteorol. 14:39–55.

Husain, I., and D. Aspinall. 1970. Water stress and apical morphogenesis in barley. Ann. Bot. 34:393–407.

Inuyama, S., J. T. Musick, and D. A. Dusek. 1976. Effect of plant water deficit at various growth stages on growth, grain yield, and leaf water potential of irrigated grain sorghum. Proc. Crop Sci. Soc. Japan 45:298–307.

Johnson, D. A., and K. H. Asay. 1978. A technique for assessing seedling emergence under drought stress. Crop Sci. 18:520–522.

Johnson, W. C. 1967. Diurnal variation in growth rate of grain sorghum. Agron. J. 59:41–44.

Jones, M. M., and N. C. Turner. 1978. Osmotic adjustment in leaves of sorghum in response to water deficits. Plant Physiol. 61:122–126.

Jordan, W. R. 1970. Growth of cotton seedlings in relation to maximum daily leaf-water potential. Agron. J. 62:699–701.

Jordan, W. R. 1979. Influence of edaphic parameters on flowering, fruiting, and cutout. A. Role of plant water deficit. p. 256–261. *In* J. M. Brown (ed.) Proc. Beltwide Cotton Prod. Res. Conf., Phoenix, Ariz. 7–11 Jan. 1979. National Cotton Council, Memphis, Tenn.

Jordan, W. R., K. W. Brown, and J. C. Thomas. 1975. Leaf age as a determinant in stomatal control of water loss from cotton during water stress. Plant Physiol. 56:595–599.

Jordan, W. R., and F. R. Miller. 1980. Genetic variability in sorghum root systems: implications for drought tolerance. p. 383–399. *In* N. C. Turner and P. J. Kramer (ed.) Adaptation of plants to water and high temperature stress. Wiley-Interscience, New York.

Jordan, W. R., F. R. Miller, and D. E. Morris. 1979. Genetic variation in root and shoot growth of sorghum in hydroponics. Crop Sci. 19:468–472.

Kanemasu, E. T., A.-J. Chen, W. C. Powers, and I. D. Teare. 1974. Stomatal resistance as an indicator of water stress. Kansas Acad. Sci. 76:159-166.

Knittle, K. H., and J. S. Burris. 1979. Effect of downward force on soybean hypocotyl growth. Crop Sci. 19:47-51.

Kozlowski, T. T. 1976. Water supply and leaf shedding. p. 191-231. *In* T. T. Kozlowski (ed.) Water deficits and plant growth. Vol. IV. Academic Press, New York.

Legg, B. J., W. Day, D. W. Lawlor, and K. J. Parkinson. 1979. The effects of drought on barley growth: models and measurements showing the relative importance of leaf area and photosynthetic rate. J. Agric. Sci. 92:703-716.

Lindstrom, M. J., R. I. Papendick, and F. E. Koehler. 1976. A model to predict winter wheat emergence as affected by soil temperature, water potential, and depth of planting. Agron. J. 68:137-140.

Loomis, W. E. 1934. Daily growth of maize. Am. J. Bot. 21:1-6.

Lyles, L., and C. D. Fanning. 1964. Effects of presoaking, moisture, tension, and soil salinity on the emergence of grain sorghum. Agron. J. 56:518-520.

Mathers, A. C., and N. H. Welch. 1964. Pans in the Southern Great Plains soils. II. Effect of duration of radial root restriction on cotton growth and yield. Agron. J. 56:313-315.

Mauney, J. R. 1979. Production of fruiting points. p. 256-261. *In* J. M. Brown (ed.) Proc. Beltwide Cotton Prod. Res. Conf., Phoenix, Ariz. 7-11 Jan. 1979. National Cotton Council, Memphis, Tenn.

McCree, K. J. 1974. Changes in stomatal response characteristics of grain sorghum produced by water stress during growth. Crop Sci. 14:273-278.

McCree, K. J., and S. D. Davis. 1974. Effect of water stress and temperature on leaf size and on size and number of epidermal cells in grain sorghum. Crop Sci. 14:751-755.

McMichael, B. L., and J. D. Hesketh. 1978. Some adaptive responses of field grown cotton to plant-water stress. p. 56. *In* Proc. Beltwide Cotton Prod. Res. Conf., Dallas, Tex. 9-11 Jan. 1978. National Cotton Council, Memphis, Tenn.

McMichael, B. L., W. R. Jordan, and R. D. Powell. 1972. Abscission processes in cotton: induction by plant water deficit. Agron. J. 65:602-604.

Meyer, R. F., and J. S. Boyer. 1972. Sensitivity of cell division and cell elongation to low water potentials in soybean hypocotyles. Planta 108:77-87.

Millar, B. C., and O. T. Denmead. 1976. Water relations of wheat leaves in the field. Agron. J. 68:303-307.

Monteith, J. L. 1977. Climate and the efficiency of crop production in Britain. Philos. Trans. R. Soc. London Ser. B. 281:277-294.

Morgan, J. M. 1977. Differences in osmoregulation between wheat genotypes. Nature 270: 234-235.

Moss, G. I., and L. A. Downey. 1971. Influence of drought stress on female gametophyte development in corn (*Zea mays* L.) and subsequent grain yield. Crop Sci. 11:368-372.

Nicholls, P. B., and L. H. May. 1963. Studies on the growth of the barley apex. I. Interrelationships between primordium formation, apex length, and spikelet development. Aust. J. Biol. Sci. 16:561-571.

Nobel, P. S. 1974. Introduction to biophysical plant physiology. W. H. Freeman Co., San Francisco, Calif.

O'Brien, L. 1979. Genetic variability of root growth in wheat (*Triticum aestivum* L.). Aust. J. Agric. Res. 30:587-595.

Oppenheimer, H. R. 1960. Adaptation to drought: xerophytism. p. 105-138. *In* Plant-water relationships for arid and semi-arid conditions. Reviews of research. UNESCO, Paris.

O'Toole, J. C., and T. T. Chang. 1979. Drought resistance in cereals—rice: a case study. p. 447-464. *In* H. Mussell and R. C. Staples (ed.) Stress physiology in crop plants. Wiley-Interscience, New York.

O'Toole, J. C., and R. T. Cruz. 1980. Response of leaf water potential stomatal resistance and leaf rolling to water stress. Plant Physiol. 65:428-432.

O'Toole, J. C., R. T. Cruz, and T. N. Singh. 1979. Leaf rolling and transpiration. Plant Sci. Lett. 16:111-114.

Owen, P. C. 1952. The relation of germination of wheat to water potential. J. Exp. Bot. 3:188-203.

Parker, J. J., Jr., and H. M. Taylor. 1965. Soil strength and seedling emergence relations. I. Soil type, moisture tension, temperature, and planting depth effects. Agron. J. 57:289-291.

Passioura, J. B. 1972. The effect of root geometry on the yield of wheat growing on stored water. Aust. J. Agric. Res. 3:745–752.

Passioura, J. B. 1976. Physiology of grain yield in wheat growing on stored water. Aust. J. Plant Physiol. 3:559–565.

Passioura, J. B. 1977. Grain yield, harvest index, and water use of wheat. J. Aust. Inst. Agric. Sci. 43:117–120.

Penning de Vries, F. W. T. 1975. The cost of maintenance processes in plant cells. Ann. Bot. 39:77–92.

Philip, J. R. 1957. Evaporation and moisture and heat fields in the soil. J. Meteorol. 14:354–366.

Phillips, R. E. 1968. Water diffusivity of germinating soybean, corn, and cotton seed. Agron. J. 60:568–571.

Rawson, H. M. 1979. Vertical wilting and photosynthesis, transpiration, and water use efficiency of sunflower leaves. Aust. J. Plant Physiol. 6:109–120.

Rawson, H. M., A. K. Bagga, and P. N. Bremner. 1977. Aspects of adaptation by wheat and barley to soil moisture deficits. Aust. J. Plant Physiol. 4:389–401.

Reginato, R. J. 1975. Sampling soil-water distribution in the surface centimeter of a field soil. Soil Sci. 120:292–294.

Ripley, E. A., and R. E. Redmann. 1976. Grassland. p. 349–398. *In* J. L. Monteith (ed.) Vegetation and the atmosphere. Vol. II: Case studies. Academic Press, London.

Ritchie, J. T., and E. Burnett. 1971. Dryland evaporative flux in a subhumid climate: II. Plant influences. Agron. J. 63:56–62.

Salter, P. J., and J. E. Goode. 1967. Crop responses to water at different stages of growth. Commonw. Agric. Bureau, Farnham Royal, Bucks, England. p. 246.

Scholander, P. F., H. T. Hammel, E. D. Bradstreet, and E. A. Hemmingsen. 1965. Sap pressure in vascular plants. Science 148:339–346.

Sedgley, R. H. 1963. The importance of liquid seed contact during the germination of Medicago tribuloides. Aust. J. Agric. Res. 14:646–654.

Sepaskhah, A. R., and E. R. Ardelsani. 1978. Effect of soil matric potential and seeding depth on emergence of barley. Agron. J. 70:728–731.

Shackel, K. A., and A. E. Hall. 1979. Reversible leaflet movements in relation to drought adaptation of cowpeas, *Vigna unguiculata* (L.) Walp. Aust. J. Plant Physiol. 6:265–276.

Sharp, R. E., and W. J. Davies. 1979. Solute regulation and growth by roots and shoots of water-stressed maize plants. Planta 147:43–49.

Slatyer, R. O. 1974. Effects of water stress on plant morphogenesis. USDA Misc. Publ. no. 1271. U.S. Government Printing Office, Washington, D.C. p. 3–13.

Slavik, B. 1974. Methods of studying plant-water relations. Ecol. Stud. 9. Springer-Verlag, New York.

Stoy, V. 1965. Photosynthesis, respiration and carbohydrate accumulation in spring wheat in relation to yield. Physiol. Plant Supp. IV:1–125.

Taylor, H. M. 1974. Root behavior as affected by soil structure and strength. p. 271–291. *In* E. W. Carson (ed.) The plant root and its environment. University of Virginia Press, Charlottesville.

Taylor, H. M., E. Burnett, and G. D. Booth. 1978. Taproot elongation rates of soybeans. Z. Acker Pflanzenbau 146:33–39.

Taylor, H. M., and W. R. Gardner. 1963. Penetration of cotton seedling taproots as influenced by bulk density, moisture content, and strength of soil. Soil Sci. 96:153–156.

Taylor, H. M., A. C. Mathers, and F. B. Latspeich. 1964. Pans in the Southern Great Plains soils. I. Why root-restricting pans occur. Agron. J. 56:328–332.

Taylor, H. M., J. J. Parker, Jr., and G. M. Roberson. 1966. Soil strength and seedling emergence relations: II. A generalized relation for Gramineae. Agron. J. 58:393–395.

Taylor, H. M., and L. F. Ratliff. 1969. Root elongation rates of cotton and peanuts as a function of soil strength and soil water content. Soil Sci. 108:113–119.

Taylor, S. A. 1968. Terminology in plant and soil water relationships. p. 49–72. *In* T. T. Kozlowski (ed.) Water deficits and plant growth. Vol. I. Academic Press, New York.

Thomas, J. C., K. W. Brown, and W. R. Jordan. 1976. Stomatal response to leaf-water potential as affected by preconditioning water stress in the field. Agron. J. 68:707–708.

Townley-Smith, T. F., and E. A. Hurd. 1979. Testing and selecting for drought resistance in wheat. p. 447–464. *In* H. Mussell and R. C. Staples (ed.) Stress physiology in crop plants. Wiley-Interscience, New York.

Turner, N. C. 1974a. Stomatal behavior and water status of maize, sorghum, and tobacco under field conditions. II. At low soil water potential. Plant Physiol. 53:360–365.

Turner, N. C. 1974b. Stomatal responses to light and water under field conditions. R. Soc. N.Z. Bull. 12:432–433.

Turner, N. C. 1979. Drought resistance and adaptation to water deficits in crop plants. p. 341–372. *In* H. Mussell and R. C. Staples (ed.) Stress physiology in crop plants. Wiley-Interscience, New York.

Wanjura, D. F., and D. R. Buxton. 1972a. Water uptake and radicle emergence of cottonseed as affected by soil moisture and temperature. Agron. J. 64:427–431.

Wanjura, D. F., and D. R. Buxton. 1972b. Hypocotyl and radicle elongation of cotton as affected by soil environment. Agron. J. 64:431–434.

Watts, W. R. 1974. Leaf extension in *Zea mays*. III. Field measurements of leaf extension in response to temperature and leaf water potential. J. Exp. Bot. 25:1085–1096.

Wenkert, W., E. R. Lemon, and T. R. Sinclair. 1978. Leaf elongation and turgor pressure in field-grown soybean. Agron. J. 70:761–764.

Wiegand, C. L. 1962. Drying patterns of a sandy clay loam in relation to optimum depths of seeding. Agron. J. 54:473–476.

Wilke, O., and D. T. Rosenow. 1976. Observations on head blasting in water-stressed sorghum. Texas Agric. Exp. Stn. Prog. Rep. no. 3387. College Station.

Williams, R. F. 1966. The physiology of growth in the wheat plant. III. Growth of the primary shoot and inflorescence. Aust. J. Biol. Sci. 19:949–966.

Wilson, D. R., C. H. M. van Bavel, and K. J. McCree. 1980. The carbon balance of water-deficient grain sorghum plants. Crop Sci. 20:153–159.

Wilson, J. H. H., and J. C. S. Allison. 1978. Effects of water stress on the growth of maize (*Zea mays* L.). Rhod. J. Agric. Res. 16:175–192.

Wright, D. L., R. E. Blazer, and J. M. Woodruff. 1978. Seedling emergence as related to temperature and moisture tension. Agron. J. 70:709–712.

Yoshida, S. 1972. Physiological aspects of grain yield. Annu. Rev. Plant Physiol. 23:437–464.

CHAPTER 7B

Whole-Plant Responses to Water Deficits: Carbon Assimilation and Utilization

DANIEL R. KRIEG

Texas Tech University and Texas Agricultural Experiment Station
Lubbock, Texas

I. INTRODUCTION

The devastating effects of water stress on plant productivity have been recognized since the beginning of agriculture. Since yield and dry matter accumulation are usually highly related, considerable effort has been expended in developing a better understanding of carbon metabolism in response to water stress. The importance of leaf area development and duration in determining the productive capacity of plants in water-limited environments has been emphasized by Jordan (Chapter 7A). The factors associated with determining the rate of net carbon assimilation and the utilization of the assimilate are equally important. This section will summarize some of the limitations in the carbon assimilation process of whole plants as related to water stress intensity.

II. ASSIMILATION OF PHOTOSYNTHATE

Green leaves, and to a limited extent stems and fruit, serve as the primary sources for assimilation of reduced carbon through the process of photosynthesis. Photosynthetic rates are usually reduced as plants begin to experience abnormal water deficits during the course of the day. Several recent reviews have addressed the photosynthetic rate response to water stress (Boyer, 1976a, 1976b; Hsaio, 1973; Lawlor, 1979). Depending upon plant species, growth environment, plant age or stage of development, and rate of stress development, one can attribute reductions in photosynthetic rate of water-stressed leaves solely to stomatal conductance changes, solely to nonstomatal components, or to both stomatal and nonstomatal factors.

Limitations to Efficient Water Use in Crop Production.

A. Stomatal Control of Photosynthesis

The literature is replete with examples of stomatal control of photosynthetic rates of leaves. Although a linear correlation usually exists between leaf conductance and photosynthetic rate, the cause-and-effect relationships have not been clearly defined. In field-grown plants where the rate of stress development is gradual, it is becoming increasingly apparent that the relationship is even more questionable. Many plant species are capable of osmotic adjustment as leaf water potentials decline due to soil water deficits. The resulting turgor maintenance may alter the leaf conductance–photosynthetic rate relationship. Leaf conductance responses through turgor maintenance via osmotic adjustment are a function of plant species, leaf age, rate of stress development, prior stress history, and evaporative demand (Ackerson et al., 1977a, 1977b, 1980; Brown et al., 1976; Cutler & Rains, 1977; Jones & Rawson, 1979; Jones & Turner, 1978; Thomas et al., 1976; Turner, 1979). Radin (Chapter 6B) and Turner and Jones (1980) have reviewed and defined osmotic adjustment and its implications with respect to physiological responses. In general, turgor maintenance through osmotic adjustment does not necessarily result in complete maintenance of physiological activity. Photosynthetic rates often change independently from stomatal conductance within a certain range of plant water deficit. Osmotic adjustment capability is apparently limited (Turner & Jones, 1980), and thus as stress intensifies positive turgor pressure cannot be maintained indefinitely and stomatal conductance decreases to low values as turgor pressure appraoches zero.

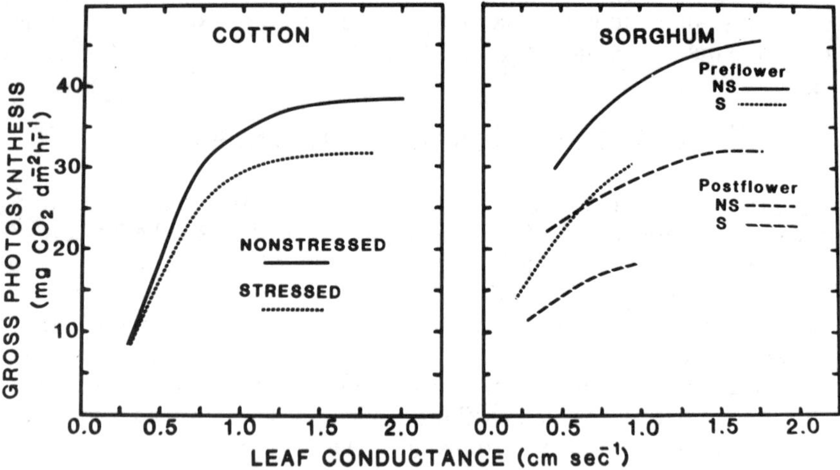

Fig. 1—Leaf conductance–gross photosynthesis relationships for cotton and sorghum at two growth stages. Data were obtained from plants grown under field conditions throughout the course of the growing season and represent results from a wide range of genotypes and soil water conditions (R. B. Hutmacher and D. R. Krieg, unpublished data).

The correlations between leaf conductance and gross photosynthetic rates of single leaves of cotton and sorghum are depicted in Fig. 1. The data are separated into pre- and post-flowering responses since we have observed differences due to growth stage in our previous efforts (Ackerson et al., 1977b, 1980). These data were taken across a number of genotypes, soil water availabilities, and times of day. The correlations were best defined using a quadratic equation for both species. As conductance increased photosynthetic activity became constant, suggesting nonstomatal limitations. Leaf-age effects are apparent in sorghum where older leaves have reduced photosynthetic rates per unit conductance compared with younger leaves. In both species, the coefficient of determination (r^2) was <0.6, indicating a high degree of variability in the photosynthetic rate–leaf conductance correlation. In sorghum, the photosynthetic rate was largely responsible for the variation, whereas in cotton, leaf conductance was most variable. These results, utilizing field-grown plants under semiarid environmental conditions, suggest that stomatal conductance may control photosynthetic rates up to a certain point beyond which nonstomatal factors exert major influence.

B. Nonstomatal Control of Photosynthesis

Pearcy (Chapter 6C) has reviewed the approaches to investigating nonstomatal inhibitions of photosynthesis and the results obtained with a variety of plant species in contrasting environments. Reductions in photosynthetic rates due to chloroplastic factors could involve both the light reaction and the enzymatic phases of CO_2 reduction. Boyer and co-workers (as reviewed by Boyer, 1976a, 1976b) have provided evidence that during mild water stress inhibition of photosynthesis was associated primarily with reduced Hill reaction activity and electron transport. As stress intensified photophosphorylation activity was reduced, and possibly became rate-limiting.

The biochemical reactions associated with CO_2 reduction are also subject to direct control by water stress, which could indirectly result in the observed effects on the light reactions. Ribulose bisphosphate carboxylase-oxygenase (RuBPCase) is usually considered to exert primary control on the CO_2-fixation process. Reductions in RuBPCase activity in water-stressed plants have been reported for a number of different species (Huffaker et al., 1970; Johnson et al., 1974; O'Toole et al., 1976). However, in most cases the reduced enzyme activities measured in vitro were not sufficient to account for the total photosynthetic rate reductions observed on a whole-leaf basis. These results suggest that either additional mesophyll resistance components are involved or that the measured enzyme activities do not truly reflect in vivo activities. The ratio of "active" to "total" RuBPCase is a function of the stromal environment especially as related to pH, Mg^{2+}, CO_2, and RuBP concentrations. The stromal environment in water-stressed leaves has not been defined, therefore in vitro assay conditions probably do not accurately reflect in vivo activities.

Since RuBPCase functions in a dual capacity, also serving as an oxygenase, its response to the environment is extremely critical to whole-plant carbon assimilation. Lawlor (1979) has recently indicated that the ratio of carboxylase to oxygenase activities decreased with increasing water stress, possibly due to a reduced CO_2/O_2 ratio caused by increased stomatal resistance to influx of CO_2. However, since evaporative cooling is reduced by increased stomatal resistance, the elevated leaf temperature may also be the cause of the ratio change. As temperature increases, CO_2 solubility decreases at a greater rate than O_2 solubility, therefore the CO_2/O_2 ratio at the site of carboxylation may be altered. The resulting CO_2/O_2 ratio would favor oxygenase activity relative to carboxylase activity and thus enhance photorespiration. Our own results with field-grown cotton suggest that the ratio of gross to net photosynthesis is not drastically altered by water stress and the main effect is on total CO_2 assimilation rates (Fig. 2). The rates of each process are reduced by stress but the ratio of gross photosynthesis/photorespiration is unchanged, suggesting a reduction in total activity (ratio of active enzyme/total enzyme) rather than a change in the carboxylase/oxygenase ratio. The carboxylase/oxygenase ratio is temperature-dependent, but appeared to be only slightly affected by water stress in this instance.

Osmond et al. (1980) have recently proposed that through the combined efforts of the stomatal and chloroplastic activities (nonstomatal), the internal CO_2 concentration remains essentially constant as stress develops (except where severe stress develops very rapidly). The internal CO_2 concentration maintained through both stomatal and nonstomatal adjustments is

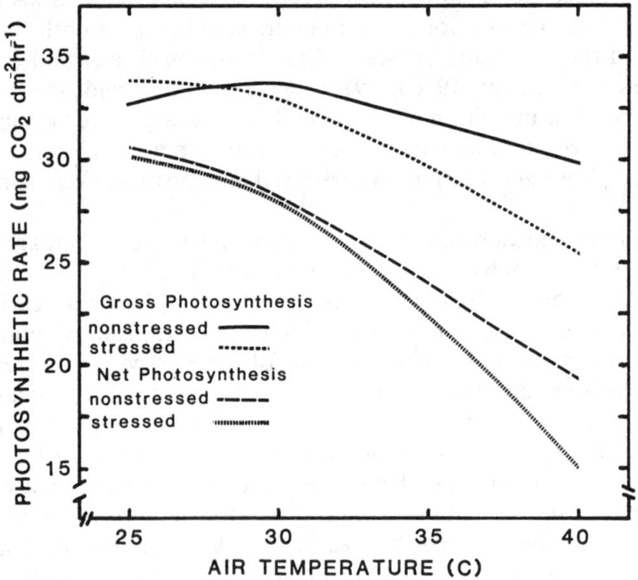

Fig. 2—Gross and net CO_2-exchange rates for field-grown cotton as functions of air temperature and plant water stress. Data were obtained from plants grown under field conditions throughout the course of the growing season for one genotype (S. W. Perry and D. R. Krieg, unpublished data).

presumably a level that is optimal for carbon assimilation in terms of transpirational water loss at the prevailing leaf temperature and evaporative demand. Photorespiration in C_3 plants may serve to maintain a level of CO_2 in the chloroplast that is adequate to dissipate energy produced in the light reactions and to prevent photoinhibition of the chloroplast system under high irradiance conditions. The mechanism involved in prevention of photoinhibition in C_4 plants is unclear.

III. UTILIZATION OF PHOTOSYNTHATE

Green plants can be divided into sources and sinks with respect to photosynthate production and accumulation. Fully expanded leaves are the major sources with some minor contributions from green stem and fruit tissue when present. Sinks include all nonphotosynthetic tissue and expanding leaves until they become fully autotrophic ($>60\%$ of maximum size).

A. Utilization by Source

Each source must retain a certain percentage of daily production for maintenance. Using fully expanded cotton leaves (a C_3 type) as an example, approximately 30% of the daily photosynthate is retained when water supplies are optimal (Table 1). Expressed as mg CH_2O g^{-1} dry weight of leaf tissue, this retention translates to 5–9% of the dry weight of the leaf, depending upon age and source-sink relationships. In sorghum (a C_4 type), retention represented 15–20% of the daily production and 3–5% when expressed on a dry-weight basis. Water stress greatly increased the retention percentage, when expressed on the basis of daily production, primarily due to reduced production. When stress was relatively mild, the retention percentage also increased on a dry-weight basis, suggesting that maintenance costs or assimilate storage may have been altered. This trend is most evident

Table 1—Source activity of cotton and sorghum as affected by water deficit during the fruit-development period.

	Cotton		Sorghum	
	Nonstressed†	Stressed†	Nonstressed†	Stressed†
Leaf area (dm^2 $leaf^{-1}$)	0.80	0.65	2.57	2.48
Avg daily net photosynthetic rate (mg CO_2 dm^{-2} h^{-1})	25	22	45	29
Daily assimilate produced by source leaf (mg CH_2O $leaf^{-1}$)	150	107	590	370
Daily assimilate retained by source leaf				
%	29	40	15	28
mg CH_2O g dry wt^{-1}	75	78	49	58

† Midday leaf water potentials of nonstressed leaves were −15 bars and of stressed leaves were −20 to −22 bars.

when fruit production is occurring in both cotton and sorghum. Wilson et al. (1980) have examined the carbon balance of young sorghum plants under water stress and have indicated that as stress intensified, the maintenance costs decreased by 50%. No explanation for the decrease was proposed. One could speculate that as the plant begins to experience water stress (especially when coupled with a high heat load) the rate of protein turnover would increase initially and then would decline as the stress became very severe. Tissue age and state of metabolic activity would influence the stress intensity required to alter maintenance costs.

B. Translocation

The rate of transport of assimilate from the source to the sink is affected by the rate of assimilate production in the source, rates of loading and unloading of the sieve-tube elements, velocity of flow in the phloem, and the rate of incorporation by the sink (Begg & Turner, 1976). The factors previously discussed that affect assimilate production directly affect the rate of transport of the assimilate from the source, since movement of photosynthate from the mesophyll cells to the conducting tissue is largely a diffusion process dependent upon concentration gradients. Phloem loading is defined as the process in which the major translocated species are transported into and concentrated in the conducting elements of the phloem of a translocation source (Geiger, 1976). Since phloem loading is an active, energy-requiring process, it represents one potential rate-limiting step in the translocation system. Membrane-transport capacity may be altered by stress due to conformational changes in membrane structure and to membrane-composition changes with prolonged water stress (Lee-Stadelmann & Stadelmann, 1976). It is unclear whether proposed membrane changes could impose limitations to assimilate utilization.

The increase in solute concentration in the phloem at the loading site results in water movement into the phloem, creating a hydrostatic pressure gradient that causes movement to occur. The total water potential of the tissue and the osmotic potential generated by phloem-loading determine the magnitude of the hydrostatic pressure gradient and thus determine the rate of movement by mass flow (Wardlaw, 1974). Phloem translocation should not be severely affected by water stress as long as a pressure gradient exists between source and sink. It is generally observed that the translocation process is more tolerant of desiccation than in source or sink activity (Wardlaw, 1976; McPherson & Boyer, 1977; Sung & Krieg, 1979) as indicated in Fig. 3.

C. Partitioning of Assimilate

The direction and rate of assimilate flow is a function of the hydrostatic pressure gradient developed and maintained between a source and sink (Crafts & Crisp, 1971). The pressure gradient maintained is a function

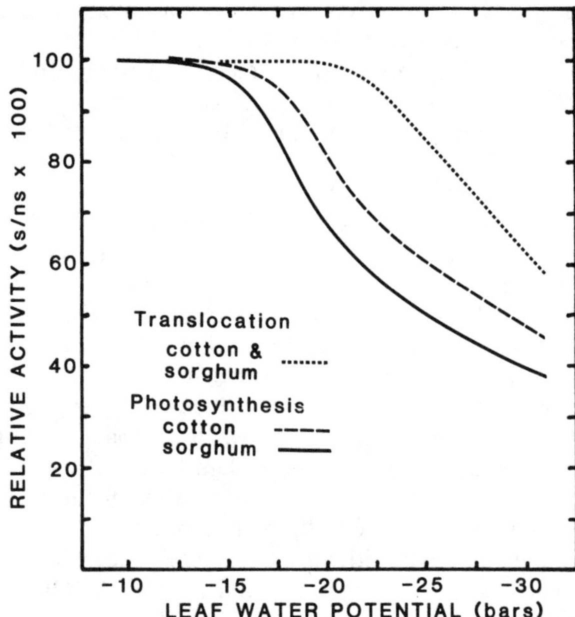

Fig. 3—Relative sensitivity of photosynthesis and rate of ^{14}C export from individual leaves of cotton and sorghum as function of leaf water potential. The response of ^{14}C export rate to decreasing leaf water potential was similar when expressed in this manner. Data derived from Karami et al. (1980) and Sung and Krieg (1979).

of source activity (photosynthetic rate, phloem-loading rate, and xylem water potential), distance between the source and sink, and sink activity (phloem-unloading rate, sink size, and strength). Water stress can alter the partitioning pattern primarily through reductions in the strength of individual sinks (Wardlaw, 1976), reflecting differential sensitivities to water stress. The rate of translocation is altered by water stress through effects on the source's capacity to supply assimilate or the sink's ability to utilize the assimilate, either of which affects the steepness of the pressure gradient.

In actively expanding tissue, the growth processes of cell division and cell expansion are very sensitive to water stress (Hsaio, 1973; McCree & Davis, 1974; Hsiao & Bradford, Chapter 6A, this book). During rapid vegetative growth prior to flowering, leaf area increases, inflorescence or floral bud differentiation and development occurs, and the root system expands rapidly. All three systems represent potentially strong sinks. If water stress develops during this growth stage, the size of the developing leaves and reproductive structures are usually reduced to a greater extent than the root system (Begg & Turner, 1976). In determinate plants, leaf expansion is complete by the time of flowering. Once seed development begins it usually becomes the strongest sink. Root growth is usually observed to be reduced during the phase of rapid dry matter accumulation by the grain and to resume in some species after grain maturity. The developing cereal seed is apparently relatively insensitive to water stress, as compared with the photosynthetic processes and source activity. Under water-deficit conditions, the seed-filling process occurs by drawing upon soluble carbon and nitrogen reserves in lower leaves and stems (Begg & Turner, 1976; Boyer & McPherson, 1975; McPherson & Boyer, 1977), and by becoming a stronger source itself, if it has photosynthetic capability (Pasternak & Wilson, 1976; Johnson et al., 1974). McPherson and Boyer (1977) have suggested that, in cereals, grain yield is more highly correlated with total seasonal photosynthetic activity than that produced only during the grain-filling period.

Indeterminate plants continue to produce new leaves simultaneous with fruit production initiated from axillary buds. As the fruit load increases, leaf initiation usually slows or ceases, suggesting source-sink control (Mauney, 1979). Whether the cessation of leaf initiation is due to limited assimilate, or to hormonal changes associated with the fruit load, is still a subject to debate. Water stress not only reduces the size of the expanding leaves, but also hastens the cessation of leaf initiation, resulting in a considerably reduced source size on a whole-plant basis. Although source size and activity are greatly reduced, it is not clear whether the primary effects of water stress are on the source or the sink, especially the fruit. In cotton, there are some assimilate-partitioning differences related to fruit production, which would suggest source limitations (Krieg & Sung, 1979). The main stem leaves contribute more assimilate to the fruit on attending sympodial branches under stressed conditions. Within the sympodial branch large differences in partitioning of assimilate from individual source leaves are observed and are dependent upon sink age. Fruits that were between 15 and 30 days old were extremely strong sinks and detracted from younger or older sinks on the same branch. Young fruits (< 10 days old) usually abscised under stress, possibly reflecting lack of sufficient assimilate for rapid growth due to excessive competition from stronger sinks on the same branch. However, fruits of this age have extremely high concentrations of growth-retarding hormones, which may be responsible for the reduced sink strength and abscission under stress (Guinn, 1979; McMichael et al., 1972). Within the cotton fruit itself some assimilate partitioning differences are also observed (Sung, 1978). The seeds are the major sink at all fruit ages but their strength decreased under stress, possibly indicating some alteration in seed metabolism. No definitive explanation has emerged as to the major effect of water stress on source-sink relations in indeterminate plants, especially those producing high concentrations of oil and protein in the seed.

D. Sink Activity

The daily carbon accumulated through the photosynthetic process is utilized by the plant for both growth and maintenance purposes. Daily dry matter accumulation by a specific sink or for a whole plant can be described as:

$$\mathrm{DMP} = a(P) - b(\mathrm{DM}), \quad [1]$$

where

DMP = dry matter production per day,
a = coefficient of conversion of assimilate to final product,
P = daily amount of assimilate available in g of CH_2O,
b = maintenance coefficient, and
DM = existing dry weight.

On a whole-plant basis, a averages about 0.70 (McCree & Van Bavel, 1977). The type of final product being assimilated greatly influences a, with lipids having a conversion efficiency of about 0.3, and carbohydrates, about 0.87 (Penning de Vries, 1975). The maintenance coefficient ranges from 0.03 to 0.07 and is more related to the protein concentration of the tissue than to the dry weight.

Dry matter production rates under water-stress conditions are a function of source activity (size and strength) and sink activity (size and strength). The effect of water stress on activity of individual sinks, especially fruit or seed, is largely unknown at this time. However, the conversion efficiency (a) of the assimilate to final product is apparently unaffected by stress (Wilson et al., 1980).

IV. WHOLE-PLANT CARBON ASSIMILATION

As water stress develops, reductions occur in both leaf area and photosynthetic rate, reducing the total assimilating capacity of the plant (Hsaio et al., 1976). The extent of the leaf area changes are related to growth habit and stage of development. Determinate species complete leaf development prior to the onset of seed formation. Stress prior to flowering reduces the area of the developing leaves, whereas stress after flowering increases lower leaf senescence, reducing whole-plant leaf area. In most determinate species, the size of the inflorescence (potential number of seed) is reduced proportionately with leaf area when stress exists prior to flowering. Since seed number is a major yield determinate, this reduction represents an irreversible yield loss unless seed weight increases can compensate (Evans & Wardlaw, 1976). Stress during the seed-filling stage usually results in seed weight reductions, but the extent of the damage is not nearly as great as are seed number reductions.

In indeterminate plants, water stress not only reduces blade size but also has an effect on reducing leaf number by causing the apical meristem to go dormant prior to its nonstressed counterpart.

The relative distribution of various aged leaves also plays a major role in yield determination, since photosynthetic activity is inversely related to leaf age after maximum size has been attained (Jordan et al., 1975; Karami et al., 1980). In determinate plants, seed production occurs with a constantly aging leaf system in which the photosynthetic rate declines rather rapidly. Water stress hastens the rate of deterioration. In an indeterminate plant, a wider range of leaf ages exist during fruit production, providing a higher photosynthetic potential during part of the total fruit-development period. The contribution of the younger, more active leaves to fruit development may be small since they are close to an active sink, namely, new leaf production.

In essentially every case, whole-plant carbon assimilation capacity declines as stress develops. The resultant effect on yield depends upon the relative contribution of newly assimilated carbon, compared with utilization of preanthesis accumulated dry matter for fruit production. Large species differences exist in these responses.

V. FUTURE RESEARCH NEEDS

Several aspects of carbon metabolism as related to water-use efficiency need clarification. Species differences are known to exist in water-use efficiency due to the type of photosynthetic system present (C_3 or C_4). However, do real genotypic differences exist in photosynthesis/transpiration ratios within a species? Or is photosynthesis so closely coupled with stomatal conductance that significant improvements are impossible? Photosynthetic rate differences for single leaves have been reported in certain species among select genotypes. However, no definitive information exists concerning genotypic differences in photosynthesis/transpiration ratios for single leaves or whole plants.

Based upon our data, it is apparent that leaf conductance and photosynthesis are correlated at low conductance levels, but uncoupled at high conductance values. Additionally, large genotypic differences existed in both sorghum and cotton, offering the possibility for genetically modifying water-use efficiency for high-evaporative-demand environments. Results from low-evaporative-demand environments indicate a high dependence of photosynthesis on leaf conductance, suggesting little possibility of improvement of water-use efficiency.

Under mild or moderate stress, many plant species are able to maintain high stomatal conductance through turgor maintenance via osmotic adjustment. Does turgor maintenance result in the same effect on physiological and biochemical systems as it does on gas exchange? Most results indicate that growth and development are reduced in stressed plants independently of turgor maintenance via osmotic adjustment. The question then arises as to the worth of osmotic adjustment in a crop improvement program.

Another major concern to many physiologists and geneticists is the effect of photorespiration on whole-plant water-use efficiency. A major concern is the genetic vs. environmental control of RuBPCarboxylase-oxygenase activity. Considerable evidence exists to suggest that RuBPCase is subject to genetic modification. However, few reports indicate a direct control of CO_2-fixation rates by RuBPCase activity. Only limited information exists to describe the environmental influences on the ratio of carboxylase to oxygenase. However, major influences are due to CO_2/O_2 ratios, which are environmentally controlled and not subject to genetic manipulation.

In conclusion, much remains to be accomplished in terms of increasing photosynthesis and crop productivity to meet future demands. Since much of the available land resources for increasing productivity are subject to water deficits, major research efforts must be directed toward developing a better understanding of the relative order of photosynthetic limitations under stress. By understanding the order of limitations in crop growth and development and determining the degree of genetic variability in existing germplasm for each process, we can begin to develop higher yielding, more stress-tolerant genotypes.

LITERATURE CITED

Ackerson, R. C., D. R. Krieg, C. L. Haring, and N. Chang. 1977a. Effect of plant water status on stomatal activity, photosynthesis and nitrate reductase activity of field grown cotton. Crop Sci. 17:81–84.

Ackerson, R. C., D. R. Krieg, T. D. Miller, and R. E. Zartman. 1977b. Water relations of field grown cotton and sorghum: temporal and diurnal changes in leaf water, osmotic and turgor potentials. Crop Sci. 17:76–80.

Ackerson, R. C., D. R. Krieg, and F. J. M. Sung. 1980. Leaf conductance and osmoregulation of field grown sorghum genotypes. Crop Sci. 20:10–14.

Begg, J. E., and N. C. Turner. 1976. Crop water deficits. Adv. Agron. 28:161–217.

Boyer, J. S. 1976a. Photosynthesis at low water potentials. Philos. Trans. R. Soc. London Ser. B. 273:501–512.

Boyer, J. S. 1976b. Water deficits and photosynthesis. p. 153–190. *In* T. T. Kozlowski (ed.) Water deficits and plant growth. Vol. IV. Academic Press, New York.

Boyer, J. S., and H. G. McPherson. 1975. Physiology of water deficits in cereal crops. Adv. Agron. 27:1–23.

Brown, K. W., W. R. Jordan, and J. C. Thomas. 1976. Water stress induced alterations of the stomatal response to decreases in leaf water potential. Physiol. Plant 37:1–5.

Crafts, A. S., and C. E. Crisp. 1971. Phloem transport in plants. W. H. Freeman and Co., San Francisco.

Cutler, J. M., and D. W. Rains. 1977. Effects of irrigation history on responses of cotton to subsequent water stress. Crop Sci. 17:329–335.

Evans, L. T., and I. F. Wardlaw. 1976. Aspects of the comparative physiology of grain yield in cereals. Adv. Agron. 28:301–359.

Geiger, D. R. 1976. Phloem loading in source leaves. p. 167–184. *In* I. F. Wardlaw and J. B. Passioura (ed.) Transport and transfer processes in plants. Academic Press, New York.

Guinn, G. 1979. Hormonal relations in flowering, fruiting and cutout. p. 265–276. *In* J. M. Brown (ed.) Proc. Beltwide Cotton Prod. Res. Conf., Phoenix, Ariz. 7–11 Jan. 1979. National Cotton Council, Memphis, Tenn.

Hsiao, T. C. 1973. Plant responses to water stress. Annu. Rev. Plant Physiol. 24:519–570.

Hsiao, T. C., E. Fereres, E. Acevedo, and D. W. Henderson. 1976. Water stress and dynamics of growth and yield of crop plants. p. 282–305. *In* O. L. Lange et al. (ed.) Water and plant life. Springer-Verlag, New York.

Huffaker, R. C., T. Radin, G. E. Kleinkopf, and E. S. Cox. 1970. Effects of mild water stress on enzymes of nitrate assimilation and the carboxylative phase of photosynthesis in barley. Crop Sci. 10:471–474.

Johnson, R. R., N. M. Frey, and D. N. Moss. 1974. Effect of water stress on photosynthesis and transpiration of flag leaves and spikes of barley wheat. Crop Sci. 14:728–731.

Jones, M. M., and H. M. Rawson. 1979. Influence of rate of development of leaf water deficit upon photosynthesis, leaf conductance, water use efficiency and osmotic potential in sorghum. Physiol. Plant. 45:103–111.

Jones, M. M., and N. C. Turner. 1978. Osmotic adjustment in leaves of sorghum in response to water deficits. Plant Physiol. 61:122–126.

Jordan, W. R., K. W. Brown, and J. C. Thomas. 1975. Leaf age as a determinant in stomatal control of water loss from cotton during water stress. Plant Physiol. 56:595–599.

Karami, E., D. R. Krieg, and J.E. Quisenberry. 1980. Water relations and carbon-14 assimilation of cotton with different leaf morphology. Crop Sci. 20:421–426.

Krieg, D. R., and F. J. M. Sung. 1979. Source-sink relations of cotton as affected by water stress during boll development. J. M. Brown (ed.) Proc. Beltwide Cotton Prod. Res. Conf., Phoenix, Ariz. 7–11 Jan. 1979. National Cotton Council, Memphis, Tenn.

Lawlor, D. W. 1979. Effects of water and heat stress on carbon metabolism of plants with C_3 and C_4 photosynthesis. p. 303–326. *In* H. Mussell and R. C. Staples (ed.) Stress physiology in crop plants. John Wiley & Sons, New York.

Lee-Stadelmann, O. Y., and E. J. Stadelmann. 1976. Cell permeability and water stress. p. 268–280. *In* O. L. Lange (ed.) Water and plant life. Springer-Verlag, New York.

Mauney, J. R. 1979. Production of fruiting points. p. 256–261. *In* J. M. Brown (ed.) Proc. Beltwide Cotton Prod. Res. Conf., Phoenix, Ariz. 7–11 Jan. 1979. National Cotton Council, Memphis, Tenn.

McCree, K. J., and S. D. Davis. 1974. Effect of water stress and temperature on leaf size and on size and number of epidermal cells in grain sorghum. Crop Sci. 14:751–755.

McCree, K. J., and C. H. M. van Bavel. 1977. Respiration and crop production: case study with two crops under water stress. p. 117–216. *In* J. J. Landsberg and C. V. Cutting (ed.) Environmental effects on crop physiology. Academic Press, New York.

McMichael, B. L., W. R. Jordan, and R. D. Powell. 1972. Abscission processes in cotton: induction by plant water deficit. Agron. J. 65:602–604.

McPherson, H. G., and J. S. Boyer. 1977. Regulation of grain yield by photosynthesis in maize subjected to water deficiency. Agron. J. 59:714–718.

Osmond, C. B., K. Winter, and S. B. Powles. 1980. Adaptive significance of carbon dioxide recycling during photosynthesis in water stressed plants. p. 139–154. *In* N. C. Turner and P. J. Kramer (ed.) Adaptation of plants to water and high temperature stress. John Wiley & Sons, New York.

O'Toole, J. C., R. K. Crookston, K. J. Trehorne, and J. L. Ozbun. 1976. Mesophyll resistance and carboxylase activity. A comparison under water stress conditions. Plant Physiol. 57: 465–468.

Pasternak, D., and G. L. Wilson. 1976. Photosynthesis and transpiration in the heads of droughted grain sorghum. Aust. J. of Exp. Agric. and Anim. Husb. 16:272–275.

Penning de Vries, F. W. T. 1975. Use of assimilates in higher plants. p. 459–480. *In* J. P. Cooper (ed.) Photosynthesis and productivity in different environments. Cambridge University Press, Cambridge, England.

Sung, F. J. M. 1978. Source-sink relationships of sorghum and cotton as affected by water stress. Ph.D. Diss. Texas Tech University, Lubbock. (Diss. Abstr. 7904974; Vol. 39).

Sung, F. J. M., and D. R. Krieg. 1979. Relative sensitivity of photosynthetic assimilation and translocation of ^{14}C to water stress. Plant Physiol. 64:852–856.

Thomas, J. C., K. W. Brown, and W. R. Jordan. 1976. Stomatal response to leaf-water potential as affected by preconditioning water stress in the field. Agron. J. 68:707–708.

Turner, N. C. 1979. Drought resistance and adaptation to water deficits in crop plants. p. 343–372. *In* H. Mussell and R. C. Staples (ed.) Stress physiology in crop plants. John Wiley & Sons, New York.

Turner, N. C., and M. M. Jones. 1980. Turgor maintenance by osmotic adjustment: a review and evaluation. p. 87–104. *In* N. C. Turner and P. J. Kramer (ed.) Adaptation of plants to water and high temperature stress. John Wiley & Sons, New York.

Wardlaw, I. F. 1974. Phloem transport: physical, chemical or impossible. Annu. Rev. Plant Physiol. 25:515–535.

Wardlaw, I. F. 1976. Assimilate partitioning: cause and effect. p. 381–392. *In* I. F. Wardlaw and J. B. Passioura (ed.) Transport and transfer processes in plants. Academic Press, New York.

Wilson, D. R., C. H. M. van Bavel, and K. H. McCree. 1980. The carbon balance of water-deficient grain sorghum plants. Crop Sci. 20:153–159.

CHAPTER 7C

Whole-Plant Response to Water Deficits: Water Deficits and the Nitrogen Economy

ANDREW D. HANSON AND WILLIAM D. HITZ

Michigan State University
East Lansing, Michigan

I. INTRODUCTION

The growth of many well-watered annual seed crops can be divided into two phases with respect to the N economy: a first phase dominated by acquisition of reduced N, and a second phase dominated by reallocation of r e-duced N from senescing leaves and stems to the seeds. These two phases can be discerned both in cereals growing on NO_3^--N, and in nodulated legumes, and correspond roughly to the periods of vegetative and reproductive growth. Data that show the two phases very clearly are given in Fig. 1 for oats, a temperate-zone cereal, and cowpea, a tropical legume. Recognizing that the two phases can often overlap, with net N acquisition continuing during seed-filling, it is convenient to separate the effects of water deficit upon the crop N economy according to whether the deficit occurs during what would be the acquisition phase or the reallocation phase in well-watered conditions. Inasmuch as water deficit can simultaneously reduce N acquisition and advance the onset of foliar senescence and N reallocation, these two sets of effects are obviously interdependent.

Using as examples annual seed crops, we discuss below some research bearing on the consequences of water deficit for N acquisition and reallocation. The consequences of stress for N acquisition are, in general, presently somewhat better described than those for reallocation. We emphasize that N acquisition and reallocation are—in reality—inseparable from carbon metabolism and from growth and development as a whole.

II. ACQUISITION OF REDUCED NITROGEN

Whereas it is clear from short-term experiments that the potentials for nitrate reduction and for N_2 fixation are both sensitive to water stress (Fig. 2), for long-term growth under water stress the situation is not so clear-cut. In principle, the acquisition of reduced N could be affected (i) to a greater

Limitations to Efficient Water Use in Crop Production.

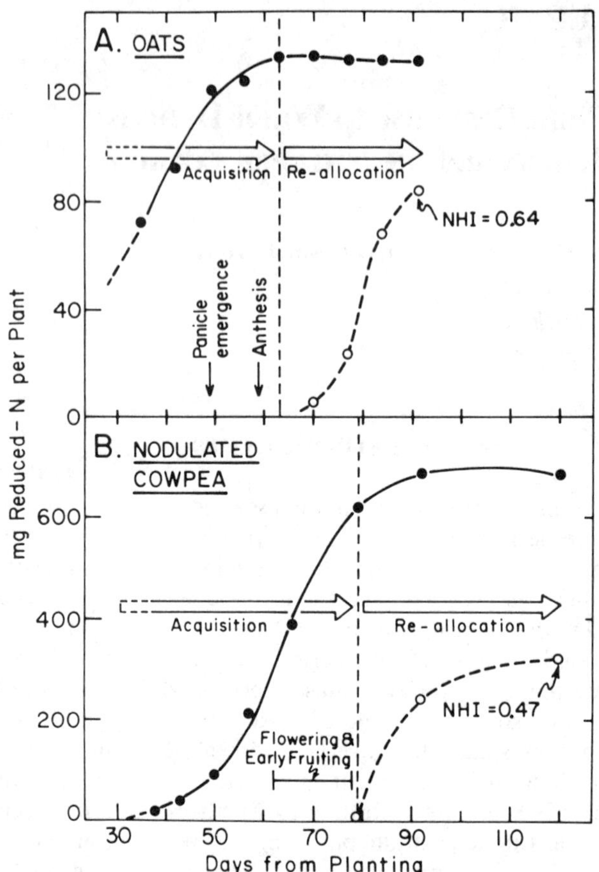

Fig. 1—Accumulation of reduced N in the aboveground parts of: (*A*) oats (cv. Orbit) grown in soil and supplied NO_3^--N and (*B*) nodulated cowpeas (cv. Caloona) grown in N-free mineral nutrient medium. ●———●: total reduced-N in aboveground parts; ○----○: reduced-N in seeds. The broken vertical lines separate the phases of N-acquisition and re-allocation. Data for oats are from Cataldo et al. (1975), and data for cowpea from Herridge and Pate (1977).

extent than growth, resulting in a decreased N concentration in the plant, or (ii) to a lesser extent, resulting in an increased N concentration (Viets, 1972). In case (i), but not (ii), growth-inhibition by water stress might be caused in part by a stress-induced N deficiency. There are interesting reports that case (i) sometimes holds for symbiotic N_2 fixation, and case (ii) for NO_3^- uptake and reduction.

In plants whose N source is soil NO_3^-, N concentration in dry matter is often increased by water stress provided that roots still have access to soil water containing NO_3^- (Viets, 1972; Arnon, 1972), whereas a decrease in the N concentration may occur in water-stressed legumes (Sprent, 1976). Some data illustrating these contrasting effects are collected in Table 1. Unfortunately, such data for NO_3^--grown plants are not definitive because

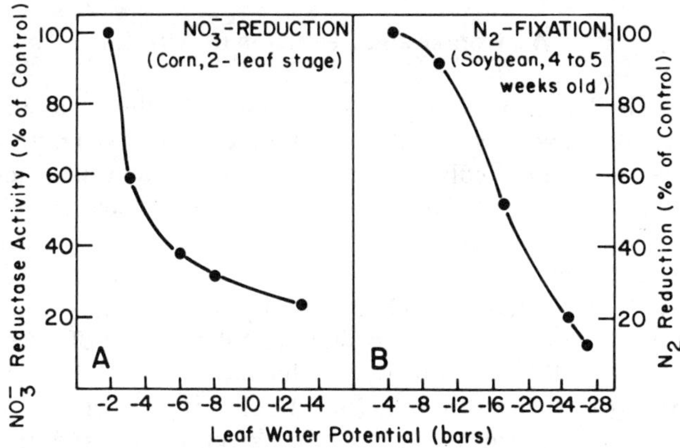

Fig. 2—Effect of short-term (4–5 days) water deficit on NO_3^- reduction by corn seedlings (*A*), and on N_2-fixation by vegetative soybean plants (*B*). Data are expressed as percent of the activity of well-watered control plants. Nitrate reductase data are from Morilla et al. (1973), and are for extracted enzyme; N_2-reduction data are from Huang et al. (1975), and are estimates based on the C_2H_2-reduction assay.

NO_3^- can accumulate in water-stressed crops, particularly in stem tissue (Hanway & Englehorn, 1958), and it is usually not clear that NO_3^--N was completely excluded from the reduced-N (Kjeldahl) determinations.

If the acquisition of reduced N by nodulated legumes is indeed more sensitive to water stress than is dry matter production, then can N_2 fixation itself restrict growth under water stress? If so, can adaptation of these crops to drought-prone environments be improved by breeding and management practices aiming to stabilize N_2 fixation during water stress? Such strategies would obviously not be suited for cereal improvement, if N-acquisition does not limit the growth of these crops during water stress.

Table 1—Effect of water stress on concentration of N in shoot dry matter harvested from plants grown either on NO_3^- fertilized soil or in N-free media.

Species†	Growth conditions/ principal N source	Percent change in N concentration of shoot dry matter‡	References
Sudangrass	Field/NO_3^-	+28	Bennett et al. (1964)
Millet	Field/NO_3^-	+28	Bennett et al. (1964)
Sorghum	Field/NO_3^-	+22	Bennett et al. (1964)
Alfalfa	Pot/NO_3^-	+ 3	Bourget & Carson (1962)
Oats	Pot/NO_3^-	+22	Bourget & Carson (1962)
Soybean	Field/N_2	−14	Hunt et al. (1979)
Soybean	Pot/N_2	−12	Kuo & Boersma (1971)
White clover	Pot/N_2	−31	Engin & Sprent (1973)
Field pea	Pot/N_2	−11	Minchin & Pate (1975)

† In all cases except oats, leaf or shoot tissue was harvested before seed-filling; oat shoots (grain + straw) were harvested at maturity.

‡ Concentrations of N were determined by Kjeldahl analysis. Percent change = [(N concentration of shoot dry matter of stressed plants/N concentration of shoot dry matter of well-watered plants) × 100] − 100.

A. Uptake and Reduction of Nitrate

The assimilation of NO_3^- in water-stressed seedlings has been studied extensively. As shown by Fig. 2A, nitrate reductase (NR) activity in corn seedling leaves is markedly decreased by moderate water deficit (e.g., Morilla et al., 1973). Similar decreases have been noted in other cereals (Huffaker et al., 1970; Rajagopal et al., 1977). In general, the sensitivity of NR to stress is similar to that of protein synthesis and of extension growth (Hsiao, 1973). The decreased NR activity is most probably linked to a decrease in the size of a specific and small metabolic pool of NO_3^-, which is concerned with NR induction, rather than to any overall decline in leaf NO_3^- content (Shaner & Boyer, 1976). The size of the metabolic NO_3^- pool is apparently determined by the NO_3^- flux from root to shoot; this flux is normally lowered during water stress (Shaner & Boyer, 1976). In summary, work with seedlings exposed to fairly brief episodes of water stress suggests that the uptake and reduction components of NO_3^- assimilation show coordinated responses to stress, which keep NO_3^- assimilation in balance with growth as a whole. It should be noted that this picture is not consistent with some field observations; balanced responses of NO_3^- assimilation and growth should tend to prevent the accumulations of both unreduced NO_3^- and reduced N, which have often been encountered in field-grown crops subjected to long-term drought stress. Well-watered corn plants can compensate for a restricted N supply during grain-filling by utilizing NO_3^--N accumulated in the stem (Friedrich et al., 1979), and the capacity to accumulate NO_3^- and reduced-N in the stem shows genetic variation (Chevalier & Schrader, 1977) in well-watered conditions. Do similar considerations apply to the capacity to accumulate N during, and utilize N after, water stress? If so, genetic improvement of NO_3^- uptake and N utilization in dry environments may be possible.

B. Symbiotic Nitrogen Fixation

The inhibition of N_2 fixation in soybean illustrated by Fig. 2B is likely to be due to a combination of (i) direct effects of water stress on the nodule itself (Pankhurst & Sprent, 1975), and (ii) indirect nutritional effects stemming from a smaller supply of photosynthate from the host plant (Huang et al., 1975). Because direct effects on the nodule, such as structural and chemical changes that reduce the rate of O_2 diffusion, have been reviewed recently (Sprent, 1976), we will focus on indirect effects.

The supply of photosynthate (sucrose) to the nodule is partitioned approximately equally between (i) carbon for energy requirements and (ii) carbon skeletons for export of reduced N (e.g., Herridge & Pate, 1977; Minchin & Pate, 1973). A drop in photosynthate supply may thus inhibit N_2 fixation either by decreasing energy availability and/or by causing the reduced products of N_2 fixation to accumulate in the nodule. The latter circumstance could result in a feedback inhibition of N_2 fixation (Pate et al., 1969). Each of these effects is consistent with the observed decreases in N concentration of water-stressed legumes (Table 1); their relative importance is not known. If the feedback effect is really an important one, is there a possibility of improving drought resistance by genetically modifying the symbiont and/or host legume to lessen the feedback effect on the whole N_2-fixation process?

III. REALLOCATION OF REDUCED NITROGEN

When seed-filling begins, the reduced N acquired during vegetative growth is mainly in leaf (and stem) proteins, especially ribulose 1,5 bisphosphate carboxylase (RuBPCase) (e.g., Dalling et al., 1976; Herridge & Pate, 1977; Huffaker & Peterson, 1974). The N stored in these proteins is withdrawn as leaves senesce and is translocated, principally in the phloem, to the seeds to meet the massive N-demand for seed growth (Fig. 1). This reduced N reallocation can be subdivided into components:

1) Turnover and net breakdown of leaf proteins.
2) Metabolism in the leaf of amino- and amide-N released by protein breakdown.
3) Phloem-transport of soluble N compounds from the leaf to the seed.
4) Amino acid interconversions and protein synthesis in the seed.

Rather little is known about these four components in either well-watered or drought-stressed crops. However, it is clear that these components must remain highly coordinated with each other, and with rates of current photosynthesis and carbohydrate remobilization, for successful seed-filling in the presence or absence of water stress. Coordination is mandatory because seed growth demands continuous, simultaneous supplies of C and N, with C/N ratios falling within limits set by the seed protein content of the genotype (e.g., Carasco et al., 1978; Dalling et al., 1976; Pate et al., 1979). The coordination is highly effective until a very late stage of senescence in well-watered crops, because in both cereals (Desai & Bhatia, 1978) and legumes (Jeppson et al., 1978) the nitrogen harvest index (NHI) can exceed 70%. General agronomic considerations show that the components also remain coordinated during senescence provoked by late-season (reallocation-phase) water stresses. Although seed yield is usually reduced by such stresses, with much of the yield reduction accounted for by a fall in harvest index (HI) (e.g., Donald & Hamblin, 1976), N movement to, and protein storage in, the seed is proportionately less affected because late-season water stress tends to increase the seed protein percentage in both cereals and legumes (e.g., Evans et al., 1975; Jurgens et al., 1978; Shibles et al., 1975). It follows that NHI falls less than HI in response to late-season stress; this is supported by high NHI values reported for dryland wheat (up to 61%) and triticale (up to 58%) (Lal et al., 1978).

In view of the above, and because (i) the protein-synthesizing machinery in the developing seeds remains active even during a drought-stress severe enough to kill most of the leaves (e.g., Jurgens et al., 1978), and (ii) N and C translocation to the seed can continue even when the supply of current photosynthate is cut drastically by water stress (e.g., Johnson & Moss, 1976; Jurgens et al., 1978; Tully & Hanson, 1979), the following speculation on the coordination of N reallocation seems reasonable. Could both the capacity of the seed to utilize imported N (component 4) and the capacity of the phloem to transport N (component 3) remain relatively high during stress? The overall rate, extent, and coordination of N reallocation during drought might then be governed principally by components 1 and/or 2 (i.e., the processes supplying N from the leaves). Selected aspects of components 1 and 2 are discussed below.

A. Turnover and Net Breakdown of Leaf Proteins (Component 1)

Because data available for both unstressed and water-stressed leaves are mainly for quite young plants, extrapolation to leaves of mature crop canopies is problematic. There are a few studies on foliar protein catabolism during grain maturation, unfortunately only in the absence of water stress (e.g., Dalling et al., 1976; Feller et al., 1977; Nair et al., 1978). Based on limited information, it seems that in mature leaves of well-watered plants, where total protein content remains steady, the rate of protein turnover corresponds to breakdown and resynthesis of perhaps 10% of total protein each day (Huffaker & Peterson, 1974; Penning de Vries, 1975). However, among the various leaf proteins, RuBPCase—the key enzyme in CO_2 fixation and the major soluble protein of the leaf—may be remarkably stable, turning over far more slowly than other proteins in the absence of senescence or stress (Huffaker & Peterson, 1974). During senescence induced by darkening or detaching leaves, there is a rapid net breakdown of total soluble protein, generally accompanied by some increased protease activities and (in attached leaves) by net N export. The initial loss of soluble protein has been shown in cereals to be almost entirely due to loss of RuBPCase (e.g., Peterson & Huffaker, 1975; Wittenbach, 1978). It thus appears that RuBPCase protein, while synthesized and degraded much more slowly than other proteins in mature leaves, may be rapidly and preferentially broken down when senescence begins. In wheat seedlings, the initial stage (rapid RuBPCase loss) of dark-induced senescence lasts about 2 days and is reversible; return to light results in restoration of soluble protein level, chlorophyll content, and photosynthetic capability (Wittenbach, 1977). After 2 days, an irreversible stage of senescence begins, when non-RuBPCase protein is rapidly lost and a return to the light fails to prevent further net protein breakdown and N export (Wittenbach, 1977, 1978).

Both sudden and slowly developing drought probably result in a pattern of senescence with most of the above features. As illustrated for barley in Fig. 3, there seems to be an initial stage (essentially reversible by rewater-

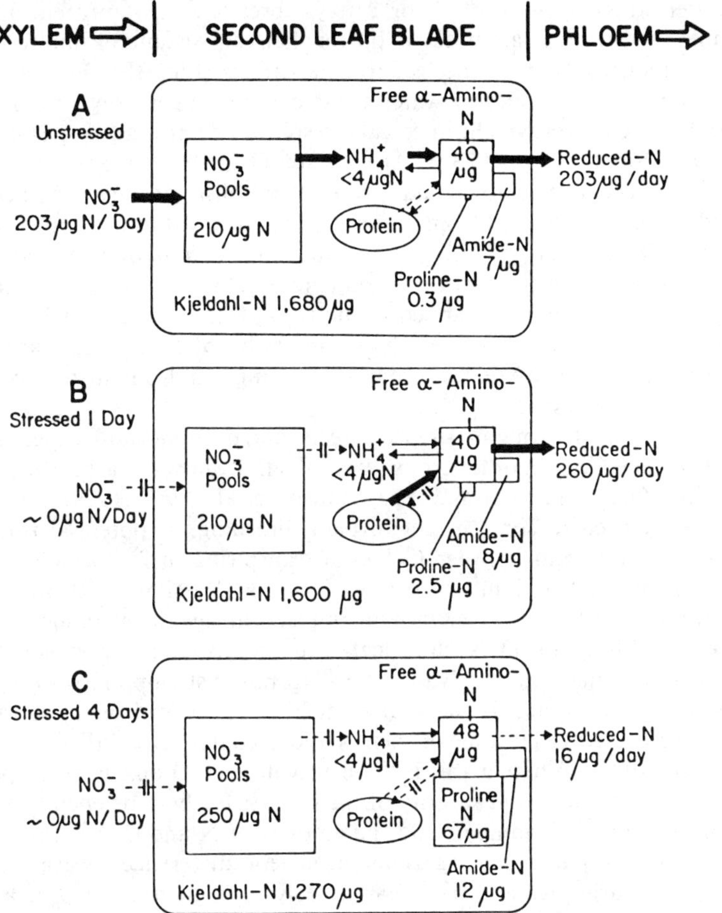

Fig. 3—Nitrogen budgets for the second leaf blade of Proctor barley plants at the three-leaf stage. (*A*) well-watered plants; (*B*) plants wilted for 1 day (ψ_{leaf} lowered by about 10 bars), with the second leaf blade still able to recover completely upon rewatering; (*C*) plants wilted for 4 days (ψ_{leaf} lowered by >30 bars), with leaf-firing extending from the tip to about the midpoint of the second leaf blade. Heavy solid arrows indicate problem major paths of N flow, light broken arrows indicate flows that are probably small, light solid arrows designate flows of uncertain size, and double bars on arrows show processes known to be inhibited during water stress. Square pool areas are proportional to their N contents. Values of α-amino-N exclude proline-N. Large boxes containing tl e various pools represent second leaf blades, which import NO_3^- from the xylem at the left, and export reduced N via the phloem at the right. (After Tully et al., 1979).

ing, Fig. 3B) during which total protein synthesis is suppressed, net breakdown of soluble protein begins, and net N export occurs (e.g., Hsiao, 1973; McMichael & Elmore, 1977; Singh et al., 1973a, 1973b; Tully et al., 1979). Certain protease activities may rise (Soong & Hageman, 1977). Because RuBPCase activity is reduced by moderate water stress (e.g., Huffaker et al., 1970; Johnson et al., 1974; O'Toole et al., 1976), the initial protein loss may involve preferential degradation of RuBPCase protein. During episodes of water stress lasting only 1 or 2 days, breakdown of protein appears capable of serving as an effective buffer that can sustain an uninterrupted flow of reduced N from the leaf to sink organs (Fig. 3B). Following the "reversible" stage, relief of water stress can no longer prevent continued degradative changes which, in cereals, result in leaf-firing (e.g., Hanson et al., 1977; Singh et al., 1973b). At some point in this "irreversible" stage (Fig. 3C) protein breakdown and further metabolism of the amino acids released become uncoupled from N export, as phloem translocation eventually ceases (Tully et al., 1979). There results an accumulation of free amino acid N especially in the form of proline (see below), which, together with unhydrolyzed protein, is trapped in the firing-leaf tissue and is lost to the plant (Hanson et al., 1977). Both accumulation of free proline and poor retrieval of foliar N are features that may distinguish drought-induced from other types of senescence.

Leaf-firing in drought-stressed cereals, and drought-induced leaf senescence and shedding in other crops, are traits that show intraspecific genetic variation, (e.g., Chang et al., 1974; Hanson et al., 1977, 1979) and which are already used to some extent in plant breeding. A potential trade-off exists with such traits between (i) loss of transpiring surface as a long-term means of regulating plant and soil water status (Passioura, 1976), and (ii) loss of photosynthetic tissue, with incomplete salvage of the tissue N and C contents. Although traits such as leaf-firing are usually viewed as deleterious, it may sometimes be advantageous, from the standpoint of long-term growth and seed yield, to allow leaves to fire (or to shed them) so rapidly during water stress that some of their N and C content is sacrificed (Hanson & Nelsen, 1979). Whether this is the case will depend on the crop species; the timing, duration, and severity of the stresses likely to be encountered in the production environment; and the amount of N and C that can be salvaged. By combining investigations on genetic differences in leaf protein catabolism during water stress—particularly on regulation of protease activity—with soil and climate data, is it possible to define ideotypes helpful to breeders seeking to exploit genetic variability in leaf-firing to improve dryland crops? Genetic differences in foliar senescence and N reallocation patterns have already been used successfully in crop breeding for non-water-stress environments (e.g., Helsel & Frey, 1978).

B. Metabolism in the Leaf of Reduced Nitrogen Released by Protein Breakdown (Component 2)

1. THE END-PRODUCTS EXPORTED

Major end-products of leaf protein catabolism are the soluble reduced-N compounds that enter the phloem and move to the developing fruits and seeds; little is known about such compounds in crops, particularly in relation to water stress. Although fruits and seeds are generally capable of NO_3^--reduction, NH_3^--assimilation, and amino acid interconversions (e.g.,

Donovan & Lee, 1978; Duffus & Rosie, 1978), the relative amounts of amino acids supplied to the seed may influence seed protein content and composition (Holowach et al., 1979) as well as the metabolic cost of storage protein synthesis (Tully & Hanson, 1979).

From ^{14}C-labeling and phloem-exudation studies with nonstressed legume and cereal crops, it appears that the amino acids carrying most of the N are species-dependent, and that for each species there is a fairly small number of major N-carriers whose relative contributions to N movement remain similar in various environments and as N withdrawal from leaves progresses (e.g., Housley et al., 1979; Pate et al., 1977; Tully & Hanson, 1979). It appears also that the molar ratio of amino acids/sucrose in the phloem of both cereals and legumes can approach or exceed 1:1 (Pate et al., 1977; Tully & Hanson, 1979); amino acid supply to the phloem in these crops clearly contributes substantially to the turgor pressure gradient required for the phloem-transport process.

The effect of water stress on the phloem amino acids in seedling leaves and flag leaves of barley has been investigated with phloem exudation and with $^{13}NH_3$ (Hanson & Tully, 1979; Tully & Hanson, 1979). In the absence of stress, major amino acids were glutamine, glutamate, serine, alanine, and aspartate. No drastic changes occurred in stressed leaves, despite the fact that water stress provoked rapid protein catabolism (Fig. 3; Tully et al., 1979). Stress increased the proportion of serine, and resulted in the appearance of small amounts of proline and γ-aminobutyric acid. There is thus as yet no evidence that hydrolysates of leaf proteins enter the phloem as such during leaf senescence in the presence (or absence) of drought. It follows that the metabolic interconversion pathways required to maintain production of the rather constant and narrow range of phloem amino acids remain fully functional during stress, at least for as long as translocation continues. Is it possible that such interconversions themselves are mandatory for phloem-transport of N, and are perhaps required also to generate solutes to drive the overall phloem-transport process?

2. LOSS OF VOLATILE NITROGEN COMPOUNDS

Although extensive protein catabolism takes place in wilted leaves, there is very little loss of volatile N under rather mild growth-chamber conditions (Tully et al., 1979), and NH_3 pools do not increase markedly in size (Fig. 3; Blum & Ebercon, 1976), presumably because the pathway for NH_3 assimilation remains highly active (Hanson & Tully, 1979). It has, however, been recently shown that senescing leaves of well-watered corn plants can release significant amounts of NH_3 at high light intensities (Farquhar et al., 1979), and that leaves of several field-grown crops can lose appreciable quantities of reduced-N compounds in transpiration vapors, particularly at high temperatures (Stutte & Weiland, 1978). Is retention of reduced N less efficient in crop canopies senescing in response to drought under high radiation levels in the field?

3. PROLINE ACCUMULATION

Proline accumulation is a characteristic biochemical feature of drought-induced leaf senescence in essentially all crops (see Hanson & Nelsen, 1979, for examples). Rapid proline accumulation in attached leaves does not usually begin until leaves are quite severely wilted, by which time net protein breakdown and net N export may have been in progress for some time (e.g., Blum & Ebercon, 1976; Routley, 1966; Tully et al., 1979; Waldren et al., 1974). As shown in Fig. 3C, although proline eventually becomes a major species among the free amino acids in severely wilted barley leaves, it accounts for only about 5% of the total N retained by such leaves.

Whereas the biochemical causes for proline accumulation are quite well-understood (Stewart & Hanson, 1980), the physiological significance of proline accumulation is unclear. The suggestions that proline is of major adaptive significance during water stress, and that proline-accumulating potential could serve as a metabolic index of drought-resistance in cereals (Singh et al., 1972, 1973b) have not been supported by a series of physiological and genetic experiments with barley (Hanson et al., 1977, 1979). Work with other crops, including rice (IRRI, 1975), sorghum (Blum & Ebercon, 1976), and *Saccharum* spp. (Moore & Maretzki, 1978), has also failed to show any clear link between proline accumulation and drought resistance. Extensive proline accumulation is therefore probably best viewed as a symptom of severe internal water deficit, and *not* as an adaptive metabolic response to water stress; selection for proline-accumulating potential in a crop breeding program would most likely achieve no genetic advance towards adaptation to drought (Hanson et al., 1977, 1979; Stewart & Hanson, 1980).

ACKNOWLEDGMENTS

The authors' research referred to in this article was carried out under Contract no. EY-76-C-02-1338, from the U.S. Department of Energy.

LITERATURE CITED

Arnon, I. 1972. Crop production in dry regions Vol. I. Barnes and Noble, New York. p. 209–211.

Bennett, O. L., B. D. Doss, D. A. Ashley, V. J. Kilmer, and E. C. Richardson. 1964. Effects of soil moisture regime on yield, nutrient content and evaporatranspiration for three annual forage species. Agron. J. 56:195–198.

Blum, A., and A. Ebercon. 1976. Genotypic responses in sorghum to drought stress. III. Free proline accumulation and drought resistance. Crop Sci. 16:428–431.

Bourget, S. J., and R. B. Carson. 1962. Effect of soil moisture stress on yield, water-use efficiency and mineral composition of oats and alfalfa grown at two fertility levels. Can. J. Soil Sci. 42:7–11.

Carasco, J. F., R. Croy, E. Derbyshire, and D. Boulter. 1978. The isolation and characterization of the major polypeptides of the seed globulin of cow pea (*Vigna unguiculata* L. Walp) and their sequential synthesis in developing seeds. J. Exp. Bot. 29:309–323.

Cataldo, D. A., L. E. Schrader, D. M. Peterson, and D. Smith. 1975. Factors affecting seed protein concentration in oats. I. Metabolism and distribution of N and carbohydrate in two cultivars that differ in groat protein concentration. Crop Sci. 15:19–23.

Chang, T. T., G. C. Loresto, and O. Tagumpay. 1974. Screening rice germplasm for drought resistance. Sabrao J. 6:9–16.

Chevalier, P., and L. E. Schrader. 1977. Genotypic differences in nitrate absorption and partitioning of N among plant parts in maize. Crop Sci. 17:897–901.

Dalling, M. J., G. Boland, and J. H. Wilson. 1976. Relation between acid proteinase activity and redistribution of nitrogen during grain development in wheat. Aust. J. Plant Physiol. 3:721–730.

Desai, R. M., and C. R. Bhatia. 1978. Nitrogen uptake and nitrogen harvest index in durum wheat cultivars varying in their grain protein concentration. Euphytica 27:561–566.

Donald, C. M., and J. Hamblin. 1976. The biological yield and harvest index of cereals as agronomic and plant breeding criteria. Adv. Agron. 28:361–405.

Donovan, G. R., and J. W. Lee. 1978. Effect of the nitrogen source on grain development in detached wheat heads in liquid culture. Aust. J. Plant Physiol. 5:81–87.

Duffus, C. M., and R. Rosie. 1978. Metabolism of ammonium ion and glutamate in relation to nitrogen supply and utilization during grain development in barley. Plant Physiol. 61: 570–574.

Engin, M., and J. I. Sprent. 1973. Effects of water stress on growth and nitrogen-fixing activity of *Trifolium repens.* New Phytol. 72:117–126.

Evans, L. T., I. F. Wardlaw, and R. A. Fischer. 1975. Wheat. p. 101–149. *In* L. T. Evans (ed.) Crop physiology. Cambridge University Press, London.

Farquhar, G. D., R. Wetselaar, and P. M. Firth. 1979. Ammonia volatilization from senescing leaves of maize. Science 203:1257–1258.

Feller, U. K., T.-S. T.Soong, and R. H. Hageman. 1977. Leaf proteolytic activities and senescence during grain development of field-grown corn (*Zea mays* L.). Plant Physiol. 59: 290–294.

Friedrich, J. W., L. E. Schrader, and E. V. Nordheim. 1979. N Deprivation in maize during grain-filling. I. Accumulation of dry matter, nitrate-N, and sulfate-S. Agron. J. 71:461–465.

Hanson, A. D., and C. E. Nelsen. 1979. Water: adaptation of crops to drought-prone environments. p. 77–152. *In* P. S. Carlson (ed.) Crop improvement. Academic Press, New York.

Hanson, A. D., C. E. Nelsen, and E. H. Everson. 1977. Evaluation of free proline as an index of drought resistance using two contrasting barley cultivars. Crop Sci. 17:720–726.

Hanson, A. D., C. E. Nelsen, A. R. Pedersen, and E. H. Everson. 1979. Capacity for proline accumulation during water stress in barley and its implications for breeding for drought-resistance. Crop Sci. 19:489–493.

Hanson, A. D., and R. E. Tully. 1979. The amino acids translocated from turgid and water-stressed barley leaves. II. Studies with ^{15}N and ^{14}C. Plant Physiol. 64:467–471.

Hanway, J. J., and A. J. Englehorn. 1958. Nitrate accumulation in some Iowa crop plants. Agron. J. 50:331–334.

Helsel, D. B., and K. J. Frey. 1978. Grain yield variations in oats associated with differences in leaf area duration among oat lines. Crop Sci. 18:765–769.

Herridge, D. F., and J. S. Pate. 1977. Utilization of net photosynthate for nitrogen fixation and protein production in an annual legume. Plant Physiol. 60:759–764.

Holowach, L. P., J. T. Madison, R. N. Beachy, and J. F. Thompson. 1979. The effect of added methionine on the growth and protein composition of soybean cotyledons. Plant Physiol. 63(suppl.):26.

Housley, T. L., L. E. Schrader, M. Miller, and T. L. Setter. 1979. Partitioning of ^{14}C-photosynthate, and long distance translocation of amino acids in preflowering and flowering, nodulated and non-nodulated soybeans. Plant Physiol. 64:94–98.

Hsiao, T. C. 1973. Plant responses to water stress. Annu. Rev. Plant Physiol. 24:519–570.

Huang, L., J. S. Boyer, and L. N. Vanderhoef. 1975. Limitation of acetylene reduction (nitrogen fixation) by photosynthesis in soybean having low water potentials. Plant Physiol. 56: 228–232.

Huffaker, R. C., and L. W. Peterson. 1974. Protein turnover in plants and possible means of its regulation. Annu. Rev. Plant Physiol. 25:363–392.

Huffaker, R. C., T. Radin, G. E. Kleinkopf, and E. L. Cox. 1970. Effects of mild water stress on enzymes of nitrate assimilation and of the carboxylative phase of photosynthesis in barley. Crop Sci. 10:471–474.

Hunt, P. G., T. A. Matheny, D. C. Reicosky, and R. B. Cambell. 1979. Nitrogen accumulation and distribution in irrigated and nonirrigated soybeans. Agron. Abstr. 1979:205.

International Rice Research Institute. 1975. Annual report for 1974. IRRI, Los Banos, Philippines. p. 173–175.

Jeppson, R. G., R. R. Johnson, and H. H. Hadley. 1978. Variation in mobilization of plant nitrogen to the grain in nodulating and nonnodulating soybean genotypes. Crop Sci. 18:1058–1062.

Johnson, R. R., N. M. Frey, and D. N. Moss. 1974. Effect of water stress on photosynthesis and transpiration of flag leaves and spikes of barley and wheat. Crop Sci. 14:728–731.

Johnson, R. R., and D. N. Moss. 1976. Effect of water stress on $^{14}CO_2$ fixation and translocation in wheat during grain filling. Crop Sci. 16:697–701.

Jurgens, S. K., R. R. Johnson, and J. S. Boyer. 1978. Dry matter production and translocation in maize subjected to drought during grain fill. Agron. J. 70:678–682.

Kuo, T., and L. Boersma. 1971. Soil water suction and root temperature effects on nitrogen fixation in soybeans. Agron. J. 63:901–904.

Lal, P., G. G. Reddy, and M. S. Modi. 1978. Accumulation and redistribution pattern of dry matter and N in triticale and wheat varieties under water stress condition. Agron. J. 70: 623–626.

McMichael, B. L., and C. D. Elmore. 1977. Proline accumulation in water stressed cotton leaves. Crop Sci. 17:905–908.

Minchin, F. R., and J. S. Pate. 1973. The carbon balance of a legume and the functional economy of its root nodules. J. Exp. Bot. 24:259–271.

Minchin, F. R., and J. S. Pate. 1975. Effect of water, aeration and salt regime on nitrogen fixation in a nodulated legume—definition of an optimum root environment. J. Exp. Bot. 26:60–69.

Moore, P. H., and A. Maretzki. 1978. Biochemical response of *Saccharum* sp. to drought. Plant Physiol. 61(suppl.):81.

Morilla, C. A., J. S. Boyer, and R. H. Hageman. 1973. Nitrate reductase activity and polyribosomal content of corn (*Zea mays* L.) having low leaf water potentials. Plant Physiol. 51:817–824.

Nair, T. V. R., H. L. Grover, and Y. P. Abrol. 1978. Nitrogen metabolism of the upper three leaf blades of wheat at different soil nitrogen levels. II. Protease activity and mobilization of reduced nitrogen to the developing grains. Physiol. Plant 42:293–300.

O'Toole, J. C., R. K. Crookston, K. J. Treharne, and J. L. Ozbun. 1976. Mesophyll resistance and carboxylase activity. A comparison under water stress conditions. Plant Physiol. 57: 465–468.

Pankhurst, C. E., and J. I. Sprent. 1975. Effects of water stress on the respiratory and nitrogen-fixing activity of soybean root nodules. J. Exp. Bot. 26:287–304.

Passioura, J. B. 1976. The control of water flow through plants. p. 373–380. *In* I. F. Wardlaw and J. B. Passioura (ed.) Transport and transfer processes in plants. Academic Press, New York.

Pate, J. S., B. E. S. Gunning, and L. G. Briarty. 1969. Ultrastructure and functioning of the transport system of a leguminous root nodule. Planta 85:11–34.

Pate, J. S., D. B. Layzell, and D. L. McNeil. 1979. Modeling the transport and utilization of carbon and nitrogen in a nodulated legume. Plant Physiol. 63:730–737.

Pate, J. S., P. J. Sharkey, and C. A. Atkins. 1977. Nutrition of a developing legume fruit. Functional economy in terms of carbon, nitrogen, water. Plant Physiol. 59:506–510.

Penning de Vries, F. W. T. 1975. The cost of maintenance processes in plant cells. Ann. Bot. 39:77–92.

Peterson, L. W., and R. C. Huffaker. 1975. Loss of ribulose 1,5-diphosphate carboxylase and increase in proteolytic activity during senescence of detached primary barley leaves. Plant Physiol. 55:1009–1015.

Rajagopal, V., V. Balasubramanian, and S. K. Sinha. 1977. Diurnal fluctuations in relative water content, nitrate reductase and proline content in water-stressed and non-stressed wheat. Physiol. Plant 40:69–71.

Routley, D. G. 1966. Proline accumulation in wilted ladino clover leaves. Crop Sci. 6:358–361.

Shaner, D. L., and J. S. Boyer. 1976. Nitrate reductase activity in maize (*Zea mays* L.) leaves. Regulation by nitrate flux at low leaf water potential. Plant Physiol. 58:505–509.

Shibles, R., I. C. Anderson, and A. H. Gibson. 1975. Soybean. p. 151–189. *In* L. T. Evans (ed.) Crop physiology. Cambridge University Press, London.

Singh, T. N., D. Aspinall, and L. G. Paleg. 1972. Proline accumulation and varietal adaptability to drought in barley: a potential metabolic measure of drought resistance. Nature New Biol. 236:188–190.

Singh, T. N., L. G. Paleg, and D. Aspinall. 1973a. Stress metabolism. I. Nitrogen metabolism and growth in the barley plant during water stress. Aust. J. Biol. Sci. 26:45–56.

Singh, T. N., L. G. Paleg, and D. Aspinall. 1973b. Stress metabolism. III. Variations in response to water deficit in the barley plant. Aust. J. Biol. Sci. 26:65–76.

Sprent, J. I. 1976. Water deficits and nitrogen-fixing root nodules. p. 291–311. *In* T. T. Kozlowski (ed.) Water deficits and plant growth. Vol. IV. Academic Press, New York.

Stewart, C. R., and A. D. Hanson. 1980. Proline accumulation as a metabolic response to water stress. p. 173–189. *In* N. C. Turner and P. J. Kramer (ed.) Adaptation of plants to water and high temperature stress. Wiley-Interscience, New York.

Stutte, C. A., and R. T. Weiland. 1978. Gaseous nitrogen loss and transpiration of several crop and weed species. Crop Sci. 18:887–889.

Tully, R. E., and A. D. Hanson. 1979. The amino acids translocated from turgid and water-stressed barley leaves. I. Phloem exudation studies. Plant Physiol. 64:460–466.

Tully, R. E., A. D. Hanson, and C. E. Nelsen. 1979. Proline accumulation in water-stressed barley leaves in relation to translocation and the nitrogen budget. Plant Physiol. 63:518–523.

Viets, F. G. 1972. Water deficits and nutrient availability. p. 217–236. *In* T. T. Kozlowski (ed.) Water deficits and plant growth Vol. III. Academic Press, New York.

Waldren, R. P., I. D. Teare, and S. W. Ehler. 1974. Changes in free proline concentration in sorghum and soybean plants under field conditions. Crop Sci. 14:447–450.

Wittenbach, V. A. 1977. Induced senescence of intact wheat seedlings and its reversibility. Plant Physiol. 59:1039–1042.

Wittenbach, V. A. 1978. Breakdown of ribulose bisphosphate carboxylase and change in proteolytic activity during dark-induced senescence of wheat seedlings. Plant Physiol. 62: 604–608.

CHAPTER 8A

Crop Manipulations for Efficient Use of Water: An Overview

ROBERT S. LOOMIS

University of California
Davis, California

I. INTRODUCTION

All farming systems reflect in some way the fact that photosynthetic productivity involves concomitant water loss through transpiration by plants. Farming practices must also contend with losses of water through runoff, deep percolation, and evaporation from the soil surface. Where water supply is limiting, several basic strategies are relied on to bring crops to maturity within the available supply: one is to ensure that a large proportion of the available water goes to transpiration, a second is to achieve a high level of production per unit transpiration, and a third involves achieving a balance between seasonal use of water and seasonal supply.

Those general strategies have led to a wide variety of specific farming practices. Crop manipulations involve variations in choice of species and cultivar, timing of events, plant density, fertility status, provision for irrigation, and other factors. The choices are influenced by economic considerations, the relative importance of animals in the larger system, access of farmers to technological and weather information and inputs such as fertilizer, and the tolerance of the system to risk. Some solutions, such as the use of drought-resistant, cool-season cereals and fallow, are used widely, while others are unique to specific situations.

The many approaches to efficiency in water use can be rationalized if we combine our knowledge of the relation between transpiration and production (Tanner & Sinclair, Chapter 1A) with ideas on risk management and crop adaptation.

II. RELATION BETWEEN PRODUCTION AND WATER USE

Photosynthetic production (the uptake and reduction of CO_2 and transpirational water loss both depend on diffusive exchange with the atmosphere through leaf stomates and on the energy environment. The ratio

Limitations to Efficient Water Use in Crop Production.

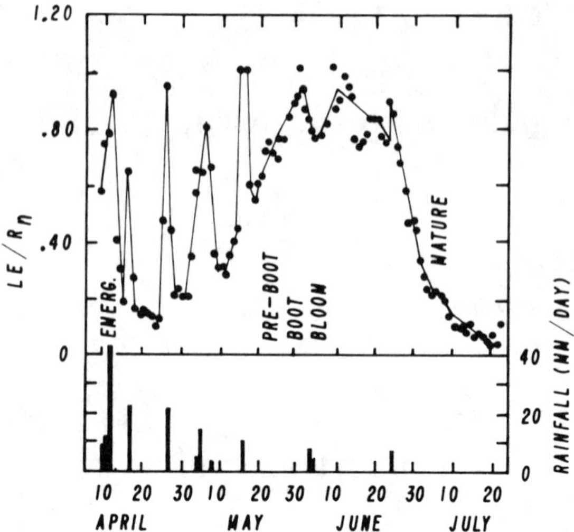

Fig. 1—Season course of ET as a fraction of net radiation (R_N) (expressed as latent heat, *LE*, to R_N) for grain sorghum, and rainfall, at Temple, Tex., in 1969 (after Ritchie, 1971). The early season peaks represent high rates of *E* following rainfall; *T* dominates in the late season.

of dry matter production (*Y*) to the amount of water transpired by a crop (*T*) is termed "water-use efficiency" (*Y/T* or WUE; kg dry matter kg^{-1} water, or kg mm^{-1}). Total evapotranspiration (ET) also includes evaporation from the soil surface, *E*, as is illustrated in Fig. 1. The distinction between *E* and *T* is important since they are subject to separate manipulation (see Ritchie, Chapter 1B). Many of those manipulations work through changes in the energy environment.

We can use the saturation vapor pressure deficit ($e^* - e$; Chapter 1) as

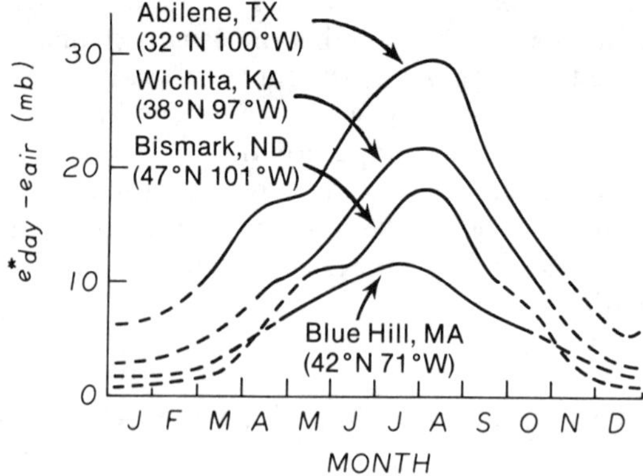

Fig. 2—Annual variation in evaporative demand (effective daytime saturation deficit, $e^*_{day} - e_{air}$) in several environments, calculated as in Chapter 1A from long-term climate data. The solid portion of each curve represents the frost-free season. Blue Hill is in a humid, temperate region; the other stations are from subhumid zones in the Great Plains and show characteristic late-summer peaks. Not shown are Davis, Calif. (Mediterranean climate), which is much like Wichita, and Nashville, Tenn. (more humid), which follows Wichita in spring and fall but has a Blue Hill-type summer maximum of 15 mbars in early July.

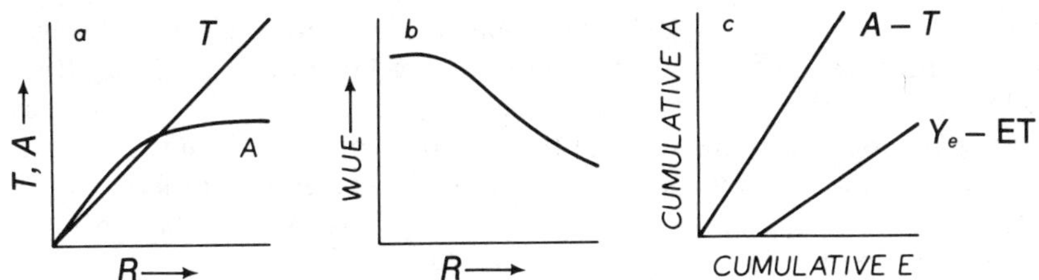

Fig. 3—Generalized relations between production and evaporation: (*a*) rates of net photosynthesis (A) and transpiration (T) as a function of daily irradiance (R) (after de Wit, 1958); (*b*) ratio of A/T (WUE) corresponding to (*a*); and (*c*) relation of cumulative A (yield of biomass, Y) to T, and agronomic yield (Y_e) to ET.

a simple index of evaporative demand and potential evapotranspiration (ET_0). Variations in $(e^* - e)$ over the annual cycle are illustrated in Fig. 2 for several locations. Showing its dependence upon both radiation and temperature, $(e^* - e)$ is at a minimum in winter and reaches a peak in late summer. The patterns are also influenced by cloudiness and advection. It is clear that there might be an advantage for efficient water use from spring cropping at these locations—$(e^* - e)$ is lower and radiation levels are higher than in late summer and fall. But $(e^* - e)$ is low in the spring because of lower temperature, which could be limiting to the choice of crop and to its rate of growth.

An analogous value for potential plant production, Y_o, can be calculated theoretically for canopies not limited by water or nutrients, or developed from field measurements of actual assimilation rates (de Wit, 1958; Loomis & Williams, 1963; Loomis & Gerakis, 1975; Monteith, 1972, 1978; de Wit et al., 1979). Crop photosynthesis rates show a tendency to saturate with increasing irradiance (e.g., Thomas & Hill, 1949), while ET increases more or less linearly. These general findings led de Wit (1958) to propose a relationship similar to that illustrated in Fig. 3A. The Y/T ratio is maximum when crop photosynthesis is not light-saturated (Fig. 3B), but declines more or less linearly after photosynthesis has reached a maximum rate. de Wit and others (e.g., Hanks et al., 1969) have concluded that cumulative Y and T are related. In environments with moderate radiation (e.g., temperate zone),

$$Y = kT, \qquad [1]$$

where k is WUE; and in high-radiation environments,

$$Y = m(T/T_o). \qquad [2]$$

The crop constants k and m vary strongly among species, with differences in photosynthetic mechanism (C_3, C_4, or CAM) being most important, and with moderate moisture stress (van Keulen, 1975; Lof, 1976). When considered on a seasonal basis, the lower relative contribution of Y and T during times of stress is very small compared with Y and T accumulated when soil moisture is adequate. This is important since it means that seasonal k and m tend to vary little with moisture supply. For practical purposes, Fischer and Turner (1978) concluded that a drop of water, whenever used, could be equated to an increment of production, a conclusion that they found to hold for natural vegetation composed of xerophytes as well as for crop plants. The insensitivity of the Y/T relation to environmental variables such as soil fertility and moderate drought suggests that there may be little that we can do about the basic WUE of crop plants other than through a choice of species and radiation levels (season). Since E can be a large fraction, perhaps 20–30%, of ET in an average situation (Chapter 1B), our attention must also focus on practices that minimize E relative to T (Chapter 10).

Other measures of efficiency are also useful. We might, for example, consider only the aboveground biomass or only economic yield (Y_e) and we might wish to base the calculation on ET rather than T. It is important that the method of calculation be defined. For example, the exclusion of roots from biomass has little influence on WUE calculations when moisture supply is adequate because the root mass is then generally only a small fraction of biomass. Under more arid conditions (or with young plants), however, the root fraction might increase to 30% or more of the total biomass (Fischer & Turner, 1978; Connor, 1975).

Figure 3C illustrates the difference between two views of efficiency in the use of water: Y/T shows the total biomass production relative to the water actually used by the plant, and Y_e/ET shows the agronomic yield of the system relative to total water use (refer to Chapter 11A for a more complete discussion). Cumulative Y_e and ET, like Y and ET, are generally related in a linear manner. The difference between those two lines in Fig. 3C represents leaks in the system (water lost to E, low harvest index, etc.). Those leaks need to be identified and made the target of crop manipulations aimed at improving efficiency. For example, an increase in the harvest index will reduce the difference. In the past, American agronomists focused mainly on improvements in Y_e as a way to improve WUE. In our semi-intensive agricultural systems, crop photosynthesis and economic yields have been limited by weed competition, diseases and pests, inferior genotypes, or nutrient deficiencies, and it has been relatively easy to increase economic yield and thus WUE. Pendleton (1966), in an earlier review of this subject, gave his attention entirely to that approach. But where the crops and cropping practices have been brought to a high standard, progress becomes much more difficult. Emphasis then shifts to the approach outlined by Robins (1967) of achieving maximum use of the seasonal supply of water while managing the risks associated with having seasonal T close to seasonal supply.

III. RISK MANAGEMENT

Farming risk increases as one pushes toward maximum use of water. The supply of water (W) can be defined from the hydrological budget as equal to precipitation (P) plus irrigation (I) and the change in storage (S), less runoff (R) and drainage (D): $W = P + I + S - R - D$. The storage term relates only to the "plant-available" portion of soil moisture. Except with irrigation, W is always variable because of variations in P and R. If the supply could be predicted each year, the cropping system could be varied accordingly. But, except in Mediterranean climates, where we at least know that summer rainfall can be taken as zero, we are unable to make accurate seasonal forecasts prior to planting. We are left with the storage term (which can be measured prior to planting) and with historic rainfall records and experience for assessment of an average degree of risk. Improved weather forecasting and specialized agricultural weather information would have a powerful influence on efficiency of water use by favoring timely changes in farm management plans (Burrows et al., 1976).

Risk varies with the cropping system, and the level of risk tolerable to the farmer will vary with economic factors (e.g., average costs and returns). The tolerable level of risk is also moderated by insurance programs, farm credit, and research and extension. Competitive forces encourage the extension into marginal environments (e.g., corn in eastern Nebraska) of crops that provide high returns (Loomis, 1976). Those situations provide a measure of the currently acceptable level of risk.

A moisture shortfall where the cropping system is matched, optimistically, to 120% of the mean of the annually available moisture has a result quite different from that where the system is matched conservatively to only 50% of the long-term mean supply. In this sense, we could design farming systems that would only rarely encounter drought risks. In practice, dryland farming techniques show as much attention to perpetuation of the farm (controlling risk) as they do to maximizing seasonal water use. That generally takes the form of choosing "conservative" systems that will use less than the expected supply or that, like forages, will vary in production according to supply; management efforts can then be directed toward maximizing the yield of that system. That approach can lead to a high efficiency per unit of water used, though not necessarily per unit of water received. The alternative plan of using an "optimistic" crop, but with minimum management to limit costs, has a low survival value for the farmer. During the drought years of the mid-1970's, corn growers in parts of South Dakota and Nebraska needed "disaster" relief for several successive years, leading one meteorologist to "ask whether corn production is a reasonable practice for that environment."

We are concerned here with crop management practices that limit risk through efficiency in the use of water. But we should be alert to the variety of options for controlling risk that exist at the farm and regional levels. Grazing systems reveal a number of those options. The high value of the animals, the relative slowness with which a herd can be increased after a decline in numbers, and the poor prices received through liquidation in times of regional shortages of grazing, place premiums on buffering feed supplies. The basic practice on semiarid lands involves a low stocking rate and, on the average, an underutilization of the forage. Stocking rate and average utilization can be increased if grass hay from wetter meadows can be saved in high-production years. Other practices include the culture of irrigated forages on arable lands (a characteristic feature in the American West) and the diversion of crop land to forages (e.g., sudangrass) in dry years.

In addition, crops such as wheat can be grazed if maintenance of the herd has higher value than grain production. Dual-purpose crops (grazing and grain) provide weatherproofing and, over years, serve to increase the average efficiency of water use on the main grazing lands. Industrialized farming systems tend to depend instead on transport to exchange feed and animals with other regions. The American practice of "finishing" beef animals in feedlots allows additional flexibility through variations in the time when animals are transferred from pastures to feedlots. Confined feeding is the general practice in many parts of the world, although with poultry and swine rather than beef animals. Poultry and swine have a high reproductive capacity and use more of the same foods as humans. Animal populations can be increased quickly after wet years (abundant food) and consumed during dry years. The average efficiency of water use by the whole farming system can be increased by such means.

Cropping practices favoring efficient use of water are influenced also by two other factors. One is the relative contribution to seasonal ET of stored moisture and precipitation (or irrigation) during the growing season. Important strategies where production is dependent upon infrequent irrigation or in-season rainfall are plant survival during intermittent drought, and timing so that the most sensitive stages of plant development are completed when drought is least likely. In Mediterranean climates, dryland production, achieved after the end of the rainy season, is entirely from stored moisture. In that circumstance, a tendency for plant dormancy under a mild stress for water could be disastrous to final yield since no rains would occur. The second factor derives from whether the economic yield consists simply of the products of vegetative growth (such as with forages) or of fruit or tubers produced at the end of the growing season after a phasic change from vegetative growth. With phasic plants, more attention must be given to sensitive stages in development and to the distribution of water supply and use between growth phases.

IV. CROP CHARACTERISTICS

A. Growth Patterns

The manner in which a crop plant distributes new assimilates in growth and development has important implications for water management. While water loss can be made to balance with uptake during short-term deficits by stomatal closure, the basic adjustments must come through changes in vegetative growth to bring a balance in the relative capacities for water uptake and loss (see Chapter 7A). That basically takes the form of altering the ratio between the surface areas of roots and leaves and the effective rhizosphere and canopy resistances. That ratio is adaptive, favoring high proportions of roots under high evaporative demand and low soil moisture. Brouwer (1963) gave the term "functional balance" to the way in which those adjustments occur. When water is nonlimiting, new assimilates are monopolized in shoot growth, where the sink capacity for growth generally exceeds photosynthate supply, and roots may grow very little. As leaf area and capacity for transpiration increase, the root surface may then be inadequate. Since moisture stress is generally more restrictive to shoot growth than to photosynthesis and root growth, roots continue growing, and, by iteration, a functional balance is maintained. Or, leaf area may be reduced through early senescence. The functional balance causes root/shoot ratios to vary widely—with wheat, roots may be 10% of biomass when soil moisture is not limiting, or over 30% under drought conditions.

A root length density (L_v in cm root cm^{-3} soil; not including root hairs) of < 1.0 can be adequate for complete extraction of available soil moisture (Chapter 2A), although many crops carry rooting density to much greater values of 2 or 3. That "excess" of roots is yet to be explained. It may reflect a high sensitivity of shoot growth to moisture stress and a frequent occurrence of photosynthate supply exceeding shoot sink capacity, or the fact that older roots become suberized and thus less active in moisture uptake. Natural selection gives heavy emphasis to competitive ability; a large L_v is perhaps an advantage for rapid extraction of moisture (in contrast to slow but complete). That would be perpetuated in plant breeding, since most methods for artificial selection also emphasize competitive ability in mixtures rather than yield in monoculture (Spitters, 1980). However, even if such roots are "excess," they are not a large fraction of biomass except under semiarid conditions. Functional balances between root and shoot growth also occur with factors other than water. A large root fraction may also reflect poor supply and distribution of nutrients within the soil.

Partitioning influences are also important with respect to biomass composition. This is understood most easily through a balance between growth and differentiation activities (Loomis, 1932, 1953). With moisture not limiting, photosynthate supplies are used principally in growth and the new biomass consists largely of proteins and cellulose, with a minimum of cell differentiation. Under mild stress, differentiation is favored and the biomass has a higher content of lignin, wax, and other materials. Such conditions favor a higher latex content in guayule tissues (Hunter & Kelley, 1946) and higher nicotine content in tobacco leaves (Van Bavel, 1953). Some of the differentiation responses are adaptive in that they contribute to stress resistance, and some responses, such as an increase in the lignin content of forages, are highly undesirable.

The most important aspect of partitioning is concerned with reproductive behavior. We distinguish first between crops such as forages and sugar cane grown for their vegetative matter and those grown for reproductive products. With vegetative crops, we are concerned more with seasonal yield (and WUE) and less with whether it accumulates intermittently during the season. Reproduction is always preceded by a vegetative period and begins later in the season, after a phasic change to flowering. Juvenile periods also are evident before root thickening or tuberization in crops such as sugar beet, manioc, and potato. Moisture stress can have quite different effects on economic yield, depending upon whether it occurs before, during, or after the phasic change. It also makes a considerable difference whether flowering is determinate (as in wheat and sunflower) or indeterminate (as in legumes). The determinate plant has only a single chance for successful reproduction, while the indeterminate plant may overcome early failures through later flowering, provided the season is long enough and the moisture sufficient. The determinate pattern works well where flowering can be matched to a period of the year with a good probability for adequate rainfall or stored moisture; indeterminate flowering could be superior where moisture supply during flowering is uncertain or total seasonal supply is highly variable.

In many indeterminately flowering plants, vegetative growth may tend to dominate over reproduction. In that case, flowers may fail to set and develop fruit when moisture and nutrients are adequate for vegetative growth. Cotton, for example, is generally more successful in fruit production (and thus harvest index and economic yield) when vegetative growth is restricted by mild deficiencies of nutrients or moisture (Grimes et al., 1969). (Additional material on reproductive behavior is presented in Chapter 7A.)

B. Growth Duration

Phasic development patterns relate closely to the duration of the crop. Variations in growth duration are one of the most obvious means for matching seasonal T to moisture supply. With a crop such as wheat, early maturing varieties generally have a strong advantage in yield over late varieties where moisture is limiting (Laing & Fischer, 1977). The maturity class might be chosen so that moisture will be limiting to yield only occasionally (e.g., 1 year in 10). That would assure a low risk and a high efficiency for water actually used by the crop. But it would also leave excess moisture in the soil in most years, so that efficiency in use of the available water would be variable and low. The long-term mean yield, while of low variation, would be smaller than it might be with a more opportunistic species or variety.

Smith and Ritchie (USDA-ARS, Temple, Tex., personal communication) used a validated growth model of wheat to simulate the influence of maturity date and moisture supply on the yield of winter wheat under rainfed conditions at Altus, Okla. (34°N 99°W). On the basis of a 1 September sowing date, grain yields were predicted with rainfall and temperature data for 55 years (1922–1976). Figure 4 illustrates the cumulative frequency distribution of simulated yields. With a "variety" having 2 March as the mean floral initiation date, 50% of the yields fell below 1.5 t ha^{-1}, and only about 20% of the cases exceeded 4.5 t ha^{-1}. That variety averaged 2.28 t ha^{-1} over the 55 years; a variety with floral initiation 4 days later (6 March as the mean date), and thus of greater potential yield, actually averaged 10% less yield, 2.04 t ha^{-1}, primarily due to moisture limitation.

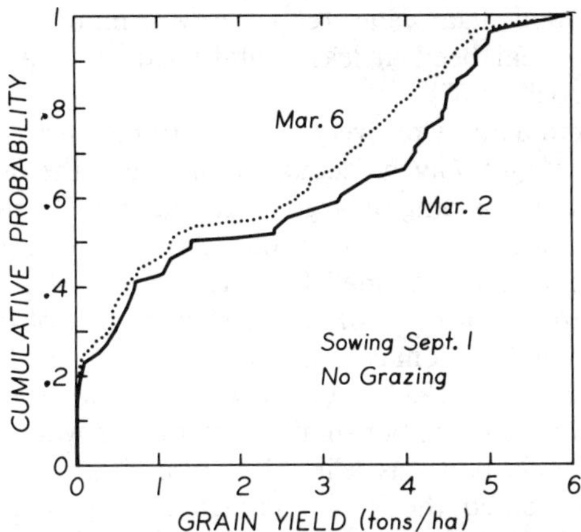

Fig. 4—Cumulative frequency of wheat yields with variable rainfall over 55 years at Altus, Okla. Simulated by R. G. Smith and J. T. Ritchie (personal communication) for two varieties differing in mean floral initiation date (2 or 6 March). The early variety is predicted to give an average 0.24 tons of grain ha^{-1} $year^{-1}$ greater yield over years.

Where moisture supply is variable, one can use an indeterminately flowering plant so that growth and flowering periods will continue in wet years. That is common with summer crops such as melons, cotton, and beans, and it can be accomplished also with sorghum, where, unlike wheat, later tillers continue to develop until terminated by drought or low temperatures. Such plants have excellent plasticity to length of growing season (e.g., to "days of transpiration"), but they also have disadvantages. With fresh-market crops, repeated hand-labor harvests are required, and with field crops, accumulated mature fruit may be subject to loss or damage before an end-of-season harvest. In either case, harvest index tends to be low because of vegetative growth and immature fruit that do not contribute to economic yield. In plants grown for secondary biochemical products (e.g., latex in guayule) or storage in primary vegetative storage organs (sugar in

Table 1—Dryland sorghum performance as a function of soil moisture supply at planting in 2 years differing in early and late-season rainfall (from Bond et al., 1964).

	Year									
	1958					1959				
	Rainfall: planting to boot stage (mm)									
	265					122				
	Rainfall: boot stage to maturity (mm)									
	51					111				
	Soil moisture storage (mm)									
Performance	71	91	137	180	Mean	33	86	112	155	Mean
Biomass yield (kg ha^{-1})	4,390	5,190	5,960	7,410	5,740	1,740	3,430	4,420	5,110	3,680
Grain yield (kg ha^{-1})	980	1,350	1,920	2,650	1,723	210	1,230	1,660	2,470	1,390
Harvest index	0.22	0.26	0.32	0.36	0.30	0.12	0.36	0.38	0.48	0.38

sugar beet roots and sugar cane stems), harvest index is also increased by stress. But with sugar beet, at least, total yield of sugar is not increased (Loomis & Haddock, 1967).

Plant size and activity are very difficult to manage during preanthesis and postanthesis stages. Given adequate moisture in the early season, plant growth may be rapid, leading to larger LAI and reproductive potential than can be supported later. Conversely, drought in the early season may lead to a reproductive potential too small for effective use of a large late-season supply of moisture. Some of those effects show in data of Bond et al. (1964) for sorghum (Table 1). Biomass and grain yields were strongly dependent upon moisture supply. In 1958, with adequate moisture to heading followed by drought, a large biomass but small harvest index was obtained; in 1959, with an early drought that was relieved by rainfall after heading, nearly as much grain was obtained, though from much less biomass.

Passioura (1976, 1977) has given particular attention to the problem of the use of stored moisture for extending growth duration. He grew wheat in containers under controlled environments with limiting supplies of moisture, also suggested by de Wit (1958) and van Keulen (1975), involving metered gradually over the life of the plants. Biomass production, and thus WUE, was independent of the method of irrigaiton, whereas grain yield and harvest index depended strongly on the amount of water available to the plants after anthesis. Those results lead to a strategy for use of stored moisture, also suggested by de Wit (1958) and van Keulen (1975), involving limitations to water use during early season. That might be accomplished by grazing early in the season to reduce leaf area, by use of genotypes that are slow to develop depth and density of rooting, or by delaying the application of limiting nutrients. Seaton and Landsberg (1978) expressed the same idea in terms of increased root-system resistance to water flow during early season.

In some situations with a deep profile of high water-holding capacity, yet not so heavy-textured that rooting depth is restricted, large amounts of stored moisture can be available. That, of course, depends on moisture in the lower profile being restored through heavy out-of-season precipitation, by an extended fallow period (e.g., crop only in alternate years), or by lateral movement of underground water from adjacent watersheds. Acevedo and Fereres (Hsiao et al., 1976; Faci & Fereres, 1979), working with a very deep and favorable soil profile and a winter-rainfall climate at Davis, Calif., were consistently successful in raising 120-day summer crops of corn and grain sorghum entirely from stored moisture. Yields were high, approaching 60–70% of the level obtained in irrigated crops with moisture not limiting. Moisture was extracted from depths >3 m, suggesting that genetic selections for rapid extension of roots in depth would be useful for such conditions (in contrast to the limited-rooting concept suggested by Passiouri for less favorable circumstances). Extensive variability in rooting habit apparently exists among cultivars of grain sorghum (Chapter 7A). Hurd (1974) also emphasized increased rooting depth as an appropriate trait for wheat in the northern Great Plains (additional discussion in Chapter 3A).

Given sufficient profile replenishment, deep rooting can be a successful strategy for a long-season crop every year. Or it can be used occasionally as a "scavenger" approach in which deep- and shallow-rooted crops are alternated in the rotation.

C. Cool-Season vs. Warm-Season Cropping, and Photosynthetic Types

One of the most striking features of Fig. 2 is the strong annual variation in evaporative demand. For a given supply of water, concentration of the cropping season around the winter minimum will allow free transpiration for a longer period than is possible in summer. At Davis, Calif., for example, with ET_0 near 3 mm day^{-1} in winter and 8 mm day^{-1} in summer, 300 mm of water would support 100 days of evapotranspiration in winter but only about 38 days in midsummer.

Thus, the choice between cool-season crops (such as small grains) and warm-season crops is a fundamental issue. In the Mediterranean climate of California, mild temperatures occur through most of the winter, and wheat and barley dominate as crops in dryland agriculture. Those are generally "spring" types (without vernalization requirement for flowering), but they are normally planted in late fall, at the beginning of the rainy season. In the Great Plains, where the winters are more severe, the same general strategy is used for cool-season cropping: in the southern areas, such as Kansas, winter wheat dominates (fall-sown, winter-vernalized, followed by growth and maturation in the spring). Farther north in the Canadian provinces, spring plantings must be employed. That delays establishment of the crop until later in the spring and places the principal period of growth and development in midsummer. But at high latitudes, summer evaporation is sharply reduced (Fig. 2), making possible the successful culture of small grains in low-rainfall environments.

The crop's response to temperature is extremely important for cool-season cropping. Frost tolerance and an ability for rapid growth at low temperatures are essential. Given those traits, efficiency in the use of water depends on the principles revealed in Fig. 3. Evaporative demand is low in the winter months for several reasons, including the lower radiation levels, so that photosynthetic rates are also lower. But if assimilation shows a strongly curvilinear response with radiation level and an optimum exists in WUE (Fig. 3B), there can be a marked gain in WUE through cropping in winter or spring rather than summer. The same reasoning applies to other means of reducing radiation, such as shelters and reflectant coatings (as discussed later).

The possibilities for summer cropping suffer mainly from the fewer number of days of full transpiration for which the water supply is sufficient. That disadvantage is offset in part by the higher rates of photosynthesis and growth that most plant species display in warm climates. Equally significant is the possibility for using a species with the C_4-type of photosynthetic pathway. Corn, sorghum, and other C_4 crops show less leveling of photosynthetic rate with increasing radiation, and thus a less pronounced optimum in the WUE-radiation relation. Most C_4 species seem susceptible to injury by chilling (exposure to temperatures below ca. 10°C) (Loomis & Gerakis, 1975) and can be grown only in the warm season. Also, C_4 plants are no more efficient in photosynthesis under low radiation than plants with C_3 metabolism (Ehleringer & Bjorkman, 1976).

Sorghum, for example, has always been an important summer crop of the semiarid zone, and scientific reports of its superior efficiency in the use of water date at least to Schroder (1895, as cited by Maximov, 1929). That observation was confirmed and extended with corn, millet, and other species in extensive experiments by Briggs and Shantz (1914) and Shantz and Piemeisel (1927). When their data are grouped into C_3 and C_4 plants, mean WUE values for the groups were 1.6 and 3.3 mg dry matter g^{-1} H_2O (Downes, 1969). Those efficiencies are rather low, but the striking advantage of C_4 plants for warm-season cropping has been amply confirmed in other experiments (Fischer & Turner, 1978).

The explanation follows in part from the resistance models of diffusive exchange of water vapor and CO_2 by leaves, summarized in Eq. [6a], [6b], and [6c] of Chapter 1A. Looking again at Eq. [6c], r'_m represents the resistance to CO_2 transport through the mesophyll cells and in carboxylation. This definition is based on the apparent photosynthesis rate and thus includes leaf respiration and photorespiration in the r'_m term. That number is usually smaller for C_4 than for C_3 plants, and the ratio of resistances to water vapor to those for CO_2 ($\Sigma r_w/\Sigma r_c$) is thus a larger number, and WUE is greater. Stated another way, C_4 plants can increase their resistances to water loss ($r_b + r_s$) and still maintain photosynthesis rates equal to or greater than those for C_3 plants.

D. Other Crop Traits

We have seen that efficient use of water implies that we maximize the fraction of the available water that goes to transpiration (T). We have also seen that drought escape (e.g., very early maturity) can be a successful strategy for dryland agriculture; however, that strategy may leave a large fraction of water to E or to drainage in high-rainfall years, lending a low efficiency to the system. Almost by definition, efficient use of water involves a strong depletion of soil moisture by the end of a season and occasionally in midseason. Necessary plant traits, then, are tolerance (i.e., an ability to function well) and avoidance of low plant water potentials. Since tolerance generally develops slowly by acclimatization to decreasing soil water potentials, the rate of "hardening" may be as important as the degree of tolerance ultimately achieved.

However, we are likely to find that little is to be gained from research aimed at increasing the tolerance of crop species to drought. A doubling of the limiting soil water potentials to which plants are tolerant, for example, would add little to seasonal moisture supplies. So, unless tolerance includes an improved WUE, little is gained in terms of production. Fischer and Turner (1978) found that ordinary crop plants and xerophytes with extreme tolerances to drought were similar in WUE. Thus, tolerance may have more to do with survival than with production.

Most crop species include germplasm drawn from extreme environments and should possess high tolerance. Hardened cotton plants can have open stomates even at leaf water potentials as low as −30 bars (Jordan & Ritchie, 1971; Cutler et al., 1977), but production rates then are low. The problem is mainly one of identifying tolerance traits and making sure that they are incorporated in new varieties. Fischer and Maurer (1978) present a very nice example of tolerance identification in cereal grains with detailed measurements of plants grown under various irrigation regimes in a rain-free environment. Also, we can employ management practices that permit hardening through a slow development of stress. Examples include soil preparation that favors deep rooting (e.g., avoids compaction), lower degree of cover (as discussed later, this provides more soil volume per unit leaf area, thereby buffering the rate of soil moisture depletion), and restriction of cropping to heavier soils with a greater water-holding capacity. Those methods also emphasize the avoidance aspects of crop resistance. We might also make genetic selections for more rapid hardening. But that could be complex—it might require a more rapid rate of development with implications to other aspects of growth. In addition, a drought-anticipating plant that hardens rapidly in response to a few hours or days of adverse weather might make little progress towards yield accumulation over a season. Hall et al. (1979) have carefully analyzed many of these problems.

The functional balance of root and shoot growth provides an effective mechanism for avoidance of plant moisture deficits provided soil water is depleted slowly. Stomatal closure serves as another principal mechanism. Temporary closure during periods of peak evaporative demand not only reduces water loss but, according to the extent that light is saturating to photosynthesis (Fig. 3), might also increase WUE. As illustrated with Eq. [6c] (Chapter 1A), an increase in leaf resistance would improve the momentary WUE. While water is saved, total photosynthesis is reduced. However, if maintenance respiration for the accumulated biomass continues at normal rates (or indeed is enhanced because of higher crop temperatures in the absence of T), then WUE calculated on accumulated biomass could be less (Hunt & Loomis, 1979). WUE based on measurements of Y/T is increased only slightly by drought (de Wit, 1958; Fischer & Turner, 1978), so respiration and stomatal influences are either small or counterbalancing (Passioura, 1976). However, WUE based on Y/ET is generally increased by drought because of reduced E (Viets, 1966), while WUE based on Y_e/ET or T is generally less because of adverse effects on reproductive growth.

Another characteristic of stomates may be important. In some species (both C_3 and C_4), and except in extreme drought or low light, stomates may operate more to regulate the CO_2 concentration within the leaf, $[CO_2]_{int}$, at a constant level than to control water loss (Raschke, 1975; Goudriaan & van Laar, 1978). In such cases, stomates are restrictive when photosynthesis is low (and thus not depleting $[CO_2]_{int}$), and less restrictive when photosynthesis is rapid. In essence, $[CO_2]_{int}$ is kept relatively constant. de Wit et al. (1979) argue that such behavior is favorable to a high WUE. Production rates are lower than without such control, but the number of days of transpiration can be increased. That phenomenon seems to offer a promising area for further research. Stomata of some species are known to respond also to humidity (Hall & Hoffman, 1976) as well as to CO_2, light, and water stress. Cowan and Farquhar (1977) suggest that stomates may respond complexly to those factors in a way that optimizes Y/T over the diurnal period. Aspects of that theory seem incompatible with our knowledge of competition and natural selection, but some experimental support has been found (A. E. Hall, personal communication). We must keep in mind that the relative constancy of Y/T in field measurements is due to the dominance of periods with high transpiration and assimilation (van Keulen, 1975). As noted before, transpiration and photosynthesis rates are low during drought, so those periods contribute little to seasonal totals of Y and T.

V. CROP MANIPULATIONS

We have seen that efficiency in the use of water depends on the relative rates of assimilation and evapotranspiration as influenced by genotype and environment. Those traits are subject to optimization through variations in cropping practice and through special treatments that might be applied to the crop plant community. The diversity of management practices taken in specific combinations is very great, and we must confine our discussion to only a small number of the major considerations.

A. Health and Welfare

Our analysis of the dependence of efficiency in water use on assimilation rate leads directly to the axiom that healthy crops will tend to make better use of available water than crops that are subject to diseases, pests, weed invasions, or stress from salinity or nutrient deficiency. The range of responses to disease is wide since those factors can enhance transpiration losses through external mycelium, block xylem transport, or devastate the root system. It is also possible that the water-supply function can be modified by mycorhizal associations. The impact of insect pests is also variable. Grazing activities reduce leaf area (biomass), but also reduce water loss, leaving WUE largely unaffected. Insects that destroy root systems, however, may cause loss of the entire crop.

Plant species that appear as weeds may be quite similar to crop species in WUE for biomass production, but economic production per total unit of water use can only be decreased by the presence of weeds. Exceptions exist where weeds develop late in the crop cycle and the crop aftermath is subject to grazing. In that case, weeds (or a sown forage) serve as a flexible means of extending the growing season to the limit of available water.

Salinity becomes a special problem wherever ET exceeds precipitation. Salinity has the general effect of reducing the availability of soil moisture and that, like mild drought, acts to increase plant water stress and stomatal closure, perhaps with some improvement in WUE. However, the overall effect is to reduce the efficiency of irrigated agriculture because a portion of the applied water must be passed through the soil profile to drainage as a means of preventing salt accumulation (Reeve & Fireman, 1967). The need for drainage is a function of the salt content of the applied water and thus is particularly high where the water supply is derived from semiarid watersheds (e.g., the Colorado Basin) (see Chapter 2C for more detail).

The following discussion assumes that pests and diseases are under reasonable control, since there is little point in discussing careful strategies for the efficient use of water if crop production is limited by other factors.

B. Fallow Cropping

Available soil moisture at planting is influenced by previous cropping and by vegetation cover and precipitation during the intervening fallow period. The fallow period can be left unmanaged, or it can be designed to meet certain objectives, including moisture storage, nutrient cycling, weed control, or maintenance of soil structure. An intervening green manure crop, for example, would contribute to all but moisture storage. Nutrient cycling is generally facilitated since residue decay, nutrient accretion from rainfall, biological N fixation and soil weathering take place during the fallow interval. Exceptions occur with P, which may be fixed into insoluble forms during fallow. Shifting agriculture in the humid tropics may involve forest regrowth between crop cycles for nutrient accumulation; lightly exploited legume stands and cover crops serve a similar purpose in humid temperate zones. In semiarid zones, moisture storage becomes a principal objective. The fallow is then kept relatively free of vegetation to eliminate *T*. Where winters are severe, as in the Great Plains of North America, cropping in alternate years and varying crop duration (e.g., alternating corn and wheat) are the only practical means of varying the length of the fallow period. With longer growing seasons, other options exist. In Mediterranean climates with mild winters, combinations of winter and summer crops such as wheat–pulse–fallow provide systems that are intermediate in moisture use compared with continuous wheat and wheat–fallow.

Hall and Dancette (1978) provided an elegant analysis of the effects of variations in growth duration of the crops included in such rotations. As moisture supply becomes limiting, the hydrological balance can be maintained by increasing the proportion of short-cycle (early maturing) crops in the rotation.

Alternate-year cropping using unvegetated fallow is an extensive practice in the Great Plains (USDA, 1974). Nutrients, especially N, and water are both limiting, though it is not always clear which factor is dominant. During the fallow period, "available" N is generally increased by residue

Table 2—Mean long-term yields of wheat with continuous cropping and crop-fallow systems, without supplemental fertilizer (from USDA, 1974).

Location	Yield period	Yield	
		After fallow	After wheat
		kg ha^{-1}	
Northern			
Harve, Mont.	1917–1951	2,100	540
Dickinson, N.D.	1908–1951	1,400	780
Newell, S.D.	1909–1948	1,420	910
Central			
Akron, Colo.	1911–1970	1,420	500
North Platte, Nebr.	1912–1967	2,140	830
Colby, Kans.	1915–1963	1,320	620
Southern			
Bushland, Tex.	1942–1970	1,010	640

decay, but some is also lost through leaching, denitrification, and soil erosion. With weedy fallow those losses are less, but moisture storage may also be less.

Fallow can have dramatic effects on yield (Table 2). The general conclusions about wheat-fallow rotations in the American Great Plains are that fallow practices strongly reduce the risk of crop failure and the relative variability of yield over years (USDA, 1974). This is illustrated in Table 3 with data from North Platte (Smika, 1970), where the coefficient of variation for continuous wheat was 74%, but only 33% for fallow wheat. Yields <400 kg ha^{-1} (failure) were obtained in 10 of the 27 years of continuous wheat, but never after fallow. Risk and variability aside, fallow is clearly a superior practice when wheat yields after fallow average more than twice as much as wheat after wheat (as shown for Akron, Colo.). Fallow was also advantageous at Mandan, N. Dak., where the ratio was only 1.5, because in that system, land was abundant and grain prices low, while labor and capital for crop production were scarce. Thus, the relative usefulness of fallow, as with other cropping practices, is determined economically. It is a luxury that a Bedouin subsistance farmer with 200 mm of rainfall and a yield ratio of <2 cannot afford.

Before widespread use of N fertilizer, the advantages of fallow were generally greater in cool (northern) than in warm (southern) climates. That situation seems to have been changed in recent years. The experiments illustrated in Table 3 reveal in the Montana example that the advantage of fallow may be reduced or even eliminated through use of N fertilizers. Such results are most common in the more humid margins of the fallow-crop region. Within the low-rainfall areas, the advantage of fallow has generally been increased through N applications or, as in the North Platte example, reduced only slightly. A correct interpretation of fallow benefits as being due to moisture or nutrients depends upon extending the experiments for several years with nutrient additions and on gathering additional data relating to plant nutrient status and soil moisture. Nitrogen fertilization may stimulate rooting and, thus, a better recovery of the available moisture, or it

Table 3—Influence of applied N on yields of wheat and barley in crop-fallow and continuous-cropping systems.

Crop system†	Yield: N applied (kg ha^{-1}) 0	34 or 45	C.V.
	kg ha^{-1} ± SD		%
Continuous wheat	730 (±540)	950 (±700)	74
Wheat—fallow	2,490 (±820)	2,650 (±880)	33

Crop system‡	Yield: N applied (kg ha^{-1}) 0	67	134
	kg ha^{-1}		
Continuous barley	910	1,940	2,260
Barley—fallow	2,100	3,170	3,120

† Yields of wheat from continuous cropping and wheat-fallow systems were compared without added N in each of the 27 years from 1941 to 1967. The same comparison was made in 19 of the years with 34 or 45 kg N ha^{-1} added (Smika, 1970). Standard deviations and C.V. calculated here were unaffected by N level.

‡ Mean yieldsof barley grain from replicate trials at three locations in Montana in 1969 and 1970 (USDA, 1974, p. 27).

may be that the N supply really was more strongly limiting than moisture. In the northern areas, where the fallow response is strongly reduced by fertilizer, it is not uncommon to find the wheat-fallow rotations replaced by wheat alternated with a summer crop such as sorghum, maize, or sunflower. As intertilled (or herbicide-treated) crops, they provide good weed control and leave the land in reasonable condition for wheat.

In general, fallow is an inefficient means for saving moisture. As little as 10–20% of the fallow-year precipitation is retained in the soil. Most of the water is lost as soil evaporation and, in uncultivated fallow, as transpiration from weeds. Thus, the basic WUE is always low. Other hazards include wind and water erosion of the fallow ground. American research has given major emphasis to those problems, with attention to practices such as retaining a standing stubble for snow entrapment and erosion protection. While efficiency is low, the dramatic responses shown in Table 2 are the basis for agriculture on $> 10 \times 10^6$ ha of land in the American Great Plains. Additional discussion on this topic is presented in Chapter 10A.

C. Plant Cover

We saw earlier that E from moist soils is energy-dependent and approaches ET_0 (Fig. 1). But, as the soil dries, E declines rapidly since soil conductivity for movement of water from lower depths in the profile declines to a very low value. The rate is also influenced by the structure of the soil-air interface. A tillage mulch is effective in reducing water loss from some soils, but most soils are "self-mulching" to some extent on drying, and additional tillage then has little effect. Plant mulch or (especially in pots) gravel mulch can reduce *E* effectively.

The main point, however, is that, after the soil surface dries, *E* is less than *T*. So, by reducing plant cover below 1.0, ET per unit area of land can be reduced. That can be done by having fewer plants per unit area, or less leaf area per plant. Since an LAI of 2–2.5 generally provides full cover of the land surface with randomly displayed leaves (and perhaps 2.5–3 with row display), LAI must be kept at a low value. A greater proportion of water is then lost as *E*, so efficiency of water use declines, but more water is available for *T* per unit cover and crop success is improved, particularly for phasic crops or long-season crops. We can also think of this as permitting more land area (precipitation catchment) and more soil volume (stored moisture) per unit of cover. The result is that crop transpiration is extended in time, and soil moisture depletion (and thus also the onset of stress) is delayed. As noted earlier, this strategy can provide an important buffering period for acclimatization or until the next rainfall.

A relationship between *T* and leaf area is illustrated in Fig. 5. The curvilinear response is explained by an increase in mutual shading and other microclimatic changes as LAI increases and cover approaches 1.0. A high *T* per unit leaf area at low LAI results directly from more solar radiation per leaf and, indirectly, from net radiation partitioning on the uncovered soil surface. The exposed soil surface is heated by solar radiation resulting in local heating of the air and greater emission of thermal radiation, both of which are intercepted by adjacent plants, thereby increasing *T*. The net effect is that cover must be reduced considerably below 1.0 in order to reduce *T* significantly.

A common method for reducing cover, in addition to grazing, is to plant the crop at less than the standard density employed when moisture is not limiting. That practice is general in semiarid regions. Extreme examples

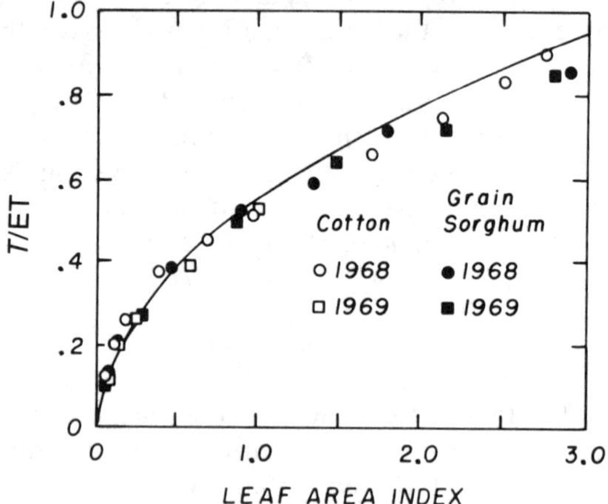

Fig. 5—Relative plant transpiration (*T*/ET) as a function of leaf area index in cotton and grain sorghum, soil water not limiting (after Ritchie & Burnett, 1971).

are found with perennial plants such as agaves, figs, and olives, where rooting may be extended over a large area of soil catchment relative to the area of green cover. Spacing of annual plants, however, is limited by the lateral extent of rooting that can be achieved in a single season. In principle, plant spacing can be varied in a way that influences the time when stored moisture is used. Roots can reach all of the soil mass earlier in the season with a regular spacing than with a clumped pattern. Where plants are closely spaced within rows (or hills), but widely spaced between rows, roots may reach the interrow soil mass much later in the season. That dependence on root extension rate has the effect of spreading the use of stored moisture over a longer period and, with a phasic plant, of conserving moisture until after anthesis. That is the theory—though experimental results are not always definitive (Blum & Naveh, 1976; Bond et al., 1964; Alessi & Power, 1976; Welch et al., 1966). An alternative is to shape the land in microwatersheds (Fuehring, 1975). In that case, rainfall flows to the rooted area and the need for a laterally extended root system is reduced. Whether crop success is improved by such arrangements depends upon moisture supply, depth and lateral extent of rooting, nutrient supply, and other factors. Alessi and Power's (1976) results are instructive because they indicate, albeit at a low level of significance, influences of row spacing, population, and maturity class.

Variations in plant spacings offer opportunities for microadjustments in plant/resource relations. Note that the principles are similar to those used for adjustments with larger scales of space or time. Examples are crop-fallow and early maturing crop systems, which represent reductions in cover over longer times, and runoff–run-on strip-cropping (part of the land is a permanent watershed to adjacent strips, which are always cropped), which offers a macroadjustment in cover over larger spaces. A useful generalization emerges: where soil resources are limiting (in this case, water), nonuniform treatment of the land or the crop can be an advantage. Where soil resources are nonlimiting, however, uniform cropping will provide the greatest efficiency in light interception and photosynthesis.

Reductions in cover can extend the days of transpiration because E from dry soil is considerably less than ET of a full-cover crop. As cover declines, however, the fraction of water lost as E increases (Fig. 1 and 5). That is a problem during fallow, during the early season in full-cover crops, and throughout the season in low-cover crops (see additional discussion by Ritchie, Chapter 1B). In the last situation there might be some advantage from selection of genotypes with an aggressive (large L_v), shallow rooting habit, which could rapidly deplete moisture near the surface after each wetting.

D. Mixed Cropping

There are several situations where it may be advantageous to crop the land with several species or cultivars sown in mixtures, in alternate rows, or in some other pattern. The most common example in North America is the use of mixtures of grasses and legumes in sown pastures. The practice is an economical way of achieving several advantages, including extension of the season for green forage through use of early- and late-season materials, "balancing" of protein and carbohydrate sources in animal rations, insurance of plant cover over a variable microenvironment, and provision of a biological source of N to grasses grown in low-N environments. Intercropping in rows is common in tropical regions, where labor is abundant, the man/land ratio is large, and the overlapping of several short-season species has the possibility of occupying the land more fully for the entire growing season. Mixed cropping is also employed in primitive systems where soil nutrients are strongly limiting. In that case, plant growth is poor, little competition occurs for light, and neighbor effects are small or absent. Thus, what occupies that aerial space makes little difference, little technological skill is required, and some insurance is gained that the remaining species can compensate if one of the species fails.

Mixtures and intercropping, however, have attendant problems not found in successive monocultures. The advantages of crop rotation for control of diseases, pests, and weeds are lost, and management (timing, density, nutrition, harvest method, etc.) cannot be optimum for each component of the mixture. In addition, water use may be less efficient. Overlapping intercrops, for example, may require reduction in cover for the second crop to be successfully established within the first. The fraction of water lost to E is thereby increased. Difficulties arise also from differences in species' reactions to variations in moisture supply, particularly in their resistances to drought. Competitive relations that may be reasonably balanced at one moisture level can be quite different at another.

A common assumption about mixtures is that different species will complement each other through use of different components of the environment. That notion rests on observations in nature that different animals may use different food sources, or that sun and shade plants may form a compatible forest community. A priori reasoning along those lines can lead one to the idea that combinations of plants with deep and shallow rooting will be more exploitive of a soil for nutrients and water than a single species.

Gause (1934) embodied those ideas into a theory that competition is more severe between like things than between unlike things. With annual crop plants, however, that is generally not the case (de Wit, 1960) and mixtures seldom exceed the yield of the best monoculture (Trenbath, 1974). The reason is that plant species compete for the *same* basic resources of light, CO_2, water, and nutrients. Thus, niche differentiation can occur only for the portion of the space within which those resources are found, and not for *different* resources. If the species differ in traits such as height or rate of growth, there will be winners and losers. One has to say that, for losers, competition is more severe between unlike things. If the species differ in physiological traits, it should be possible to judge one to be superior in a particular environment. Mixture of superior with less able species leads to a deterioration of community productivity. Thus, the photosynthetic rate of a full canopy of leaves of high physiological capacity and arranged in an optimal way can be maximized with a single species, and it cannot be improved by introducing a species with inferior traits. Similarly, complementation of species with deep and shallow rooting cannot exceed in nutrient and moisture extraction a single species capable of exploring the entire profile. Agriculturally, an inadequacy in rooting or other trait must be viewed as a genetic defect subject to improvement through a change in species or cultivar (Hurd, 1974; de Wit et al., 1979; Chapter 8C, this book).

In an evolutionary sense, natural selection gives little merit to conservation of common resources. Resources used slowly or not at all simply go to increasing the reproductive yield of competing genotypes. Selection thus tends to emphasize competitive ability and rapid use of resources (Spitters, 1980). In managed monocultures, however, the possibility exists of selecting for genotypes that have low responses to competition (Donald, 1968). In that case, conservation can benefit the individual and the crop since all competing plants are doing the same thing; competition is equal, with no winners or losers. Such behavior applies in the use of stored moisture, outlined earlier, where a slow root development in early season would reserve moisture for postanthesis development, thus favoring a high economic yield per unit of water.

Enthusiasm for mixed cropping also drew support from theory by Margelef (1969) and others that "diversity" lends "stability" to plant communities. There are many definitions of diversity and stability, but that theory fails in extensions to agriculture, where stability (continued success of farming) is provided more by managerial decisions on safe crops and safe methods. Systems with over 10,000 years of continuous cropping to rice in Southeast Asia and over 100 years to corn in Iowa must be considered as reasonably stable. The theory also fails with natural vegetation (Goodman, 1975). Intermediate successional stages in vegetation are highly diverse but also highly unstable as compared with climax stands of less diversity.

Decisions on the use of mixtures depends in part on climate variability. Where patterns of moisture supply and use are reasonably stable, one can choose a single species that will give the highest degrees of safety and economic yield. Where the climate is highly uncertain, a mixture could be superior over a poor choice for monoculture, particularly where length of season (in days of transpiration) and uncertainty of stand establishment are principal variables. For example, long-season species or cultivars could provide the farmer with additional yield in high-rainfall years. Their presence in a mixture, however, must reduce yield in low-rainfall years. We might approach that situation better with a monoculture of an indeterminate species, or with improvements in weather forecasting. We still need good research involving comparisons of farming with mixtures vs. diverse monocultures under such conditions. In experiments to date, either the mixtures or the comparison monocultures generally involve the same variety and the same low density as is necessary in the mixture, rather than optimum practices for monoculture. Experiments with forages generally have much more validity. In that case, pure grass is usually superior when N is not limiting; with a low supply of N, a pure stand of legumes is usually superior in yield.

E. Nutrient Status

Nutrients are commonly or potentially limiting in all agricultural systems since the export of products may carry nutrients away from the farm more rapidly than they are replenished by rainfall and other natural sources. Nutrient deficiencies reduce plant growth and the expansion of cover, and also photosynthesis. Both effects can reduce the efficiency of water use even though actual WUE is lowered only by extreme deficiencies (de Wit, 1958).

In terms of biomass production, the effects of nutrient deficiency are generally similar to those of incomplete cover. Special cases occur, however, since nutrient supply can modify the course of growth and development. Improved nutrition may alter rooting habit, thus changing access to stored moisture. The effect is generally to increase rooting intensity. That may not change the measured efficiencies, although it can increase biomass production significantly. Where the seasonal supply of moisture is less-than-adequate in days of transpiration to maturity, timing of fertilization can have strong effects on reproductive yields. Early applications expand cover and the use of stored moisture during the vegetative phase. Unless the crop also responds with an accelerated development, serious moisture deficits can develop late in the season. In contrast, fertilization after flowering may have favorable effects, such as increasing access to stored moisture and delaying the senescence of old leaves. As noted earlier, reduced or delayed fertilization may be a desirable practice with phasic crops grown under limiting moisture. Interactions of water supply with efficient nutrient use are discussed in Chapter 10B.

F. Modifications of Evapotranspiration Rates

Thus far we have emphasized modifiations of the cropping system that would balance water supply and ET by increasing the water supply (fallow, irrigation) or by reducing ET (early maturing and cool-season crops and incomplete cover). We also can seek to slow water loss.

Losses through E were seen to be a special problem, and attention has been given to minimizing E with soil covers of plant residues, plastic, and asphalt. Such practices can reduce E, although side problems in disease control may increase. In general, such mulches are expensive in terms of energy, equipment, and labor, and thus far are seldom cost-effective. Exceptions occur where the mulch comes as a residue or standing stubble from the previous crop.

In irrigated crops, efficiency of water use can be strongly affected by the method, amount, and timing of irrigation. For example, where cover is small, trickle irrigation or subirrigation can reduce E below that with sprinklers, furrows, or flooding, which generally wet most of the soil surface. Subirrigation requires special soil conditions and control of the water table, and trickle irrigation (like artificial mulches) is expensive for annual crops. Unfavorable patterns of salt accumulation are common with both methods, requiring periodic leaching by rain or uniform surface irrigation. In addition, the surface soil dries, and nutrients, which are more abundant there than lower in the root zone, are then poorly available (Henderson et al., 1968). With closely spaced annual crops, significant savings of water are achieved only during the early season, when cover is low. Even then, we have the problem of energy from net radiation exchange at the dry soil surface being transferred to the adjacent plants. Savings can be more dramatic in young orchards, where cover is very low for several years. In that case, however, cover during those years can be increased by intercropping with annual plants.

The simplest strategy for efficient water use in irrigated agriculture is generally intensive cropping. By "intensive," I mean a system that comes rapidly to complete cover and extends the cover for the full growing season without limitations by nutrients, diseases, or pests. In that way, E is brought to a low fraction of ET without resort to expensive preventives of evaporation. Underlying the intensive approach is a question that researchers in the western United States faced during the late 1800's: given a limited supply of water, is it better to grow crops on a small area without limitation by water or on a larger area with water strongly limiting? Those "economy-of-water" experiments took the form of applying 1 m of water ot 1, 2, or 4 ha sown to various crops. Although the experiments did not detect subtle differences due to mild plant stress, deep rooting and so forth, they did provide the general answer that best economy was obtained through practices that minimized severe stress. Those results and earlier European findings on transpiration ratios also made it clear that there might be differences in WUE among species—a question that Briggs and Shantz (1914) then attempted to answer in their famous experiments.

Irrigation practices in arid regions must always provide an excess of moisture for leaching and drainage of accumulated salt (Reeve & Fireman, 1967). Leaching is required only occasionally, however, rather than at each irrigation, opening the possibility for "deficit irrigation." Deficit irrigation is based on supplying slightly less water than the potential ET of the crop. That can be done by applying only some fraction of weather-based estimates of ET_0 or lengthening the irrigation cycles to greater degrees of soil moisture depletion based on soil or plant indicators. The practice restricts water use through mild stress, which can be done with some crops without greatly affecting production. More important, perhaps, it brings drainage under better control. Irrigation timing is discussed in detail in Chapter 9A.

Extensive attention has been given to foliage treatments such as chemical sprays that might reduce transpiration. Such treatments can be used at any stage in the soil-plant-atmosphere continuum but are best directed to leaves where transpiration can be slowed without inducing plant stress. Three approaches merit brief discussion: (i) reflectant coatings on leaves to modify the net radiation balance; (ii) "antitranspirant" materials, which either add an additional diffusion barrier over the stomates or chemically regulate stomatal resistance; and (iii) changes in the aerial environment above the plants to modify radiation or increase aerodynamic resistance to water loss.

Waggoner (1966) and Gale and Hagen (1966) reviewed early work on antitranspirants and reflective coatings. Progress since those thorough analyses has been mostly in our understanding of growth-regulator influences on stomates (Raschke, 1975; Vaadia, 1976). Commercial practices in this area are limited.

Reflective coatings of materials such as kaolinite act by reducing the absorption of radiant energy. That procedure can be quite effective in reducing water loss and is practiced with some orchard crops. Water-use efficiency, however, is generally not improved because the absorbance of green leaves is strongest only in the wave bands that are effective also in photosynthesis. Thus, of the incoming short-wave solar radiation, there is little "excess" absorption that adds to evaporation but not to photosynthesis. Most reflectants, like kaolinite, have broad reflectance over the whole short-wave spectrum. The best means for increasing WUE might be selective reflection of the high-energy photons of blue light (good quantum efficiency in photosynthesis but poor energy efficiency).

Reflective and impermeable film coatings both present difficult problems in coverage. A thin coating on all leaves exposed to radiation requires perhaps 20 kg material ha^{-1}, and there is then the problem of maintaining it in the face of dew and light rains but removing it when moisture becomes sufficient again. The film coatings seek to restrict diffusion, essentially by an increase in boundary resistance, r_b. Materials such as polyethylene appear attractive because they have much greater conductivity to CO_2 than to water. But photosynthetic reductions, coverage, and removal are still problems. All leaves must be treated to some level of coverage since they act as parallel conduits for water loss. Full coverage might require 40–50 kg ha^{-1}.

Stomatal-control agents seem more promising in that much smaller amounts of material are needed. Phenyl mercuric acetate and growth-regulators such as abscisic acid can be used to induce stomatal closure. Despite the enthusiasm among researchers for such materials, they may have limited potential in agriculture. We again have difficulties in managing the extent (the parallel-conduit problem) and duration of their effects. More fundamentally, however, they accomplish only what plants do naturally. Thus, their potential lies with doing it in advance of stress, to slow water loss and thus conserve stored moisture for later use during a temporary drought. As we have seen, delays in rooting by fertilization and other means can also solve that problem, although without the degree of timing control that may eventually be possible with stomatal control. A potential danger of early stomatal control is that, through avoidance of water deficits, hardening is also prevented. After the control ends, the plants would then be more susceptible to a continuing drought. Equation [6c] in Chapter 1A again serves to assess the benefits from stomatal control for WUE (with the same caveats as before regarding gas exchange vs. dry matter gain).

The gain in WUE is potentially large for plants with large internal resistances to CO_2 (i.e., C_3 plants). Such control could extend the season for C_3 plants under limiting moisture. However, if we account for the greater losses of water in *E* and of dry matter in maintenance respiration over that longer season, the real gains might be small and still less than the production of an unregulated C_4 plant except in a cold climate unsuited for chilling-sensitive C_4 plants. The practical use of any coating or antitranspirant treatment for protection against short-term drought may eventually depend on our ability in long-range weather forecasting. Definite knowledge that a drought will continue may be the only way of justifying the expense and managerial skill required for such treatments.

Another approach to the problem is to modify the aerial environment above the crop through various types of shelter (Barfield & Gerber, 1979). An extreme example would be to induce the formation of cloud cover over the crop (e.g., by jet contrail development). Other approaches with glass or plastic houses and screen shelters are common in horticulture. Screen shelters, for example, provide important changes in the plant environment that lead to lower water use and increased WUE (Allen, 1975). The most extensive use of that approach in North America is with shade-grown tobacco. More interesting and practical for extensive agriculture are windbreak shelters, which alter the aerodynamic transport between the crop and the bulk atmosphere. The movement of water vapor away from the crop can be represented by a flux equation:

$$\mathrm{ET} = (\mathrm{dH_2O}/\mathrm{d}z)\, K_{\mathrm{H_2O}}, \quad [3]$$

where the flux, ET, is dependent upon the height gradient of water vapor, $\mathrm{dH_2O}/\mathrm{d}z$, and the "eddy diffusivity" coefficient of turbulent air for water vapor, $K_{\mathrm{H_2O}}$. This coefficient represents the effectiveness of the air in carrying water vapor. The reciprocal ($1/K_{\mathrm{H_2O}}$) can be visualized as an "air resistance" above the crop, in analogy with the resistance approach used in Eq. [11] (Chapter 1A). If wind speed is reduced, $K_{\mathrm{H_2O}}$ is decreased (resistance increased). Brown and Rosenberg (1971) used this approach to reduce ET and increase WUE and yield in a sugar beet crop with rows of taller maize plants spaced every 15 m within the beet crop. The porous maize canopy, like a snow fence, maintained a slower moving layer of air near the ground. Windbreaks have been used on a larger scale, as with the rows of eucalyptus trees found in California. The outcomes of such efforts are variable: humidity can build within the crop, increasing the gradient and thus restoring the ET rate; the increased resistance to the downward CO_2 flux may diminish photosynthesis to a greater extent than transpiration; or the loss of yield through the land area and water assigned to the windbreak crop may exceed the benefits. On the whole, however, this seems a promising area for continued research.

VI. SUMMARY

Relatively few principles underlie the farming practices that provide for efficient use of water. A maximum proportion of water should be expended in *T*, with minimum losses to *E*, drainage, and runoff. Basically, that involves extending complete green cover as long as practical to minimize *E*, maintaining soil characteristics favorable to rapid infiltration, and cropping at a time and with a rooting intensity appropriate to the interception of drainage. It also involves a careful match between *T* and water supply. Some of the means for doing that are summarized in Fig. 6. Cool-season cropping, incomplete cover, and early maturity are shown to be effective. A second principle involves achievement of a favorable ratio of CO_2 assimilation to *T* (i.e., a high WUE). That is mainly a matter of species selection, although time of year for cropping, time of day for stomatal closure, nutrient status, and other factors are also important. A third principle is that partitioning of assimilates should favor crop safety (e.g., appropriate timing and extent of root growth and flowering) and a high level of economic yield.

In the broad sense, partitioning involves all aspects of growth and development including hardening, functional balance, and other adjustments to changing water status. The latitude found here for successful water-use

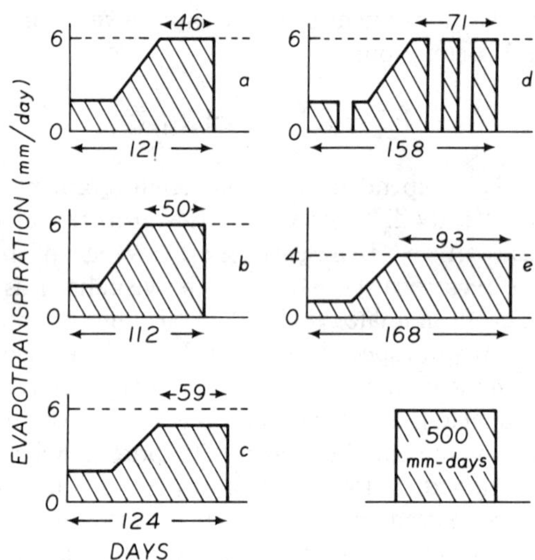

Fig. 6—Illustration of crop manipulation tactics designed to match seasonal ET to an available moisture supply of 500 mm. ET_0 (----) is shown at 4 mm day^{-1} (cool season) or 6 mm day^{-1} (warm season). ET is plotted against crop duration with the area under the relation equal to 500 mm. Crop duration and days of maximum ET are given numerically. Taking (*a*) as the performance of a standard annual crop during the warm season (ET = 6 mm day^{-1}), full-cover ET is extended by an early maturing variety having a shortened juvenile period (*b*) or by drought tolerance that allows the crop to exist on periodic rains (*d*). But the most effective means for extending growth duration are to limit cover to < 1.0 (*c*) or to grow the crop during the cool season (*e*) (ET_0 = 4 mm day^{-1}).

tactics is very broad, and our scientific knowledge is still quite small. It may be that existing genetic populations represent a rather complete collection of possible solutions to partitioning problems. It is more likely that they do not. One point of my discussion on mixtures is that natural and most artificial selections emphasize competitive ability rather than agricultural success. Success in the present instance is defined as dependably high yields with limiting water. I do not expect that we will find some means for a large jump in photosynthesis rate and thus in WUE. In fact, we may find it quite difficult to improve on basic WUE already known for C_3, C_4, and CAM plants. It is also doubtful that we would gain much from a greater drought-tolerance than our plants now possess. Even so, we can hope for more efficient partitioning patterns, particularly in timing and extent of root and leaf growth, chemical composition, and economic yield, so that the basic WUE is achieved more consistently and with a larger fraction of the water supply.

We have emphasized the behavior of crops in semiarid and arid regions where efficient use of water is always a problem and cropping practices are adjusted accordingly. In more humid areas, drought occurs less frequently and the present attitude of farmers and researchers generally follows a "dryland perspective." Drought is accepted as a risk, and farming practices give little attention to water use. We may nevertheless be approaching a time when we can no longer afford occasional crop losses. A better approach to farming in such regions may be found in intensification through arrangements for supplemental irrigation and other practices common to the more arid zones.

ACKNOWLEDGMENTS

I greatly appreciate the valuable contributions to this paper made through discussion and comment by R. Benjamin, E. Fereres, C. T. de Wit, H. van Keulen, and J. T. Ritchie.

LITERATURE CITED

Alessi, J., and J. F. Power. 1976. Water use in dryland corn as affected by maturity class and plant spacing. Agron. J. 68:547–550.

Allen, H. 1975. Shade-cloth microclimate of soybeans. Agron. J. 67:175–181.

Barfield, B. J., and J. F. Gerber (ed.). 1979. Modification of the aerial environment of plants. Am. Soc. of Agric. Eng. Monogr. no. 2. St. Joseph, Mich.

Blum, A., and M. Naveh. 1976. Improved water-use efficiency in dryland grain sorghum promoted by plant competition. Agron. J. 68:111–116.

Bond, J. J., T. J. Army, and O. R. Lehman. 1964. Row spacing, plant populations and moisture supply as factors in dryland grain production. Agron. J. 56:3–6.

Briggs, L. J., and H. L. Shantz. 1914. Relative water requirement of plants. J. Agric. Res. 3: 1–64.

Brouwer, R. 1963. Some aspects of the equilibrium between overground and underground plant parts. Jaarb. Inst. Biol. Scheik. Onderz (IBS) 1963:31–39.

Brown, K. W., and N. J. Rosenburg. 1971. Shelter effects on microclimate, growth and water use by irrigated sugar beets in the Great Plains. Agric. Meteorol. 9:241-263.

Burrows, W. C., L. Hahn, R. E. Jensen, R. S. Loomis, and W. Oschwald. 1976. Crop and livestock management. p. 79-105. *In* Climate and food. Natl. Res. Council Rep. Natl. Academy of Sciences, Washington, D.C.

Connor, D. J. 1975. Growth, water relations and yield of wheat. Aust. J. Plant Physiol. 2:353-366.

Cowan, I. R., and G. D. Farquhar. 1977. Stomatal function in relation to leaf metabolism and environment. Symp. Soc. Exp. Biol. 31:471-505.

Cutler, J. M., D. W. Rains, and R. S. Loomis. 1977. Role of changes in solute concentration in maintaining favorable water balance in field-grown cotton. Agron. J. 69:773-779.

de Wit, C. T. 1958. Transpiration and crop yields. Versl. Landbouwkd. Onderz. (Agric. Res. Rep.) no. 64.6. s'Gravenhage, The Netherlands.

de Wit, C. T. 1960. On competition. Versl. Landbouwkd. Onderz. no. 66.8. Pudoc, Wageningen, The Netherlands.

de Wit, C. T., H. H. van Laar, and H. van Keulen. 1979. Physiological potential of crop production. p. 47-82. *In* J. Sneep and A. J. T. Hendriksen (ed.) Plant breeding perspectives. Pudoc, Wageningen, The Netherlands.

Donald, C. M. 1968. The breeding of crop ideotypes. Euphytica 17:385-403.

Downes, R. W. 1969. Differences in transpiration rates between tropical and temperate grasses under controlled conditions. Planta 88:261-273.

Ehleringer, J., and O. Bjorkman. 1977. Quantum yields for CO_2 uptake in C3 and C4 plants. Plant Physiol. 59:86-90.

Faci, J. M., and E. Fereres. 1980. Responses of grain sorghum to variable water supply under two irrigation frequencies. Irrig. Sci. 1:149-159.

Fischer, R. A., and R. Maurer. 1978. Drought resistance in spring wheat cultivars. I. Grain yield responses. Aust. J. Agric. Res. 29:897-912.

Fischer, R. A., and N. C. Turner. 1978. Plant productivity in the arid and semiarid zones. Annu. Rev. Plant Physiol. 29:277-317.

Fuehring, H. D. 1975. Yield of dryland grain sorghum as affected by antitranspirant, nitrogen, and contributing watershed. Agron. J. 67:255-257.

Gale, J., and R. M. Hagan. 1966. Plant antitranspirants. Annu. Rev. Plant Physiol. 17:269-282.

Gause, G. F. 1934. The struggle for existence. Williams and Wilkins, Baltimore, Md.

Goodman, D. 1975. The theory of diversity-stability relationships in ecology. Quart. Rev. Biol. 50:237-266.

Goudriaan, J., and H. H. van Laar. 1978. Relations in leaf resistance, CO_2-concentration and CO_2 assimilation in maize, beans, lalang grass and sunflower. Photosynthetica 12:241-249.

Grimes, D. W., H. Yamada, and W. L. Dickens. 1969. Functions for cotton (*Gossypium hirsutum* L.) production from irrigation and nitrogen fertilization variables: I. Yield and evapotranspiration. Agron. J. 61:769-773.

Hall, A. E., and C. Dancette. 1978. Analysis of fallow-farming systems in semi-arid Africa using a model to stimulate the hydrologic budget. Agron. J. 70:816-823.

Hall, A. E., K. W. Foster, and J. G. Waines. 1979. Crop adaptation to semi-arid environments. p. 148-179. *In* A. E. Hall et al. (ed.) Agriculture in semi-arid environments. Ecol. Stud. Vol. 34. Springer-Verlag, New York.

Hall, A. E., and G. J. Hoffman. 1976. Leaf conductance response to humidity and water transport in plants. Agron. J. 68:876-881.

Hanks, R. J., H. R. Gardner, and H. R. Florian. 1969. Plant growth–evapotranspiration relations for several crops in the Central Great Plains. Agron. J. 61:30-34.

Henderson, D. W., F. J. Hills, R. S. Loomis, and E. F. Nourse. 1968. Soil moisture conditions, nutrient uptake, and growth of sugar beets as related to method of irrigation of an organic soil. J. Am. Soc. Sugar Beet Technol. 15:35-48.

Hsiao, T. C., E. Fereres, E. Acevedo, and D. W. Henderson. 1976. Water stress and dynamics of growth and yield of crop plants. p. 281-305. *In* Water and plant life. Ecol. Stud. Vol. 19. Springer-Verlag, Berlin.

Hunt, W. F., and R. S. Loomis. 1979. Respiration modelling and hypothesis testing with a dynamic model of sugar beet growth. Ann. Bot. 44:5-17.

Hunter, A. S., and O. J. Kelley. 1946. The growth and rubber content of guayule as affected by variations in soil moisture stresses. Agron. J. 38:118-134.

Hurd, E. A. 1974. Phenotype and drought resistance in wheat. Agric. Meteorol. 14:39–55.

Jordan, W. R., and J. T. Ritchie. 1971. Influence of soil water stress on evaporation, root absorption, and internal water status of cotton. Plant Physiol. 48:783–788.

Laing, D. R., and R. A. Fischer. 1977. Adaptation of semidwarf wheat cultivars to rainfed conditions. Euphytica 26:129–139.

Lof, H. 1976. Water use efficiency and competition between arid zone annuals especially the grasses *Phalaris minor* and *Hordeum murinum*. Agric. Res. Rep. no. 853. Pudoc, Wageningen, The Netherlands.

Loomis, R. S. 1976. Agricultural ecosystems. Sci. Am. 235(3):99–105.

Loomis, R. S., and P. A. Gerakis. 1975. Agricultural productivity. p. 145–172. *In* J. P. Cooper (ed.) Photosynthesis and productivity in different environments. Cambridge University Press, Cambridge, U.K.

Loomis, R. S., and J. G. Haddock. 1967. Sugar beets. p. 640–648. *In* R. M. Hagan et al. (ed.) Irrigation of agricultural lands. Am. Soc. of Agron., Madison, Wis.

Loomis, R. S., and W. A. Williams. 1963. Maximum crop productivity: an estimate. Crop Sci 3:67–72.

Loomis, W. E. 1932. Growth-differentiation balance vs. carbohydrate-nitrogen rate. Proc. Am. Soc. Hort. Sci. 29:240–245.

Margelef, R. 1969. Diversity and stability: a practical proposal and a model of interdependence. Brookhaven Symp. Biol. 22:25–37.

Maximov, N. A. 1929. The plant in relation to water. George Allen and Unwin, Ltd., London.

Monteith, J. L. 1972. Solar radiation and productivity in tropical ecosystems. J. Appl. Ecol. 9:747–766.

Monteith, J. L. 1978. Reassessment of maximum growth rates for C3 and C4 crops. Exp. Agric. 14:1–5.

Passioura, J. B. 1976. Physiology of grain yield in wheat growing on stored moisture. Aust. J. Plant Physiol. 3:559–565.

Passioura, J. B. 1977. Grain, yield, harvest index, and water use of wheat. J. Aust. Inst. Agric. Sci. 43:117–120.

Pendleton, J. W. 1966. Increasing water use efficiency by crop management. p. 231–258. *In* W. H. Pierre et al. (ed.) Plant environment and efficient water use. Am. Soc. of Agron., Madison, Wis.

Raschke, K. 1975. Stomatal action. Annu. Rev. Plant Physiol. 26:309–340.

Reeve, R. C., and M. Fireman. 1967. Salt problems in relation to irrigation. p. 988–1008. *In* R. M. Hagan et al. (ed.) Irrigation of agricultural lands. Am. Soc. of Agron., Madison, Wis.

Ritchie, J. T. 1971. Dryland evaporative flux in a subhumid climate. I. Micrometeorological influences. Agron. J. 63:51–55.

Ritchie, J. T., and E. Burnett. 1971. Dryland evaporative flux in a subhumid climate. II. Plant influences. Agron. J. 63:56–62.

Robins, J. S. 1967. Reducing irrigation requirements. p. 1143–1158. *In* R. M. Hagan et al. (ed.) Irrigation of agricultural lands. Am. Soc. of Agron., Madison, Wis.

Seaton, K. A., and J. J. Landsberg. 1978. Resistance to water movement through wheat root systems. Aust. J. Agric. Res. 29:913–924.

Shantz, H. L., and L. N. Piemeiscl. 1927. The water requirement of plants at Akron, Colorado. J. Agric. Res. 34:1093–1190.

Smika, D. E. 1970. Summer fallow for dryland winter wheat in the semiarid Great Plains. Agron. J. 62:15–17.

Spitters, C. J. T. 1980. Competition and its consequences for selection in barley breeding. Agric. Res. Rep. no. 893. Pudoc, Wageningen, The Netherlands.

Thomas, M. D., and G. R. Hill. 1949. Photosynthesis under field conditions. p. 19–52. *In* J. Franck and W. E. Loomis (ed.) Photosynthesis in plants. Iowa State University Press, Ames.

Trenbath, B. R. 1974. Biomass productivity of mixtures. Adv. Agron. 26:177–210.

U.S. Department of Agriculture. 1974. Summer fallow in the western United States. Conserv. Res. Rep. no. 17, USDA-ARS. U.S. Government Printing Office, Washington, D.C.

Vaadia, Y. 1976. Plant hormones and water stress. Philos. Trans. R. Soc. London Ser. B. 273: 513–522.

Van Bavel, C. H. M. 1953. Chemical composition of tobacco leaves as affected by soil moisture conditions. Agron. J. 45:611–614.

van Keulen, H. 1975. Simulation of water use and herbage growth in arid regions. Pudoc, Wageningen, The Netherlands.

Viets, F. G., Jr. 1966. Increasing water use efficiency by soil management. p. 259–274. *In* W. H. Pierre et al. (ed.) Plant environment and efficient water use. Am. Soc. of Agron., Madison, Wis.

Waggoner, P. E. 1966. Decreasing transpiration and the effect upon growth. p. 49–72. *In* W. H. Pierre et al. (ed.) Plant environment and efficient water use. Am. Soc. of Agron., Madison, Wis.

Welch, N. H., E. Burnett, and H. V. Eck. 1966. Effect of row spacing, plant population, and nitrogen fertilization on dryland grain sorghum production. Agron. J. 58:160–163.

CHAPTER 8B

Crop Manipulation for Efficient Use of Water: Inadvertent and Intended Manipulations of Crop Water Use

GARY H. HEICHEL

Agricultural Research Service, USDA; and University of Minnesota
St. Paul, Minnesota

I. INTRODUCTION

Changes in crop water use may occur through inadvertent modification of the atmosphere, or by intentional modification of plant traits. This section briefly discusses how crop water use might respond to atmospheric enrichment with CO_2, and how manipulations of water use by plant breeding might be achieved.

Our current framework of crop manipulations to achieve efficient use of water has been developed by decades of research on crop growth and agronomic practices. Despite our confidence in these principles, we have little understanding of whether they will serve agriculturalists as reliably in the future as they have in the past. One reason for this uncertainty is the inadvertent and irreversible changes that humankind is making in the earth's atmosphere.

II. CARBON DIOXIDE ENRICHMENT, PLANT GROWTH, AND WATER-USE EFFICIENCY

The CO_2 concentration of the atmosphere has risen about 1.2 ppm annually for the past 20 years because of increased combustion of fossil fuels and accelerated deforestation of the tropics. It is now conservatively predicted (Kellogg, 1979; Rotty, 1979) that the atmospheric CO_2 concentration in 50 years will be 50–100% greater than the present value of 330 ppm, and the consequences of this enrichment on growth and water use over the life cycle of the crop are difficult to anticipate. Studies of short-term CO_2 enrichment of the atmosphere around individual plants or leaves show that a doubling or tripling above ambient levels stimulates CO_2 exchange and rates of growth and development (Ford & Thorne, 1967; Akita & Moss, 1973;

Limitations to Efficient Water Use in Crop Production.

Heichel & Jaynes, 1974; Hardy & Havelka, 1977), exerts an antitranspirant effect upon stomata (Moss et al., 1961; Pallas, 1965; Akita & Moss, 1972), modifies the partitioning of assimilate between roots and shoots (Ford & Thorne, 1967), lessens the differences in photosynthetic CO_2 exchange between C_3 and C_4 species (Akita & Moss, 1972, 1973), increases tillering, branching, and leaf area (Ford & Thorne, 1967; Heichel & Jaynes, 1974; Gifford, 1977), modifies crop maturity and senescence (Chang, 1975; Hardy & Havelka, 1977), and increases water-use efficiency (Gifford, 1979; Carlson & Bazzaz, 1980). Numerous differences in physiological responses among species (Akita & Moss, 1973) and among cultivars within species (Heichel & Musgrave, 1969) have been observed.

Future increases in atmospheric CO_2 concentration are virtually guaranteed, so there is every reason to believe that important modifications of crop growth and water use will ensue. Furthermore, concurrent but gradual changes in temperature patterns among major agricultural regions may modify areas of plant adaptation and the concepts of cool- vs. warm-season cropping for efficient water use.

It is therefore likely that the principles that presently guide the development of practices to assure the efficient use of water may be somewhat inadequate to cope with future water-use patterns resulting from atmospheric CO_2 enrichment. We must anticipate and implement the kinds of research needed for plant breeders, physiologists, and agronomists to develop cultivars and cropping systems to meet the needs resulting from this inadvertent atmospheric modification. There has been only one long-term field experiment on the response of crop production to CO_2 enrichment (Hardy & Havelka, 1977), and none on crop water use. Except for a few recent experiments in growth cabinets and glasshouses (Gifford, 1977, 1979; Carlson & Bazzaz, 1980), we have virtually no analytical understanding of how CO_2 enrichment over the entire life cycle mediates the growth, development, and water use of crop plants. Growth and water use of a crop stand under CO_2 enrichment may differ markedly from that of single plants (Gifford, 1979). Furthermore, some species display photosynthetic acclimation to CO_2 enrichment (Raper & Peedin, 1978). These recent findings illustrate the problem of trying to anticipate the effects of future atmospheric CO_2 enrichment on efficiency of water use with present knowledge from short-term studies in artificial environments. These problems will likely become more acute unless appropriate research is undertaken.

III. BREEDING CROPS FOR HIGHER WATER-USE EFFICIENCIES: A PERSPECTIVE

Whether the objective of research is to meet the challenge of inadvertent climate modification or to develop improved cultivars with attributes for more efficient use of water, there is considerable potential for intentional modification of crop water use through plant breeding. Despite this potential, there has been a paucity of achievement. Therefore, I offer a perspective on how to expedite the task.

Because of its complexity, breeding crops for more efficient water use might be most effectively approached by a team research effort. The success of such an effort to intentionally modify the use of water by crops depends upon several factors: identification in target species of genetic variability for characters that are associated with water use; an understanding of the climate of the area or the region for which the species is being developed; the joining of a plant breeder-geneticist, a plant or crop physiologist, and a soil-crop management scientist (as a minimum) with support personnel into a multidisciplinary team with a common purpose; and a commitment of financial resources for several years to permit the unhindered pursuit of objectives.

Considerable thought should be given to constituting the membership of a research team. Because so much of the success of a team effort depends upon interpersonal relationships among its members, an effective group will seldom be constituted by the unilateral actions of an administrator distant from the research. A better approach would be to identify a coordinator who would eventually be a team member, and then allow that person and his close colleagues to recruit others to the team. In this way the goals of the team develop with the addition of new members, each with unique competence and perspective.

Developing plants with improved water-use characteristics will be a long-term effort that in its early stages is probably poorly suited to rapid experimental progress or rapid publication. Therefore, prospective team members should be sought from among established scientists with strong credentials who might endure the challenges of slow progress and slow publication better than junior scientists with prescribed criteria for success and advancement. A long period of experience with the candidate species is also desirable, and this is also another strong reason for selection of established scientists for team membership.

Once the team is constituted, all members should share equally in establishing research objectives and in accepting responsibility for various aspects of the program. It is especially important that breeding for attributes related to efficient use of water be a major portion of the position description of the plant breeder, and not just an added assignment to a program that would probably already be overburdened with breeding for the traditional objectives of yield, quality, and pest resistance. The latter objectives are important, of course, but the new objective will be far more difficult because the breeding procedures are ill-defined (see Chapter 8C).

Before beginning the program, the climatic and soil characteristics of the area for which the crop is being selected must be understood (Chapter 8A). The seasonal progression of temperature, the distribution and intensity of rainfall, and the availability of soil moisture will largely govern which attributes of the plant might be beneficially altered to improve efficiency of water use. These attributes might include: depth of rooting early in the season to promote seedling survival, rapid canopy cover to minimize soil evaporation, deep rooting to exploit stored soil moisture during long droughts, stomata that close quickly at threshold water potentials, waxy leaves to increase reflection of radiation, reduced stomatal density to limit transpiration, leaf-rolling to reduce light interception, survival of protracted drought, premature leaf senescence, or maintenance of extension growth under severe stress. Selection within a specific environment will result in a narrowly adapted cultivar that will probably not be well-suited to other environments. Ideotypes for particular environments may be developed.

Familiarity with the crop and with the seasonal availability of water will reveal characteristics that might be manipulated for improved efficiency of water use (for more complete discussions see Chapters 7A and 8A). Breeding programs can only be attempted with characters or attributes known to be under genetic control and for which some information on heritability is available. It is important here to note that merely showing that a group of cultivars differs in expression of a trait does not adequately demonstrate genetic control because the cultivar differences may be variations in response to environment.

Relatively few attributes associated with efficiency of water use are known to be under genetic control. Stomatal size, density, and behavior (Heichel, 1971; Miskin et al., 1972; Henzell et al., 1976; Roark & Quisenberry, 1977), and depth or pattern of rooting (Hurd, 1971; Raper & Barber, 1970) are heritable characters in a number of species, as are the more well-known attributes of leaf dimension, leaf angle, and crop maturity. A number of the traits mentioned above have been associated with crop water use (Blum, 1979; Jones, 1979; Townley-Smith & Hurd, 1979), but techniques have not been tested on sufficient plant material to demonstrate genetic variability. The view that many physiological screening procedures are too slow or laborious for application to breeding programs has probably unduly hampered their application. A slow screening procedure can often be hastened with additional technical support for processing samples. Additionally, the screening procedure should be executed to reveal ranges in genetic expression in a test population rather than the detailed understanding of the functioning of a specific plant process that the physiologist might desire.

Physiologists have frequently compared divergent cultivars or divergent species within a genus to establish differences in physiological or morphological traits that they then suggest for inclusion in a plant breeding program. This is operationally a difficult situation because the tested plant material is often agronomically unadapted to the desired geographical area and the plant breeder is faced with the task of minimizing the introduction of undesirable correlated traits while attempting the successful transfer of the desired trait into the agronomically useful population. The inevitable cotransfer of undesirable traits into the adapted population reduces its usefulness for cultivar development until the undesirable traits have largely been eliminated. Besides being a very laborious procedure, it could be exceptionally difficult if the traits for efficient water use were not independent of desirable agronomic traits.

A more effective procedure might be for the plant breeder, physiologist, and management specialist to identify adapted cultivars with apparent differences in efficiency of water use. They should simultaneously subject these cultivars to intense physiological analyses to reveal how the adapted cultivars have achieved contrasting performance, and make plant selections. These selections would be used to develop an adapted population with a broad genetic base and one that is known to contain several types of traits influencing efficient water use. The physiological dissection of the adapted cultivars will be particularly useful in clarifying to team members how certain physiological or morphological traits might be associated and how they could best be selected from the broad base population.

Selection from this population for elite materials with improved water-use characteristics should be based upon selection for several traits simultaneously. The combination of several individual traits will improve the likelihood of isolating superior genotypes. Because the population is already agronomically adapted, it should require much less refinement after selection than a population that has been perturbed by the introduction of widely divergent germplasm.

IV. SUMMARY

Breeding crop plants for more efficient water-use characteristics and adapting these plants to cropping systems will require more coordination of effort than is usually possible among individual scientists working on different species at diverse locations. Development of interdisciplinary research teams with a long-term mandate to discover new principles and to develop new techniques should expedite the task. I have suggested a protocol that is proving successful in other team research on physiological and morphological traits.

LITERATURE CITED

Akita, S., and D. N. Moss. 1972. Differential stomatal response between C_3 and C_4 species to atmospheric CO_2 concentration and light. Crop Sci. 12:789–793.

Akita, S., and D. N. Moss. 1973. Photosynthetic responses to CO_2 and light by maize and wheat leaves adjusted for constant stomatal aperture. Crop Sci. 13:234–237.

Blum, A. 1979. Genetic improvement of drought resistance in crop plants: a case for sorghum. p. 429–445. *In* H. Mussell and R. C. Staples (ed.) Stress physiology in crop plants. Wiley-Interscience, New York.

Carlson, R. W., and F. A. Bazzaz. 1980. The effects of elevated CO_2 concentrations on growth, photosynthesis, transpiration, and water use efficiency of plants. p. 609–623. *In* J. J. Singh and A. Deepak (ed.) Environmental and climatic impact of coal utilization. Academic Press, New York.

Chang, C. W. 1975. Carbon dioxide and senescence in cotton plants. Plant Physiol. 55:515–519.

Ford, M. A., and G. N. Thorne. 1967. Effect of carbon dioxide concentration on growth of sugarbeet, barley, kale, and maize. Ann. Bot. 31:629–644.

Gifford, R. M. 1977. Growth pattern, carbon dioxide exchange and dry weight distribution in wheat growing under different photosynthetic environments. Aust. J. Plant Physiol. 4:99–110.

Gifford, R. M. 1979. Growth and yield of CO_2-enriched wheat under water-limited conditions. Aust. J. Plant Physiol. 6:367–378.

Hardy, R. W. F., and U. D. Havelka. 1977. Possible routes to increasing the conversion of solar energy to food and feed by grain legumes and cereal grains (crop production): CO_2 and N_2 fixation, foliar fertilization, and assimilate partitioning. p. 299–322. *In* A. Mitsui et al. (ed.) Biological solar energy conversion. Academic Press, New York.

Heichel, G. H. 1971. Genetic control of epidermal cell and stomatal frequency in maize. Crop Sci. 11:830–832.

Heichel, G. H., and R. A. Jaynes. 1974. Stimulating emergence and growth of *Kalmia* genotypes with CO_2. Hortic. Sci. 9:60–62.

Heichel, G. H., and R. B. Musgrave. 1969. Relation of CO_2 compensation concentration to apparent photosynthesis in maize. Plant Physiol. 44:1724–1728.

Henzell, R. G., K. J. McCree, C. H. M. van Bavel, and K. F. Schertz. 1976. Sorghum genotype variation in stomatal sensitivity to leaf water deficit. Crop Sci. 16:660–662.

Hurd, E. A. 1971. Can we breed for drought resistance? p. 77–88. *In* K. L. Larson and J. D. Eastin (ed.) Drought injury and resistance in crops. CSSA Spec. Publ. no. 2. Crop Sci. Soc. of Am., Madison, Wis.

Jones, H. G. 1979. Stomatal behavior and breeding for drought resistance. p. 408–428. *In* H. Mussell and R. C. Staples (ed.) Stress physiology in crop plants. Wiley-Interscience, New York.

Kellogg, W. W. 1979. Influences of mankind on climate. Annu. Rev. Earth Planet. Sci. 7:63–92.

Miskin, K. E., D. C. Rasmusson, and D. N. Moss. 1972. Inheritance and physiological effects of stomatal frequency in barley. Crop Sci. 12:780–783.

Moss, D. N., R. B. Musgrave, and E. R. Lemon. 1961. Photosynthesis under field conditions. IV. Some effects of light, carbon dioxide, temperature, and soil moisture on photosynthesis, respiration, and transpiration of corn. Crop Sci. 1:83–87.

Pallas, J. E., Jr. 1965. Transpiration and stomatal opening with changes in carbon dioxide content of the air. Science 147:171–173.

Raper, C. D., and S. A. Barber. 1970. Rooting systems of soybeans. I. Differences in root morphology among varieties. Agron. J. 62:581–584.

Raper, C. D., Jr., and G. F. Peedin. 1978. Photosynthetic rate during steady-state growth as influenced by carbon-dioxide concentration. Bot. Gaz. 139:147–149.

Roark, B., and J. E. Quisenberry. 1977. Environmental and genetic components of stomatal behavior in two genotypes of upland cotton. Plant Physiol. 59:354–356.

Rotty, R. M. 1979. Present and future production of CO_2 from fossil fuels—a global appraisal. p. 36–43. *In* W. P. Elliott and L. Machta (ed.) Workshop on the global effects of carbon dioxide from fossil fuels. CONF-770385. U.S. Dep. of Energy, Washington, D.C.

Townley-Smith, T. F., and E. A. Hurd. 1979. Testing and selecting for drought resistance in wheat. p. 447–463. *In* H. Mussell and R. C. Staples (ed.) Stress physiology in crop plants. Wiley-Interscience, New York.

CHAPTER 8C

Crop Manipulation for Efficient Use of Water: Constraints and Potential Techniques in Breeding for Efficient Water Use

R. W. ZOBEL

Agricultural Research Service, USDA, and Cornell University
Ithaca, New York

I. INTRODUCTION

Improved efficiency of water use can be obtained by (i) modifying cultural practices, such as land preparation, irrigation, cultivation, and crop selection, or (ii) genetic modification of the plant. Modifying cultural practices generally provides short-term solutions (from several weeks to several seasons), while genetic modification is a more long-term solution.

In this book the authors discuss soil, plant, and atmospheric effects, and their interactions in terms of efficiency of water use. Several authors (especially Loomis, Jordan, and Taylor) discuss plant characteristics that impact directly on water-use efficiency. They also attempt to provide some insight into the directions plant breeders will have to emphasize if any improvement is to be obtained. Heichel (Chapter 8B) addressed the subject of constraints on breeding for efficiency of water use in terms of the physiological parameters involved. On the other hand, the short discussion presented here will consider some of the techniques available and their applicability to physiological breeding in general, and to breeding for efficiency of water use in particular.

Genetic modification of a plant to improve water use depends upon three research steps: (i) identifying and characterizing the aspects that need change, (ii) identifying and characterizing plant material capable of filling the needs determined in step (i), and (iii) manipulation of genes to produce an adapted cultivar that has the required characteristics and fills other specific needs.

II. IDENTIFICATION OF CHARACTERS TO BE MODIFIED

Improved water use by plants can be explored at seven levels, based on genetic complexity (Fig. 1). None of these levels are mutually exclusive. In

Limitations to Efficient Water Use in Crop Production.

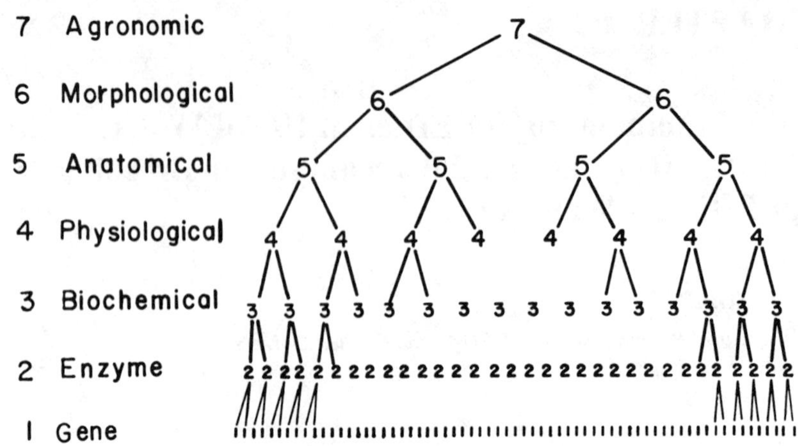

Fig. 1—Hierarchical pattern of the seven levels of characterization.

fact, a character at any one level consists of specific combinations of characters from succeedingly lower levels. Therefore, in terms of the lowest level, the gene, higher levels are increasingly more complex, with agronomic characters being the most complex. Since plant breeding is essentially manipulation of genes, agronomic characters are certainly the most difficult to breed for. In practice, however, this does not always appear to be the case.

The process of exploring any desirable attribute such as efficiency of water use often starts at the highest level, in this case agronomic aspects. Agronomic characteristics, such as planting density and timing of canopy closure, are determined directly or indirectly by morphological characteristics, and these in turn by anatomical characteristics, and so on to the genetic level. Each character has its own unique heritable components, and therefore is, to some extent, amenable to breeding procedures.

Consider, for example, planting density as an example of an agronomic (level 7) character that impacts heavily on water use. Planting density can be changed without genetic modification of the plant, but if it was desirable to improve efficiency of water use without a change in density per se, either other agronomic characters would need to be changed or plant morphology would need to be modified. At least two morphological (level 6) aspects of the plant are impacted at specific planting densities: leaf performance through mutual shading, and root performance through competition for water and minerals. These two characters can be genetically modified to improve plant responses to planting density. Many potential approaches are possible to improve root performance; two are to alter numbers or sizes (diameters) of roots (Passioura, 1972). Changes in root diameter bring about anatomical (level 5) changes in the relationship of internal tissues such as cortex thickness vs. stele diameter. A decrease in stele diameter without a reduction in cortex thickness will result in restriction of water movement from soil to shoot, while the reverse, reduction of cortex without a reduction in stele diameter, could result in an increase in water flow. The first of these two cases can be accomplished by a reduction in average xylem-vessel diameter or in xylem-vessel numbers. Both of these anatomical changes will alter the axial resistance to water flow (Chapter 3A). Xylem-vessel numbers are closely correlated with physiological (level 4) status of the root apex, specifically hormonal balance (Torrey, 1957). Several hormones interact to produce the final hormonal balance resulting in a given number of xylem vessels. Two of these hormones, cytokinin and auxin, are primary controlling agents, and are in turn controlled through the biochemical (level 3) pathways that regulate their production, activity, and half-lives. Indole acetic acid (IAA), for instance, is synthesized via several biochemical steps from the amino acid tryptophan. The last step in the IAA synthesis pathway is conditioned by the enzyme Indole–acetaldehyde oxidase, (IAAld oxidase) (level 2). IAAld oxidase is a gene (level 1) product. Some genes regulate the synthesis of enzymes and others the activity of genes that code for enzymes.

Although some enzymes result from combining the products from several genes, IAAld oxidase is a product of a single gene. The component amino acid structure of IAAld oxidase, and therefore its activity and mode of action, can be modified by modifying the specific nucleotide sequence of its encoding gene. This one-to-one relationship is the simplest of all to modify. This example demonstrates that a level 7 character is the product of many different genes. A crude minimum estimate of the number of genes involved in this example would be 2^6 or 64 (levels 1–6, with at least two characters added at each succeeding level; this estimate is certainly conservative, Fig. 1).

This example was selected because it demonstrates an impact on water-use efficiency and because it could be traced through all seven levels—we do not yet, however, know precisely which gene codes for IAAld oxidase. It is possible to breed for any one of the characters identified at each level by selecting for the appropriate modification of that character and then using conventional or, for that matter, nonstandard (e.g., recombinant DNA) breeding techniques to put that character in the appropriate genetic background. A character at a higher level (e.g., level 7 or level 6) is more difficult to characterize in terms of which gene or genes are modified than one at a lower level (2 or 1). These observations have some important impacts on ability to breed for any given character, and on selecting an appropriate breeding technique; these will be discussed later.

Many plant characteristics have been described as affecting efficiency of water use (Table 1 lists some of these characters and their levels of genetic complexity). If we use the conservative estimate of two characters of a lower complexity being added for each character at a given level, an estimate of the minimum number of genes involved in a characteristic is obtained (Table 1). For any of these characters the actual number of genes could be as high as 10 times this estimate. No two characters affecting the same characteristic have the same relative effect on that characteristic; one will tend to be predominant over the other. With more than two characters, a hierarchy

Table 1—Characters with potential for altering efficiency of water use.

Character†	Gene complexity level	Estimated minimum number of genes
Timing of flowering	4	8
Stomatal density	6	32
Stomatal response	4	8
Cuticle waxiness	3	4
Xylem diameter	5	16
Stele diameter	5	16
Leaf area	6	32
Duration of seed development	4	8
Membrane thermostability	3	4
Proline content	3	4
Osmoregulation	4	8
Abscisic acid regulation	4	8
Root geometry	6	32
Growth stability	7	64
Drought tolerance	7	64
Drought avoidance	7	64
Growth cycle	7	64
Growth stability	7	64
Cropping patterns	7	64
Heat tolerance	7	64

† This is not an exhaustive list, it just samples some of the characters mentioned in other chapters in this book.

of predominance is obtained. This hierarchical pattern is determined by the relationships between the genes of the total complement and the characters that make up a given character. Since all of the characteristics of efficient water use involve more than two genes, they result from hierarchical interactions.

To breed or select for differences in a characteristic, genetic differences have to be identified. This does not mean adding new genes, or discarding old ones, but identifying different forms (alleles) of the genes already conditioning that characteristic. Different alleles of the same gene have a hierarchy of dominance in relation to each other, as well as in their relation to the other genes conditioning a given character. Thus, replacing one allele with another will have more or less effect on the final character, dependent on the levels in the heirarchy that the two alleles have in relation to each other. For this reason it is difficult to predict the final effect of replacing one allele with another. The hierarchial response to such a change will determine the amount of effort necessary to complete such a replacement and produce a cultivar containing the new allele.

When a desired change can be accomplished by the replacement of a single gene with its allele, analysis of the F_2 generation (if the analysis is nondestructive) or the F_3 generation (if it is destructive) will identify plants with the modified characteristic (those containing the new allele). If the gene conditions a predominant character, and there is little genetic difference between the two parents, as few as a dozen plants (F_2) or lines (F_3) need be scored to acquire the character. If, on the other hand, the parents differ significantly and/or the character is not a predominant one, up to several

Table 2—Smallest perfect population necessary to identify a specific recessive genotype.

Number of genes with two alleles	Smallest perfect population
1	4
2	16
3	64
4	256
5	1,024
10	1,084,576
20	1.1×10^{12}

thousand F_2 plants or F_3 lines will have to be analyzed to identify one or more with the desired character in a usable plant phenotype. An estimate of the difficulty involved here can be made with the information from Table 2. The smallest possible F_2 population needed for the case where the parents differ by only the alleles in question would be four plants. Where the difference between parents is 10 genes, the smallest possible population is 1,084,576 plants, etc. Although it is possible to quantify the population size necessary to find the one plant desired when the genes are at the top of the dominance hierarchy, it becomes increasingly difficult to provide any estimate of population size with genes of succeeding lower hierarchical status. Ease of characterization (cost, time requirement, and labor required to analyze a single plant), variability due to measurement, destructiveness of the analytic procedure, and genetic nature of a given characteristic all determine the suitability of a given character for modification.

Initial studies to determine the genetic nature of some of the characters from Table 1 have been undertaken (membrane thermostability, Martineau et al., 1979; stomatal density, Dos Santos, 1978[1]; timing of flowering, Masaya, 1978[2]; leaf area, Woodworth, 1932). A measure of the difficulty described in the preceding discussion can be seen by reviewing the works of Masaya (1978)[2], and Martineau et al. (1979). Masaya studied two close ly related dry bean cultivars to determine the inheritance of photoperiod sensitivity (earliness of flowering). He discovered that two genes, one for lateness vs. earliness, and one for short duration vs. long duration of flowering, conditioned the differences between the two cutlivars. The gene for lateness was dominant over the earliness gene, and the two alleles for duration of flowering were co-dominant. This precision was possible because the two cultivars were closely related. Martineau et al. studied the genetic relation between a line with thermostable membranes and one with thermolabile membranes. The progeny distribution of the F_2 population demonstrated the probable presence of a dominant gene for tolerance, but also indicated that there were many other genes with lower predomiance segregating in the population. This second result is typical of genetic analy-

[1] E. O. Dos Santos, 1978. Germplasm evaluation and inheritance of stomata density and other plant characteristics in soybean (*Glycine max* (L.) Merrill). Ph.D. Thesis. Cornell University, Ithaca, N.Y.

[2] P. N. Masaya, 1978. Genetic and environmental control of flowering in *Phaseolus vulgaris* L. Ph.D. Thesis. Cornell University, Ithaca, N.Y.

ses for characters of presumed agronomic advantage. Generally cultivars that are geneologically far apart have F_2 population distributions that approach a normal distribution, thus obscuring the genetic nature of a character.

III. IDENTIFYING VARIABILITY

Breeding for, or against, some characteristic presumes the existence of genetic variability for that character. Identification of variability is therefore a critical first step after establishing that a given character needs to be genetically modified. There are essentially two sources of variability—natural and induced. These are in reality the two extremes, and variability exists or occurs in a virtual continuum between them. Ideally, variability for a character will be found in a closely related cultivated variety. If it is simply inherited, breeding will be rapid. A method of establishing simply inherited characters in a closely related line is to use mutagenesis on the line in which the modification is desired, and select for the desired characteristic. One of the major drawbacks to mutagenesis is the increased number of simultaneously occurring undesirable mutants, which then need to be genetically removed. As the level of mutagensis increases, the likelihood of finding a specific character increases, but so does the number of mutations per individual. Mutagenesis therefore involves a tradeoff between the number of individuals to be screened vs. the number of mutations per individual.

An alternative to mutagenesis is the use of tissue-culture procedures. Screening plants for a given characteristic can involve millions of plants and becomes more prohibitive as the complexity of the characteristic or its method of identification increases. Tissue-culture techniques allow millions of individual genetic units to be grown and characterized in a small area, over a very short period. Because of their natures, even the most complex identification (selection) criteria can be easily adopted for use with tissue cultures. In addition, the large numbers and rapid life cycles allow selection for spontaneous mutations (occurrance of spontaneous mutations is on the order of one random mutation per 10^2 individuals or one specific mutation per 10^8 individuals). Mutagenesis can also be used to increase the likelihood of finding a given character, but the same drawbacks as discussed for whole plants obtain.

Screening and selection via tissue culture has several requirements that can severely limit its usefulness: (i) the character must be a level 1, 2, 3 (or 4) character, and (ii) the character must be expressed by cells that are not at the same level of differentiation or organization as those in the intact plant where the character is ultimately to be expressed. The first requirement addresses the observation that with only a few exceptions (e.g., tracheid development and adventitious rooting) anatomical and morphological characters are the result of very precise patterns of differentiation and organization. Cell-membrane thermostability might be amenable to tissue-culture screening and selection, since the cells in tissue culture have completely developed cell membranes. If cell cultures are exposed to a period of heat that is just sufficient to inhibit normal membrane function, only mutants with higher thermostability will survive, and therefore be able to be selected. This type of technique has been successfully carried out with resistance to herbicides and many other factors (Chaleff, 1981).

A refinement of tissue-culture techniques—molecular engineering/recombinant DNA—increases the potential of tissue culture for developing materials for use in a breeding program. The most dramatic possibility is the potential to move a specific gene from one species to another in essentially a single step. It may be possible with these techniques to transplant genes from bacteria, for instance, into plants. There are other aspects of these techniques that will be discussed later, but the concept of moving or transplanting a single gene has certain requirements that severely limit its usefulness in the present context.

As discussed earlier, efficient water use is not conditioned by a single gene, but an integrated complex of genes. Any modification of one characteristic is likely to result in a series of changes in the other characteristics that interact to give the final expression of efficient water use. To change one or a few genes with some assurance of success would require much greater knowledge about the interactions of those genes with the others involved.

Natural variability sometimes exists as a single gene difference that arose spontaneously in a cultivar of importance. More frequently it exists as part of a complex of different genes in related cultivars and even wild lines. Obviously the greater the genetic or evolutionary distance between the cultivar to be improved and the cultivar containing the desired character, the greater the difficulty of transferring that character. Genetic distance is defined as the number of genes that differ between two cultivars. Sometimes the desired character exists in exotic lines (other species, genera, or even families) and the effort required to manipulate it can be enormous (see the work of Sears, 1956). Some new techniques have potential to ease the effort required, and these will be discussed later. Getting the desired character into a cultivated line is only the beginning. The differences between the cultivated line to be improved, and the line containing the desired character determine the number of extraneous genes that need to be selected out before a final product is ready for release to the farmer. Characters coming from great distances often are closely associated with genes which in the new genetic backgrounds are deleterious. If the association is genetically close (i.e., if the genes involved are in close physical proximity to each other on a chromosome) a great effort is required to separate them. More distant material has a greater likelihood of genetic linkage of deleterious genes to desirable ones. These observations explain the heavy reliance on closely related plant material for all but the most extreme needs.

IV. MANIPULATION

There are many possibilities for manipulating characters (genes) once they are identified and the needed variation located. In the case of induced variation, where the variant is induced in the desired cultivar, no additional breeding is necessary except for removal of additional induced variability that is not desired.

Normal plant-breeding techniques are suitable for most sources of variability once the character is transferred to the desired background. Briefly, this involves a process of crossing the line containing the desired character to the cultivar to be improved, and then in succeeding generations selecting for the new character while maintaining the advantageous characteristics of the original cultivar. This procedure is most successful with easily selected, simply inherited characters. Complex characters or characters requiring complicated or tedious selection procedures are handled only with difficulty.

Some new techniques provide assistance in the breeding process. Plant materials that are amenable to haploid-induction techniques can be quickly brought to homozygosity through doubled haploids. This allows selection on true-breeding individual plants followed by backcrossing and further selection. This procedure accelerates the elimination of deleterious genes and offers a greater chance of keeping complexly inherited characters together.

Characters conditioned by recessive genes can only be observed when the genes are homozygous or hemizygous. Hemizygous is where there is only one copy (allele) of the gene rather than the normal two copies (alleles). Haploids or doubled haploids provide expression of recessive alleles without competition with dominant alleles. Twenty-five percent of the individuals in an F_2 population will carry both alleles of a given gene as the recessive allele. In populations derived from F_1 pollen and haploid techniques, 50% of the individuals are hemizygous or homozygous for the recessive allele. When the desired character is conditioned by a recessive gene, haploid procedures can provide improved manipulation. The alternative is extensive progeny testing until later stages in the breeding process when the desired gene can be brought to and maintained in a homozygous state.

Protoplast fusion and recombinant DNA techniques are two procedures that can provide unique inputs to plant breeding. When working with plant materials that do not easily crossbreed, plant breeders have had to resort to excising immature hybrid embryos and culturing them in vitro. Often the desired variability exists in plant species that are sexually incompatible with the species to be improved. When protoplast-fusion techniques are perfected it may be possible to fuse the protoplasts of the two plants (the cultivar to be improved and the cultivar containing the desired character), regenerate a plant, and, through gene elimination similar to that used by Sears (1956) (or use of haploid protoplasts and backcrossing the resulting diploid hybrid), develop a plant which, through standard breeding techniques, can be used to develop the desired improved variety. Unfortunately, at this point in time protoplast-fusion techniques have not been satisfactorily perfected for routine use. At best they will only allow access to germplasm not accessable by normal sexual means. Standard breeding techniques will still be required to produce the final improved cultivar.

Stomatal response to water potential changes is an area in which efficiency of water use can often be improved. Genes conditioning reduced water loss from tomato leaves have been found in a wild relative of tomato (*Solanum pennelli*) (Rick, 1973). This species is cross-compatible with cultivated tomato, but the transfer of desired characteristics is long and tedious. After five generations of backcrossing and selection for cultivated tomato characteristics, abnormal genetic ratios are still found (Rick, 1969). This drastically slows progress in breeding and must be taken into consideration when characters from wild relatives or more distant material are considered for use.

Recombinant DNA techniques are, however a different story. In essence, when perfected, they will allow the transfer of a single gene or several contiguous genes from one organism—plant, animal, or bacteria—to a specific host plant with little additional modification to the recipient plant. This technique has the advantages of induced mutations in that the cultivar to be improved is not otherwise modified. The major difficulty with recombinant DNA techniques is that they are limited to genes that can be characterized in vitro and whose effects can be characterized in the recipient cultivar. As discussed earlier, recombinant DNA techniques operate at levels 1 and 2, where the complexity of interactions and effects on the intact plant are difficult to predict. Of the characters listed in Table 2, proline content and cuticle wax have some potential for manipulation via recombinant DNA techniques. Membrane thermostability is a level 3 character, but it acts at level 5, the membrane, and is therefore not amenable to in vitro characterization. Before proline content and cuticle waxiness can be manipulated by these techniques, much more information about the interactions of these characters and a plant's physiology and biochemistry is needed. There will be a long period of research both in recombinant DNA techniques and plant molecular biology before any real use can be made of this technique for improving efficiency of water use. The technique is never likely to be useful with complexly inherited characters or characters not amendable to level 1 or level 2 characterization.

By their very natures, level 3, 4, and 5 characters involve characterization techniques that are usually too tedious to be routinely employed in a breeding program, and too genetically complex to be amenable to those new techniques capable of moving or modifying single-gene characters. Many characters at levels 2 and 6 also are difficult to manipulate without extensive effort in terms of time, manpower, and facilities. Standard breeding procedures involve one or more techniques to improve a cultivar through selection—usually human-mediated selection. Some of these techniques are:

pedigree, single-seed descent, bulk, double- and single-cross hybrids, and backcross. All involve selection for a specific character(s) at one or more points during the process. These are all more-or-less labor-intensive and in routine use are generally restricted to level 6 and 7 characters.

Animal breeders often use a different system, since their material is even more difficult to bring to homozygosity than corn, for instance (see Falconer, 1960, for discussion). One approach is to determine a desired characteristic (say at level 4), characterize its correlation to other characters that are at levels 6 or 7, and then place selection pressure on the correlated character. This technique is especially useful with complex characters, since the selection is not on any one or several of the characters, but rather on the result of their interactions. To some extent this is exactly what selection of level 6 and 7 characters represents, since most of these characters are the products of a series of level 5 characters, which are in turn the products of an even greater series of level 4 characters. There are, of course, many level 6 and 7 characters that jump directly to levels 1 or 2 and are conditioned by single genes, e.g., flower color or seed oil quality. As breeding pressure has increased, these simpler characters have been thoroughly exploited, leaving the more complex characters for future treatment.

The process of breeding for correlated characters rather than directly for the character desired (for convenience this can be called associative breeding) can easily be used to make more thorough use of level 3, 4, and 5 characters in breeding programs without broad-spectrum teams of scientists and funding (see Chapters 3C and 8B). As an example, let us examine associative breeding for one of the characters in our example of levels of characterization. If a series of random crosses are made and the resulting F_2 progeny are placed in a soil susceptible to drought conditions, survival would be dramatically reduced over that in an irrigated or nondroughty field. There would also be little likelihood that the survivors would convey any increased ability to avoid or tolerate drought conditions. From Passioura's work (1972) it is known that under fixed soil water conditions (as exist in the western United States), a reduced number of roots attached directly to the shoot will result in slower water uptake, delaying the onset of severe drought. If the reduction in water uptake rate is balanced correctly, as in his experiments, more water will be available during seed development, thus increasing yield. Since the total root mass in each treatments was similar, the delay in drought appears to be due to restriction of water movement to the shoot rather than to reduced uptake capacity. Reducing the number of roots attached to the shoot is level 6 manipulation and marginally amenable to selection in the field. Reduction of stele diameter might accomplish the same thing, but stele diameter is a level 5 character and would be prohibitively expensive to select for in a breeding program. If, however, several different sources for narrow stele are intercrossed with advanced breeding lines and cultivars, successful combinations should correlate closely with increased yield when grown under fixed soil water conditions. Combining this technique with that of Suneson (1956), where all the lines are intercrossed and grown for several generations in bulk, "natural selection'' will rapidly select out the best combination of genes under the specific field conditions.

This associative breeding approach will be the most powerful with self-pollinating species because of their tendency toward homozygosity. With care the procedure should be suitable with crosspollinating species where the genes in the pool have a chance to recombine many times, resulting in the most suitable combinations possible under those growing conditions. With self-pollinated crops, optimum performance would be found with intercrossing the better lines after several generations. Very complex characters can be manipulated in this way, since the ultimate character screened for its yield, and selection pressure is provided by environment. Any character (levels 1, 2, 3, 4, or 5) affecting a level 7 or 6 character could easily be bred for by ''associative breeding,'' simply through selection for the correlated (associated) level 6 or 7 character. Of course the simplest technique is the one just described: placing environmental pressure on the level 6 or 7 character and then screening for yield.

V. SUMMARY

Investigation of a phenomena like efficient plant water use can be carried out on any of seven different levels: (7) agronomic, (6) morphological, (5) anatomical, (4) physiological, (3) biochemical, (2) enzymatic, and (1) genetic. Characterization of desired changes in plants will be made at one of these seven levels. Standard plant breeding procedures can manipulate characters without extensive commitments of resources at levels 6 and 7 only. Recombinant DNA techniques operate most effectively on those characters that can be characterized at level 1 (the gene). Tissue-culture techniques require that characters be expressed in undifferentiated cells and therefore are restricted to levels 1-4, and to single genes. Mutagenesis can be used at any level where a single predominant gene conditions the character, while protoplast fusion and standard hybridization work effectively at all levels of characterization and genetic complexity. On the other hand, tissue culture and mutagenesis are generally restricted to direct modifications of the cultivar to be improved, while hybridization, protoplast fusion, and recombinant DNA techniques are able to introduce characters from greater phylogenetic distances. Unfortunately these procedures are not all currently available for most crops. Regardless of the technique used, most modifications of the genome of a cultivar must be followed by a period of breeding to refine the ''fit'' of the introduced character into an agronomically acceptable background. This is most easily accomplished with level 6 and 7 characters under current breeding procedures. Associative breeding, where a desired level 2, 3, 4, or 5 character is incorporated into a series of cultivars and allowed to become adopted through natural-selection pressure on its associated level 6 and 7 character(s), allows breeding for those characters that would normally require prohibitive levels of manpower and resources.

LITERATURE CITED

Chaleff, R. 1981. Genetics of higher plants: applications of cell culture. Cambridge University Press, New York.

Falconer, D. S. 1960. Quantitative genetics. The Ronald Press Company, New York. Chapt. 19:312–329.

Martineau, J. R., J. H. Williams, and J. E. Specht. 1979. Temperature tolerance in soybean. II. Evaluation of segregating populations for membrane thermostability. Crop Sci. 19:79–81.

Passioura, J. B. 1972. The effect of root geometry on the yield of wheat growing on stored water. Aust. J. Agric. Res. 23:745–752.

Rick, C. M., Jr. 1969. Controlled introgression of chromosomes of *Solanum pennellii* into *Lycopersicon esculentum*: segregation and recombination. Genetics 62:753–768.

Rick, C. M., Jr. 1973. Potential genetic resources in tomato species: clues from observations in native habitats. p. 255–269. *In* A. M. Srb. (ed.) Genes, enzymes, and populations. Plenum Press, New York.

Sears, E. R. 1956. The transfer of leaf-rust resistance from *Aegelops umbellulata* to wheat. Brookhaven Symp. Biol. 9:1–22.

Suneson, C. A. 1956. An evolutionary plant breeding method. Agron. J. 48:188–190.

Torrey, J. G. 1957. Auxin control of vascular pattern formation in regenerating pea root meristems grown *in vitro*. Am. J. Bot. 44:859–870.

Woodworth, C. M. 1932. Genetics and breeding in the improvement of the soybean. Illinois Agric. Exp. Stn. Bull. no. 384.

CHAPTER 9A

Yield and Water-Use Relationships: An Overview

R. J. HANKS
Utah State University
Logan, Utah

I. INTRODUCTION

The water requirement of crops has been a subject of much study in the past (Jensen, 1973; Doorenbos & Pruitt, 1975). The design of irrigation systems, land allocation, etc., depends on a knowledge or estimate of the amount of water required to produce a crop. Most of these "water-requirement" studies have concentrated on the climatic factors that influence water use for maximum crop production. Thus crop yield relations have not been considered in any detail. Recent emphasis on yield–water-use studies have been conducted because of the increasing scarcity of water for irrigation. Soil water storage and rainfall relations have been considered to a lesser extent. Where water is scarce and yield is limited, the storage of water by the soil is a much more important factor.

Considerable research was conducted in the early part of this century to determine the yield–water-use relations of a large group of crops (reviewed by de Wit [1958] and discussed in some detail in Chapter 1). Out of this work came the concept that there was a close relation of yield (aboveground dry matter) to water use or transpiration (i.e., the ratio of transpiration to yield was near a constant value). In this chapter the attempts to use the information on yield and water-use relationships to make predictions of crop yield are reviewed. The general approach is to first estimate crop transpiration and then from this estimate to calculate crop yield.

II. EARLY STUDIES

Of critical importance in these predictions is, of course, the relationship between yield and transpiration. As reviewed in Chapter 1, substantial data were collected on this relationship at the beginning of this century. However, many of these measurements were made in small containers under rather unusual conditions, especially since evaporation from the soil

Limitations to Efficient Water Use in Crop Production.

was eliminated, so few attempts were made to make practical use of this vast amount of information. Many researchers have criticized this type of research as being too simplistic because of the neglect of many factors that influenced either yield or water use.

The practical implications of this early work were not fully investigated until mid-1900's, when de Wit (1958) re-examined the old data on transpiration ratio (sometimes called water requirement). He concluded that there was a nearly unique yield–water-use relation for limited climatic regions provided a climatic correlation was imposed. He concluded that for semiarid climates, like those where Briggs and Shantz (1914) and similar data were collected, an equation relating yield to water use is:

$$Y = mT/E_o, \quad [1]$$

where Y is dry matter yield, T is transpiration, m is a crop factor dependent only on variety and species, and E_o is free-water (potential) evaporation. Many factors were tested to determine which best accounted for the climatic

Table 1—Dry matter yield (Y), transpiration (T), and pan evaporation (E) relations for the data of Shantz and Piemeisel (1927).

Crop	Factor	1911	1912	1913	1914	1915	1916	1917	Avg.
Alfalfa	Y (g)	196	338	292	306	217	123	183	236
(ADIE-23)	T (kg)	209	222	243	272	151	129	150	202
	E (cm d^{-1})	0.62	0.48	0.55	0.54	0.43	0.60	0.55	0.54
	m† (kg ha^{-1} d^{-1})	58	73	66	61	62	57	67	63
Oats	Y	174	396	276	264	332	103	171	245
(Swedish	T	107	168	170	158	149	90	109	147
select	E	0.62	0.48	0.55	0.54	0.43	0.60	0.55	0.54
CI:134)	m	101	113	89	90	96	69	86	90
Barley	Y	136	297	--	182	262	104	140	187
(Hannshen	T	72	132	--	91	106	69	73	--
CI531)	E	0.62	0.48	0.55	0.54	0.43	0.60	0.55	0.54
	m	117	108	--	108	104	90	105	106
Wheat	Y	94	274	280	306	238	92	154	205
(Kubanka	T	44	108	139	159	96	58	73	99
C.Z. 1440)	E	0.62	0.48	0.55	0.54	0.43	0.60	0.55	0.54
	m	132	122	111	104	106	94	117	112
Corn	Y	161	432	356	304	112	181	337	269
(North-	T	59	121	142	112	28	90	117	96
western	E	0.62	0.48	0.55	0.54	0.43	0.60	0.55	0.54
dent)	m	168	171	138	147	170	121	159	151
Millet	Y	202	243	211	301	256	171	229	230
(Kursk)	T	58	45	60	89	52	63	65	63
	E	0.62	0.48	0.55	0.54	0.43	0.60	0.55	0.54
	m	216	257	192	183	212	163	194	198
Sorghum	Y	308	487	606	457	203	193	276	361
(Minnesota	T	92	116	181	130	41	57	75	97
amber)	E	0.62	0.48	0.53	0.54	0.43	0.60	0.55	0.54
	m	208	200	185	190	212	203	202	200

† $m = YE_o/T$, where E_o = evaporation from a BPI pan (E above) × 0.92.

relation. Free-water evaporation, measured as pan evaporation corrected by a pan factor, was found to give the best correlation. Table 1 shows the *m* value calculated from the data of Shantz and Piemeisel (1927) for the different crops for different years. For the data of Table 1, the coefficients of determination of *Y* vs. *T* were 0.70, 0.80, 0.95, 0.89, 0.80, 0.19, 0.94 for the same sequence of crops. Figure 1 shows a plot of *Y* vs. Y/E_0 for sorghum, wheat, and alfalfa grown in containers in different locations in the U.S. Great Plains. Also included are data from India. Thus, the correlations of *Y* vs. T/E_0 were always better than *Y* vs. *T*.

There still remained the question of how applicable these early data were to the field. de Wit (1958) considered this problem in detail. He concluded the value of *m* found in the containers generally applied to the field with an adjustment (decrease) of about 10% for ground-level plants grown in the field and another decrease (by 0.92) to adjust the evaporation pan data to a large free-water body. Using these values of *m* (207, 115, and 55 kg ha^{-1} d^{-1} for sorghum, wheat, and alfalfa, respectively), he was able to make good predictions compared with a wide variety of field data.

de Wit (1958) also examined the question of whether the value of *m* was sensitive to soil water status. He concluded that a constant value of *m* would account for water-limiting situations also. The problem of evaluating the amount of *T* when evapotranspiration, ET (both soil evaporation, *E*, and *T*), is measured in the field makes this evaluation somewhat questionable. Another field-related problem involves drainage below the root zone and/or runoff, which could cause ET measurements to be faulty.

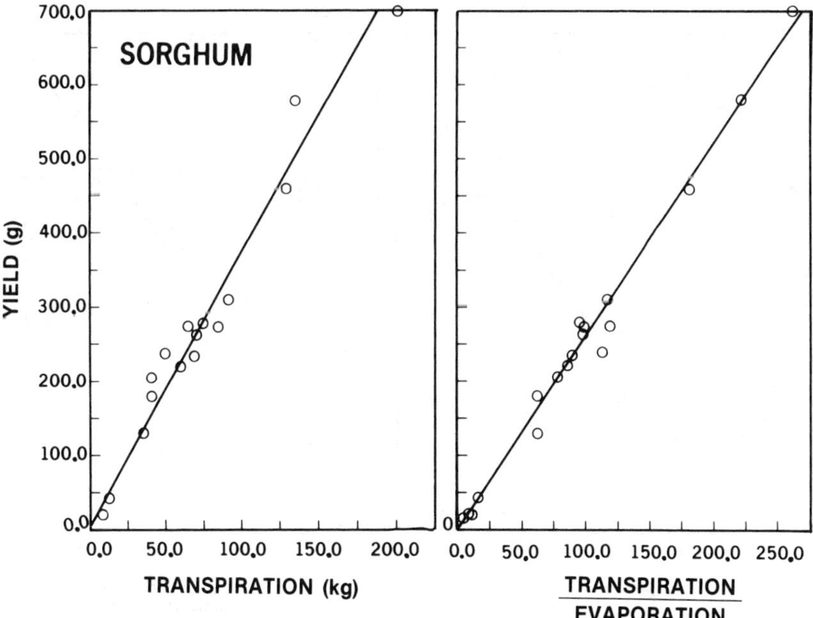

Fig. 1a—Relation of yield to transpiration and yield of the ratio of transpiration to free-water evaporation for sorghum. Data from containers.

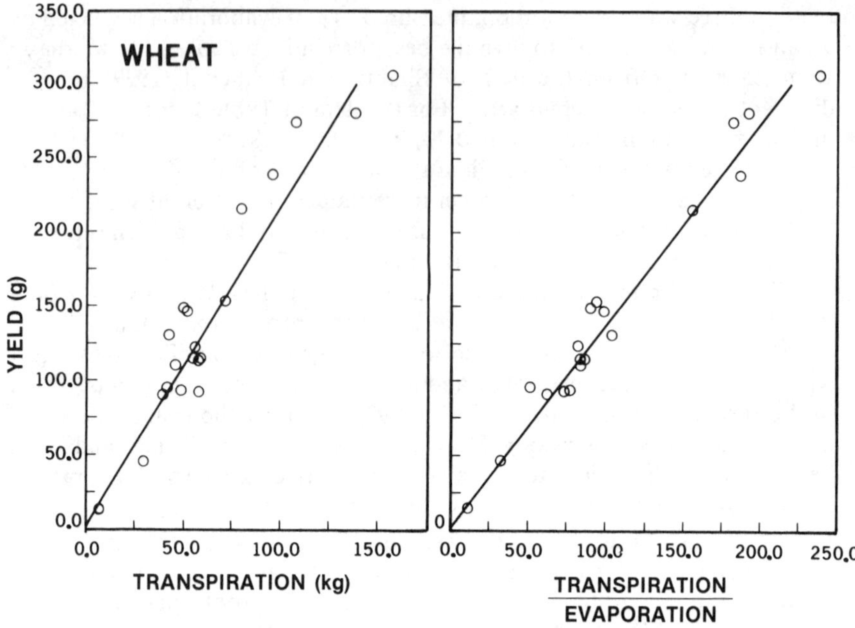

Fig. 1b—Relation of yield to transpiration and yield to the ratio of transpiration to free water evaporation for wheat. Data from containers.

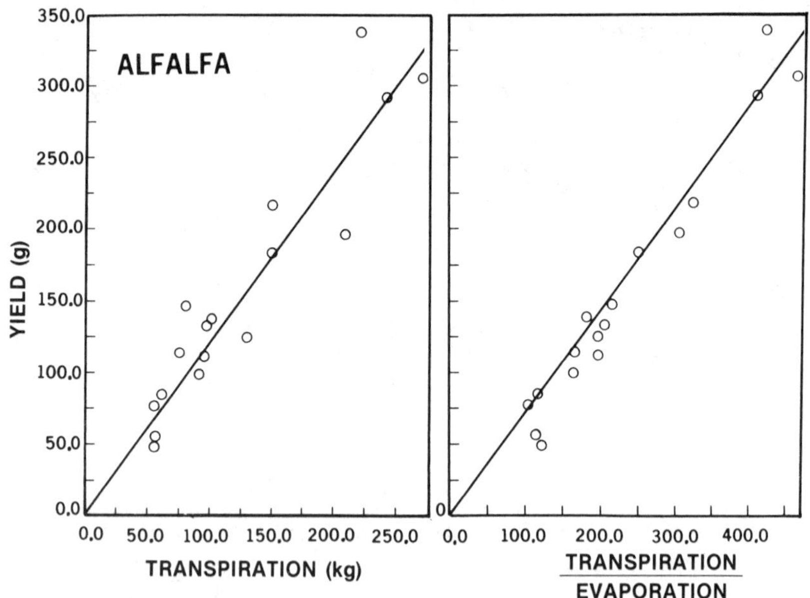

Fig. 1c—Relation of yield to transpiration, and yield to the ratio of transpiration to free water evaporation for alfalfa. Data from containers.

de Wit (1958) also concluded that for the humid parts of the world where solar radiation was limiting, the relation of Y to T was of the form:

$$Y = nT, \quad [2]$$

with the value of n being 610, 340, and 260 to dry matter ha^{-1} cm^{-1} for beets, peas, and oats, respectively. While the field data for testing were limited, there was good prediction of field yields using the appropriate n value compared with measured field yields, provided water was the main factor limiting yields. Here again field measurements of transpiration were uncertain because of many problems.

Arkley (1963) considered some of the same data and concluded the climatic correlation factor, E, could be replaced by a relative humidity term, and a single equation would hold for all climates. Tanner and Sinclair (Chapter 1) have extended this approach from a second theoretical background, which would indicate the climatic correlation factor should be vapor pressure deficit.

A recent review of plant productivity in arid and semiarid zones by Fischer and Turner (1978) concluded that the variation in water-use efficiency between communities could generally be described by Eq. [1]. If root production is included in the total dry matter production, they predicted an m factor of 110–140 kg ha^{-1} d^{-1} for wheat and many other C_3 species, and about 210 kg ha^{-1} d^{-1} for corn and many other C_4 species. The analysis of Fischer and Turner (1978) is shown in Table 2 for four plant communities. Note that if only the above ground portion of the production was included there was much more variation in the value of m. These crop

Table 2—Yield and water-use relations for four plant communities (taken from Fischer & Turner, 1978).

Community	E_o	ET	T	Y_T	Y_A	$m = \frac{Y_T E_o}{T}$	$m = \frac{Y_A E_o}{T}$
	cm d^{-1}	cm		kg ha^{-1}		kg ha^{-1} d^{-1}	
1. Shrubland *Ceretoides lanata*, Curlew Valley, Utah, 1972–1974	0.48	25.8	12.3	3,580	860	140	34
2. Annual pasture, Migda, Israel, 1971–1974	0.22	27.6	18.6	9,500	6,870	112	81
3. Wheat cv. Heron, Wagga Wagga, N.S.W., Australia, 1962	0.24	30.0	22.9	11,640	10,300	122	108
4. Corn (maize) cv. 85 RM, Mandan, N. Dak., 1968–1970	0.55	33.5	26.8	11,170	8,930	229	183

Y_T = Dry matter production of roots and shoots.
Y_A = Dry matter production of aboveground portion only.

plants had a rather small proportion of dry matter going to roots, so the value of m for the aboveground portion, as used by de Wit (1958), was only slightly lower than that for the total plant.

Hanks et al. (1969) considered the applicability of the de Wit equation (Eq. [1]) for field measurements of yield where water was limiting under dryland conditions. They concluded that the equation represented the field situation well. The values of m found (aboveground dry matter) were 125, 141, and 150 for wheat, sorghum, and millet, respectively. The wheat value agrees with that of de Wit, but the values for millet and sorghum are lower. The value of m has been shown to be quite dependent on variety, which could account for much of the difference. The variety differences have been found to be due to differences in dry matter production and not in evapotranspiration (Retta & Hanks, 1980; Sorensen et al., 1979). Here again there was uncertainty in predicting T in the field from measurements of ET.

All of the above discussion of yield is for dry matter yield only. If the yield of interest is grain, then the situation is more complicated. However, the first approximation, a constant ratio of grain to dry matter (harvest index, HI) could be used. The HI for Kubanka wheat grown in containers from Briggs and Shantz (1914), Shantz and Piemeisel (1927), and Dillman (1931) give an HI of 0.36. An HI of 0.42 was reported by Cole and Mathews (1923) for many locations in the U.S. Great Plains. Many factors, such as extreme droughts at critical periods, may cause the HI to be lower than these values. This problem will be discussed in more detail later.

It should also be pointed out that the m value found in the studies discussed is not nearly as constant if computed from each year's data. The values of m computed for each year, as shown in Table 1, show considerable variation from year to year. The general trend appears to be a low m for the hot years (1916) and a high m in the cool years (1912). When m was computed for part of the crop year, the variation, as would be expected, increased.

III. RECENT FIELD STUDIES

In recent years there have been many studies done under field conditions to evaluate the relation of water use to yield. One rather extensive study with corn was done in California, Colorado, Utah, and Arizona, using the same experiment design (Stewart et al., 1977). This study involved irrigation amount, irrigation frequency, and salinity treatments. Its main conclusion was that there was a strong relation of ET to yield. The correlation was better for dry matter than for grain. Figures 2, 3, 4, and 5 show the data for the four locations. The Arizona data appear less reliable because of the difficulty in getting corn to grow at that location on a sandy soil.

The data of Fig. 2–5 include some treatments where irrigation was withheld during pollination or earlier in the year. In spite of this, the soil water-holding capacity was sufficient to carry plants over during this period where irrigation had earlier been high. The treatments where irrigation was

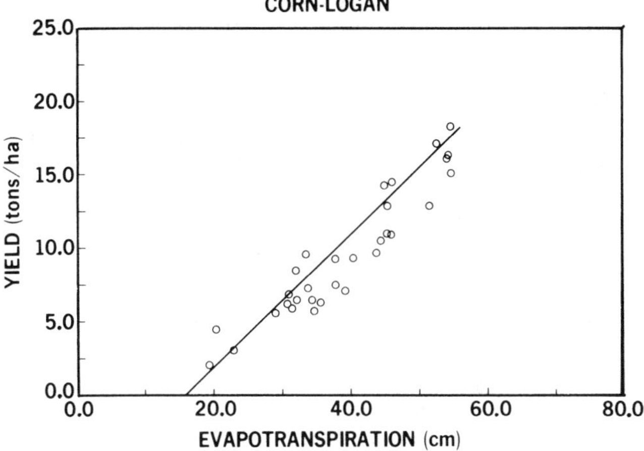

Fig. 2—Yield of corn as related to ET at Logan, Utah, in 1975.

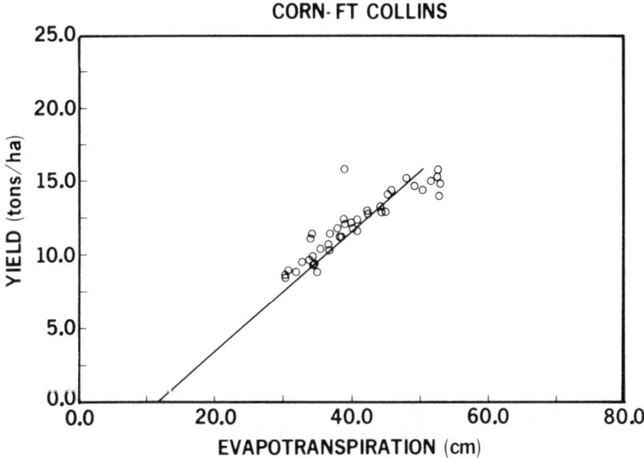

Fig. 3—Yield of corn as related to ET at Fort Collins, Colo., in 1975.

low throughout apparently induced some conditioning in the plant so that complete irrigation withdrawal did not substantially change the ET-yield relation.

Figures 2, 3, 4, and 5 also show data for soil treatments where soil salinity was imposed prior to planting. Dry matter yields on these treatments were all decreased, but ET was decreased similarly. Again the ET-yield relations appeared to be similar regardless of different irrigation frequencies or salinity treatments. The grain yield–ET relation was generally similar to the dry matter yield–ET relation except for Fort Collins, Colo., where the grain yields for the same ET were lower for the salinity treatments than for the check treatments. Grain yields at Yuma, Ariz., were nearly zero, apparently because of extremely hot conditions during a few critical

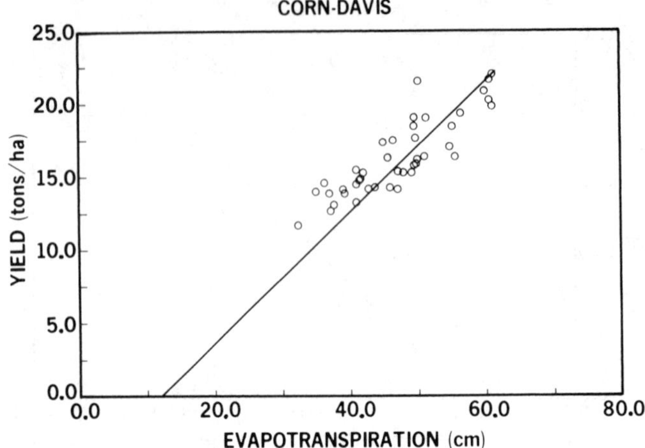

Fig. 4—Yield of corn as related to ET at Davis, Calif., in 1975.

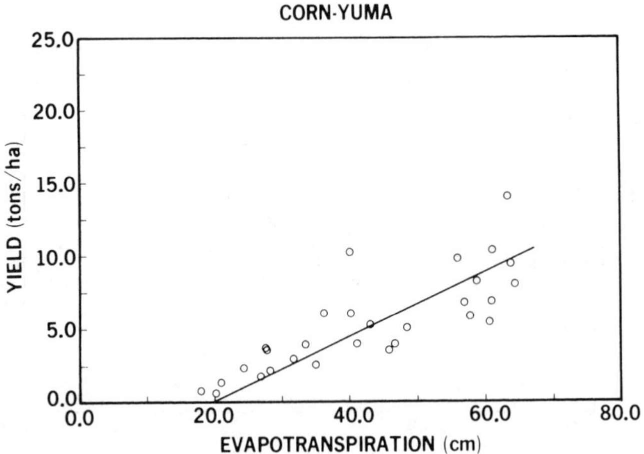

Fig. 5—Yield of corn as related to ET at Yuma, Ariz., in 1975.

days. For a given irrigation level, ET was lower in the salinity treatments because of lower soil water depletion. This is shown in Fig. 6 for Logan, Utah. Similar results were found at Davis, Calif. and Fort Collins.

These data can be used to estimate a value of m provided some method is available for estimating the proportion of ET that is T. This was done using the model of Hanks (1974) and reported in Stewart et al. (1977). Table 3 shows the computations for 1974 and 1975 using the data from California, Colorado, and Utah. The m factors found for Logan and Fort Collins were essentially the same. Where the same variety was used, PX20, the m factor found was 256 at Fort Collins and 248 kg ha^{-1} d^{-1} at Logan, which is well within errors of estimates. The m found at Davis was higher than at Logan or Fort Collins by about 70 kg ha^{-1} d^{-1}. This may be a real difference associated with a more efficient longer-season variety. All of these m values were

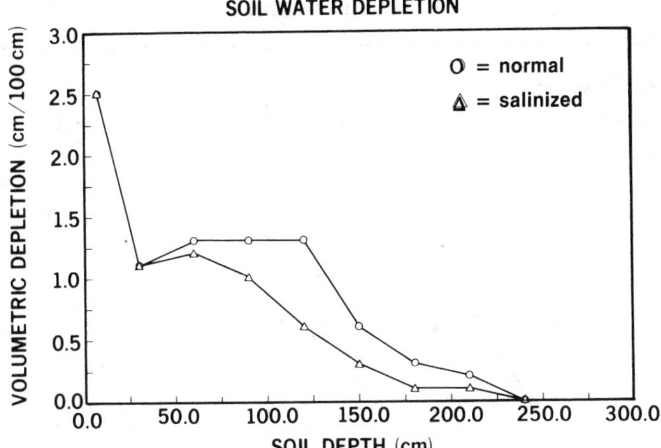

Fig. 6—Soil water depletion for normal and salinized treatment for the Logan, Utah, study of Fig. 2.

higher than those computed by the Briggs and Shantz (1914) data of Table 1. This could be attributed to improved modern-day varieties. Tanner and Sinclair (Chapter 1) have also indicated that this variation in *m* could be caused by use of an improper climatic convection factor. They show less variation if the vapor pressure deficit is used rather than free-water evaporation. Again, it should be noted that there is considerable uncertainty in this type of computation. The values of maximum evapotranspiration (ET_{max}), maximum yield (Y_{max}), and maximum transpiration (T_{max}) are all subject to considerable uncertainty in estimation. If all the data points going into Fig. 2, 3, and 4 were used to compute *m*, a much wider range of values would have been found.

Another study where different corn varieties were used has recently been completed (Sorensen et al., 1979). There was a linear relation between yield and ET for each variety and the same relation applied if relative yield was plotted against ET. Different relations were found for each year, how-

Table 3—Estimated values of the *m* factor at three locations in 1974 and 1975† for several varieties of corn.

Location	Year	Variety	ET_{max}	E	T_{max}	Y_{max}	E_o	m
			cm					kg ha⁻¹ d⁻¹
Davis, Calif.	1974	F4444	67.4	12.0	55.4	22.0	0.78	310
	1975	F4444	61.6	12.0	49.6	22.0	0.73	324
Fort Collins, Colo.	1974	PX20	52.9	14.0	38.9	17.2	0.58	256
	1975	P3955	54.5	16.0	38.5	15.8	0.58	238
Logan, Utah	1974	UH544A	64.4	17.9	46.6	17.9	0.63	241
	1975	PX20	56.2	18.1	40.2	18.1	0.55	248

† ET_{max} is maximum ET measured, E is soil evaporation, T_{max} is maximum transpiration computed as $T_{max} = ET_{max} - E$, and Y_{max} is maximum dry matter yield measured, where $ET = ET_{max}$. The value of E is estimated as the portion of ET where $Y = 0$ (the linear line relating ET to Y is extrapolated to $Y = 0$).

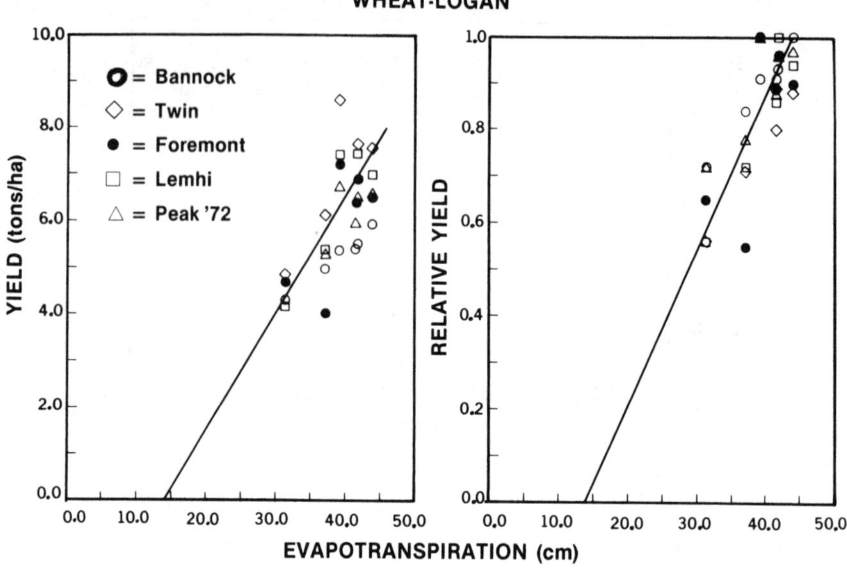

Fig. 7—Relation of dry matter and relative yield to ET for five spring wheat varieties.

ever. No measurable differences in ET were found due to variety, but differences in aboveground dry matter and grain yield were found.

Data from a number of studies and for other crops are available that show similar relations between yield and ET. Shalhevet et al. (1976) have summarized studies done in Israel over several years showing many instances of linear relations of yield and ET. Data from different locations have been "lumped" together and linear relation of relative yield vs. water applied are apparent. They found ET was a large component of water applied, but is a better indicator because it is corrected for drainage or soil water storage. However, irrigation was measured and drainage was not. The crops studied were wheat, sorghum, corn, cotton, sugar beets, peanuts, alfalfa, and tomatoes.

In addition, Rasmussen and Hanks (1978) studied the dry matter and grain-yield–water-use relations of five varieties of spring wheat where variable amounts of irrigation were applied. The actual yield vs. ET relations were different for different varieties, because some varieties seemed to respond to irrigation more than others (Fig. 7). However, a plot of relative yield vs. ET tends to lump these data from different varieties together, as also shown in Fig. 7. The estimates of ET in this study contain some uncertainty because soil water depletion was only measured on two varieties and was assumed the same for all varieties.

A similar study has just been completed testing four varieties of spring barley at two locations (two soils in 1978).[1] There were some problems with lodging at the high irrigation levels, which tended to obscure the results

[1] M. A. Ali. 1979. The yield of grain and dry matter in four varieties of barley as affected by the amounts of water available to the crop. M.S. Thesis. Utah State University, Logan.

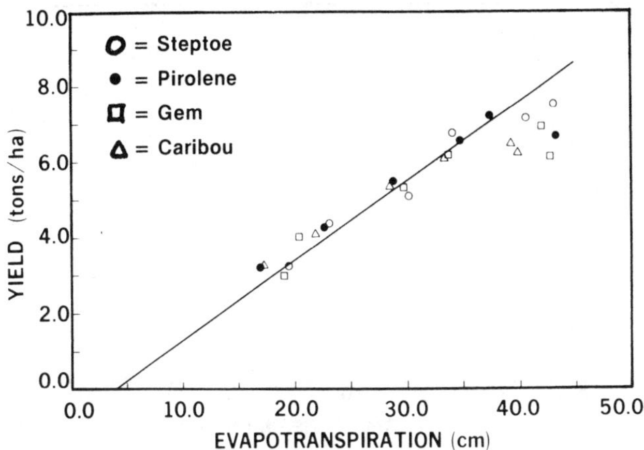

Fig. 8—Relation of yield to ET for four barley varieties.

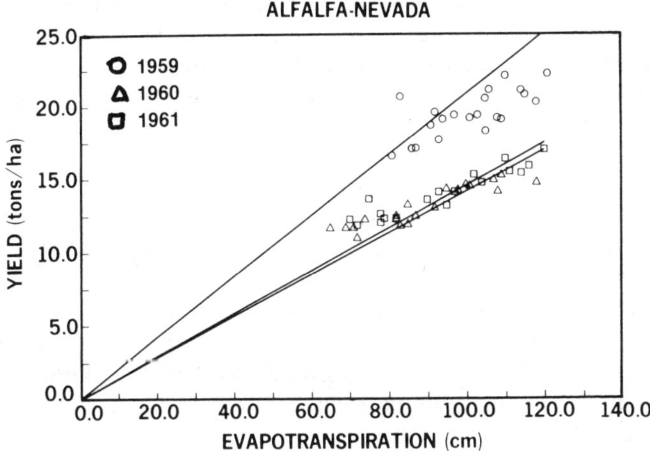

Fig. 9—Relation of yield to ET for alfalfa grown in lysimeters in Nevada.

somewhat, but the strong linear relation between yield and ET was also apparent, as shown in Fig. 8. There is some departure from linearity at high ET values due to lodging, but there was a strong linear trend that was slightly better for dry matter than for grain.

There has been a large amount of yield data collected for alfalfa. The linear relation between yield and ET is also very strong. Data were collected for several years in Nevada on yield-ET relations for alfalfa grown in lysimeters (Tovey, 1963). The treatments imposed were (i) various depths to water table and (ii) various amounts of irrigation. These data show a strong linear relation to yield but also show the relation is dependent on the particular year (Fig. 9). However, about 7 cm of water per metric ton/ha is an average value. Alfalfa covers the ground most of the year so the portion

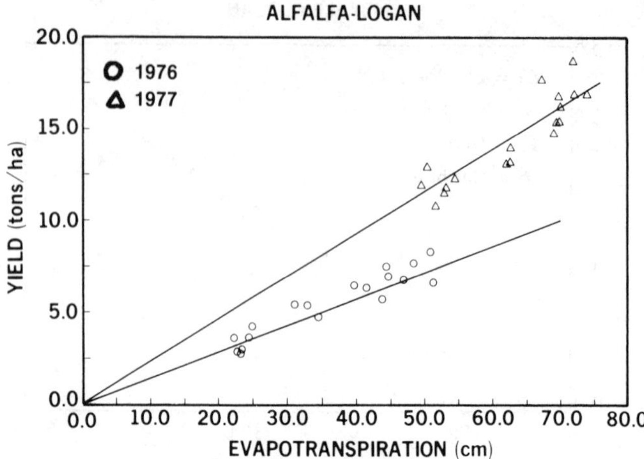

Fig. 10—Relation of yield to ET for alfalfa grown in the field at Logan, Utah in 1976 and 1977. The 1976 data is for only the second and third crop, and the 1977 data is for three crops.

of ET that is E is very small. This ratio would give an m value for alfalfa of 72 kg ha^{-1} d^{-1} ($E_o \cong 0.5$ cm d^{-1}), which is about 30% higher than computed from the Briggs and Shantz (1914) data. New, more efficient, varieties of alfalfa could be responsible for this increase, or it may be that partitioning of photosynthate between tops and roots has been changed. As discussed by Tanner and Sinclair (Chapter 1), the use of free-water evaporation instead of vapor pressure deficit might cause some of the differences.

Another recent study has been made at Logan on three alfalfa varieties (Retta & Hanks, 1980). These data are shown in Fig. 10. There was no appreciable difference in yield or ET due to variety. The yields were much lower in 1976 than in 1977 because only the second and third crop were harvested in 1976, having just been planted (Retta & Hanks, 1980). The maximum yields were about 7.5 metric tons ha^{-1} in 1976 and about 17.5 metric tons ha^{-1} in 1977. The maximum values of ET were about 50 and 74 cm for the 1976 and 1977 seasons, respectively.

IV. PREDICTING YIELD FROM WATER USE

The measured data discussed so far indicate that yields, of dry matter at least, can be approximated from measurements of predictions of ET if we can distinguish between T and ET using a de Wit-type equation. This is really only useful if a method is also available for predicting T and ET from soils, crops, and climatic data. Several simple water-balance-type methods have been proposed to do this. Most of these work well if upward soil water flow from a water table is not a problem. One such method, described by Hanks (1974), involves the following inputs:

1) *Climatic data*—values of E_o by day or by period and the amount and time of rain. The value of E_o has generally been taken as equal

to class A evaporation pan data times an appropriate pan factor. Other methods as described by Jensen (1973) could be used to supply this information.

2) *Irrigation*—the data and amount of irrigation that enters the soil.
3) *Crop*—rooting depth as related to time and information on how T_{max} is varied as a function of time. The value of T_{max} is usually related to some function of amount of cover.
4) *Soil*—the amount of availability for each soil depth and the water content at the beginning of the season.

From the above information, the computer model of Hanks (1974) computes water loss by E from the top soil layer according to the following equation:

$$E = E_{max}t^{-1/2}, \qquad [3]$$

where t is the time in days since the soil was last wet, and $E_{max} = E_o - T_{max}$.

Transpiration is computed on a day-to-day basis from the wettest layer or layers according to the following equation:

$$T = T_{max}, \text{ if SW/AW} > 0.5; \qquad [4]$$

or

$$T = T_{max}\text{SW/0.5ASW}, \quad \text{if SW/ASW} < 0.5; \qquad [5]$$

where SW is actual soil water content of the layer and ASW is the total available water possible in the layer.

When the soil is wet by rain or irrigation, the top layer is wet until SW = ASW of the top layer. If more water is applied than needed to fill the top layer, it is added to the next layer. This process continues throughout the profile and drainage occurs if the total water added exceeds soil water storage. It is assumed that upward flow from the water does not occur.

This model thus computes ET (as well as E and T separately) as related to the soil, climate, and crop information that was input at the beginning. The information can then be used to predict yields by several methods. The equation of de Wit could be used. An alternative would be to use the following equation given by Hanks (1974) for dry matter yield:

$$Y/Y_{max} = T/T_{max}. \qquad [6]$$

This equation applies to either of the two de Wit (1958) or Arkley (1963) equations, or to the Tanner and Sinclair equations (Chapter 1). This equation also assumes that the linear relation applies to only 1 year. Different relations could apply from year to year. It does not require a value of m or k, but instead requires a value for Y_{max}. The value of Y_{max} might vary from year to year, so the above equation is most useful for comparing conditions or treatments within a given year. A more complicated version of

Eq. [6] has been used by Stewart et al. (1977) to predict grain yields by dividing the season into growth stages, but in most instances Eq. [6] is as good as the more complicated version.

Another useful equation has been devised to predict yields if ET is known or can be estimated, as given by Stewart et al. (1977) as follows, for dry matter:

$$Y/Y_{max} = 1 - \beta_o ET_D = 1 - \beta_o + \beta_o ET/ET_{max}, \quad [7]$$

where β_o is the slope of relative yield vs. ET_D (ET deficit = $ET_D = 1 - ET/ET_{max}$). The only unknown in this equation besides Y_{max} is the value of β_o. Note that the ratio ET/ET_{max}, where $Y/Y_{max} = 0$, approximates soil E (Hanks, 1974). The portion of ET_{max} that is T_{max} is $1/\beta_o$. Thus a β_o of 1.0 would mean no evaporation from the soil and a β_o of 1.5 would mean that one-third of ET_{max} was E and two-thirds T_{max}. The problem with this equation at present is that the value of β_o must be determined from field measurements. Table 4 shows the results of the four-state study where this equation was applied. The values of β_o given in Stewart et al. (1977) are 1.0 or less, which cannot be physically correct. This is probably caused by limited data covering a wide range of ET and the problem discussed earlier of uncertainties in deciding on the values of ET_{max} and Y_{max}. The value of β_o is dependent on the portion of ET_{max} that is E and T_{max}, so it will be dependent on the frequency of wetting when crop cover is small. For alfalfa, a value of 1.05 would be reasonable for β_o because there would be very little soil evaporation.

Equation [7] is, however, quite useful because of its simplicity. It is not difficult in most instances to estimate a reasonable value for β_o for prediction purposes. In many situations β_o is close to the same from year to year for the same crop.

Note also that the value of the de Wit m can be determined as follows:

$$m = Y_{max} E_o \beta_o / ET_{max}. \quad [8]$$

Thus, if Y_{max} = 20,000 kg ha^{-1}, and β_o = 1.25, ET_{max} = 60 cm, and E_o = 0.50 cm d^{-1}, m would be 208 kg ha^{-1}. The β_o values shown in Table 4 were

Table 4—Values of β_o, Y_{max}, E_o, and ET_{max} and m for corn grown at several locations as reported by Stewart et al. (1977). Data shown are adjustments of original published tables.

Location	Year	β_o	E_o	Y_{max}	ET_{max}	Variety	m
			cm	kg ha^{-1}	cm		
Davis, Calif.	1974	1.22	0.78	22,000	67.4	F4444	310
	1975	1.24	0.73	22,000	61.6	F4444	324
Fort Collins, Colo.	1974	1.36	0.58	17,200	52.9	PX20	256
	1975	1.42	0.58	15,800	54.5	P3955	238
Logan, Utah	1974	1.38	0.63	17,900	64.4	U4544A	241
	1975	1.40	0.55	18,100	56.2	PX20	248

computed assuming the same m values as shown in Table 3. This is the same as evaluating β_o as equal to ET_{max}/T_{max}.

Stewart et al. (1977) have developed a more complicated equation, using concepts similar to Eq. [7], to predict grain production. This equation accounts for variable water supply at different crop growth stages by evaluating ET/ET_{max} for each stage. Each stage is then weighted by a factor. The factor has been found to be complicated by the extent of stress conditioning that has occurred up to the particular stage. This more complicated equation was found to give improved predictions in only a few instances compared with the simpler Eq. [7].

Another approach to predicting yields has been developed by Kanemasu and associates (Morgan et al., 1980; Kanemasu et al., 1976, 1978) that is quite different from that of Hanks (1974) and Stewart et al. (1977). They developed a functional relationship between solar radiation, temperature, and T_{max} (in cm) for various crops as follows:

$$T_{max} = 1.74\,(1 - \tau)[\Delta/(\Delta + \gamma)][0.86\,R_s - 103.9]/590, \quad [9]$$

$$\text{for LAI} < 3.0\text{; and}$$

$$T_{max} = (\alpha - \tau)[\Delta/(\Delta + \gamma)][0.848\,R_s - 144.5]/590, \quad \text{for LAI} \geq 3.0; \quad [10]$$

where α is crop- and location- dependent (1.35, 1.28, 1.45, and 1.35 for corn, sorghum, soybeans, and winter wheat, respectively, in Kansas), Δ is the slope of the saturation vapor pressure curve, γ is the pyschrometric constant, R_s is solar radiation (cal cm^{-2} d^{-1}), and LAI is leaf area index. This approach is a modification of the method by Jury and Tanner (1975). The variable, τ, is the fraction of energy applied to the soil surface and is related to LAI by the following:

$$\tau = \exp(-0.389\,\text{LAI} + 0.15), \quad \text{for LAI} > 0.39. \quad [11]$$

If LAI < 0.39, τ is taken as 1.0.

Evaporation directly from the soil is computed as

$$E_{max} = \tau\,[\Delta/(\Delta + \gamma)]\,R_N/590, \quad [12]$$

where R_N is net radiation, which is directly measured or is estimated from measurements of R_s provided water can be transmitted to the soil surface readily. If water cannot be transported to the surface quickly enough to meet atmospheric demands, then Eq. [12] is superseded by

$$E = ct^{1/2} - c(t - 1)^{1/2}, \quad [13]$$

where c is a soil factor depending on water-transmission properties (about 0.2–0.4).

To account for the limitation on T as soil water content decreases, an equation similar to Eq. [4] or [5] is used, except that a limiting value of 0.35

is used instead of 0.5. The computer model computes a soil water balance similar to that discussed earlier by Hanks (1974). Thus, daily and seasonal measurements of T and E are made.

To compute yield, the Kanemasu model described above computes a growth response on a daily basis according to the Hanway (1963) growth response for corn. This growth response is modified according to the level of available soil water. A different growth-response curve is used for vegetative growth early in the year than for reproductive growth late in the year.

Other, more sophisticated, models have been used to predict ET and yield. Nimah and Hanks (1973) have described a more fundamental model for predicting E, T, and drainage that accounts for upward flow and does not require many of the approximations of the Hanks (1974) model. Much more soils data is required, however. Childs and Hanks (1975) modified the earlier method of Nimah and Hanks (1973) to include the effect of salinity. Both of these models use the same T/T_{max} approach to computing the influence of climate, soil, or climatic conditions on yield.

Childs et al. (1977) had used the soil water-flow portion of the Childs and Hanks (1975) model and have incorporated it into a more sophisticated model to predict corn yield through estimation of soil and leaf water potentials. Photosynthesis, respiration, and available carbohydrates in the leaf are all estimated in this model. Yield prediction using this model has generally been good.

V. APPLICATIONS

The relation of water use to yield has been widely used by many people. Engineers have long used water-requirement data to design irrigation systems. Much recent interest in yield–water-use relations comes from economists who desire to know the economic properties of given irrigation systems. In recent years many lawsuits have arisen because of disagreements over the water requirements of crops and the yield effect if the available water for irrigation is decreased.

It is not possible to answer the many yield-ET questions in a straight forward and simple manner because of the complicated interactions of soil, climate, crop, and irrigation management. The yield influence of 10 cm of irrigation applied in one location will probably not be the same as that at another location, nor will it be the same from year to year if everything else is equal. Similarly, 10 cm of irrigation water applied at one time of the year will not necessarily influence yield in the same way as 10 cm applied at a different time. A general answer can be given of the influence of irrigation on dry matter yield if the influence of irrigation on T can be deduced. For grain production it may also be necessary to know the timing of the T during the season.

The advantage of the model approach is that specific situations can be simulated. Many answers to particular questions have been published. Figure 11 shows a simulation of the influence of irrigation interval on corn

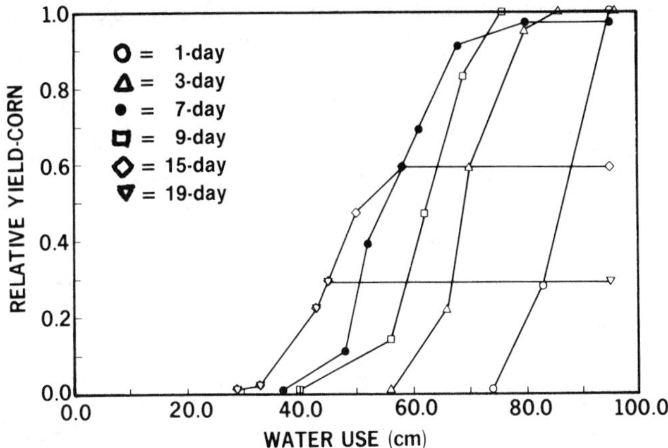

Fig. 11—Computed dry matter production of corn, as related to water use (soil water depletion + irrigation + rain) for different frequencies of irrigation assuming available soil water storage of 5 cm.

dry matter production for a certain soil and climatic situation. Where irrigation was applied frequently, yield was lower for a given water use (ET + drainage) than where irrigation was applied less frequently (up to 9 days). This difference was caused by more water being lost by E when irrigation was frequent. Figure 11 also illustrates the problem that might occur if water was applied at very long intervals. In this situation, the soil water storage was not enough to supply plant needs between irrigations, so T/T_{max} is limited and, for the same amount of water used, much of it is lost to drainage.

Hill et al. (1974) used the model of Hanks (1974) in a modified form to include a climatic prediction to predict growth stages. They tested the best time of the year to apply a limited amount of irrigation assuming a given rainfall-climatic situation like that of Mead, Nebr., in 1972. This prediction is illustrated in Table 5 for two different planting dates.

Table 5—Predicted influence of incremental application levels of irrigation water on corn yield (Mead, Nebr.; 1972 data).

Planting date	Dates of application of 2.54 cm irrigation (in order of selection)	Predicted yield (with irrigation on previously selected dates)
		% of potential
4 May	None	33
	17 July	60
	19 July	74
	3 July	85
	7 July	93
	26 May	96
18 May	None	63
	23 July	85
	15 June	92
	15 June	96

Table 6—Results of simulated irrigation schedules based on estimated crop-response model.

Simulated run number	Water applied to soil in root zone	Water pumped	Reduction of available moisture in root zone over growing season	Total water used	Predicted yield	Variable cost of irrigation	Predicted net return†
	cm				kg ha^{-1}	$ ha^{-1}	
1	30.5	50.8	21.1	71.9	7400	78.23	92.28
2	38.1	63.5	15.1	78.6	7865	97.79	109.58
3	45.7	76.2	11.8	88.0	8034	117.35	94.35
4	53.3	88.9	11.8	100.7	8223	136.91	89.06
5	61.0	101.6	11.2	112.8	8511	156.46	102.88
6	38.1	63.5	13.3	76.8	7099	97.79	48.62
7	30.5	50.8	18.1	68.9	5699	78.23	25.01
8	30.5	50.8	18.0	68.8	6026	78.23	−7.34
9	30.5	50.8	19.2	70.0	7136	78.23	71.35
10	30.5	63.5	10.5	74.0	5739	97.79	−60.39
11	38.1	63.5	11.4	74.9	6379	97.79	−9.38
12	45.7	76.2	8.1	84.3	7930	117.35	95.24

† Not including land charge.

The Kanemasu model has been used (Kanemasu et al., 1976) to predict the effect of irrigation frequency and amount on yield and net return per hectare. These results are shown in Table 6. They found the highest net return (smallest loss) was for an intermediate irrigation that produced less than the maximum yield. Similar amounts of irrigation water applied produced different returns because of different irrigation timing.

Andersen and Hanks (1981) have used the model of Childs and Hanks (1975) to predict economic relations of saline irrigation water. They put an economic value on the relation of irrigation management to the salt content of the drainage water.

LITERATURE CITED

Andersen, J. C., and R. J. Hanks. 1981. Physical and economic evaluation of irrigation return flow and salinity on a farm. p. 173–199. *In* D. Yaron (ed.) Salinity in irrigation and water resources. Marcel Dekker, New York.

Arkley, R. J. 1963. Relationships between plant growth and transpiration. Hilgardia 34:559–584.

Briggs, L. J. and H. L. Shantz. 1914. Relative water requirements of plants. J. Agric. Res. 3: 1–63.

Childs, S. W., J. R. Gilley, and W. E. Splinter. 1977. A simplified model of corn growth under moisture stress. Trans. ASAE 20:858–865.

Childs, S. W., and R. J. Hanks. 1975. Model of soil salinity effects on crop growth. Soil Sci. Soc. Am. Proc. 39:617–622.

Cole, J. S., and O. R. Mathews. 1923. Use of water by spring wheat on the Great Plains. USDA Bull. 1004. U.S. Government Printing Office, Washington, D.C.

de Wit, C. T. 1958. Transpiration and crop yields. Versl. Landbouwkd. Donderzock. no. 64.6.

Dillman, A. C. 1931. Water requirement of certain crop plants and weeks in the Northern Great Plains. J. Agric. Res. 42:197–238.

Doorenbos, J., and W. O. Pruitt. 1975. Crop water requirements. Irrigation and Drainage Paper no. 24. FAO, United Nations, Rome.

Fischer, R. A., and N. C. Turner. 1978. Plant productivity in the arid and semi-arid zones. Annu. Rev. Plant Physiol. 29:277–317.

Hanks, R. J. 1974. Model for predicting plant growth as influenced by evapotranspiration and soil water. Agron. J. 66:660–665.

Hanks, R. J., H. R. Gardner, and R. L. Florian. 1969. Plant growth–evapotranspiration relation for several crops in the Central Great Plains. Agron. J. 61:30–34.

Hanway, J. J. 1963. Growth stages of corn (*Zea mays* L.). Agron. J. 55:487–492.

Hill, R. W., R. J. Hanks, J. Keller, and V. P. Rasmussen. 1974. Predicting corn growth as affected by water management: an example. Pub. no. CUSUSWASH 211(d)-6. Utah State University, Logan. p. 18.

Jensen, M. E. (ed.). 1973. Consumptive use of water and irrigation water requirements. Am. Soc. Civ. Eng., New York.

Jury, W. A., and C. B. Tanner. 1975. Advective modifications of the Priestly and Taylor evapotranspiration formula. Agron. J. 67:840–842.

Kanemasu, E. T., V. P. Rasmussen, and J. Bagley. 1978. Estimating water requirements for corn with a "pocket calculator." Kansas Agric. Exp. Stn. Bull. no. 615, Manhattan.

Kanemasu, E. T., L. R. Stone, and W. L. Powers. 1976. Evapotranspiration model tested for soybean and sorghum. Agron. J. 68:569–572.

Kiesselbach, T. A. 1916. Transpiration as a factor in crop production. Nebr. Agric. Exp. Stn. Res. Bull. 6, Lincoln.

Morgan, T. H., A. W. Biere, and E. T. Kanemasu. 1980. A dynamic model of corn yield response to water. Water Resour. Res. 16:59–64.

Nimah, M. N., and R. J. Hanks. 1973. Model for estimatng soil water and atmospheric interrelations. Soil Sci. Soc. Am. Proc. 37:522–532.

Rasmussen, V. P., and R. J. Hanks. 1978. Model for predicting spring wheat yields with limited climatological and soil data. Agron. J. 70:940–944.

Retta, A., and R. J. Hanks. 1980. Corn and alfalfa production as influenced by limited irrigation. Irrig. Sci. 1:135–147.

Shalhevet, J., A. Mantell, A. Bielorai, and D. Shimshi. 1976. Irrigation of field and orchard crops under semi-arid conditions. Int. Irrig. Info. Ctr. Pub. no. 1. Bet Dagan, Israel.

Shantz, H. L., and L. N. Piemeisel. 1927. The water requirement of plants at Akron, Colo. J. Agric. Res. 34:1093–1190.

Sorensen, V. M., R. J. Hanks, and R. L. Cartee. 1979. Cultivation during early season and irrigation influences on corn production. Agron. J. 72:266–270.

Stewart, J. I., R. E. Danielson, R. J. Hanks, E. B. Jackson, R. M. Hagan, W. O. Pruitt, W. T. Franklin, and J. P. Riley. 1977. Optimizing crop production through control of water and salinity levels in the soil. Utah Water Lab. Pub. no. PRWG 151-1, Logan. p. 191.

Tovey, R. 1963. Consumptive use and yield of alfalfa in the presence of static water tables. Tech. Bull. no. 232. Nevada Agric. Exp. Stn., Reno.

CHAPTER 9B

Yield and Water-Use Relationships: Some Problems of Relating Grain Yield to Transpiration[1]

E. T. KANEMASU
Kansas State University
Manhattan, Kansas

I. INTRODUCTION

In Chapter 9A, Hanks adequately discusses and summarizes both early and current research on yield–water-use relations. One can conclude from his discussion that, in general, (i) dry matter production and water use are more closely correlated than grain yield and water use, (ii) improvements in the correlations are observed when transpiration is used instead of evapotranspiration, (iii) incorporating effects of water deficit at critical growth stages may improve the correlations, and (iv) economic yield–water-use relationships can change from location to location, year to year, crop to crop, and variety to variety.

Hanks argues correctly that there exists a close relation between yield (dry matter) and transpiration. This is especially true in water-limiting conditions. Obviously, aboveground dry matter (DM) production and photosynthesis are closely related; and photosynthesis (N_C) and transpiration (T) are also related, via the diffusional resistances, as discussed in Chapter 1A. Shown in Fig. 1 are the daily WUE (N_C/ET) of grain sorghum as a function of solar radiation. Photosynthesis was measured with large field chambers and ET was determined by weighing lysimeters at Manhattan, Kans. The LAI exceeded 2.8 and growth stages corresponded to 5 and 6 (Vanderlip & Reeves, 1972). Clearly, N_C/ET declined with increasing solar radiation, with considerably more scatter at low radiation, which may stem from differences in daily vapor pressure deficit. Several days of soil water deficit (10–20 August 1972), gave N_C/ET ratios that did not differ from well-watered sorghum ratios. These data substantiate some of the concepts of de Wit (1958) regarding m during water-limiting conditions.

[1]Contribution no. 80-156-A, Kansas Agricultural Experiment Station, Kansas State University, Manhattan.

Limitations to Efficient Water Use in Crop Production.

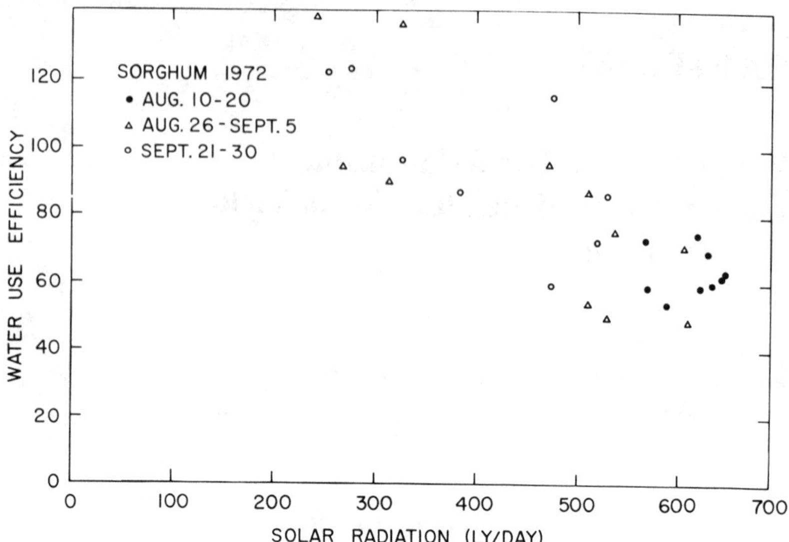

Fig. 1—Water use efficiency (N_C/ET) vs. solar radiation. All values were on a 24-h basis.

During conditions of high radiation, the ratio of N_C/T decreases with increasing energy input; therefore, T is usually normalized (e.g., by pan evaporation). Possibly, much of the observed variation in the value of m is due to the problems associated with an appropriate means of normalizing T, as discussed briefly in Chapter 1A. For example, pan evaporation does not properly weight daytime and nighttime evaporative demand for many vegetative surfaces. Additionally, during days of high air temperature and well-watered conditions, ET can exceed net radiation. These advective contributions should not be included in the analysis of m, as N_C and T would not be similarly affected. Bierhuizen and Slatyer (1965) reported that the variation in N_C/T stems primarily from differences in the mean vapor pressure deficit.

The more tenuous arguments appear to be in equating yield and N_C, especially in the case of grain yield (Y). Photosynthate is distributed to roots, stems, leaves, and head (ear)/fruit. The distributions vary with stages of development. In the case of winter wheat in Kansas, roots have nearly extended to their maximum depth by the jointing stage; subsequent photosynthate then is primarily partitioned among the aboveground organs. Yield components (spikelets and kernels) are determined at later growth stages and provide a sink for assimilates (Fig. 2). Translocation of stored assimilates to the kernel varies depending upon genetic and environmental conditions. Reports indicate 10–20% of the total photosynthate is translocated to the grain. Because of the observed linearity in kernel growth with time, it has been suggested that reduced N_C is partially compensated for by translocation (Gallagher et al., 1976). However, reduced accumulation of preanthesis DM may affect the amount of assimilate translocated.

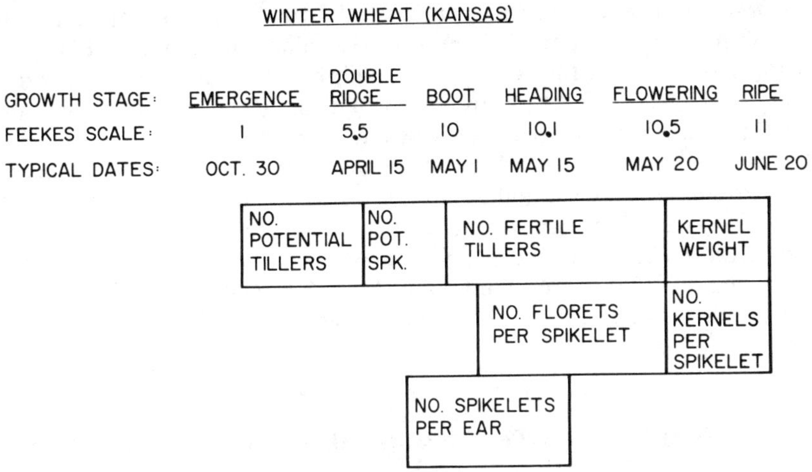

WINTER WHEAT (KANSAS)

GROWTH STAGE:	EMERGENCE	DOUBLE RIDGE	BOOT	HEADING	FLOWERING	RIPE
FEEKES SCALE:	1	5.5	10	10.1	10.5	11
TYPICAL DATES:	OCT. 30	APRIL 15	MAY 1	MAY 15	MAY 20	JUNE 20

NO. POTENTIAL TILLERS
NO. POT. SPK.
NO. FERTILE TILLERS
KERNEL WEIGHT
NO. FLORETS PER SPIKELET
NO. KERNELS PER SPIKELET
NO. SPIKELETS PER EAR

Fig. 2—Development of yield components for winter wheat in Kansas.

II. HARVEST INDEX

The harvest index (HI) is a widely cited index and provides information on the relation of economic and biological yields (Y/total DM). However, the index gives no information on how the specific yield components (e.g., head number, kernels per head, and kernel weights) performed. Hanks assumes a constant HI; however, he indicates that HI may decrease during extreme droughts.

Slabbers et al. (1979), after determining Y from DM, reported the following regression equations for sorghum and maize, respectively:

$$Y(\text{kg ha}^{-1}) = 0.58\,(\text{DM}) - 1{,}263 \qquad (r = 0.97, N = 34), \text{ and}$$

$$Y(\text{kg ha}^{-1}) = 0.49\,(\text{DM}) - 1{,}208 \qquad (r = 0.93, N = 42).$$

Independent tests indicated that their equations explained 94% of the variation in sorghum yields and 86% in maize. However, they found that water stress during sensitive periods of growth needed to be considered in "partitioning" equations. Passioura (1977) found a positive linear relationship between harvest index (Y/total DM) and percent water used after anthesis of wheat. His glasshouse harvest indices (HI) ranged from about 0.18 to about 0.55, which emphasizes the importance of the grain-filling period when almost all the photosynthate produced is going to the kernels.

Hodges (1978) and Rasmussen (1979), working in the Evapotranspiration Laboratory at Kansas State University, found large variations in HI in sorghum and winter wheat. Water-stressed and well-watered sorghum had HI values of 0.24 and 0.53, respectively. At Manhattan, Kans., during 1978, 33 wheat plots ($\overline{Y}$ = 1,875 kg ha^{-1}; C.V. = 15) had a mean HI of 0.31 (C.V. = 0.3). In the same year, samples from 75 wheat fields uniformly distributed over Kansas had a mean HI of 0.44, with a C.V. of 0.3. In a population-density and maturity-rating study for corn, DeLoughery and Crookston (1979) reported decreasing HI values (maximum HI = 0.50) with increasing moisture stress and a high positive correlation between Y and HI. However, in a nonstress environment, HI was weakly correlated with Y. They conclude that "harvest indices obtained in one environment had no predictable relationship to grain yields in a different environment."

Presumably the variability in HI is reflecting, in part, the effects of (i) water stress at various growth stages, (ii) environment on partitioning of assimilates, (iii) environment on plant development, and (iv) environment on translocation. Therefore, Y estimates from ET relationships are variety-, site-, and year-specific. Yield estimates over time and space must take another approach.

III. OTHER EVAPOTRANSPIRATION-YIELD APPROACHES

It seems reasonable physiologically to weight the normalized T (e.g., T/ET_{max}) according to growth stages (i.e., with winter wheat: emergence to jointing [stage 1], jointing to heading [stage 2], and heading to soft dough [stage 3]), and to provide preference to longer time periods within a growth stage. Rasmussen (1979), who allowed for both of those concepts in his winter wheat study, developed the following grain-yield–water-use equation:

$$Y(\text{kg ha}^{-1}) = 192.1[\Sigma(T/ET_{max})]_1^{0.172} [\Sigma(T/ET_{max})]_2^{0.104} [\Sigma(T/ET_{max})]_3^{0.646}, \quad [1]$$

where the subscripts refer to the growth stages indicated above. Daily T/ET_{max} ratios are summed during each of the stages. The value of ET_{max} is estimated by the Priestley-Taylor equation. The maximum T/ET_{max} summation would be the total number of days in each growth period. Equation [1] has the advantage that neither the potential yield nor the m factor needs to be known. In 95 observations at three sites over 3 years, Rasmussen (1979) reported a correlation coefficient of 0.68. Better results were obtained with a data subset in which water was the primary yield-limiting factor.

The inclusion of a potential yield term (Y_o) is attractive because it can account for some of the effects of crop management and soil productivity. However, it is difficult to obtain independent estimates of Y_o. Hanks indicates that Y_o may vary from one year to another; therefore, Y_o approaches are most useful when comparing treatments within a given year.

Fischer (1979) proposed a conceptual yield model for wheat that emphasizes the importance of preanthesis conditions and sink size (grain number). In general, Fischer (1979) constructs the sink and then fills it. The sink size is assumed to be dependent on the DM accumulation at anthesis, which, in turn, depends on transpiration efficiency (DM/T). He assumes that T efficiency does not change greatly with stress. This type of approach may serve some short-term needs.

IV. SUMMARY

In general, ET-*Y* (DM) relationships are attractive because of their simplicity and their routine data requirements, and may serve many current needs. However, in estimating grain yields, the ET-*Y* relationships are not unique. Because of complex interactions among development, assimilate partitioning, and environment, it is doubtful that an ET–*Y* relationship can be fruitfully extended to climatically diverse regions. However, the same complexities must also be reflected in the more physiologically based models that will necessarily use photosynthate production and distribution as well as plant development. This emphasizes the need for adequate data sets from diverse climates to quantitatively assess plant-enrivonment interactions.

LITERATURE CITED

Bierhuizen, H. F., and R. O. Slatyer. 1965. Effect of atmospheric concentration of water vapor and CO_2 in determining transpiration-photosynthesis relationships of cotton leaves. Agric. Meterol. 2:259–270.

DeLoughery, R. L., and R. Kent Crookston. 1979. Harvest index of corn affected by population density, maturity rating, and environment. Agron. J. 71:577–580.

de Wit, C. T. 1958. Transpiration and crop yields. Versl. Landbouwkd. Donderzock. no. 64.

Fischer, R. A. 1979. Growth and water limitation to dryland wheat yield in Australia: a physiological framework. J. Aust. Inst. Agric. Sci. 45:83–94.

Gallagher, J. N., P. V. Biscoe, and B. Hunter. 1976. Effects of drought on grain growth. Nature 264:541–542.

Hodges, T. 1978. Photosynthesis, growth, and yield of sorghum and winter wheat as functions of light, temperature, water, and leaf area. Ph.D. Thesis. Kansas State University, Manhattan. Microfilm no. LD2668D5—1978H63. Ann Arbor, Mich.

Passioura, J. B. 1977. Grain yield, harvest index, and water use of wheat. J. Aust. Inst. Agric. Sci. 43:117–120.

Rasmussen, V. P. 1979. Modelling winter wheat yields as affected by water relations and growth regulants. Ph.D. Thesis. Kansas State University, Manhattan. Microfilm no. LD2668D5—1979R37. Ann Arbor, Mich.

Slabbers, P. J., V. Sorbello Herendorf, and M. Stopper. 1979. Evaluation of simplified water-crop yield models. Agric. Water Manage. 2:95–129.

Vanderlip, R. L., and H. E. Reeves. 1972. Growth stages of sorghum (*Sorghum bicolor* L. Moench). Agron. J. 64:13–16.

CHAPTER 10A

Soil Management for Efficient Water Use: An Overview[1]

PAUL W. UNGER AND B. A. STEWART

Conservation and Production Research Laboratory, USDA-ARS
Bushland, Texas

I. INTRODUCTION

Annual precipitation in Nevada averages 23 cm $year^{-1}$, which is the lowest for any of the 48 contiguous states of the United States (Geraghty et al., 1973). However, even this is well above the amount of water used to produce favorable grain yields when evaporation during the growing season has been effectively controlled (Table 1). Studies by Griffin et al. (1966), Peters (1960), and Willis et al. (1963) showed that at least one covered-soil treatment resulted in less water use than the lowest statewide average annual precipitation in the United States. Grain yields generally were as good as those from treatments using more water. Evapotranspiration (ET) water-use efficiency generally was highest when evaporation of soil water was prevented or greatly reduced by the soil cover.

Precipitation, however, is highly variable in time and areal distribution, and not all of it becomes available for crop production. Besides precipitation variability, the efficiency of water use is limited also by soil factors that affect runoff and infiltration, water distribution and retention in the profile, water evaporation and percolation, and plant root growth and proliferation within the profile to absorb the soil water. To use water more efficiently for crop production, we must continue to develop new technology for effectively capturing and retaining precipitation.

II. SCOPE OF CHAPTER

In Chapter 10 we are concerned with soil management practices that affect the fate of water potentially available for crop production. The main water source is precipitation, but, where appropriate, we also consider

[1] Contribution from the Agricultural Research Service, USDA, Bushland, Texas, in cooperation with the Texas Agricultural Experiment Station and Texas A&M University, College Station, Texas.

Limitations to Efficient Water Use in Crop Production.

Table 1—Effect of evaporation control on grain yield, water use, and water-use efficiency (from Griffin et al., 1977; Harrold et al., 1959; Peters, 1960; Willis et al., 1963).

Location	Treatment	Plant population	Yield	Water used	Water-use efficiency
		1,000 ha^{-1}	kg ha^{-1}	cm	kg ha-cm^{-1}
Goodwell, Okla.					
	Grain sorghum, 1963				
	Conventional irrigation		5,800 c†	46 b	126 b
	100% cover, plastic sheet, no water added after planting		6,300 bc	18 c	350 a
	No cover, soil maintained at 50% available water		8,300 a	130 a	64 c
	80% cover, plastic sheet, on ridges between rows, irrigated at the first sign of stress		7,200 ab	51 b	141 b
	Grain sorghum, 1964				
	No cover, soil maintained at 50% available water		7,200 a	76 a	95 b
	80% cover, plastic sheet, on ridges between rows, irrigated at first sign of stress		5,600 b	38 b	147 a
	80% cover, sprayed asphalt, on ridges between rows, irrigated at first sign of stress		5,200 b	38 b	137 a
	57% cover, asphalt coated paper, on ridge between rows, irrigated at first sign of stress		5,600 b	38 b	147 a
	Corn, 1957				
Coshocton, Ohio	Covered		7,800	30.9	252
	Uncovered		9,300	57.1	163
Urbana, Ill.	Covered	20	6,800	13.0	523
	Covered + irrigation	20	7,300	18.5	395
	Natural rainfall	20	6,300	31.2	202
	Natural rainfall + irrigation	20	6,300	33.8	186
	Covered	40	9,300	15.2	612
	Covered + irrigation	40	9,700	20.6	471
	Natural rainfall	40	10,300	34.3	300
	Natural rainfall + irrigation	40	9,200	35.6	258
Mandan, N. Dak.					
	Corn, 1960 (wet year)				
	Check		3,200 a	40.4	79
	Ridged, surface bare		3,350 a	37.3	90
	Ridged, surface 90% plastic covered		4,330 a	37.8	115
	Corn, 1961 (dry year)				
	Check		1,620 b	16.0	101
	Ridged, surface bare		1,390 b	12.7	109
	Ridged, surface 90% plastic covered		3,930 a	19.8	198

† Column values within a group followed by the same letter(s) are not significantly different (Duncan's Multiple Range Test, 5% level).

water from irrigation. In Chapter 10A, we discuss (i) the effects of tillage and engineering practices designed to retain water on the surface by preventing or controlling runoff, thus providing more time for infiltration; (ii) water infiltration per se as influenced by conditions at the soil surface and within the profile, by previous crops, and by crop residues; and (iii) water retention in the profile with respect to water storage volume, deep percolation, and evaporation from the surface.

In Chapter 10B, Power discusses the effects of soil fertility on efficient water use. Emphasis is on principles involved, plant growth and yield, water use, and water-use efficiency.

In Chapter 10C, Reicosky discusses soil management effects on water use, plant growth, and crop yields, with emphasis on the acid and high-strength soils of the southeastern United States. Included are descriptions of the soils, water use by various crops, and management systems for some problem soils of the Southeast.

III. WATER RETENTION ON THE SOIL SURFACE

When the precipitation or irrigation rate exceeds a soil's infiltration rate, the excess water collects in surface depressions, moves laterally across the surface as overland flow, and eventually enters surface drainage systems. This runoff potentially can be conserved for future crop production (Bertrand, 1966).

A. Tillage Effects

Tillage has long been considered an essential crop production practice, with one benefit being increased water conservation through runoff control and improved infiltration. Although tillage has also been criticized (Faulkner, 1943), it helps control or prevent runoff on some soils. The following examples illustrate the results that can be expected from various tillage-induced soil conditions.

1. SURFACE ROUGHNESS AND MICRODEPRESSIONS

To control runoff, tillage should produce a rough soil surface with numerous nonconnected depressions that temporarily store the potential runoff water. The effectiveness of a roughened surface depends, among other factors, on the intensity and amount of precipitation, and on the stability of the surface soil. These factors are interrelated, especially within a given time period. High-intensity precipitation readily disperses low-stability surface soil aggregates, thus sealing the surface, decreasing infiltration rates, and allowing water to accumulate on the surface. If precipitation is adequate, the depressions fill and become connected, thus increasing the potential for runoff. In contrast, the same amount of low-intensity precipitation generally is less destructive of surface structure and, therefore, causes less runoff.

Because the most favorable soil conditions for runoff control and for seedbeds differ, Larson (1962, 1964) established criteria for row (seedling environment) and interrow (water management) zones for optimizing plant growth and water storage, respectively. Minimum tillage systems, such as plow-plant, wheel-track planting, lister planting, and strip tillage, effectively create and maintain favorable row and interrow zones (Bertrand, 1966). Primary tillage with a moldboard, chisel, disk, or sweep plow also creates surface depressions that temporarily store water. Secondary tillage generally decreases depression storage (Hays, 1961). However, if previous rainfall caused a surface crust, loosening the crust by cultivation decreased runoff (Meyer & Mannering, 1961). Surface-depression storage after rotary tillage generally was slight, but exceeded 7.5 cm of water with lister tillage on the contour on land where the slope was <2%. Where slopes were greater, depression storage was less (Larson, 1964). A technique for determining depression storage was developed by Burwell et al. (1966), who measured surface roughness with a microrelief meter. In their study, runoff control was proportional to tillage-induced surface roughness (Table 2).

2. LOOSENED TILLAGE LAYERS

Runoff control by water storage in loosened tillage layers is related to a soil's air-filled porosity. Loosening a soil increases the total porosity and the thickness of the plowed layer. The following example is adapted from Larson (1962): If a 17.8-cm-thick layer of soil having a 1.4 g cm^{-3} bulk density is loosened by tillage to a 1.0 g cm^{-3} bulk density, total porosity increases from about 47 to 62% and thickness of the layer increases to 24.9 cm. Potential water storage in the plow layer at saturation is 8.4 cm before plowing and 15.5 cm after plowing. Based on a field capacity of 25% by weight, the plow-layer storage potential (between field capacity and saturation) is about 2.3 cm before and 9.4 cm after plowing. The extra storage volume is temporary, however, because the loosened layer generally settles when water is added. Also, loosening a soil by tillage may cause rapid drying of the exposed soil.

Table 2—Effect of tillage-induced plow-layer porosity and surface roughness on cumulative infiltration of simulated rainfall (from Burwell et al., 1966).

	Surface conditions		Cumulative infiltration§		
Tillage treatment†	Pore space‡	Roughness	To initial runoff	To 2.5 cm runoff	To 5.0 cm runoff
	cm				
Untilled	8.1	0.8	0.9	2.1	2.4
Plow	13.7	5.0	17.1	21.7	23.0
Plow-disk-harrow	12.4	2.5	5.3	7.3	8.4
Cultivated	9.7	2.9	5.7	8.3	9.1
Rotovated	11.7	1.5	2.4	3.8	4.1

† Plowing and rotovating performed to a 15-cm depth; cultivating to a 7.5-cm depth on otherwise untilled soil.
‡ Measured to the tillage depth.
§ Water applied at a 12.7-cm h^{-1} rate.

Various implements effectively increase soil porosity. An additional increase occurs when crop residues are incorporated with soil by tillage. In contrast, removing surface residues may adversely affect infiltration, as will be shown in later sections. As for depression storage, secondary tillage generally decreases temporary storage of water in the plow layer.

Burwell et al. (1966) evaluated the effects of tillage-induced porosity and surface roughness on infiltration of simulated rainfall (Table 2). Cumulative infiltration approached the total pore-space and surface-roughness retention volumes for the plow treatment before runoff began and exceeded the total volumes before 2.5 cm of runoff occurred. For other treatments, the potential storage volume was not filled, even though 5.0 cm of runoff occurred. Cumulative infiltration before initial runoff was more closely related to surface roughness than to total pore space (Fig. 1). Apparently, the smoother surfaces with treatments other than plowing resulted in more rapid surface soil dispersion and sealing, which reduced infiltration into the tillage-loosened soil.

3. CONTOUR AND GRADED FURROWS

Tillage near or on the contour associated with ridge terraces has long been used to control runoff on relatively steep land in humid regions. Although designed to orderly convey excess water from fields with minimum erosion and little profile recharge (Bertrand, 1966), graded-furrowing and

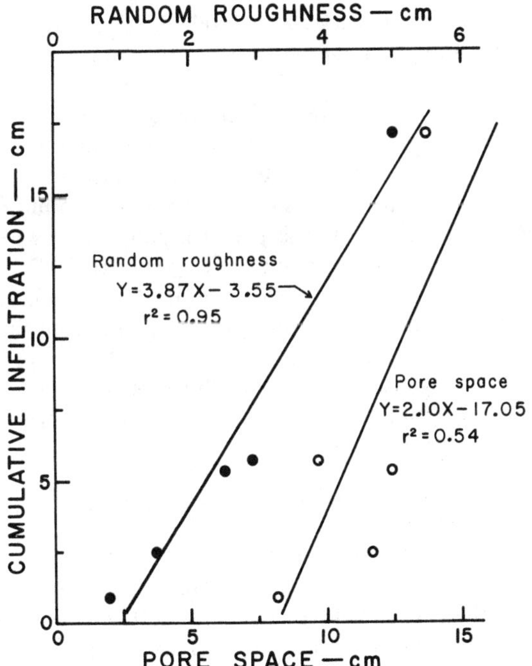

Fig. 1—Effect of soil random roughness and pore space on cumulative infiltration (from Burwell et al., 1966).

Table 3—Runoff and sediment yield from corn watersheds at Coshocton, Ohio, during a severe rainstorm on 5 July 1969 (from Harrold & Edwards, 1972).

Tillage	Slope	Rainfall	Runoff	Sediment yield
	%	cm		kg ha^{-1}
Plowed, clean-tilled sloping rows	6.6	14.0	11.2	50,700
Plowed, clean-tilled contour rows	5.8	14.0	5.8	7,200
No-tillage, contour rows	20.7	12.9	6.4	70

contouring affect runoff from cropland. At Temple, Tex., runoff during a 32-month period totaled 18.7 cm from a graded-furrow watershed and 23.6 cm from a terraced watershed (Richardson, 1973). Less runoff from the graded-furrow watershed probably resulted from more uniform distribution of the excess water over the entire field. On the terraced watershed, the excess water concentrated in terrace channels and ran off the field more rapidly.

Harrold and Edwards (1972) measured rainfall, runoff, and sediment yields for three watersheds cropped to corn during a storm near Coshocton, Ohio, that had an expected recurrence frequency of more than 100 years. Over 12.9 cm of rain fell in 7 hours. Rainfall was identical and slopes were similar for clean-tilled watersheds having sloping (upslope and downslope) or contour rows (Table 3). Runoff and sediment yield from the contoured watershed were only 52 and 14%, respectively, of that from the sloping-row watershed. No-tillage with contour rows resulted in 57% of the runoff and 0.1% of the sediment yield as compared with that from the sloping-row watershed, even though the slope of the no-tillage watershed was much greater.

In water-deficit farming areas, contour-furrowing for row crops is an effective runoff control and water conservation practice (Dickson et al., 1940). Contouring with terracing often reduces runoff more than terracing alone because the contour furrows hold the water on a major portion of a field. When a lister or disk bedder is used for contour plowing, each furrow serves as a miniature terrace. At Spur, Tex., runoff for 27 years averaged 7.0, 5.0, and 0.0 cm and cotton yields averaged 130, 160, and 210 kg ha^{-1} from areas with sloping rows, contour rows, and contour rows supplemented with closed-end level terraces, respectively (Fisher & Burnett, 1953). As compared with row crops, winter wheat responded little to contour tillage and planting because flat-tillage methods (sweep or oneway plowing) were used (Finnell, 1944).

4. RESIDUES—CLEAN vs. CONSERVATION TILLAGE

Conservation tillage rates among the top advances in crop production practices in recent agricultural history. Among the many benefits ascribed to conservation tillage is decreased runoff, which decreases soil erosion and increases water conservation. The large decreases in runoff result from surface residues, which are an integral part of the systems (Wittmuss et al., 1973). Surface residues dissipate the energy of falling raindrops, thus pro-

Table 4—Effect of tillage on runoff and soil losses from bare-fallow, plowed, and no-tillage areas in Nigeria due to rainfall† (from Rockwood & Lal, 1974).

Slope	Bare-fallow		Plowed		No-tillage	
	Runoff	Soil loss	Runoff	Soil loss	Runoff	Soil loss
%		tons ha^{-1}	%	tons ha^{-1}	%	tons ha^{-1}
1	18.8	0.2	8.3	0.04	1.2	0.0007
5	20.2	3.6	8.8	2.16	1.8	0.0007
10	17.5	12.5	9.2	0.39‡	2.1	0.0047
15	21.5	16.0	13.3	3.92	2.2	0.0015

† Total rainfall was 44.2 mm.
‡ Probably an error in original publication.

Table 5—Cropping effects on rainfall runoff from three soils in Puerto Rico (from Barnett et al., 1972).

Cropping and treatment†	Soil	Slope	Storm‡ 1	Storm‡ 2	Storm‡ 3	Runoff 1	Runoff 2	Runoff 3
		%	cm					
Fallow	Humatas clay	35	6.4	13.0	19.4	0.0	7.2	7.2
	Juncos silty clay	32	6.5	12.4	18.9	4.4	13.0	17.4
	Pandura sandy loam	26	6.6	13.0	19.7	0.5	5.8	6.3
Tobacco								
Conventional tillage	Humatas clay	38	6.0	11.9	18.0	1.4	8.2	9.6
	Juncos silty clay	33	6.6	12.6	19.2	4.0	11.6	15.6
	Pandura sandy loam	28	7.2	12.9	20.1	0.2	4.8	5.1
Mulch tillage	Humatas clay	38	6.0	11.9	18.0	1.4	9.9	11.2
	Juncos silty clay	34	6.2	12.3	18.6	3.6	11.5	15.0
	Pandura sandy loam	26	6.5	13.1	19.6	0.0	2.9	2.9
Grass strips	Humatas clay	37	6.4	13.0	19.4	1.1	10.6	11.7
	Juncos silty clay	32	6.2	12.6	18.8	3.8	11.0	14.8
	Pandura sandy loam	29	6.8	12.5	19.3	0.1	3.2	3.2
Pangolagrass								
Full sod	Humatas clay	39	6.0	12.7	18.7	0.1	0.7	0.7
Tops removed	Humatas clay	46	6.0	11.9	17.9	1.7	9.1	10.8

† Conventional tillage was plowing, smoothing, and planting on contour; mulch tillage was planting in sod; grass strips were one row wide between three clean-tilled rows; and pangolagrass conditions were as shown.
‡ Rainfall was applied at 6.35 cm h^{-1} for 60 min (storm 1), and 12.7 cm h^{-1} for 60 min (storm 2) beginning 10 min after storm 1. Results for storm 3 are the sums of those for storms 1 and 2.

tecting surface soil from dispersion, decreasing surface sealing, and increasing infiltration.

Surface mulches decreased runoff for such widely different conditions as the high-rainfall areas of the tropics and the southeastern United States and the low-rainfall area of the U.S. Great Plains. In the tropics (Nigeria), Rockwood and Lal (1974) measured much lower runoff and soil losses from no-tillage areas than from bare-fallow and plowed areas (Table 4). In Puerto Rico, Barnett et al. (1972) applied artificial rainfall and measured runoff from three soils having slopes from 26 to 46% (Table 5). Runoff was

Table 6—Soil management effect on runoff from topsoil and exposed subsoil (from Batchelder & Jones, 1972).

		Soil condition					
		Topsoil		Subsoil		Mulched subsoil	
Dates	Factor	I_1†	I_0†	I_1	I_0	I_1	I_0
		cm					
9 May–20 Oct. 1966	Rainfall	48.7	48.7	48.7	48.7	48.7	48.7
	Rainfall runoff	11.8	12.6	11.3	15.8	10.4	2.7
	Irrigation	34.7	0	34.7	0	34.7	0
	Irrigation runoff	5.5	0	2.2	0	3.4	0
24 May–31 Oct. 1967	Rainfall	37.5	37.5	37.5	37.5	37.5	37.5
	Rainfall runoff	9.5	11.5	7.9	9.1	1.9	0.3
	Irrigation	19.8	0	17.5	0	12.5	0
	Irrigation runoff	5.9	0	2.2	0	0.3	0
10 May–25 Oct. 1968	Rainfall	49.9	49.9	49.9	49.9	49.9	49.9
	Rainfall runoff	12.4	16.6	10.9	13.7	1.2	0.5
	Irrigation	17.1	0	14.2	0	6.2	0
	Irrigation runoff	3.7	0	2.0	0	0.1	0

† I_1 and I_0 refer to irrigated and nonirrigated plots, respectively.

not greatly affected by cropping treatment on Humatas clay (kaolinitic, isohyperthermic Typic Tropohumult), apparently because runoff was more influenced by water conductivity through this fine-textured soil than by the potential infiltration rate. On Pandura sandy loam (mixed, isohyperthermic, shallow Typic Eutropept), mulch tillage and grass strips greatly reduced runoff. According to Barnett et al. (1972), the high runoff for Juncos silty clay (fine, montmorillonitic, isohyperthermic Vertic Eutropept) is misleading because the rainfall entered the soil but returned as interflow at the lower end of the plots. The imposed rainstorms had simulated recurrence frequencies of 2 years (storm 1), 50–100 years (storm 2), and 150–500 years (storm 3).

Batchelder and Jones (1972) measured rainfall and irrigation-water runoff from Groseclose clay loam (clayey, mixed, mesic Typic Hapludult) in Virginia from 1966 to 1968 (Table 6). Runoff was lowest from the mulched subsoil, except in 1966 when runoff of sprinkler-applied water was higher from mulched than from unmulched subsoil. The higher runoff in 1966 indicated that the mulched subsoil plots were excessively irrigated. In subsequent years, the mulch effectively controlled runoff. In 1967 and 1968, water applied to different treatments was not uniform because corn used less water on the subsoil and mulched subsoil plots; however, yields were not significantly affected by reducing the irrigation amount. Except on the mulched plots, rainfall runoff was greater from nonirrigated than from irrigated plots because the irrigated plots remained wetter and thus a hard, dry crust did not form as it did on nonirrigated plots. Also, more rainfall was intercepted by plants on irrigated plots because of greater plant growth and leaf areas.

Table 7—Tillage effects on runoff and crop yields in Wisconsin (from Hays, 1961).

Soil	Period	Crop and seedbed preparation	Runoff	Yield
			cm	kg ha^{-1}
Fayette silt loam, 15% slope	1955–1959	Corn after hay		
		Normal	2.1	6,110
		Wheel-track	0.9	5,440
		Mulch	1.1	5,950
Fayette silt loam, LaCrosse	1954–1959	Corn in corn-grain-hay rotation	1.5	7,530
		Corn after corn and corn stover mulch	0.3	7,880
Miami silt loam, Madison	1955–1959	Corn in corn-grain-hay rotation	2.0	4,930
		Corn after corn and corn stover mulch	0.2	4,770
Fayette silt loam, 15% slope	1957–1959	Small grain	9.0	5,720
		Fall plowed	7.2	5,240
		Field cultivated		
		Corn stover mulched + field cultivated	4.8	5,540

Mulches greatly decreased runoff in several seedbed-preparation studies by Taylor et al. (1964) and Hays (1961) in Wisconsin (Table 7). For corn after hay, wheel-track planting or mulch planting (field cultivator) decreased runoff about 50% as compared with normal seedbed peparation. Corn stover mulch in corn after corn decreased runoff more than corn in a rotation on soils at LaCrosse and Madison. For small grain, runoff with field cultivation was lower than with fall plowing. Even less runoff occurred with mulching and field cultivation. In all studies, yields were not greatly affected by seedbed-preparation techniques, but soil loss was greatly reduced by runoff control.

In Ohio, runoff during 1964–1966 totaled 0.5 and 1.9 cm from no- and conventional-tillage areas cropped to corn. Soil losses from the respective areas were 130 and 6,500 kg ha^{-1} (Harrold et al., 1967). Also in Ohio, Triplett et al. (1968) evaluated infiltration of simulated rainfall into plowed and unplowed (no-tillage) Wooster silt loam (fine, mixed, mesic Typic Fragiudalf). Some no-tillage areas had corn stover on the surface (Table 8). Infiltration was significantly greater with 80% surface cover than with other treatments, even though soil density was as great or greater and air-filled pore space was as low or lower than for other treatments. Greater infiltration with high residue levels resulted primarily from improved soil structural stability rather than from physical protection of the soil surface.

The value of residues was also shown by Mannering and Meyer (1961), who studied runoff and soil losses on Warsaw loam (fine over sandy or sandy-skeletal, mixed, mesic Typic Argiudoll) and Russell silt loam (fine, mixed, mesic Typic Hapludalf) in Indiana. Cornstalk management treatments were (i) check—stalks as left by cornpicker, (ii) stalks shredded, and (iii) stalks shredded and disked once. Runoff with check and shredded-corn-

Table 8—Treatment effects on infiltration at the end of 3 years of different surface-residue treatments in Ohio (from Triplett et al., 1968).

Treatment†	Total infiltration in 1 hour‡	
	Initial run	Wet run
	cm	
Plowed—bare	1.80 b§	1.04 c§
No-tillage—bare	1.22 b	0.64 c
No-tillage—40% residue	2.34 b	1.35 bc
No-tillage—80% residue	4.39 a	3.48 a
No-tillage—40% residue, residue removed	1.63 b	1.04 c
No-tillage—80% residue, residue removed	4.17 a	2.49 ab

† Bare—residue removed each year; 40 and 80% residue—normal and twice-normal residue maintained on surface; residue removed—all residues removed before making infiltration measurements.
‡ Water applied for 60 min at rates of 10.2–11.9 cm h^{-1}.
§ Values within columns followed by the same letter(s) are not significantly different (Duncan's Multiple Range Test, 5% level).

stalk treatments was similar, but shredding the cornstalks decreased soil losses to about one-half those of the check. Soil losses with the shredding-plus-disking treatment were intermediate. Although runoff was not affected, the study showed that increasing the surface coverage by shredding was an effective soil-loss control practice.

In South Dakota, Onstad (1972) measured runoff and soil losses from 1965 to 1970 from plots on Egan and Wentworth silty clay loams (fine, mixed, mesic Udic Haplustolls) having 5.8% slopes. Although average runoff was relatively low as compared with that at locations with more rainfall, the results show the value of surface residues and improved tillage practices for decreasing runoff and soil losses (Table 9).

Under graded-furrow conditions on Pullman clay loam (fine, mixed, thermic Torrertic Paleustoll) at Bushland, Tex., Allen et al. (1975) evaluated rainfall and irrigation-water runoff from clean- and no-tillage areas cropped continuously to grain sorghum. For four irrigations in 1971 totaling 36.4 cm, runoff totaled 9.7 and 4.1 cm from clean- and no-tillage areas, respectively. For the fifth irrigation when 6.3 cm of rain fell immediately after applying 8.2 cm of water, runoff from the respective areas totaled 5.8

Table 9—Average rainfall, runoff, and soil loss on tillage plots at Madison, S. Dak. (from Onstad, 1972).

Tillage practice	Average values (1965–1970)		
	Rainfall	Runoff	Soil loss
	cm		tons ha^{-1}
Fallow	42.0	5.0 a†	17.5 a†
Conventional	42.0	2.9 b	6.0 b
Mulch	42.0	2.4 c	3.7 c
Till-plant (with slope)	42.0	2.1 c	3.5 c
Till-plant (on contour)	42.0	1.0 d	0.9 d

† Values within a column followed by the same letter are not significantly different (Duncan's Multiple Range Test, 5% level).

Table 10—Water loss by runoff and soil loss in runoff during initial, wet, and very wet runs in a laboratory study. Rain intensity was 6.4 cm h^{-1} (from Lattanzi et al., 1974).

Rain applied	Mulch rate	Water loss S†= 2%	Water loss S= 6%	Water loss S= 12%	Water loss S= 20%	Soil loss S= 2%	Soil loss S= 6%	Soil loss S= 12%	Soil loss S= 20%
	tons ha^{-1}	kg m^{-2}				g m^{-2}			
Initial run, 60 min	0	49.3	52.7	53.0	49.2	950	1,230	1,870	2,140
	0.5	50.9	52.4	51.1	52.4	600	750	970	1,250
	2	43.5	43.7	44.0	44.1	240	260	310	490
	8	2.5	4.3	4.1	5.7	7	10	6	2
Wet run,‡ 30 min	0	28.6	28.9	28.2	27.9	410	550	770	820
	0.5	29.9	29.8	29.5	30.0	260	370	480	520
	2	27.7	28.1	27.2	27.3	90	130	150	210
	8	2.5	5.0	9.3	7.1	2	5	3	8
Very wet run,‡ 30 min	0	28.4	29.3	29.1	28.1	400	560	720	810
	0.5	30.8	30.8	30.1	30.6	260	380	480	540
	2	28.9	28.1	28.2	29.3	80	120	150	220
	8	4.0	6.4	12.5	9.1	1	5	3	5

† S = slope steepness.
‡ The wet and dry wet runs were separated by 15 min without rain and were applied 1 day after the initial run.

and 4.6 cm. Less runoff with no-tillage was attributed to slower water advance due to residues in the furrows and increased infiltration, which allowed deeper penetration of the water than in clean-tillage furrows. The extra water increased plant growth, but grain yields were not increased because uncontrolled volunteer sorghum plants resulted in plant populations above the optimum for grain production.

Lattanzi et al. (1974) determined the influences of wheat straw mulch rates and slopes on water and soil losses from Russell silt loam under laboratory conditions. Runoff was independent of slope but was slightly decreased by the 2-ton ha^{-1} mulch rate and greatly decreased by the 8-ton ha^{-1} rate (Table 10). At 40 min, surface sealing by raindrop impact had decreased infiltration rates for treatments with 0–2 tons ha^{-1} of mulch to only 20% of those with 8 tons ha^{-1} of mulch. Interrill erosion with 0.5 and 2 tons ha^{-1} of mulch was decreased about 40 and 80%, respectively, as compared with no mulch. Soil loss decreased more than runoff, which shows the value of small amounts of surface residue for controlling erosion, provided rills are not formed beneath the residues. The value of small residue amounts for decreasing soil losses was also shown by Kramer and Meyer (1969).

B. Engineering—Structure Effects

Land-forming practices are widely used to control runoff from excess rainfall. An extreme example is the bench leveling of entire steep hillsides in some Asiatic countries. In the United States, less-intensive land-forming practices have received considerable attention.

1. TERRACES

Precipitation amount and distribution strongly influence the type of terrace used in a particular region. In higher-precipitation regions where erosion is a major problem, excess water is channeled from fields by graded terraces, which drain the water at nonerosive velocities into grassed waterways or other grassy areas. In lower-precipitation areas, where water conservation for subsequent crop use is desirable, level, blocked-end level, conservation bench, and level bench terraces have received considerable attention.

Because graded terraces are designed primarily to orderly convey excess water from fields, runoff may be higher from terraced land than from land where other runoff control practices, such as graded furrows, are used (Richardson, 1973). However, in the long term, graded terraces may still result in more efficient water use by maintaining a productive soil through improved erosion control. Controlling erosion provides better soil conditions for crops on the terraced area and better quality runoff water for downstream users.

When designed for water conservation, terraces often are level and may have closed ends. Graded and level terraces were evaluated for water storage and crop yields at several locations in the Southern Great Plains. Runoff was lower and crop yields were higher on graded-terrace fields than on nonterraced fields (Dickson et al., 1940; Finnell, 1944). Generally, runoff was lower and yields were higher with level terraces, especially with closed ends and contour furrows between terraces (Burnett & Fisher, 1956; Dickson et al., 1940; Fisher & Burnett, 1953). When Foard silt loam (fine, montmorillonitic, thermic Typic Natrustoll) in Oklahoma was cropped to wheat, runoff was stored mostly in channels of closed-end level terraces. During dry seasons, wheat yields were higher in the channels, but during wet seasons, yields in the channels often were lower than on areas between terraces. Water had to be drained from terrace channels during wet periods to prevent damage to wheat (Harper, 1941).

On Pullman clay loam, wheat and grain sorghum yields from 1949 to 1960 were similar with graded and closed-end level terraces in a wheat-sorghum-fallow (WSF) system. Hauser et al. (1962), therefore, suggested using open-end level terraces, which would avoid the need for high terrace ridges to store large volumes of runoff or to drain water from terrace channels during wet periods.

For conserving potential runoff water by storing it in soil over larger areas and for later crop use, bench terraces with and without contributing watersheds were evaluated at several Great Plains locations. At Bushland, Tex., bench terraces with contributing watersheds (Conservation Bench Terraces—CBT) were constructed in 1955 on land with 1.0–1.8% slope. The watershed/bench ratio was 2:1. The benches were continuously cropped to grain sorghum and the watersheds were in a WSF sequence. The CBT system controlled runoff, largely prevented water erosion, and distributed runoff for storage in soil (Hauser, 1968; Zingg & Hauser, 1959). Water storage on level benches of the CBT system was greater than on level benches without a watershed, and about the same as on level-terraced fields after 11 months of fallow. Yields were about 1.5 times greater with the CBT system than with level terraces because all of the level-terraced area was cropped in a WSF system, whereas the benches of the CBT system were continuously cropped to grain sorghum. Total yield was highest for a system of bench terraces without a watershed because these terraces were annually cropped to sorghum. Further evaluation of the systems from 1959 to 1972 substantiated earlier results (Jones & Hauser, 1975). Again, overall yield was highest on level benches without a watershed, but this system increased the probability of low yields in dry years due to low soil water storage. The major advantage of the CBT system over level benches without a watershed was that only one-third of the area was leveled for the CBT system. Higher yields on benches without watersheds did not adequately offset the additional construction costs when compared with yields and construction costs for the CBT system (Jones & Shipley, 1975).

The CBT systems at Big Spring, Tex., were on Amarillo fine sandy loam (mixed, thermic Aridic Paleustalf) with a 1.3–1.9% slope and watershed/bench ratios of 2:1, 4:1, or 6:1 (Armbrust & Welch, 1966). A level bench without watershed and a check treatment were also evaluated. The CBT systems did not increase cotton or grain sorghum yields. The soil had a high infiltration rate and an available water storage capacity of only 10.2 cm to a 1.2-m depth. Because runoff was limited to large, high-intensity rains or rains that followed previous rains within 1 or 2 days, the limited capacity of the reservoir resulted in the impounded runoff being lost mostly through deep percolation.

The CBT systems at Akron, Colo., were on Rago silt loam (fine, montmorillonitic, mesic Pachic Argiustoll) on 1% slope. The benches were 30 m wide and had watershed/bench ratios of 0:1, 1:1, 2:1, or 3:1. The benches captured runoff from the watersheds and increased soil water contents by 1.8–4.3 cm over the water contents resulting from fallow during a period of below-normal precipitation. Water storage in benches during 7 months between crops with annual cropping was almost the same as storage during 19 months of fallow on the watershed where a sorghum-fallow cropping system was used. Grain sorghum grown annually on the benches yielded 450 kg ha^{-1} more grain than sorghum after fallow. The 2:1 watershed/bench ratio resulted in highest grain sorghum yield (Mickelson, 1968).

At Hays, Kans., the CBT terraces, which had 0:1, 1:1, or 2:1 watershed/bench ratios, were constructed on a slowly permeable Crete soil (fine, montmorillonitic, mesic Pachic Argiustoll) in 1958. Because of high rainfall and runoff, the benches were drained in 1959; in 1960, runoff was slight. In both years, grain sorghum yielded more on the annually cropped benches than on the watersheds, which were fallowed for 12 months before planting the sorghum (Hauser & Cox, 1962).

At Mandan, N. Dak., water storage in soil and alfalfa yields were more influenced by bench location for snow collection than by the watershed/

bench ratio in a CBT system. Overwinter water storage for 5 years averaged 3.6 cm on the watershed and ranged from 12.2 to 23.1 cm on the benches. Alfalfa yields averaged 3.4 tons ha^{-1} on the watershed and ranged from 7.2 to 9.6 tons ha^{-1} on the benches (Haas & Willis, 1971).

2. FURROW DIKES

Basin listing or furrow diking was first used in the Great Plains in the 1930's, but by 1950 it was little used because of operational, weed-control, and erosion problems. In addition, yield increases were slight (Daniel, 1951; Locke & Mathews, 1953) and stubble mulching, terracing, and other conservation practices became more popular and easier to manage (Clark & Hudspeth, 1976).

With improved technology, Clark and Hudsepth (1976) overcame most problems causing early failures with furrow diking. Long-term data for Bushland, Tex., showed that runoff was highest during May and June, just before or soon after planting summer row crops. Having the dikes in place during the growing season for row crops instead of during fallow after wheat when the potential for runoff is low, as was done previously, allows the extra water to be used almost immediately and, therefore, minimizes evaporation.

In early diking studies, weeds were controlled by cultivation. Now herbicides are used, and many fields are not cultivated after planting. When cultivation is necessary, dikes can be removed by a device mounted on the front of the tractor, the crop cultivated, and the dikes re-established, all in one operation (Lyle & Dixon, 1977).

Clark and Hudspeth (1976) eliminated dike over-topping and potential erosion by using large furrows to increase furrow-storage capacity. Also, the crop used stored water between rains, thus increasing subsequent storage capacity. For grain sorghum on Pullman clay loam, furrow diking increased water content throughout the growing season over the water content of areas with open furrows and over flat-planted areas. In 1975, grain yields at Bushland were 2,920, 2,580, and 2,470 kg ha^{-1} on diked-furrow, open-furrow, and flat-planted areas, respectively. On Amarillo fine sandy loam with 0.2% slope at Lubbock, runoff occurred from open furrows, but not from diked furrows. Cotton lint yields were 220 and 280 kg ha^{-1} with open and diked furrows, respectively.

Aarstad and Miller (1973) in Washington evaluated furrow diking for decreasing runoff from high-frequency, center-pivot sprinkler irrigations. Runoff was negligible with the diking treatment (Table 11). Because runoff was lower than with the control, irrigation frequency was reduced for the diking and hay-application treatments. In all cases, except for sugarbeets under Circle 1 (center-pivot area), yields were significantly lower with the control than with the other treatments. Since equipment is available to easily build and eliminate the dikes from row-cropped land, furrow diking seems to be an improved, practical method for holding potential runoff on the field where the water falls.

Table 11—Average runoff as percent of water applied and crop yields in Washington (from Aarstad & Miller, 1973).

Crop and soil	Treatment	No. of measurements	Runoff	Yield
			%	tons ha^{-1}
Circle 1—sugarbeets—3% slope—silt loam	Control	28	11	80.2
	60-cm long basins	2	1	83.8
	11. tons ha^{-1} hay (incorporated)	8	2	82.9
	4.5 tons ha^{-1} hay (mulch)	8	5	85.6
Circle 2—potatoes—3% slope—silt loam	Control	26	16	54.9
	60-cm long basins	3	1	68.5†
	11.2 tons ha^{-1} hay (incorporated)	11	2	74.8†
	4.5 tons ha^{-1} hay (mulch)	11	2	70.3†
Circle 3—potatoes—7% slope—loam	Control	24	17	57.1
	60-cm long basins	5	1	62.5†
	11.2 tons ha^{-1} hay (incorporated)	12	1	64.4†
	4.5 tons ha^{-1} hay (mulch)	12	3	62.9†
Circle 4—sugarbeets—7% slope—loam	Control	22	41	72.6
	60-cm long basins	8	1	93.4†
	11.2 tons ha^{-1} hay (incorporated)	15	6	86.7†
	4.5 tons ha^{-1} hay (mulch)	19	30	85.8†

† Significantly higher than the control at the 5% level of probability.

3. MICROWATERSHEDS AND VERTICAL MULCHES

Although microwatersheds and vertical mulches serve two distinct purposes, they often are used in a combination system. Microwatersheds increase runoff from a portion of the field and concentrate the water on a relatively small area to increase depth of water penetration. Vertical mulching, by providing a residue-filled soil slot open to the surface, results in rapid channeling of water into soil.

Under laboratory conditions, vertical mulches on a level surface saved 30–40% more of the applied rather than did a furrow treatment. Storage efficiency with the vertical mulch decreased 17% when the surrounding soil surface was wetted during water application. A microwatershed with a vertical mulch caused 7–10% more of the applied water to be stored than did a microwatershed without a mulch. Depth of water penetration and dry surface soil adjacent to the mulch appeared to be factors that decreased evaporation from vertical-mulched soil (Fairbourn & Gardner, 1972).

Under field conditions at Akron, Colo., a vertical mulch saved 16% more water than a nonmulched microwatershed and 41% more than the control treatment. Check dams across the furrows were an important feature of the microwatershed systems. Grain sorghum yields with the vertical-mulch treatment were 37–150% higher than with the control treatment (Fairbourn & Gardner, 1974).

Parr (1959) compared vertical mulching (without a microwatershed) and subsoiling on Crosby silt loam (fine, mixed, mesic Aeric Ochraqualf) in Indiana. Besides decreasing soil bulk density and increasing aggregation, vertical mulching as compared with subsoiling generally increased soil water content. Kingsley and Shubeck (1964) reported similar results for Aberdeen (fine, montmorillonitic Glossic Udic Natriboroll) and Exline (fine, montmorillonitic Leptic Natriboroll) silty clay loams and Tetonka silt loam (fine, montmorillonitic, mesic Argiaquic Argialboll) in South Dakota.

Wendt (1973) obtained no benefit from a vertical mulch on Olton clay loam (fine, mixed, thermic Aridic Paleustoll) at Lubbock in 1970 because no runoff occurred. In 1971, rainfall in late August, September, and October (15.9, 12.9, and 5.1 cm, respectively) increased water contents at the 30- to 90-cm depth in vertical-mulched plots. The water contents remained higher until sorghum was killed by frost. Yields were 2,090, 2,490, and 3,110 kg ha^{-1} on check, vertical mulch, and vertical mulch with oil (sprayed on soil between rows) plots, respectively, and the differences were statistically significant.

Heilman and Gonzalez (1973) trenched Harlingen clay (very-fine, montmorillonitic, hyperthermic Entic Chromustert) in Texas to 61- or 102-cm depths and backfilled the trenches with soil or vermiculite. Trenching increased water infiltration rates, decreased soil bulk density in the trenches, increased rooting depth of cotton, and decreased soil salinity. Cotton lint yields were significantly influenced by trenching in 2 of the 3 years (Table 12).

Hauser and Taylor (1964) compared chiseling at 2-m intervals to a 60-cm depth, with and without vertical mulching, and disk plowing to a 60-cm depth with conventional moldboard plowing to a 20-cm depth on Pullman clay loam. Infiltration rates were 1.5 and 1.9 times greater with chiseling plus vertical mulching and disk plowing, respectively, than with conventional tillage during the 3-year test. Chiseling increased infiltration only the first year because chisel spacing was too wide to effectively loosen the soil.

Table 12—Cotton lint yields in Texas for 1970, 1971, and 1972 (from Heilman & Gonzalez, 1973).

Treatment	Year			Avg.
	1971	1972	1973	
	kg ha^{-1}			
Check	590 a†	820 d	580 b	660
61-cm trench, soil-backfilled	620 a	1,000 c	670 b	760
61-cm trench, vermiculite-backfilled	620 a	1,160 a	740 a	840
102-cm trench, soil-backfilled	520 a	1,020 b	660 b	730

† Column values followed by the same letter are not significantly different at the 5% level (Duncan's Multiple Range Test).

Grain sorghum yields were significantly higher with deep tillage than with other treatments in only 1 year when the amount of water applied to the level-border plots was inadequate for high yields. The effectiveness of the vertical-mulch treatment was low because the area was disked after the mulch was installed; hence, the mulch did not extend to the surface. Also, limited lateral water movement from the trench indicated that the 2-m spacing was too wide.

4. WATER HARVESTING, RUNOFF FARMING, AND WATER SPREADING

Water harvesting implies treating watersheds to enhance runoff. Nearly complete recovery of rainfall is possible with applications of asphalt emulsions, aluminum foil, butyl rubber, or plastic film, but these materials are costly and easily damaged by livestock and wild animals. Less effective runoff inducement was obtained by smoothing land surfaces, removing rocks, and spraying soil with water repellents (Bertrand, 1966).

Contour-ridging to prevent runoff and covering the ridges between 1-m spaced rows with black plastic film increased water conservation and resulted in an average corn grain yield of 4,130 kg ha^{-1} at Mandan, N. Dak. Yields with the conventional check averaged 2,410 kg ha^{-1} (Willis, 1962; Willis et al., 1963). The yield increase was attributed to better utilization of small rains, lower evaporation, and higher soil temperatures in the spring.

Mickelson et al. (1965) and Mickelson (1966) at Akron, Colo., constructed level basins in or adjacent to natural waterways to intercept runoff water from the waterways. Watershed/basin ratios ranged from 3:1 to 56:1. Runoff above a predetermined amount in one basin flowed through or bypassed the basin to fill the next basin at a lower elevation. Available soil water contents at sorghum planting for 1962–1964 averaged 10.2, 18.9, and 19.7 cm on continuously cropped (nonlevel), after fallow (nonlevel), and continuously cropped (level basin) areas, respectively. For the 3 years, sorghum grain yields on the respective areas averaged 350, 1,320, and 3,030 kg ha^{-1} (Mickelson, 1966). The major yield increase with level basins over that for the fallow area apparently resulted from runoff captured during the growing season, because soil water content at planting was similar for these two treatments.

IV. WATER INFILTRATION

Water infiltration and runoff control are virtually inseparable. Infiltration, therefore, was involved in most of the studies discussed in the previous section. Infiltration is influenced by soil conditions at the surface, in the tillage layer, and within the profile. Surface- and tillage-layer factors affecting infiltration include soil aggregation, surface roughness and porosity, surfactants and water repellents, and crop residues. Profile factors include depth to slowly permeable or impervious layers; soil texture and type of clay; and previous crops, especially with regard to rooting types and depths and residual soil water content.

A. Aggregation

1. NATURAL AGGREGATION

Soils of humid tropic and subtropic regions have generally high infiltration rates, even on steep slopes, due to natural aggregation of mineral origin (Table 4). Intensive weathering has destroyed most original minerals and has caused great losses of silica and basic cations. The remaining substances—Fe, Al, and some silica (SiO_2)—have recrystalized to form hydrous oxides of Fe, Al, and sometimes Ti, plus some kaolinite clay. Many particles in tropic soils are the size of sand grains and consist mainly of altered minerals cemented by Fe (Donahue et al., 1977). Because of the stable particles, infiltration rates for tropic and subtropic soils often are similar to those for deep sands and the rates remain high during prolonged rainfall.

Organic substances that contribute to soil aggregation are derived from plant materials, either after alteration by soil animals, bacteria, fungi, etc., or directly from the plants themselves. Earthworms improve soil structure, which increases infiltration (Hopp & Slater, 1961). While feeding on organic materials and burrowing in soils, earthworms secrete gelatinous substances that coat and stabilize soil aggregates. Water-stable aggregates result also from water-insoluble gummy substances secreted by bacteria, fungi, and actinomycetes (Donahue et al., 1977).

Earthworm activity and intensive soil tillage are not very compatible (Gantzer & Blake, 1978). Hence, little earthworm activity occurs in many intensively cultivated soils. Although cultivation decreases earthworm activity, bacteria and other microorganisms are active in cultivated soils as they feed on decaying plant roots and other plant parts returned to the soil by tillage. Besides enhancing aggregation and thus water infiltration, the activity of soil organisms also influences soil productivity through its effect on plant nutrients. For maximum earthworm activity, no-tillage is desirable. Where the soil is tilled, enough crop residues should be returned to the soil to provide abundant food for soil organisms, thus providing the potential for soil aggregation and increased water infiltration.

Plants directly influence soil aggregation through exudates from roots, leaves, and stems, and through leachates from weathering and decaying plant materials, which bind soil particles together; through plant canopies, which protect surface aggregates against breakdown due to raindrop impact; and through root action in soil, which promotes aggregate formation. If aggregates formed through these processes are subsequently maintained on the surface, water infiltration will be higher than in intensively cultivated, poorly aggregated soils (Donahue et al., 1977).

2. INDUCED AGGREGATION

Two of the most widely used soil conditioners for increasing aggregate stability were hydrolyzed polyacrylnitril (HPAN) and a modified vinyl

Table 13—Rainfall and runoff from 1952 to 1958 at Lincoln, Nebr. (from Weakly, 1960).

Year	Total annual rainfall	Treatments				
			HPAN† (kg ha^{-1})			Straw (kg ha^{-1})
		None	1,120	2,240	4,480	5,600
		Runoff, cm				
1952	80.1	12.2	10.7	7.8	7.3	1.0
1953	46.7	3.5	3.1	1.5	0.5	0.8
1954	77.6	22.9	18.1	15.9	12.4	2.6
1957	91.6	9.9	9.3	8.3	5.9	--
1958	90.4	12.3	12.0	10.8	12.7	4.1

† Hydrolyzed polyacrylnitril.

acetate-maleic acid compound (VAMA). When applied to soil, these substances resulted in larger aggregates. In general, VAMA was more effective than HPAN when applied at the same rate (Martin, 1953). Although not economical, except possibly for some high-value crops, the substances improved aggregation, which increased infiltration and decreased runoff as shown by Weakly (1960) for Sharpsburg silty clay loam (fine, montmorillonitic, mixed Typic Argiudoll) with 8.5% slope at Lincoln, Nebr. (Table 13). The HPAN decreased runoff and erosion, but a straw mulch was more effective than any of the HPAN treatments.

B. Surfactants and Water Repellents

Some soils are difficult to wet because of their natural conditions, the crops grown on them, or treatments applied. Examples are the loess soils of Israel, sandy soils under citrus trees in Florida, and soil made water-repellent by wildfires in California. Water repellency is desirable for water harvesting from soils, such as those in Israel (Evanari et al., 1968), but it decreases water storage in soil for crop use and may result in severe erosion.

Resistance to wetting of sandy surface soils under citrus trees in Florida caused trees in older groves to decline and decreased irrigation efficiency (Jamison, 1943, 1947). Resistance to wetting, which was greatest under old trees and least beyond the area where water dripped from leaves, was related to colloidal organic matter in the surface soil. Where declining trees were replaced by young trees and the soil was stirred and wetted, the soil regained its wettability. Relative wetting rates were 1.0 and 0.08 near reset trees and under declining trees, respectively. Cultivation before irrigation also increased irrigation efficiency and soil wetting. Soil wetting was even more effective when wetting agents or suitable clays were used in combination with irrigation and cultivation.

Water repellency of sands decreased germination and yield of barley in Australia (Bond, 1972). It also caused difficulty in golf-green management in Ohio (Miller & Wilkinson, 1977), where a coating of fulvic acid on sand grains was believed responsible for the water repellency.

Osborn et al. (1964) recognized soil wettability as a major contributor to severe erosion where a wildfire burned the moderately dense chaparral vegetation. Other studies (Debano et al., 1976; Savage et al., 1972) showed that soil water content at burning and burn temperature contributed to the amount and depth of deposition of hydrophobic substances in soil. Water repellency was mainly in the upper 1.5 cm of wet sand, but occurred to a 4-cm depth in dry sand. The amount of water-repellent substances produced increased as fire temperature increased. Wetting agents decreased runoff and erosion from watersheds burned by a wildfire, as compared with that in untreated areas (Osborn et al., 1964).

C. Deep Plowing and Profile Modification

When subsurface soil conditions restrict or prevent water infiltration and storage and plant root growth to the soil water, deep tillage and profile modification may result in more efficient use of water. Soil improvement by deep tillage and profile modification usually involves loosening such layers as fragipans, hardpans, plowpans, claypans, high-clay horizons, and saline-sodic horizons.

The benefits to be expected from deep tillage and profile modification depend on such factors as the severity and nature of the problem, climatic conditions, management, and the crop grown (Unger et al., 1981).

1. FRAGIPANS, HARDPANS, AND PLOWPANS

Bradford and Blanchar (1977) modified Hobson silt loam (fine, siliceous, mesic Typic Fragiudalf) to a depth of 152 cm in Missouri in 1974 by loosening the profile with a trenching machine and adding lime, fertilizer, and sawdust. This soil has a dense, sandy fragipan at a depth of about 50–70 cm. The treatments greatly reduced soil bulk density, increased saturated hydraulic conductivity, caused more uniform distribution of sand, silt, clay, and soil pores throughout the profile, and increased available water storage and root proliferation to greater depths. Consequently, grain sorghum yields were significantly increased. Re-excavation in 1975 of a pit in Hobson silt loam that was excavated and refilled in 1959 with the same soil showed little indication of fragipan reformation during 16 years. In contrast, Fritton and Olson (1972) in New York found that modification of Erie channery silt (coarse, loamy, mixed, mesic Aeric Fragiaqualf) by mechanical disturbance alone (trenching for a pipeline) resulted in the re-establishment of dense fragipan layers within 11 years.

Van Doren and Haynes (1961) subsoiled and chiseled Ohio soils having fragipans at the 36- to 90-cm depths. Subsoiling or chiseling to 38- or 46-cm depths, respectively, did not increase corn, oats, and clover yields because not enough of the fragipan was loosened to adequately increase water storage and plant rooting depth.

At some locations in California, severe flooding occurs if rainfall continues after soil above the hardpan is wetted. To alleviate these problems, the indurated hardpan is ripped to depths of 1.5–2.1 m at 0.9- to 1.2-m spacings. After ripping, water and root penetration are vastly improved and the results are long-lasting. Ripping is accomplished by up to four tractors operating as one unit having a combined power rating of over 1,100 × 10^3 W (1,500 hp) (Calif. Builder & Eng., 1975).

In Louisiana, some soils near the Mississippi River have natural hardpans about 21–45 cm below the surface. Plants on these soils experience water stress well before those on surrounding soil. Deep tillage increased rainwater infiltration, which more readily refilled the soil water-storage reservoir. Cotton lint yields were substantially increased by deep tillage in dry years, but not when rainfall was adequate (Saveson et al., 1961). Also in Louisiana, Patrick et al. (1959) and Saveson and Lund (1958) obtained yield increases with 30-cm-deep moldboard plowing or 35-cm-deep chiseling on soils that had high-density or high-strength layers below normal plow depth.

Compaction by tillage machinery has formed plowpans that impede water infiltration and root penetration in some soils. Taylor and Gardner (1963) showed that soil strength, rather than bulk density or other physical factors of soil, controls the penetration of roots in soil. On a sandy clay loam soil at Big Spring, Tex., cotton yield decreased as soil strength increased. Few roots penetrated soil with strength above 25 bars at field capacity (Taylor et al., 1964). Both soil water content and bulk density affect soil strength.

Plowpans more often develop on fine sandy loams or other nonswelling soils than on swelling clay soils. Plowpans on fine sandy loam soils at Woodward, Okla., decreased infiltration of water, hindered root penetration, increased erosion, and decreased crop yields, whereas the one at Mandan, N. Dak., was not severe enough to influence crop yields (Locke et al., 1960). In a field study at Woodward from 1958 to 1963, when precipitation was average, sorghum yields were not affected by the plowpan (Taylor et al., 1964).

Hobbs et al. (1961) deep-tilled eight soils in Kansas and found that deep tillage lowered bulk density and improved water permeability in dense layers. However, subsequent tillage decreased water movement into and storage in the soils. Crop yields were seldom increased by deep tillage. In Michigan, deep tillage improved crop yields when compact zones were just below the normal tillage depth (Robertson et al., 1977). In Connecticut, plowpans restricted root development of tobacco plants, but shattering the plowpan resulted in only small yield increases (De Roo, 1961).

2. DENSE CLAY HORIZONS AND CLAY PANS

Soils with thick, dense montmorillonitic clay layers restrict drainage, water penetration, soil aeration, and root development, especially for annual crops. One such soil is Houston Black clay (very-fine, thermic Udic Pellustert), which with associated soils covers about 2 million ha in central Texas. Burnett and Tackett (1968) modified this soil to 60- and 120-cm depths by rotary tillage or with a trenching machine. Mixing the profile to the 120-cm depth, as compared with mixing to the 60-cm depth or normal tillage, caused greater and more uniform aeration to a 180-cm depth, which increased root proliferation. Roots penetrated to the 150-cm depth on unmodified soil, but branching in the lower profile was greater in modified soil. Deep mixing generally increased cotton and grain sorghum yields as compared with conventional tillage. Also, improved aeration, which decreased profile CO_2 concentrations, greatly decreased cotton root rot incidence.

Low water infiltration rate after initial surface wetting and filling of shrinkage cracks, and shallow rooting of crops are major problems on Pullman clay loam and associated soils of the Southern Great Plains. Pullman clay loam has a very slowly permeable swelling clay horizon at about the 23-cm depth, the depth of primary tillage on cultivated soils. The lower boundary is variable, but is typically at about the 68-cm depth. After initial wetting, the infiltration rate decreases to about 1.3 mm h^{-1}, making it difficult to refill the soil water reservoir. Thus, the goal of deep tillage and profile-modification studies on Pullman clay loam soil has been to increase water infiltration and storage in the lower profile and to increase rooting depth.

Jensen and Sletten (1965) found that chiseling Pullman clay loam at 50- to 60-cm intervals to the 38-cm depth after harvest when the soil was dry doubled basic infiltration rates during the first irrigation as compared with conventional tillage. By the fourth year, the infiltration rate with chiseling was 20% greater than with conventional tillage. The plots had borders that kept all added water on them; therefore, chiseling increased wheat yields only about 6% and sorghum yields were not affected.

In 1966, Pullman clay loam was moldboard-plowed to 40-, 60-, and 80-cm depths. Conventional tillage was moldboard plowing to the 20-cm depth. The 40- and 60-cm depth treatments loosened parts of the slowly permeable soil layer and mixed it with surface soil. Plowing 80 cm deep mixed the slowly permeable layer with surface soil and with other more permeable soil from below the slowly permeable layer. Tillage to the 40-, 60-, or 80-cm depth, as compared with conventional tillage, increased irrigation water infiltration in 1966, but not in 1968, on the furrow-irrigated plots (Table 14). In 1968, lack of response to deep tillage resulted from surface compaction during sugarbeet harvest in 1967 while the soil was wet. The compaction was alleviated by moldboard plowing 20 cm deep after the 1968 crop. With one growing-season irrigation, grain sorghum yields in 1966 and 1968 averaged 1,990 kg ha^{-1} more with 80-cm-deep tillage than with conventional tillage. With two irrigations, yields were highest with 40-cm-deep tillage (Table 14) (Schneider & Mathers, 1970). Greater water storage and subsequent depletion permitted sorghum on deep-tilled plots to produce satisfactory grain yields with limited growing-season irrigations. Where water for irrigation is limited, as in parts of the Southern Great Plains, deep tillage increases early season water storage and decreases the demand for irrigation water during critical growth periods of grain sor-

Table 14—Effect of deep tillage of Pullman clay loam in Texas on irrigation water infiltration and grain sorghum yields (from Schneider & Mathers, 1970).

Tillage depth	Infiltration†		Average grain yield	
	1966	1968	One irrigation	Two irrigations
cm	cm		kg ha^{-1}	
20	19.8	15.2	3,830	5,620
40	22.1	14.7	4,870	7,100
60	29.7	14.7	5,280	6,600
80	29.5	14.7	5,820	6,200

† Based on average infiltration for treatment receiving two growing-season irrigations.

Table 15—Effect of deep tillage of Pullman clay loam in Texas on sugarbeet root and sugar yields (from Mathers et al., 1971).

Tillage depth	Yields†	
	Root	Sugar
cm	tons ha^{-1}	kg ha^{-1}
20	41.3	6,900
40	49.6	8,280
60	45.1	7,350
80	44.9	7,480

† Based on the average for 0, 112, and 224 kg ha^{-1} N treatments.

ghum. For greatest ET water-use efficiency by grain sorghum, Musick and Dusek (1971, 1975) showed that the slowly permeable horizon of Pullman clay loam should not be completely penetrated by deep tillage. Complete penetration allowed excessive infiltration and deep percolation of irrigation water.

In 1967, adequately irrigated sugarbeets yielded significantly more roots and sugar with 40-cm-deep tillage than with conventional tillage (Table 15). Tillage to 60- and 80-cm depths did not further increase yields (Mathers et al., 1971).

In 1964, Eck and Taylor (1969) modified Pullman clay loam to 90- and 150-cm depths with a wheel-type ditching machine. Modification treatments significantly increased sorghum grain yields with limited irrigation but not with full irrigation. With modification, more water was stored deeper in the soil. Greater water storage, coupled with depletion to a depth of about 240 cm on modified plots as compared with about a 150 cm depth on unmodified plots, increased ET water-use efficiencies for grain and for total dry matter production. A subsequent study on these plots showed that modification increased alfalfa yields and ET water-use efficiencies. Because water infiltrated more rapidly and ponding was not a problem, modification also simplified irrigation management for the alfalfa by permitting one large water application between harvests, whereas two smaller irrigations were required on unmodified soil to obtain similar yields (Eck et al., 1977). Water infiltration rates due to modification were still greater in 1976, 12 years after the profile was modified.

The profile-modification treatments significantly decreased soil bulk density and resistance to penetration (measured with a penetrometer) within the modified layers (Unger, 1970). Modification also resulted in more uniform particle-size distribution and greater air-filled porosity, and it tended to increase the hydraulic conductivity of soil cores. Water retention and plant-available water were not altered by the treatments, but water infiltrated much more rapidly into modified than into unmodified field plots.

Freeman silt loam (fine, mixed, mesic Mollic Haploxeralf) in eastern Washington and northern Idaho was modified by backhoe mixing or moldboard plowing (Cary et al., 1967; Mech et al., 1967). The soil has a silt loam A horizon about 30 cm thick and a well-developed A2 horizon from the 30- to 46-cm depth overlying a dense silty clay loam B horizon. Loosening and mixing the B horizon increased root proliferation and withdrawal of water in the deeper depths. With backhoe mixing, wheat used 11.4 cm of water from the top 150 cm of soil with conventional tillage, and 14.5, 29.0, and 25.0 cm, respectively, with mixing to 46 cm, 120 cm (topsoil and subsoil removed and replaced in original position), and 120 cm (topsoil and subsoil mixed and replaced). Profile mixing to 120 cm increased wheat yields by almost one-half (2,000 vs. 2,880 kg ha^{-1}) and almost doubled 3-year average alfalfa hay yields (3,230 vs. 5,940 kg ha^{-1}). Moldboard plowing (90 cm deep) slowed water infiltration rate because it deposited the clayey subsoil on the soil surface. However, the deep-plowed soil stored 5.3 cm more water from precipitation in the surface 90 cm than did conventionally plowed soil. Mech et al. (1967) hypothesized that water entering the conventionally plowed soil may have been lost by seepage along the A2 horizon. Infiltration into the undisturbed B2 horizon, with the A and A2 horizons removed, dropped to <0.25 mm h^{-1} after it was wetted about 2.5 cm deep.

3. SALT-AFFECTED SOILS

Salt-affected soils occur in subhumid to arid regions of the United States and Canada. Although their characteristics vary, these soils generally have a dense surface layer that is high in adsorbed or exchangeable Na. Beneath this layer is a saline to moderately saline layer that may also contain gypsum ($CaSO_4$).

In a field study started on the Chilcott-Sebree soil complex in Idaho in 1959, treatments were: subsoiling 71 cm deep, moldboard plowing 61 or 76 cm deep, and a check. On Sebree soil (fine-silty, mixed, mesic Xerollic Nadurargid), the effect of added gypsum (27 tons ha^{-1}) was also evaluated (Rasmussen et al., 1964). The Sebree is a saline-sodic soil that occurs in close association with less salt-affected or nonsaline soils of the Chilcott (fine, montmorillonitic, mesic Abruptic Xerollic Durargid) and other series. Plant growth on Sebree soil is very poor because of unfavorable soil physical conditions and extremely low water infiltration during irrigations. The low infiltration results from the deteriorated soil structure caused by the high exchangeable Na, primarily in the shallow clayey subsoil. Deep tillage greatly increased the rate and depth of water infiltration in 24 hours on both soils. Subsoiling increased infiltration on Chilcott soil, but not on Sebree

Table 16—Effects of deep tillage of soils of the Sebree-Chilcott complex in Idaho on wheat grain and alfalfa yields (from Rasmussen et al., 1964).

Treatment	Yields	
	Wheat	Alfalfa
	kg ha^{-1}	tons ha^{-1}
Chillcott		
Check	1,190	9.9 bc†
Subsoiled 71 cm	1,820	10.9 abc
Plowed 61 cm	2,500	12.7 ab
Plowed 76 cm	2,080	13.2 a
Sebree		
Check	780	3.4 d
Gypsum	990	9.7 c
Subsoiled 71 cm	1,050	3.7 d
Subsoiled 71 cm + gypsum	1,810	12.4 abc
Plowed 61 cm	1,860	11.7 abc
Plowed 61 cm + gypsum	2,580	12.2 abc
Plowed 76 cm	1,970	12.2 abc
Plowed 76 cm + gypsum	2,510	12.9 a

† Alfalfa yield values for both soils followed by the same letter(s) are not significantly different at the 5% level (Duncan's Multiple Range Test).

soil, after the first year. Applying gypsum, with and without subsoiling, increased infiltration on Sebree soil. Associated with the increased infiltration were increased leaching, water storage, root penetration, and crop yields (Table 16).

The Rhoades silt loam-silty clay complex (fine, montmorillonitic Leptic Natriboroll) in western North Dakota has a shallow, dense sodic claypan. Plowing these soils 30 or 60 cm deep and adding S or gypsum increased barley yields only slightly, but greatly increased wheat yields, as compared with plowing 15 cm deep (Sandoval et al., 1972). Average barley yields were 2,230, 2,290, and 2,390 kg ha^{-1} with 15-, 30-, and 60-cm-deep plowing, respectively. Wheat yields for the respective treatments were 1,310, 1,760, and 2,110 kg ha^{-1}. The yield increases were attributed to greater soil water availability and to increased soil leaching. Sulfur and gypsum affected yields mainly with 15-cm-deep plowing, but the increases were not as large as for the plowing treatments.

Bowser and Cairns (1967) deep-tilled Duagh silt loam and silty clay loam soils in Alberta, Canada, which have a hydraulic conductivity, usually, <0.3 mm h^{-1}, an electrical conductivity of 10 mmhos cm^{-1}, and a profile that contains considerable amounts of gypsum, Na_2SO_4 and $MgSO_4$. The soil was tilled 56 cm deep with a moldboard plow in the fall of 1959. Deep tillage increased the infiltration rate and greatly decreased the total salt content of the soil. In 1966, soluble Na contents in the profile averaged 4.3 and 0.34 meq 100 g^{-1} in check and deep-tilled plots, respectively. Deep tillage had less effect on soluble Ca and Mg contents. Average yields of wheat and barley from 1960 to 1965 were similar on deep-tilled and check plots (2,320 and 2,400 kg ha^{-1}, respectively). Alfalfa and bromegrass were grown on the plots in 1966. Alfalfa rooted much deeper in deep-tilled than in check plots. Because of the more extensive root system and greater plant growth, it also extracted significantly greater amounts of water.

Robinson and Luthin (1968) chiseled and slip-plowed a nonstratified clay at 56- and 120-cm depths, respectively, in an area that had tile drains at a 1.8-m depth at 36-m intervals. Infiltration rates were 11.6, 12.0, and 14.8 mm d^{-1} for normal tillage, chiseling, and slip-plowing treatments, respectively, when water was ponded on the soil. The greater infiltration with slip-plowing resulted in the greatest decrease in electrical conductivity at the midpoint between tile drains. In the tile trench, untreated and chiseled areas had the lowest conductivities. Apparently, slip plowing opened the soil profile enough so that leaching from all points at the same depth proceeded at similar rates.

Kaddah (1976) chiseled and slip-plowed Rositas loamy fine sand (mixed, hyperthermic Typic Torripsamment), which overlies Imperial silty clay (fine, montmorillonitic (calcareous), hyperthermic Vertic Torrifluvent), 90 cm deep in one or two directions and compared this with disking 20 cm deep. Soil bulk density between plowed slits and in many slits did not differ greatly among treatments. Penetration tests showed similar soil strengths at the 0- to 30-cm depths in all plots, but lower strengths at the 30- to 90-cm depths in slip-plowed than in disked plots. Examination of plant rooting in open pits after harvest showed greater root penetration and density in deeply loosened soil than in disked soil, and along slits as compared with areas between slits. Chiseling and slip-plowing significantly increased wheat grain and straw yields as compared with disking.

D. Previous Crops

Crop rotations involving grasses and legumes have long been recognized for increasing soil aggregation and maintaining higher organic matter (OM) contents than continuous growing of row crops (Johnston et al., 1943; Mazurak et al., 1955; Van Bavel & Schaller, 1951; Wilson & Browning, 1946). For a 10-year rotation on Marshall silt loam (fine, mixed, mesic Typic Hapludoll), aggregates were largest with continuous bluegrass and were successively smaller after red clover, oats, and corn in rotation and after continuous corn (Johnston et al., 1943). The clover maintained a loose, granular structure, whereas continuous corn resulted in a cloddy soil that was difficult to manage. With continuous corn, OM content decreased from 3.39% in 1931 to 2.86% in 1942. Organic matter contents with the rotation and continuous bluegrass were similar. Associated with the larger aggregates and higher OM contents were less runoff and soil erosion. Precipitation runoff was as follows: bluegrass—2%, clover—2%, oats—10%, rotation corn—12.5%, and continuous corn—18%. Average annual soil losses, in the same order as for runoff, were 0.04, 0.2, 24, 43, and 88 tons ha^{-1}. With limited water, yields of rotation and continuous corn were similar, but with adequate water, yields were higher with rotation corn.

Similar results were reported by Van Bavel and Schaller (1951) and Wilson and Browning (1946).

When row crops replaced sod crops, aggregation and infiltration declined and soil losses generally increased (Adams, 1974; Jensen & Sletten, 1965; Mazurak & Ramig, 1963; Van Bavel & Schaller, 1951). The residual effect on aggregation increased with age of sod before plowing. Replacing grain crops with grasses increased soil aggregation and water infiltration, which generally improved with the age of sod (Mazurak & Conard, 1959; Mazurak & Ramig, 1962; Mazurak et al., 1960). Cool-season grasses, as a group, more favorably affected soil aggregation and water infiltration than warm-season grasses (Mazurak & Conard, 1959). About 4 years in sod were needed before substantial increases in water infiltration were measured (Mazurak et al., 1960).

Infiltration and subsequent soil water contents are also influenced by harvest dates, length of the nongrowing period, and rooting and water-extraction patterns of previous crops. Laws and Brawand (1969), at Renner, Tex., measured higher soil water contents, especially at increasing soil depths, when cotton followed wheat than when cotton followed cotton, clover, sorghum, or corn. The difference resulted largely from the wheat plots being fallowed from harvest in June until cotton planting the next April. The other crops were harvested later than wheat, thus there was less time for water storage between crops.

E. Crop Residues and Growing Crops

Crop residues and growing crops protect soil surfaces from dispersion due to raindrop impact and from flowing water. The latter may be a factor with irrigation or when precipitation causes runoff. By protecting soil against dispersion, the favorable surface structure is maintained, which decreases surface sealing and, therefore, permits more rapid infiltration than a dispersed, sealed surface layer.

Surface protection against dispersion is largely a function of coverage provided by surface materials. Because different materials have different densities and diameters, equal weights of them provide different amounts of surface coverage and, thus, differently influence infiltration (Van Doren & Allmaras, 1978). For example, wheat straw is about three times more effective than German millet straw and seven times more effective than grain sorghum stalks per unit weight of material for preserving infiltration. As for residues, growing crops provide protection that is related to the extent of surface coverage.

The influence of residues on infiltration is highly dependent on the surface condition when the residues are applied. A straw mulch at 4,500 kg ha^{-1} applied to a loose, porous surface decreased runoff 75% as compared with runoff from a similar but unprotected surface. On a crusted soil, the same amount of straw decreased runoff only 13% (Borst & Woodburn, 1942).

V. WATER RETENTION IN SOIL PROFILE

Frequently, the water storage capacity of soils is too limited for good plant growth to continue until the next rewetting. Effective water storage is limited by storage volume, low water-holding capacity, and high evaporation. Sometimes, the amount of water retained in soil is too high for effective crop growth. Under these conditions, crop performance is better when adapted crops are grown or if the soil is drained.

A. Increasing Storage Volume

The amount of water retained in soil is directly affected by, among other factors, the volume of soil available for storing water. Soils underlain at shallow depths by natural or tillage-induced layers that prevent or impede water infiltration greatly decrease water availability for subsequent plant growth. Soils with dense clay in the profile, such as the montmorillonitic Pullman clay loam in Texas, are slowly recharged with water. Although their potential water-holding capacities are relatively high, these soils are only filled to capacity during extensive wet periods or with frequent irrigation. To increase the volume of soil available for water storage and to recharge the soil more readily, the restricting layers must be loosened to depths compatible with water recharge frequencies, plant rooting requirements, and economics of the treatments.

High yields can be obtained on relatively shallow soils if plant water and nutrient requirements can be met. In irrigated regions, this can be done through frequent irrigation and fertilization. Under rainfed cropping, however, greater soil volumes generally are desirable for storing more water, which decreases the potential for plant water stress between rains if the plants root deeply enough to extract the deeply stored water. Deeply loosening a soil has little or no value if the crop's roots do not grow to depths greater than the restricting layer, unless the loosening increases soil aeration that is limiting root growth.

The economics of deeply loosening a soil for increasing the water storage volume are influenced by the crops grown, the water supply, and the nature of the restricting layer. As crop values and yields increase, more extensive soil loosening becomes economical. An extreme example is the ripping of indurated hardpans for crops in California (Calif. Builder & Eng., 1975).

B. Increasing Water Retention

Sandy soils have low water-holding capacity, but crop yields are often favorable on these soils because they have higher water infiltration rates and, therefore, are more readily refilled with water than are fine-textured soils. To use water more efficiently on sandy soils, treatments that increase water retention within the root zone should be beneficial.

1. ADDING FINE-TEXTURED MATERIALS

Mixing fine-textured materials from outside sources or from greater depths within the profile with sandy surface horizons can increase the water-holding capacity of the sandy surface soil. Adding materials from outside sources may be practical in limited areas, but not in large areas because of the large amount of material needed and the expenses involved in transporting the materials. Mixing finer materials from deeper in the profile with the coarser materials near the surface is, however, within the realm of practicality. The water-holding capacity of the entire profile may not be increased, but soil near the surface should hold more water, thus improving seedling establishment and early growth.

The Hezel (sandy over loamy, mixed, nonacid, mesic Xeric Torriorthent) and similar soils in central Washington have coarse, wind-deposited materials overlying high-silt, water-deposited materials. The Hezel soil is sandy loam to the 46-cm depth and silt loam at the 46- to 90-cm depth. In 1970, Miller and Aarstad (1972) moldboard-plowed the soil 100 cm deep in an attempt to increase the water-holding capacity by bringing finer material to the surface and by turning under the coarse surface material. After plowing, the texture was silt loam at the 0- to 15-cm depth and loam at the 15- to 30-cm depth. The available water-holding capacity of the surface 30 cm of soil was 36 mm before plowing and 61 mm after plowing, a 69% increase, which was largely due to the change in texture. The sand content of the surface layer was reduced from about 70 to 40 or 50%. Below the 30-cm depth, the water-holding capacities were similar for unplowed and deep-plowed soil.

Increasing the water-holding capacity and plant nutrient supply of the surface soil layer and decreasing the wind erosion potential were goals of deep-plowing studies on sandy soils at five locations in Oklahoma (Harper & Brensing, 1950). Before plowing, the surface soil was 87–95% sand and 3–4% clay. Initial subsurface soil clay contents and plowing depths were not specified, but plowing was about 25–40 cm deep. After plowing, the surface soil was 65–83% sand and 10–18% clay. Deep plowing generally increased crop yields, but the reason was not readily apparent because soil water and plant nutrient contents were not measured. Undoubtedly, surface soil water content was a factor because Unger (1975) showed that plant-available water (based on −0.33- and −15-bar metric potentials) in soil cores increased about 0.09% by weight for each percentage unit increase in soil clay content. For sieved soil, the increase was about 0.28% per percentage unit. This suggested that modification treatments which thoroughly disrupt the structure would increase the water-holding capacities of soils over those of unmodified soils. Such thorough disruption, however, is not practical under large-scale field conditions.

Sand, silt, and clay distribution in a profile after deep plowing depends largely upon their initial distribution and the type of implement used. Miller and Aarstad (1972) found water retention below the surface layer to be highly variable because this layer was not cross-tilled after deep plowing. Harper and Brensing (1950) advised against deep plowing sandy soils that had subsurface layers with very high clay contents because it places too much clay in the surface layer, which could decrease water infiltration. This condition, however, can be avoided by shallower plowing or by more thoroughly mixing the soil when plowing. For even greater homogenization, Eck and Taylor (1969) modified Pullman clay loam to 120- or 150-cm depths with a wheel-type ditching machine, which caused more uniform sand, silt, and clay distributions than in the unmodified soil (Unger, 1970).

2. ADDING ORGANIC MATERIALS

Increasing a soil's OM content has long been considered effective for increasing its available water-holding capacity. Jamison (1953) found a high positive correlation between water retention and OM content for sandy soils that are $< 15\%$ clay. The benefits resulted from diluting the droughty sandy material with material having a greater water-retention capacity. Unger (1975), using multiple regression analyses, showed that plant-available soil water (volume basis) increased about 1.8 percentage units for each percentage-unit increase in OM content for soils ranging in texture from sand to clay.

Increasing the OM content of fine-textured soils does not increase the available water-storage capacity as much as it does that of sandy soils. The benefits on fine-textured soils result from improved structure, which decreases crusting and surface sealing, permits greater water infiltration, and increases root proliferation throughout the soil profile (Jamison, 1953). Although available water-storage capacity per se is not increased, the additional OM does result in plants using the water more effectively.

Although water retention in sandy soils increases with increased OM (Jamison, 1953; Peele et al., 1948), the gain in available water retention by adding OM to sandy soils may be small (Allison, 1973). Even small gains are important, however, because the retention capacity of these soils is low. For example, a one-percentage-unit increase in OM content, which increases available water retention by 1.8 percentage units (Unger, 1975), increases the available water retained in sandy soil from 2.3 to 4.1 percentage units, or 78%. The problem, however, is in increasing the OM content enough to be of much value. To raise the soil OM content by one percentage unit to a 30-cm depth requires adding about 45 tons ha^{-1} of dry material, an amount greatly above annual crop production. Smaller additions result in smaller increases in water retention, but may be practical if the materials are grown in place. Returning most or all residues from well-managed, high-residue crops to the soil should maintain or gradually increase the soil OM content, thus increasing water storage capacity over a long period. Additional benefits may result from decomposing roots and adding manure, woody materials (bark, chips, sawdust), peat, and composts (Allison, 1973). Because not all added materials become stable OM, repeated applications are needed.

C. Decreasing Deep Percolation (Sandy Soils)

Water retention in deep sandy soils can be increased by installing barriers that restrict the downward movement of water. The theory concerning the influence of impermeable subsurface barriers on water movement in sandy soils was discussed by Erickson et al. (1968). Water near the edges of the barrier rapidly drains deeply into the soil, thus creating capillary discontinuities that further slow the drainage of low-tension water from above the barrier.

In 1966, Erickson et al. (1968) installed asphalt barriers at 55- to 60-cm depths in sandy soils at two locations in Michigan. Vegetables were then grown with and without sprinkler irrigation on the treated and control areas. During the growing season, the water content at the 30- to 60-cm depth was greater above the barriers, except for the nonirrigated plots in Montcalm County, where not enough rain fell to fill to capacity the water-storage reservoir above the barrier. The barriers had little effect on soil water content at the 0- to 15-cm depth because water readily drained below this depth. In Allegan County, 85 mm of rain fell on 15 August. After 3 days, soil at the 30- to 60-cm depth contained 25 mm more water with than without the barrier.

The differences in rainfall at the two locations resulted in different responses in vegetable yields (Table 17). Favorable rainfall in Allegan County increased cucumber and cabbage yields with the barriers, but irrigation decreased cabbage yields, possibly because of nutrient leaching. With less favorable rainfall in Montcalm County, the barriers did not increase potato and bean yields without irrigation (Erickson et al., 1968). These results showed that the extra water-retention capacity increased yields when the rainfall distribution or irrigation frequency was such that the increased storage capacity decreased drought injury to crops.

Table 17—Effects of asphalt barriers and irrigation on crop yields on sandy soil with favorable and unfavorable rainfall in Michigan (from Erickson et al., 1968).

			Treatments			
Rainfall	Location	Crop	Control	Control with irrigation	Asphalt layer	Asphalt layer with irrigation
			Yield, tons ha^{-1}			
Favorable	Allegan County, Mich.	Cucumbers	18.7 b†	19.5 b	24.9 a	25.5 a
		Cabbage	23.0 b	18.5 c	32.7 a	23.8 b
Unfavorable	Montcalm County, Mich.	Potatoes	24.3 b	37.3 a	25.6 b	38.6 a
		Beans	13.8 b	25.1 a	14.4 b	25.3 a

† Row values followed by the same letter are not significantly different at the 5% level (Duncan's Multiple Range Test).

An asphalt barrier at a 60-cm depth in Lakeland fine sand (thermic, coated Typic Quartzipsamment) in Florida decreased rooting depth of tomato and corn (Saxena et al., 1973). However, the soil water reservoir within the 60-cm root zone was about 110 mm, the same as in 90 cm of soil without asphalt. The water above the asphalt was maintained at higher potentials, which, along with increased root concentration, increased the availability of water to plants. The extra water generally increased yields (Saxena et al., 1969, 1971).

In 1971, corn on the asphalt-barrier plots yielded more fodder than on check and irrigation-treatment plots during the first 11 weeks when water stress was moderate. Then an 11-day drought caused severe plant water stress on check and asphalt-treatment plots (without irrigation), which significantly decreased final fodder and grain yields. The asphalt decreased the need for supplemental water and increased rainfall efficiency in the well-drained sandy soil (Robertson et al., 1973). The asphalt barrier still effectively increased water retention in the soil after 4 years (Saxena et al., 1973).

D. Controlling Evaporation

Evaporation accounts for the major loss of water from many cultivated soils. In the Great Plains, for example, about 60% of the 50 cm of average annual precipitation evaporates directly from soil (Bertrand, 1966). As plant canopies develop, evaporation decreases and transpiration increases.

Soil water evaporation occurs in three stages (Lemon, 1956; Chapter 2B, this volume). In the first stage, water loss is rapid and steady. It depends on the net effects of water transmission to the surface and on environmental conditions, such as wind speed, temperature, relative humidity, and radiant energy. During the second stage, the loss rate decreases rapidly as the soil water supply is depleted. Soil factors control the rate of water movement to the surface and aboveground factors have little influence. Evaporation during the third stage is extremely slow and is controlled by adsorptive forces at the liquid-solid interface.

According to Lemon (1956), the greatest potentials for decreasing soil water evaporation lie within the first two stages. Potential methods include (i) decreasing turbulent transfer of water vapor to the atmosphere, (ii) decreasing capillary continuity, and (iii) decreasing capillary flow and water-holding capacity of surface soil layers.

1. MULCHES

Materials such as crop residues, plastic films (including polyethylene, polyvinyl chloride, etc.), petroleum-based products, gravel, and soil itself have been widely studied as potential mulches for decreasing evaporation. The effect of mulches on evaporation is, however, difficult to establish because of the interacting influences of mulches on soil water infiltration, distribution, and subsequent evaporation. Consequently, higher water contents resulting from surface mulches may be due to lower evaporation,

Table 18—Effect of tillage method on average precipitation storage, sorghum grain yield, and grain-yield ET water-use efficiency for the sorghum crop in an irrigated-wheat-dryland-grain-sorghum cropping system in Texas (from Unger & Wiese, 1979).

Tillage method	Precipitation storage	Grain yield	Grain yield ET water-use efficiency
	%	kg ha^{-1}	kg ha-cm^{-1}
No-tillage	35	3,140	89
Sweep	23	2,500	77
Disk	15	1,930	66

Precipitation was: fallow—34.8 cm; growing season—26.4 cm.

but water infiltration and distribution may be involved also, especially in field studies where there is little control over soil wetting by precipitation.

In field studies in Colorado, Montana, and Nebraska, the amount of precipitation stored as soil water during fallow ranged from 16% of the total with no residues to 34% with 11 tons ha^{-1} of wheat straw on the surface (Greb et al., 1967). When Unger (1978) placed wheat straw on Pullman clay loam at Bushland, Tex., precipitation storage during fallow ranged from 23% with no mulch to 46% with 12 tons mulch ha^{-1}. Dryland grain sorghum after fallow yielded 1,780 and 3,990 kg ha^{-1} with the 0 and 12 ton ha^{-1} mulch treatments, respectively. Unger and Wiese (1979) at Bushland used no-, sweep, and disk tillage for residue management and weed control from irrigated winter wheat harvest until dryland sorghum planting 11 months later. Precipitation storage during fallow, sorghum grain yields, and grain-yield ET water-use efficiency were highest with no-tillage and lowest with disk tillage (Table 18). Although water infiltration and distribution as well as evaporation control undoubtedly were involved, these studies showed that wheat residues maintained on soil as a mulch can greatly increase precipitation storage and crop yields.

Materials such as gravel, petroleum-coated soil particles, and, to some extent, crop residues move the zone of water retention from the air-mulch interface to the mulch-soil interface and result in a condition conducive to lower evaporation than that which occurs when the water is mainly retained at the air-soil interface. Gravel retains little water; thus, the water readily enters the soil. Use of gravel mulches on large areas, however, is not practical. Crop residues absorb considerable amounts of water and, therefore, may not be as effective as gravel mulches, especially when the water is added in small amounts as by light rains. Crop residues, however, are more abundant and practical than gravel as mulches. The generally greater thickness of residue mulch than of gravel mulch tends to minimize the difference in their effectiveness.

Plastic films are highly effective for decreasing evaporation because the path of vapor flow to the air is blocked (Willis, 1962; Willis et al., 1963). Plastic films, however, cause water entry problems unless special provisions have been made for this purpose. Also, plastic films are expensive and difficult to manage. They are generally impractical, except for high-value crops.

Mulches of dry soil decreased evaporation more than gravel and corn cob mulches in one study (Benoit & Kirkham, 1963) and caused an evaporation reduction similar to straw in another study (Gerard et al., 1959). The different responses with a soil mulch may have resulted from differences in amounts of the other materials used, times and methods of mulch establishment, evaporativity, and soil type (Gill et al., 1977). For maximum effectiveness, the soil mulch must be established before the soil dries appreciably. This usually is not practical under field conditions and the soil mulch must be re-established after each rainstorm.

Mulch effectiveness for decreasing evaporation increases as mulch thickness increases (Bond & Willis, 1969; Hanks & Woodruff, 1958; Unger & Parker, 1976). Because material density greatly influences mulch thickness, low-density materials, such as wheat straw, more effectively decrease evaporation than sorghum stubble or cotton stalks, which are more dense. For similar evaporation decreases, twice as much sorghum stubble and four times as much cotton stalks were needed as compared with wheat straw on a weight basis (Unger & Parker, 1976).

Crop residue mulches effectively reduce evaporation during the first stage (Bond & Willis, 1969; Unger, 1976; Unger & Parker, 1976). However, for long-term effectiveness, enough water must be added to penetrate deeply into the soil profile or large amounts of residue must be present (Bond & Willis, 1971; Gardner & Gardner, 1969; Unger, 1976). Since the water-content difference between mulched and bare soil is mainly near the soil surface, especially soon after water additions, mulches are very useful for seedling establishment (Bertrand, 1966).

Surface mulches of crop residues and similar materials, as compared with bare soil, will result in lower soil temperatures during some seasons of the year. Factors involved are radiation reflectance and possibly soil insulation (Unger, 1978; Van Doren & Allmaras, 1978). Seemingly, the lower temperatures should result in less evaporation from the soil. However, Hanks et al. (1961) found no direct relationship between soil temperature and evaporation. Apparently, water movement within the soil after surface drying had a greater influence on evaporation than did soil temperature. Although evaporation may not be affected, mulch-related soil temperatures have a marked influence on root and top growth of plants (Allmaras et al., 1964; Allmaras & Nelson, 1973; Van Doren & Allmaras, 1978). Within limits, growth is generally more rapid in warm soil than in cold soil.

2. SURFACTANTS

Surfactants may decrease evaporation by decreasing capillary rise of water to the surface. The surfactants probably cause decreased surface tension at the solid-liquid interface, thereby decreasing capillary flow of water and causing the formation of a dry diffusion barrier (Law, 1964). Materials frequently used as surfactants are fatty alcohols and nonionic compounds.

As compared with an untreated check, a fatty alcohol applied at a rate of 450 kg ha^{-1} to washed sand decreased evaporation by 87%. When applied to Amarillo fine sandy loam, the most effective rate was 900 kg ha^{-1}, but the evaporation decrease was only 24% as compared with that from untreated soil. A nonionic material applied to the soil as a 0.1% solution in water decreased evaporation 52% (Law, 1964).

Olsen et al. (1964) found that a fatty alcohol was most effective and enduring for decreasing evaporation when applied to the surface of Weld loam (fine, montmorillonitic, mesic Aridic Paleustoll). During the first 10 days, evaporation decreased 43% as compared with that from bare soil. In another study, Olsen et al. (1962) found that adding fatty alcohols (hexadecanol and octadecanol) to Weld loam did not decrease transpiration water use per unit of biomass production by corn. However, transpiration per unit of biomass produced increased with rate of alcohol addition when mixed with soil, but not when banded. Both alcohols significantly decreased the yield of corn when mixed with soil at a rate of 25 g per 3 kg of soil. When banded, octadecanol decreased yield as compared with that from untreated soil.

Seemingly, the use of surfactants for making more efficient use of water has limited potential. Although evaporation was decreased, total water use by crops was not affected and yields were decreased. Also, the materials had limited stability in soil (Olsen et al., 1962, 1964), and thus would have to be reapplied periodically, possibly after each tillage operation.

3. OTHER METHODS

As pointed out earlier, evaporation from soil decreases as plant canopies increase. The decrease in evaporation results from decreased wind movement and air turbulance at the soil surface and from interception of radiant energy by the plants. Other materials for decreasing wind movement and air turbulence include windbreaks—such as trees, grasses, and other crops—and snow fences, etc. Their effectiveness for decreasing evaporation, however, was questioned by Bertrand (1966) because the materials or devices also served as heat traps, which increased evaporation. While windbreaks decreased evaporation (Aase & Siddoway, 1976; Pelton, 1967; Staple & Lehane, 1955), their major benefit in northern areas apparently results from trapping snow, which can increase soil water contents (Frank & Willis, 1978).

E. Drainage

A crop such as rice can tolerate saturated soils and is frequently grown partially submerged in water. For such crops, restricted percolation and drainage are desirable for decreasing the total water requirement and for minimizing weed problems. For most crops, however, poorly drained soil conditions are detrimental to good growth and yields, and drainage is a highly important management practice. Soil drainage is achieved by surface channels, tile drains, or loosening of impervious layers in the profile.

Drainage of agricultural lands has been discussed in detail in monographs edited by Luthin (1957) and van Schilfgaard (1974).

VI. SUMMARY

The time and areal distribution of precipitation is highly variable in most areas of the United States. This, coupled with soil factors, often results in inefficient use of precipitation for crop production. Soil factors involved are those that affect runoff and infiltration, water distribution and retention in the profile, water evaporation and percolation, and growth and proliferation of roots within the profile to absorb the soil water.

Alteration of precipitation patterns is largely beyond the control of man, but soil management may have a major influence on how effectively the available precipitation, or even irrigation, is used for crop production. One of the foremost requirements for more efficiently using precipitation is retaining the water where it falls so that it will infiltrate and replenish the soil water storage reservoir. Other factors include increasing the soil water storage capacity, reducing water percolation and evaporation, and altering the soil physical and chemical conditions so that plant roots will proliferate within a larger soil volume.

Runoff, infiltration, distribution, and retention of water are influenced by surface conditions and by inherent and induced profile conditions. These can be altered by tillage or modification of surface and subsurface layers, management of crop residues, and use of engineering practices. Management practices that affect the aggregation, wettability, and organic matter content of the soil also influence runoff, infiltration, distribution, and retention of water. Mulches, surfactants, and plant canopies affect soil water evaporation.

In Chapter 10A, we described and discussed practices that have increased the amount of water available for plant growth or have removed physical limitations to root growth so that potentially available soil water could be used by plants. In Chapter 10B, Power discusses the effects of soil fertility on water-use efficiency. Reicosky, in Chapter 10C, emphasizes the effects of profile modification on plant growth and yield on the acid and high-strength soils of the southeastern United States.

LITERATURE CITED

Aarstad, J. S., and D. E. Miller. 1973. Soil management to reduce runoff under center-pivot sprinkler systems. J. Soil Water Conserv. 28:171–173.

Aase, J. K., and F. H. Siddoway. 1976. Influence of tall wheatgrass wind barriers on soil drying. Agron. J. 68:627–631.

Adams, J. E. 1974. Residual effects of crop rotations on water intake, soil loss, and sorghum yield. Agron. J. 66:299–304.

Allen, R. R., J. T. Musick, and A. F. Wiese. 1975. No-till management of furrow irrigated continuous grain sorghum. Texas Agric. Exp. Stn. Prog. Rep. PR-3332C, College Station. p. 66–78.

Allison, F. E. 1973. Soil organic matter and its role in crop production. Elsevier Scientific Publishing Co., New York. p. 346–359, 417–444.

Allmaras, R. R., W. C. Burrows, and W. E. Larson. 1964. Early corn growth as affected by soil temperature. Soil Sci. Soc. Am. Proc. 28:271–275.

Allmaras, R. R., and W. W. Nelson. 1973. Corn root-configuration response to soil temperature and matric suction. Agron. J. 65:725–730.

Armbrust, D. W., and N. H. Welch. 1966. Evaluation of Zingg conservation bench terraces on Amarillo fine sandy loam soil. J. Soil Water Conserv. 21:224–226.

Barnett, A. P., J. R. Carreker, F. Abruna, W. A. Jackson, A. E. Dooley, and J. H. Holladay. 1972. Soil and nutrient losses in runoff with selected cropping treatments in tropical soils. Agron. J. 64:391–395.

Batchelder, A. R., and J. N. Jones, Jr. 1972. Soil management factors and growth of *Zea mays* L. on topsoil and exposed subsoil. Agron. J. 64:648–652.

Benoit, G. R., and D. Kirkham. 1963. The effect of soil surface conditions on evaporation of soil water. Soil Sci. Soc. Am. Proc. 27:495–498.

Bertrand, A. R. 1966. Water conservation through improved practices. p. 207–235. *In* W. H. Pierre et al. (ed.) Plant environment and efficient water use. Am. Soc. of Agron., Madison, Wis.

Bond, J. J., and W. O. Willis. 1969. Soil water evaporation: surface residue rate and placement effects. Soil Sci. Soc. Am. Proc. 33:445–448.

Bond, J. J., and W. O. Willis. 1971. Soil water evaporation: long-term drying as influenced by surface residue and evaporation potential. Soil Sci. Soc. Am. Proc. 35:984–987.

Bond, R. D. 1972. Germination and yield of barley when grown in a water-repellent sand, Agron. J. 64:402–403.

Borst, H. L., and R. Woodburn. 1942. Effect of mulches and surface conditions on the water relations of Muskingum soils. USDA Tech. Bull. 825. U.S. Government Printing Office, Washington, D.C.

Bowser, W. E., and R. R. Cairns. 1967. Some effects of deep plowing a Solonetz soil. Can. J. Soil Sci. 47:239–244.

Bradford, J. M., and R. W. Blanchar. 1977. Profile modification of a Fragiudalf to increase crop production. Soil Sci. Soc. Am. J. 41:127–131.

Burnett, E., and C. E. Fisher. 1956. Land leveling increases dryland cotton production. Texas Agric. Exp. Stn. Prog. Rep. PR-1914, College Station.

Burnett, E., and J. L. Tackett. 1968. Effect of soil profile modification on plant root development. Int. Congr. Soil Sci. Trans. 9th (Adelaide, Australia) III:329–337. American Elsevier Publishing Co., Inc., New York.

Burwell, R. E., R. R. Allmaras, and L. L. Sloneker. 1966. Structural alteration of soil surfaces by tillage and rainfall. J. Soil Water Conserv. 21:61–63.

California Builder and Engineer. 1975. Ripping the unrippable. Reprinted from California Builder and Engineer. 14 Mar. 1975.

Cary, E. E., G. M. Horner, and S. J. Mech. 1967. Relationship of tillage and fertilization to the yield of alfalfa on Freeman silt loam. Agron. J. 59:165–168.

Clark, R. N., and E. B. Hudspeth. 1976. Runoff control for summer crop production in the Southern Plains. Paper no. 76-2008. Am. Soc. Agric. Eng., St. Joseph, Mich.

Daniel, H. A. 1951. Water conservation for wheat production in Oklahoma. Soil Sci. Soc. Am. Proc. 15:408–412.

Debano, L. F., S. M. Savage, and D. A. Hamilton. 1976. The transfer of heat and hydrophobic substances during burning. Soil Sci. Soc. Am. J. 40:779–782.

De Roo, H. C. 1961. Deep tillage and root growth. Connecticut Agric. Exp. Stn. Bull. 644, Storrs.

Dickson, R. E., B. C. Langley, and C. E. Fisher. 1940. Water and soil conservation experiments at Spur, Texas. Texas Agric. Exp. Stn. Bull. no. 587, College Station.

Donahue, R. L., R. W. Miller, and J. C. Shickluna. 1977. Soils—an introduction to soils and plant growth. 4th ed. Prentice-Hall, Inc., Englewood Cliffs, N.J. p. 155–171, 302–320, 463–490.

Eck, H. V., T. Martinez, and G. C. Wilson. 1977. Alfalfa production on a profile modified slowly permeable soil. Soil Sci. Soc. Am. J. 41:1181–1186.

Eck, H. V., and H. M. Taylor. 1969. Profile modification of a slowly permeable soil. Soil Sci. Soc. Am. Proc. 33:779–783.

Erickson, A. E., C. M. Hansen, and A. J. M. Smucker. 1968. The influence of subsurface asphalt barriers on the water properties and the productivity of sand soils. Int. Congr. Soil Sci. Trans. 9th (Adelaide, Australia) I:331–337. American Elsevier Publishing Co., Inc., New York.

Evanari, M., L. Shanan, and N. H. Tadmor. 1968. "Runoff farming" in the desert. I. Experiment layout. Agron. J. 60:29–32.

Fairbourn, M. L., and H. R. Gardner. 1972. Vertical mulch effects on soil water storage. Soil Sci. Soc. Am. Proc. 36:823–827.

Fairbourn, M. L., and H. R. Gardner. 1974. Field use of microwatersheds with vertical mulch. Agron. J. 66:740–744.

Faulkner, E. H. 1943. Plowman's folly. University of Oklahoma Press, Norman.

Finnell, H. H. 1944. Water conservation in Great Plains wheat production. Texas Agric. Exp. Stn. Bull. no. 655, College Station.

Fisher, C. E., and E. Burnett. 1953. Conservation and utilization of soil moisture. Texas Agric. Exp. Stn. Bull. no. 767, College Station.

Frank, A. B., and W. O. Willis. 1978. Effect of winter and summer windbreaks on soil water gain and spring wheat yield. Soil Sci. Soc. Am. J. 42:950–953.

Fritton, D. D., and G. W. Olson. 1972. Bulk density of a fragipan soil in natural and disturbed profiles. Soil Sci. Soc. Am. Proc. 36:686–689.

Gantzer, C. J., and G. R. Blake. 1978. Physical characteristics of Le Sueur clay loam soil following no-till and conventional tillage. Agron. J. 70:853–857.

Gardner, H. R., and W. R. Gardner. 1969. Relation of water application to evaporation and storage of soil water. Soil Sci. Soc. Am. Proc. 33:192–196.

Geraghty, J. J., D. W. Miller, F. Van Der Leeden, and F. L. Troise. 1973. Water atlas of the United States. Water Infor. Ctr. Pub., Port Washington, N.Y.

Gerard, C. J., M. E. Bloodworth, and W. R. Cowley. 1959. Effect of tillage, straw mulches and grass upon soil-moisture losses and soil temperatures in the Lower Rio Grande Valley. Texas Agric. Exp. Stn. Misc. Publ. no. 382, College Station.

Gill, K. S., S. K. Jalota, S. S. Prihar, and T. N. Chaudhary. 1977. Water conservation by soil mulch in relation to soil type, time of tillage, tilth and evaporativity. J. Indian Soc. Soil Sci. 25:360–366.

Greb, B. W., D. E. Smika, and A. L. Black. 1967. Effect of straw-mulch rates on soil water storage during summer fallow in the Great Plains. Soil Sci. Soc. Am. Proc. 31:556–559.

Griffin, R. H., II., B. J. Ott, and J. F. Stone. 1966. Effect of water management and surface applied barriers on yields and moisture utilization of grain sorghum in the Southern Great Plains. Agron. J. 58:449–452.

Haas, H. J., and W. O. Willis. 1971. Water storage and alfalfa production on level benches in the Northern Plains. J. Soil Water Conserv. 26:151–154.

Hanks, R. J., S. A. Bowers, and L. D. Bark. 1961. Influence of soil surface conditions on radiation, soil temperature, and evaporation. Soil Sci. 91:233–238.

Hanks, R. J., and N. P. Woodruff. 1958. Influence of wind on water vapor transfer through soil, gravel, and straw mulches. Soil Sci. Soc. Am. J. 86:160–164.

Harper, H. J. 1941. The effect of terrace ridges on the production of winter wheat. Soil Sci. Soc. Am. Proc. 6:474–479.

Harper, J., and O. H. Brensing. 1950. Deep plowing to improve sandy land. Oklahoma Agric. Exp. Stn. Bull. no. B-362, Stillwater.

Harrold, L. L., and W. M. Edwards. 1972. A severe rainstorm test of no-till corn. J. Soil Water Conserv. 27:30.

Harrold, L. L., G. B. Triplett, Jr., and R. E. Youker. 1967. Watershed test of no-tillage corn. J. Soil Water Conserv. 22:98–100.

Hauser, V. L. 1968. Conservation bench terraces in Texas. Trans. ASAE 11:385–386, 392.

Hauser, V. L. and M. B. Cox. 1962. Evaluation of Zingg conservation bench terrace. Agric. Eng. 43:462–464, 467.

Hauser, V. L., and H. M. Taylor. 1964. Evaluation of deep-tillage treatments on a slowly permeable soil. Trans. Am. Soc. Agric. Eng. 7:135–136, 141.

Hauser, V. L., C. E. Van Doren, and J. S. Robins. 1962. A comparison of level and graded terraces in the Southern High Plains. Trans. ASAE 5:75–77.

Hays, O. E. 1961. New tillage methods reduce erosion and runoff. J. Soil Water Conserv. 16: 172–175.

Heilman, M. D., and C. L. Gonzalez. 1973. Effect of narrow trenching in Harlingen clay soil on plant growth, rooting depth, and salinity. Agron. J. 65:816–819.

Hobbs, J. A., R. B. Herring, D. E. Peaslee, W. W. Harris, and G. E. Fairbanks. 1961. Deep tillage effects on soils and crops. Agron. J. 53:313–316.

Hopp, H., and C. S. Slater. 1961. The effect of earthworms on the productivity of agricultural soil. p. 67–83. *In* R. Rodale (ed.) The challenge of earthworm reseach. The Soil and Health Foundation, Emmaus, Pa.

Jamison, V. C. 1943. The slow reversible drying of sandy surface soils beneath citrus trees in Central Florida. Soil Sci. Soc. Am. proc. 7:36–41.

Jamison, V. C. 1947. Resistance to wetting in the surface of sandy soils under citrus trees in Central Florida and its effect upon penetration and the efficiency of irrigation. Soil Sci. Soc. Am. Proc. 11:103–109.

Jamison, V. C. 1953. Changes in air-water relationships due to structural improvement of soils. Soil Sci. 76:143–151.

Jensen, M. E., and W. H. Sletten. 1965. Effects of alfalfa, crop sequence, and tillage practice on intake rates of Pullman silty clay loam and grain yields. USDA-ARS Conserv. Res. Rep. 1. U.S. Government Printing Office, Washington, D.C.

Johnston, J. R., G. M. Browning, and M. B. Russell. 1943. The effect of cropping practices on aggregation, organic matter content, and loss of soil and water in the Marshall silt loam. Soil Sci. Soc. Am. Proc. 7:105–107.

Jones, O. R., and V. L. Hauser. 1975. Runoff utilization for grain production. Water Harv. Symp. Proc., Phoenix, Ariz. March 1974. USDA-ARS W-22. U.S. Government Printing Office, Washington, D.C.

Jones, O. R., and J. L. Shipley. 1975. Economics of land leveling for dryland grain production. J. Soil Water Conserv. 30:177–181.

Kaddah, M. T. 1976. Subsoil chiseling and slip plowing effects on soil properties and wheat grown on a stratified fine sandy soil. Agron. J. 68:36–39.

Kingsley, Q. S., and F. E. Shubeck. 1964. The effects of organic trenching on runoff. J. Soil Water Conserv. 19:19–22.

Kramer, L. A., and L. D. Meyer. 1969. Small amounts of surface mulch reduce soil erosion and runoff velocity. Trans. ASAE 12:638–641, 645.

Larson, W. E. 1962. Advances in minimum tillage for corn. Reprinted from 17th Annu. Hybrid Corn Industry Res. Conf., American Seed Trade Assoc., Washington, D.C. 17:44–51.

Larson, W. E. 1964. Soil parameters for evaluating tillage needs and operations. Soil Sci. Soc. Am. Proc. 28:118–122.

Lattanzi, A. R., L. D. Meyer, and M. F. Baumgardner. 1974. Influences of mulch rate and slope steepness on interrill erosion. Soil Sci. Soc. Am. Proc. 38:946–950.

Law, J. P., Jr. 1964. The effect of fatty alcohol and a nonionic surfactant on soil moisture evaporation in a controlled environment. Soil Sci. Soc. Am. Proc. 28:695–699.

Laws, W. D., and H. Brawand. 1969. The effect of farming systems on soil moisture reserve and cotton production in the Texas Blacklands. Hoblitzelle Agric. Lab. Bull. no. 27. Texas Research Foundation, Renner.

Lemon, E. R. 1956. The potentialities for decreasing soil moisture-evaporation loss. Soil Sci. Soc. Am. Proc. 20:120–125.

Locke, L. F., H. V. Eck, B. A. Stewart, and H. J. Haas. 1960. Plowpan investigations at the Great Plains Field Stations, Woodward, Okla., and Mandan, N. Dak. USDA-ARS Prod. Res. Rep. no. 40. U.S. Government Printing Office, Washington, D.C.

Locke, L. F., and O. R. Mathews. 1953. Relation of cultural practices to winter wheat production, Southern Great Plains Field Station, Woodward Okla. USDA Circ. no. 917. U.S. Government Printing Office, Washington, D.C.

Luthin, J. N. (ed.). 1957. Drainage of agricultural lands. Agronomy 7. Am. Soc. of Agron., Madison, Wis.

Lyle, W. M., and D. R. Dixon. 1977. Basin tillage for rainfall retention. Trans. ASAE 20: 1013–1017.

Mannering, J. V., and L. D. Meyer. 1961. The effects of different methods of cornstalk residue management on runoff and erosion as evaluated by simulated rainfall. Soil Sci. Soc. Am. Proc. 25:506–510.

Martin, W. P. 1953. Status report on soil conditioning chemicals: I. Soil Sci. Soc. Am. Proc. 17:1–9.

Mathers, A. C., G. C. Wilson, A. D. Schneider, and P. Scott. 1971. Sugarbeet response to deep tillage, nitrogen, and phosphorus on Pullman clay loam. Agron. J. 63:474–477.

Mazurak, A. P., and E. C. Conard. 1959. Rates of water entry in the three great soil groups after seven years in grasses. Agron. J. 51:264–267.

Mazurak, A. P., H. R. Cosper, and H. F. Rhoades. 1955. Rate of water entry into an irrigated chestnut soil as affected by 39 years of cropping and manurial practices. Agron. J. 47: 490–493.

Mazurak, A. P., W. Kriz, and R. E. Ramig. 1960. Rates of water entry into a chernozem soil as affected by age of perennial grass sods. Agron. J. 52:35–37.

Mazurak, A. P., and R. E. Ramig. 1962. Aggregation and air-water permeabilities in a chernozem soil cropped to perennial grasses and fallow-grain. Soil Sci. 94:151–157.

Mazurak, A. P., and R. E. Ramig. 1963. Residual effects of perennial grass sod on the physical properties of a chernozem soil. Soil Sci. Soc. Am. Proc. 27:592–595.

Mech, S. J., G. M. Horner, L. M. Cox, and E. E. Cary. 1967. Soil profile modification by backhoe mixing and deep plowing. Trans. ASAE 10:775–779.

Meyer, L. D., and J. V. Mannering. 1961. Minimum tillage for corn: its effects on infiltration and erosion. Agric. Eng. 42:72–75, 86–87.

Mickelson, R. H. 1966. Level pan construction for diverting and spreading runoff. Trans. ASAE 9:568–570.

Mickelson, R. H. 1968. Conservation bench terraces in eastern Colorado. Trans. ASAE 11: 389–392.

Mickelson, R. H., M. R. Cox, and J. Musick. 1965. Runoff water-spreading on leveled cropland. J. Soil Water Conserv. 20:57–60.

Miller, D. E., and J. S. Aarstad. 1972. Effect of deep plowing on the physical characteristics of Hezel soil. Washington Agric. Exp. Stn. Circ. 556, Pullman.

Miller, R. H., and J.F. Wilkinson. 1977. Nature of the organic coating of sand grains of nonwettable golf greens. Soil Sci. Soc. Am. J. 41:1203–1204.

Musick, J. T., and D. A. Dusek. 1971. Grain sorghum response to preplant and seasonal irrigation in relation to deep plowing on Pullman clay loam. Texas Agric. Exp. Stn. Prog. Rep. PR-2952, College Station. p. 13–25.

Musick, J. T., and D. A. Dusek. 1975. Deep tillage of graded-furrow-irrigated Pullman clay loam. Trans. ASAE 18:263–269.

Olsen, S. R., F. S. Watanabe, F. E. Clark, and W. D. Kemper. 1964. Effect of hexadecanol on evaporation of water from soil. Soil Sci. 97:13–18.

Olsen, S. R., F. S. Watanabe, W. D. Kemper, and F. E. Clark. 1962. Effect of hexadecanol and octadecanol on efficiency of water use and growth of corn. Agron. J. 54:544–545.

Onstad, C. A. 1972. Soil and water losses as affected by tillage practices. Trans. ASAE 15: 287–289.

Osborn, J. F., R. E. Pelishek, J. S. Krammes, and J. Letey. 1964. Soil wettability as a factor in erodibility. Soil Sci. Soc. Am. Proc. 28:294–295.

Parr, J. F. 1959. Effects of vertical mulching and subsoiling on soil physical properties. Agron. J. 51:412–414.

Patrick, W. H., Jr., L. W. Sloane, and S. A. Phillips. 1959. Response of cotton and corn to deep placement of fertilizer and deep tillage. Soil Sci. Soc. Am. Proc. 23:307–310.

Peele, T. C., O. W. Beale, and F. F. Lesesne. 1948. Irrigation requirements of South Carolina soils. Agric. Eng. 29:157–158, 161.

Pelton, W. L. 1967. The effect of a windbreak on wind travel, evaporation, and wheat yield. Can. J. Plant Sci. 47:209–214.

Peters, D. B. 1960. Relative magnitude of evaporation and transpiration. Agron. J. 52:536–538.

Rasmussen, W. W., G. C. Lewis, and M. A. Fosberg. 1964. Improvement of the Chilcott-Sebree slick spot soils in southwestern Idaho. USDA-ARS Pub. no. 31-91. U.S. Government Printing Office, Washington, D.C.

Richardson, C. W. 1973. Runoff, erosion, and tillage efficiency on graded-furrow and terraced watersheds. J. Soil Water Conserv. 28:162–164.

Robertson, L. S., A. E. Erickson, and C. M. Hansen. 1977. Tillage systems for Michigan soils and crops. I. Deep, primary, supplemental and no-till. Michigan Agric. Exp. Stn. Bull. E. 1041, East Lansing.

Robertson, W. K., L. C. Hammond, G. K. Saxena, and H. W. Lundy. 1973. Influence of water management through irrigation and a subsurface asphalt layer on seasonal growth and nutrient uptake of corn. Agron. J. 65:866–870.

Robinson, F. E., and J. N. Luthin. 1968. Slip plowing in non-stratified clay. Calif. Agric. 22(11):8–9.

Rockwood, W. G., and R. Lal. 1974. Mulch tillage: a technique for soil and water conservation in the tropics. Span 17(2):77–79.

Sandoval, F. M., J. J. Bond, and G. A. Reichman. 1972. Deep plowing and chemical amendment effect on a sodic claypan soil. Trans. ASAE 15:681–684, 687.

Savage, S. M., J. Osborn, J. Letey, and C. Heaton. 1972. Substances contributing to fire-induced water repellency in soils. Soil Sci. Soc. Am. Proc. 36:674–678.

Saveson, I. L., and Z. F. Lund. 1958. Deep tillage improves delta "Hot Spot" areas. Crops Soils 10(5):15.

Saveson, I. L., Z. F. Lund, and L. W. Sloane. 1961. Deep-tillage investigations of compacted soil in the cotton area of Louisiana. USDA-ARS Pub. no. 41-41. U.S. Government Printing Office, Washington, D.C.

Saxena, G. K., L. C. Hammond, and H. W. Lundy. 1969. Yields and water-use efficiency of vegetables as influenced by a soil moisture barrier. Proc. Fla. State Hort. Soc. 82:168–172.

Saxena, G. K., L. C. Hammond, and H. W. Lundy. 1971. Effect of an asphalt barrier on soil water and on yields and water use by tomato and cabbage. J. Am. Soc. Hort. Sci. 96:218–222.

Saxena, G. K., L. C. Hammond, and W. K. Robertson. 1973. Effects of subsurface asphalt layers on corn and tomato root systems. Agron. J. 65:191–194.

Schneider, A. D., and A. C. Mathers. 1970. Deep plowing for increased grain sorghum yields under limited irrigation. J. Soil Water Conserv. 25:147–150.

Staple, W. J., and J. J. Lehane. 1955. The influence of field shelterbelts on wind velocity, evaporation, and moisture and crop yield. Can. J. Agric. Sci. 35:440–453.

Taylor, H. M., and H. R. Gardner. 1963. Penetration of cotton seedling taproots as influenced by bulk density, moisture content, and strength of soil. Soil Sci. 96:153–156.

Taylor, H. M., L. F. Locke, and J. E. Box. 1964. Pans in Southern Great Plains soils: III. Their effects on yield of cotton and grain sorghum. Agron. J. 56:542–545.

Taylor, R. E., O. E. Hays, C. E. Bay, and R. M. Dixon. 1964. Corn stover mulch for control of runoff and erosion on land planted to corn after corn. Soil Sci. Soc. Am. Proc. 28: 123–125.

Triplett, G. B., Jr., D. M. Van Doren, Jr., and B. L. Schmidt. 1968. Effect of corn (*Zea mays* L.) stover mulch on no-tillage corn yield and water infiltration. Agron. J. 60:236–239.

Unger, P. W. 1970. Water relations of a profile-modified slowly permeable soil. Soil Sci. Soc. Am. Proc. 34:492–495.

Unger, P. W. 1975. Relationships between water retention, texture, density and organic matter content of west and south central Texas soils. Texas Agric. Exp. Stn. Misc. Pub. MP-1192C, College Station.

Unger, P. W. 1976. Surface residue, water application, and soil texture effects on water accumulation. Soil Sci. Soc. Am. J. 40:298–300.

Unger, P. W. 1978. Straw-mulch rate effect on soil water storage and sorghum yield. Soil Sci. Soc. Am. J. 42:486–491.

Unger, P. W., H. V. Eck, and J. T. Musick. 1981. Alleviating plant water stress. p. 61–97. *In* G. F. Arkin and H. M. Taylor (ed.) Modifying the plant root environment to reduce crop stress. Am. Soc. Agric. Eng., St. Joseph, Mich.

Unger, P. W., and J. J. Parker. 1976. Evaporation reduction from soil with wheat, sorghum, and cotton residues. Soil Sci. Soc. Am. J. 40:938–942.

Unger, P. W., and A. F. Wiese. 1979. Managing irrigated winter wheat residues for water storage and subsequent dryland grain sorghum production. Soil Sci. Soc. Am. J. 43: 582–588.

Van Bavel, C. H. M., and F. W. Schaller. 1951. Soil aggregation, organic matter, and yields in a long-time experiment as affected by crop management. Soil Sci. Soc. Am. Proc. 15: 399–404.

Van Doren, D. M., Jr., and R. R. Allmaras. 1978. Effect of residue management practices on the soil physical environment, microclimate, and plant growth. p. 49–83. *In* W. R. Oschwald (ed.) Crop residue management systems. Am. Soc. of Agron. Spec. Pub. no. 31. Madison, Wis.

Van Doren, D. M., Jr., and J. L. Haynes. 1961. Subsoil tillage on fragipan soils in Ohio. Ohio Agric. Exp. Stn. Res. Circ. 95, Columbus.

van Schilfgaarde, J. (ed.). 1974. Drainage for agriculture. Agronomy 17. Am. Soc. of Agron., Madison, Wis.

Weakly, H. E. 1960. The effect of HPAN soil conditioner on runoff, erosion and soil aggregation. J. Soil Water Conserv. 15:169–171.

Wendt, C. W. 1973. Effects of a minimum tillage-vertical mulch concept on soil moisture and yield of grain sorghum at Lubbock, Texas. 1970–1971. Texas Agric. Exp. Stn. Prog. Rep. PR-3152, College Station.

Willis, W. O. 1962. Effect of partial surface covers on evaporation from soil. Soil Sci. Soc. Am. Proc. 26:598–601.

Willis, W. O., H. J. Haas, and J. S. Robins. 1963. Moisture conservation by surface or subsurface barriers and soil configuration under semi-arid conditions. Soil Sci. Soc. Am. Proc. 27:577–580.

Wilson, H. A., and G. M. Browning. 1946. Soil aggregation, yield, runoff, and erosion as affected by cropping systems. Soil Sci. Soc. Am. Proc. 10:51–57.

Wittmuss, H. D., G. B. Triplett, Jr., and B. W. Greb. 1973. Concepts of conservation tillage systems using surface mulches. p. 5–12. *In* Conservation tillage. Proc. Natl. Conf., Des Moines, Iowa. March 1973. Soil Conserv. Soc. Am., Ankeny, Iowa.

Zingg, A. W., and V. L. Hauser. 1959. Terrace benching to save potential runoff for semiarid land. Agron. J. 51:289–292.

CHAPTER 10B

Soil Management for Efficient Water Use: Soil Fertility

J. F. POWER

Agricultural Research Service, USDA; and University of Nebraska
Lincoln, Nebraska

I. INTRODUCTION

Water-use efficiency (WUE) is defined as the yield of plant product (grain, silage, forage, tuber, or other plant product of concern) produced per unit of water used. Mathematically, water-use efficiency is expressed as:

$$\mathrm{WUE} = Y/\mathrm{ET}, \qquad [1]$$

where Y is the quantity of the plant product (either biomass or marketable product) produced on a given surface area in a given time period (usually a growing season), and ET is evapotranspiration from the same surface area during the same time interval. As indicated in Chapter 1, transpiration (T) could be used instead of ET, but in almost all field situations, ET, rather than T, is measured. Thus, units are expressed as mass produced per volume of water used. Water use is restricted to that removed from the soil by evaporation and transpiration, and excludes nonproductive losses occurring by deep percolation or surface runoff.

In Chapter 10B, the effects of soil fertility upon WUE are discussed and the principles involved illustrated. Examples of how soil fertility can influence WUE are provided. The relation of soil fertility to crop yield, the numerator of Eq. [1], is discussed first, followed by the relation to ET, the denominator. The interactions between soil fertility and water availability are discussed, and examples are given of how these effects can be beneficially manipulated through soil and crop management.

II. PRINCIPLES

Viets (1962, 1967) published several excellent summaries of the relationship between soil fertility and WUE. The principles outlined apply directly to agriculture today, and the coverage in these publications is more comprehensive than that possible here.

Limitations to Efficient Water Use in Crop Production.

Data in Fig. 1 show how available water supply can affect plant growth responses to available N levels. These data from Smika et al. (1965) show how native grass production responded to N fertilization as available water supplies varied. When water supply was very limited (<25 cm), grass production was very low, and little response to N fertilization occurred at any N rate. As available water supply increased to over 40 cm, yield from unfertilized grass increased by several hundred kg ha^{-1}. However, as rate of annual N fertilization increased, the slopes of the regression curves relating yield to available water likewise increased. Thus, while N rate had very little effect on production when water supply was limited, responses of several thousand kg ha^{-1} were recorded when 40 cm of water was available. Likewise, responses to increased water were very modest with no fertilizer N, but increased greatly as the rate of N fertilization increased.

These results showed that growth responses to either variable were very limited as long as the supply of the other growth factor was also limited. However, when either the water or the N deficiency was eliminated, responses to the enhancement of the other factor were excellent.

The data in Fig. 1 illustrate the close relationship between water availability and fertilizer responses. Many other examples could be cited to illustrate this relationship. These data lead to the conclusion that a strong interaction exists between the availability of water and plant nutrition, and that changing one of these factors can greatly affect responses to the other factor. Greenwood (1976) reviewed a quantitative method for evaluating yield reduction caused by nutrient deficiencies. Power (1971) showed earlier that this concept could be applied to water deficits as well as to nutrient de-

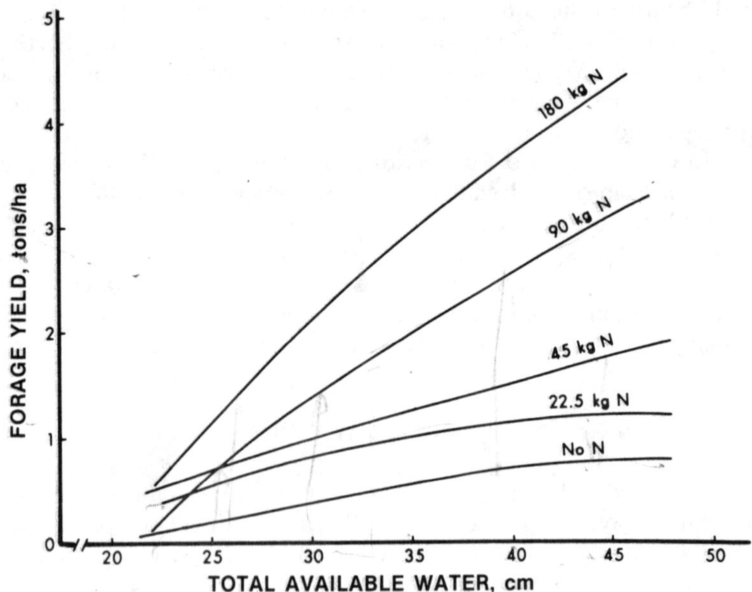

Fig. 1—Effect of water availability upon responses by native mixed prairie to N fertilization (from Smika et al., 1965).

ficits. Yield reductions arising from N and water deficits were additive. For example, June growth of smooth bromegrass, which had received 90 kg N ha^{-1}, was reduced 24% by N deficits, 32% by water deficits, and 54% when both water and N were deficient.

Increased available water supplies not only enhance fertilizer responses directly by eliminating water as a growth-limiting factor, but, in many instances, may also affect nutrient availability and efficiency of utilization. This mode of action has long been recognized by soil scientists and was documented in the early part of this century in several publications relating soil water content to soil NO_3 production (Buckman, 1910; Greaves & Carter, 1920). Numerous other publications illustrating such effects have appeared in the literature since then. Stanford and Epstein (1974) concluded that, for all practical purposes, rate of nitrification in a soil is related linearly to soil water content in the range from field capacity to wilting point. Thus, water has both direct and indirect effects on growth responses to soil fertility levels.

III. PLANT GROWTH AND YIELD

Soil fertility can affect WUE by altering plant growth and yield, the numerator in Eq. [1], because plant growth and yield can be improved by correcting those nutrient deficiencies that restrict yield. Frequently, commercial fertilizers or animal manures are used for this purpose. However, other practices—such as growth of N-fixing legumes, use of certain cropping systems and tillage practices, residue management, and similar activities—can be used beneficially.

Much of the soil fertility research conducted the last several decades has been aimed at developing techniques to correct nutrient deficiencies through fertilization, and thereby increase yields. An additional objective of more recent research has been to improve the efficiency of fertilizer use to reduce cost and protect the environment. Because the yield-response curve resulting from fertilization is usually curvilinear, yield per unit fertilizer applied (fertilizer efficiency) frequently reaches a maximum at fertilizer rates lower than those required for maximum yield per unit water used (WUE). Likewise, maximum WUE is often achieved with fertilizer rates somewhat lower than those that provide maximum yield (Anderson & Read, 1966; Bauer & Young, 1966; Doss et al., 1966; Olson et al., 1964; Viets, 1962).

Crop yield may also be increased by correcting nutrient deficiencies by means other than fertilization. Including an N-fixing legume in the cropping system is one common method, and animal manures and other organic byproducts are also frequently employed. The capability of legumes and animal manures to eliminate N deficiencies in sugar-beet production was demonstrated in a 20-year experiment by Halvorson and Hartman (1975). This experiment also illustrated the adverse effects of above-optimum rates of N fertilization on yields.

Soil and residue manipulation through tillage can also affect nutrient availability and subsequent crop yield. The direct effects of tillage upon nitrification were documented as early as 1914 in Kansas (Call, 1914) and have been shown in numerous other publications since. Tillage generally increases soil O_2 concentrations and, subsequently, activity of aerobic organisms, including nitrifiers (Campbell et al., 1976).

Residues are a major storehouse of potentially available nutrients, and residue placement is normally achieved through tillage. Consequently, residue placement may affect the distribution of available nutrients in a soil profile. This is of particular concern for no-tillage methods, wherein all residues remain at the surface. Several investigators have shown that no-tillage management can result in a large accumulation of available nutrients near the soil surface and can reduce nutrient concentrations below the 10- to 15-cm depth (Bauer & Kucera, 1969; Blevins et al., 1977; Moschler et al., 1972). This distribution could affect fertilizer requirements and final yield.

Water availability can influence soil nutrient availability and subsequent yield by two different mechanisms: (i) by influencing the rate of transformation of indigenous soil nutrients (including those in crop residues) to plant-available form, and (ii) by affecting the positional availability of nutrients from either soil or fertilizer for uptake by the plant.

The effects of water upon the cycling of nutrients through the soil-plant system have been studied in recent years (Campbell & Paul, 1978; Patrick & Reddy, 1976; Paul & Myers, 1971; Power & Legg, 1978). From these and a number of other publications, it is apparent that nutrients returned to the soil in crop residues and those immobilized in the cells of soil microorganisms are a major source of the indigenous soil nutrients that become available during a growing season (Allen et al., 1973; Campbell et al., 1973; Parr & Papendick, 1978; Shields et al., 1973, 1974; Westerman & Kurtz, 1972). Residue placement, which is normally achieved with tillage, therefore plays a major role in the cycling of these nutrients. Several workers have recently outlined many of the effects of residue placement and tillage practices on nutrient availability, water conservation, and crop yield (Campbell et al., 1976; Larsen et al., 1978; Power & Legg, 1978; Stanford et al., 1973). Generally, surface placement of residues decreases evaporation and soil temperature, and results in surface accumulation of organic matter and of nonmobile nutrients, and creates a less oxidative soil environment.

The quantity of crop residues annually returned to the soil is of major importance in determining the availability of plant nutrients. Black (1973a, 1973b) has documented these effects for a wheat-fallow rotation in Montana. These data show that, after the first few years of transition, increasing the quantity of wheat straw returned annually increased the levels of soil NO_3-N and $NaHCO_3$-soluble P. Workers in Saskatchewan have investigated the physical, chemical, and microbiological changes occurring in a soil as a result of returning residues (Campbell et al., 1976, 1977; Myers & Paul, 1971; Paul & Myers, 1971; Sadler & Stewart, 1974). Likewise, numerous other investigators have studied the role of water in these transformations (Greaves & Carter, 1920; Patrick & Reddy, 1976; Stanford & Epstein,

1974; Viets, 1967). The reduced air-filled porosity, lower temperature, and larger soil organic-matter pool resulting from surface placement of residues all favor slower transformation rates, but often greater quantities, of NO_3 production.

Removal of crop residues likewise affects soil nutrient availability and water relations. Barnhardt et al. (1978) have shown that continued removal of corn silage from an Iowa soil resulted in decreased soil organic-matter and total N content, when compared with plots where grain only was removed. In addition, removal of crop residues exposes the soil to water and wind erosion. Larson et al. (1978) calculated that removal of crop residues from Minnesota cornland could result in removal of 38–65 kg N ha^{-1} directly in the residues, plus loss of an additional 50–110 kg N ha^{-1} annually in accelerated soil erosion, making total removal from the soil-crop system of 88–175 kg N ha^{-1} $year^{-1}$. Such losses of N over a period of years would eventually reduce soil fertility levels and also reduce WUE.

The effects of water upon the positional availability of soil and fertilizer nutrients have also been studied extensively. The effects of excess water upon losses of nutrients by leaching or denitrification are well-documented. The inability of plants to remove available nutrients from dry soil has likewise been frequently documented (Campbell et al., 1977; Paul & Myers, 1971; Power et al., 1961a, 1961b). Often, the benefits ascribed to starter fertilizers banded with or close to the seed depend upon the availability of water. Nutrients applied by banding are available for almost immediate uptake by the young roots whenever there is enough water for plant growth; however, lack of adequate available water may result in germination damage from banding because of the high levels of salinity sometimes encountered in such bands.

IV. WATER USE

Water use is the denominator in calculating WUE; however, Sections A and C of Chapter 10 contain discussions of many aspects of soil water conservation, pointing out management practices that may be used to increase water conservation and improve WUE. In this Section the influence of soil fertility upon plant water use will be reviewed in more detail. Again, the reader can refer to publications by Viets (1962, 1967) for a more extensive discussion than is possible here.

Water use by plants is regulated primarily by (i) the evaporative potential of the atmosphere, (ii) the amount of water available, and (iii) certain plant characteristics. The available water supply is affected by precipitation and by irrigation practices, drainage, certain inherent soil physical properties affecting water storage and movement, and numerous cultural and management practices. In arid and semiarid regions, amount and distribution of precipitation are the major factors determining the available water supply, whereas, in many humid regions, factors such as drainage and soil physical properties are more important. For a given soil at a given location, however, the management practices used—tillage, residue placement, fertilizer practices, cropping systems, and other such factors—largely determine the amount of water available for plant growth. See Sections A and C of Chapter 10 for a more thorough discussion of factors affecting soil water availability.

Evaporation (E) potential is regulated mainly by the energy available in the atmosphere for evaporation and, to a lesser extent, by soil properties and management. Solar radiation is the ultimate energy source for E. Proper soil management can be used to control rates of E to some extent and thus reduce the amount of soil water lost. Again, specific examples of this are given in Sections A and C of Chapter 10.

Plant characteristics are important parameters in determining water use. Since most transpiration (T) occurs from leaf surfaces, some of the more important plant characteristics that affect water use are leaf area and leaf area duration (see Chapter 1). These factors are markedly affected by soil fertility status (Campbell et al., 1977; Power & Alessi, 1978). Generally, adequate supplies of soil nutrients result in rapid expansion of leaf area, increasing the potential for water use by T. On the other hand, greater leaf area results in more rapid ground cover and reduced penetration of radiant energy to the soil surface for E. If the soil surface is moist, then these two processes—increased potential for transpiration and reduced evaporation—may offset each other, providing very little change in total water use resulting from fertilization. If the soil surface is drying during most of the growing season, then all available water in the root zone is generally used anyway (Anderson & Read, 1966; Doss et al., 1966; Smika et al., 1965; Viets, 1967).

Numerous publications have shown that soil fertility status may alter plant rooting habits (Anderson & Read, 1966; Bond et al., 1971; Campbell et al., 1977; Power et al., 1961b; Welch et al., 1966; Wight & Black, 1978). Plant rooting patterns can markedly affect water use, particularly in well-drained soils (Campbell et al., 1977; Olson et al., 1964). Characteristics of major concern include rooting depth, root density, and rate of root penetration (Taylor & Klepper, 1973).

Frequently, plants grown with an adequate nutrient supply extend roots to greater depths than the same plants grown in soil that is deficient in one or more nutrients (Campbell et al., 1977; Campbell & Paul, 1978; Olson et al., 1964; Power et al., 1961b; Wight & Black, 1978). This reaction to enhanced soil fertility, of course, increases the volume of soil water available to the plant, thereby increasing the potential for water use and thus reducing the probability of plant growth being restricted by intermittent periods of drought (Brown, 1971). In regions of limited precipitation, this effect may influence plant growth for only the first year or two until the water in these deeper soil layers has been used. After that, this characteristic has little influence unless soil water at the lower depths is recharged (Campbell et al., 1977; Olson et al., 1964; Wight & Black, 1978).

Maintaining adequate fertility often increases plant root density (Campbell et al., 1977). This characteristic may have little influence on water use, however, if the water can readily move to the root surface after the required minimum root density is achieved. In soils with physical properties that restrict internal water movement, root density may be more important.

The rate of root growth and penetration is another factor that soil fertility can influence. When water supply in the upper part of the soil is limited, enhanced root penetration resulting from improved fertility status may assume major importance in maintaining a favorable water status in the plant. Frequently, adequate fertilization hastens the rate of root development, thereby providing access to a larger supply of available water. The classical P-deficiency symptoms often develop in plants growing in soils that restrict root development, indicating that root growth may affect the positional availability of relatively immobile nutrients.

In most semiarid and even subhumid regions, agricultural crops frequently dry most of the soil within the zone of maximum rooting to near the wilting point during the growing season (Call, 1914; Campbell & Paul, 1978; Power et al., 1961b; Power & Alessi, 1970; Wight & Black, 1978). Thus, any practice that increases the depth of rooting, rate of root development, or root density, particularly at lower depths, can potentially increase the water available for use and can also stimulate crop growth by reducing water stress. However, under dryland agriculture, the potential depth of rooting is often not totally recharged without the use of fallow or irrigation. Consequently, under these conditions, improved soil fertility enhances water use only until the plant roots have used the water stored near the bottom of the normal root zone.

V. WATER-USE EFFICIENCY

In general, WUE improves as the availability of plant nutrients increases, as long as enough water is available to provide reasonable growth rates. When water availability is limited, maximum values of WUE are often achieved at fertility levels considerably below those that provide maximum yield.

Increasing the availability of plant nutrients somctimes increases yields as well as water use by the crop; however, the increase in water use is usually small—generally <25%. Also, water use may be increased to near maximum with less-than-adequate quantities of fertilizer for maximum yield. Increasing fertilizer application rates progressively increases yield until a maximum is reached. The yield increases obtained may be severalfold greater than the yields from the check. Thus, WUE increases with level of fertility, but often the greatest increases come from fertilizer rates below optimum for maximum yield. Many of these effects of soil fertility on the relationships between water and yield are illustrated in Fig. 1.

In discussing crop response in the WUE equation, crop quality is also a matter of concern. With many crops, it is not unusual for total crop yield to increase with increased level of available nutrients, while quality of the crop is declining. The work of Halvorson and Hartman (1975) with sugar beets is a good example. Also, in the production of malting barley, yields can be increased by increasing either the available water supply or the level of available nutrients, but both have effects upon quality. Banasik and Power (1973) found that increased water availability improved both barley yield and malting quality, while increased rates of N increased yield but often reduced quality. Thus, when yield is restricted to achieve the highest quality product, fertilizer N rates below optimum for maximum growth may be desirable. These adjustments may have a variable effect on WUE.

In many soils, factors other than water or nutrient availability limit yield. In such instances, correction of the growth-limiting factor could drastically change water and fertility requirements. In 1974, van Schaik and Cairns found that application of NH_4 fertilizers at rates of 400–800 kg N ha^{-1} reclaimed some sodic-claypan soils of eastern Alberta. Ammonium ions replaced Na on the exchange sites, improving physical structure and, subsequently, available water supply and plant rooting patterns. Sodic-claypan soils that contain gypsum in the profile can be greatly improved by deep plowing, which improves water infiltration and plant growth (Sandoval et al., 1972). In Section C of Chapter 10, improvement of the physical structure in Coastal Plains soils is also discussed. Such practices greatly alter soil water regimes and plant-growth characteristics, which, of course, are reflected in WUE.

Soil temperature can also alter relationships between soil fertility levels and water use. Below-optimum soil temperatures reduced growth rate of barley without greatly altering P metabolism (Power et al., 1964); however, at above-optimum soil temperatures, each unit of P absorbed by the plant was less efficient in stimulating plant growth. Also, higher soil temperatures increased water losses by *T*, all of which reduced WUE. Campbell and coworkers (1973) found that freezing-thawing cycles were lethal to some soil organisms, resulting in rapid mineralization of N in those cells. As discussed earlier, the enhanced N availability that results may alter WUE.

VI. CONCLUSIONS

This review points out the general relationships between water and soil fertility and their effects on WUE. These relationships are not simple. Myriad factors can affect availability of both water and plant nutrients, and these factors may or may not be related. Thus, general summaries are difficult to make because there can be so many exceptions to each general statement. It should be kept in mind, however, that water use by plants is controlled mainly by the abiotic properties of the environment in which they grow, especially by radiant energy. Crop yield, on the other hand, is the integrated product not only of the abiotic environment, but also of the numerous biologically mediated chemical reactions that occur in the soil and plant. Soil fertility is a major factor in controlling these reactions, of course. The effects of soil fertility on WUE are both complex and numerous; however, these effects cannot be ignored in development of management systems to utilize efficiently the water that is available.

LITERATURE CITED

Allen, A. L., F. J. Stevenson, and L. T. Kurtz. 1973. Chemical distribution of residual fertilizer nitrogen in soil as revealed by nitrogen-15 studies. J. Environ. Qual. 2:120–124.

Anderson, C. D., and D. W. L. Read. 1966. Water-use efficiency of some varieties of wheat, oats, barley, and flax grown in the greenhouse. Can. J. Plant Sci. 46:375–378.

Banasik, O. J., and J. F. Power. 1973. Effect of water and nitrogen supply on malting quality of barley. Brew. Dig. 48(2):58–62.

Barnhardt, S. L., W. D. Shrader, and J. R. Webb. 1978. Comparison of soil properties under continuous corn grain and silage cropping systems. Agron. J. 70:835–837.

Bauer, A., and H. L. Kucera. 1969. Influence of tillage method on incorporation of fertilizer phosphorus. N. Dak. Agric. Exp. Stn. Farm Res. 26(5):4–6. Fargo.

Bauer, A., and R. A. Young. 1966. Fertilized wheat uses water more efficiently. N. Dak. Agric. Exp. Stn. Farm Res. 24(3):4–11. Fargo.

Black, A. L. 1973a. Crop residue, soil water, and soil fertility related to spring wheat production and quality after fallow. Soil Sci. Soc. Am. Proc. 37:754–758.

Black, A. L. 1973b. Soil properties associated with crop residue management in a wheat-fallow rotation. Soil Sci. Soc. Am. Proc. 37:943–946.

Blevins, R. L., G. W. Thomas, and P. L. Cornelius. 1977. Influence of no-tillage and nitrogen fertilization on certain soil properties after 5 years of continuous corn. Agron. J. 69:383–386.

Bond, J. J., J. F. Power, and W. O. Willis. 1971. Soil water extraction of N-fertilized spring wheat. Agron. J. 63:280–283.

Brown, P. L. 1971. Water use and soil water depletion by dryland winter wheat as affected by nitrogen fertilization. Agron. J. 63:43–46.

Buckman, H. O. 1910. Moisture and nitrate relations in dry-land agriculture. Proc. Am. Soc. Agron. 2:121–138.

Call, L. E. 1914. The effect of different methods of preparing a seed bed for winter wheat upon yield, soil moisture, and nitrates. J. Am. Soc. Agron. 6:249–259.

Campbell, C. A., V. O. Biederbeck, and F. G. Warder. 1973. Influence of simulated fall and spring conditions on the soil system: III. Effect of method of simulating spring temperatures on ammonification, nitrification, and microbial populations. Soil Sci. Soc. Am. Proc. 37:382–386.

Campbell, C. A., D. R. Cameron, W. Nicholiachuk, and H. R. Davidson. 1977. Effects of fertilizer N and soil moisture on growth, N content, and moisture use by spring wheat. Can. J. Soil Sci. 57:289–310.

Campbell, C. A., and E. A. Paul. 1978. Effects of fertilizer N and soil moisture on mineralization, N recovery, and A-values, under spring wheat grown in small lysimeters. Can. J. Soil Sci. 58:39–51.

Campbell, C. A., E. A. Paul, and W. B. McGill. 1976. Effect of cultivation and cropping on the amounts and forms of soil N. p. 9–101. *In* W. A. Rice (ed.) Proc. Western Can. Nitrogen Symp., Calgary, Alberta. 20–21 Jan. 1976. Alberta Agric., Edmonton.

Doss, B. D., D. A. Ashley, and O. L. Bennett. 1966. Interaction of soil moisture, nitrogen, and clipping frequency on yield and nitrogen content of coastal bermudagrass. Agron. J. 58:510–512.

Greaves, J. E., and E. G. Carter. 1920. Influence of moisture on the bacterial activities of the soil. Soil Sci. 10:361–387.

Greenwood, E. A. N. 1976. Nitrogen stress in plants. Adv. Agron. 28:1–35.

Halvorson, A. D., and G. P. Hartman. 1975. Long-term nitrogen rates and sources influence sugarbeet yield and quality. Agron. J. 67:389–393.

Larson, W. E., R. F. Holt, and C. W. Carlson. 1978. Residues for soil conservation. p. 1–15. *In* W. R. Oschwald (ed.) Crop residue management system. ASA Spec. Publ. 31. Am. Soc. of Agron., Madison, Wis.

Moschler, W. W., G. M. Shear, D. C. Martens, G. D. Jones, and R. R. Wilmouth. 1972. Comparative yield and fertilizer efficiency for no-tillage and conventionally tilled corn. Agron. J. 64:229–231.

Myers, R. J. K., and E. A. Paul. 1971. Plant uptake and immobilization of ^{15}N-labelled ammonium nitrate in a field experiment with wheat. p. 55–64. *In* Nitrogen-15 in soil-plant studies. Int. Atomic Energy Agency, Vienna.

Olson, R. A., C. A. Thompson, and P. H. Grabouski. 1964. Water requirement of grain crops as modified by fertilizer use. Agron. J. 56:427–432.

Parr, J. F., and R. I. Papendick. 1978. Factors affecting the decomposition of crop residues by microorganisms. p. 101–129. *In* W. R. Oschwald (ed.) Crop residue management systems. ASA Spec. Publ. 31. Am. Soc. of Agron., Madison, Wis.

Patrick, W. H., Jr., and K. R. Reddy. 1976. Fate of fertilizer nitrogen in a flooded rice soil. Soil Sci. Soc. Am. J. 40:678–681.

Paul, E. A., and R. J. K. Myers. 1971. Effect of soil moisture stress on uptake and recovery of tagged nitrogen by wheat. Can. J. Soil Sci. 51:37–43.

Power, J. F. 1971. The evaluation of water and nitrogen stresses on bromegrass growth. Agron. J. 63:726–728.

Power, J. F., and J. Alessi. 1970. Effect of nitrogen source and phosphorus on crested wheatgrass growth and water use. J. Range Manage. 23:175–178.

Power, J. F., and J. Alessi. 1978. Tiller development and yield of standard and semidwarf spring wheat varieties as affected by nitrogen fertilization. J. Agric. Sci., Camb. 90:97–108.

Power, J. F., P. L. Brown, T. J. Army, and M. G. Klages. 1961a. Phosphorus responses by dryland spring wheat as influenced by moisture supplies. Agron. J. 53:106–108.

Power, J. F., D. L. Grunes, and G. A. Reichman. 1961b. The influence of phosphorus fertilization and moisture on growth and nutrient absorption by spring wheat: I. Plant growth, N uptake, and moisture use. Soil Sci. Soc. Am. Proc. 25:207–210.

Power, J. F., D. L. Grunes, G. A. Reichman, and W. O. Willis. 1964. Soil temperature effects on phosphorus availability. Agron. J. 56:545–548.

Power, J. F., and J. O. Legg. 1978. Effect of crop residues on the chemical environment and nutrient availability. p. 95–100. *In* W. R. Oschwald (ed.) Crop residue management systems. ASA Spec. Publ. 31. Am. Soc. of Agron., Madison, Wis.

Sadler, J. M., and J. W. B. Stewart. 1974. Residual fertilizer phosphorus in western Canadian soils: a review. Sask. Inst. Pedology Pub. no. R136. Saskatoon, Saskatchewan.

Sandoval, F. M., J. J. Bond, and G. A. Reichman. 1972. Deep plowing and chemical amendment effect on a sodic claypan soil. Trans. ASAE 15:681–687.

Shields, J. A., E. A. Paul, and W. E. Lowe. 1973. Turnover of microbial tissue in soil under field conditions. Soil Biol. Biochem. 5:753–764.

Shields, J. A., E. A. Paul, and W. E. Lowe. 1974. Factors influencing the stability of labelled microbial material. Soil Biol. Biochem. 6:31–37.

Smika, D. E., H. J. Haas, and J. F. Power. 1965. Effects of moisture and nitrogen fertilizer on growth and water use by native grass. Agron. J. 57:483–486.

Stanford, G., O. L. Bennett, and J. F. Power. 1973. Conservation tillage practices and nutrient availability. p. 54–62. *In* Proc. Conservation Tillage Symp., Des Moines, Iowa. 28–30 Mar. 1973. Soil Conserv. Soc. Am., Ankeny, Iowa.

Stanford, G., and E. Epstein. 1974. Nitrogen mineralization–water relations in soils. Soil Sci. Soc. Am. Proc. 38:103–107.

Taylor, H. M., and B. Klepper. 1973. Rooting density and water extraction patterns for corn (*Zea mays* L.). Agron. J. 65:965–968.

van Schaik, J. C., and R. R. Cairns. 1974. Influence of ammonium on the behavior of clay particles in a sodic soil and bentonite. Can. J. Soil Sci. 54:39–44.

Viets, F. G., Jr. 1962. Fertilizers and the efficient use of water. Adv. Agron. 14:223–264.

Viets, F. G., Jr. 1967. Nutrient availability in relation to soil water. *In* R. M. Hagan et al. (ed.) Irrigation of agricultural lands. Agronomy 11:458–471. Am. Soc. of Agron., Madison, Wis.

Welch, N. H., E. Burnett, and H. V. Eck. 1966. Effect of row spacing, plant population, and nitrogen fertilization on dryland grain sorghum production. Agron. J. 58:160–163.

Westerman, R. L., and L. T. Kurtz. 1972. Residual effects of ^{15}N-labeled fertilizers in a field study. Soil Sci. Soc. Am. Proc. 36:91–94.

Wight, J. R., and A. L. Black. 1978. Soil water use and recharge in a fertilized mixed prairie plant community. J. Range Manage. 31:280–282.

CHAPTER 10C

Soil Management for Efficient Water Use: Soil-Profile Modification Effects on Plant Growth and Yield in the Southeastern United States[1]

D. C. REICOSKY

North Central Soil Conservation Research Laboratory, USDA-ARS
Morris, Minnesota

I. INTRODUCTION

Since soils in the southeastern United States developed under wide ranges of rainfall amounts, parent materials, lengths of growing seasons, and kinds of vegetation, their varieties in this region are endless. This discussion will be limited to that part of the southeastern United States that lies east of the Mississippi River and in the Atlantic Coastal Plains. This region has most of the agricultural soils of southeastern United States. The Coastal Plains belt is about 240 km wide along the Atlantic seaboard and about 640 km wide in the Gulf of Mexico. The topography ranges from plateaus at low elevations to broad, swampy flats (Buol, 1973).

II. SOIL AND CLIMATIC FACTORS

The abundance of rainfall for plant growth, high temperatures, and soil water leaching are the main factors controlling the development of these soils. Mean annual temperature of the area ranges from 15 to 20°C and mean precipitation ranges from 100 to 120 cm year^{-1}, with erratic distribution during the growing season. Total evaporation potential ranges from 120 to 150 cm year^{-1} and water deficits occur in some areas, particularly those with sandy soils. The long, hot summers and short, mild winters facilitate rapid rates of chemical, physiological, and biological weathering. Climatic conditions favor high rates of microbial activity, even during the cold season; therefore organic matter is rapidly oxidized, producing much

[1]Contribution from the North Central Soil Conservation Research Laboratory, USDA-ARS, Morris, Minn.; in cooperation with the Minnesota Agricultural Experiment Station, St. Paul, Journal Series 10795.

Limitations to Efficient Water Use in Crop Production.

CO_2 that combines with water to form carbonic acid (H_2CO_3), and providing favorable conditions for acid leaching.

In early spring, the soil water content is generally high because of winter rains. Most spring rainfall is lost as runoff, resulting in a high erosion potential. During the growing season, recharge is often incomplete because high-intensity thunderstorms on unprotected soil surfaces cause significant amounts of runoff. Thus, most of the growing-season rainfall is not readily stored or available for crop growth.

In addition to limited recharge during the growing season, the low soil water storage capacity of many sandy soils in the Coastal Plains adds to the water deficit. Campbell et al. (1974) characterized water retention curves for Varina sandy loam (clayey, kaolinitic, thermic Plinthic Paleudults) in the South Carolina Coastal Plains. Nearly 42% of the total pore volume in the Ap horizon was drained between −0.01 and −0.1 bar, as compared with 17% in the more dense A2 horizon. Water held in the B horizon between the −0.01- to −1.0-bar range was released more gradually than that from the surface layers. In general, the surface horizon of sandy loam or loamy sand soils releases most of the water at high matric potentials within 1 or 2 days after a rain, and relatively little is released below the −1.0-bar range. Crops grown on the soils become water-stressed within 3–7 days after rainfall, and for maximum crop production frequent water replacement is necessary.

Water availability to the plant is not solely determined by soil water-retention characteristics, but it also depends on root development and soil water transmission. Particle-size analysis indicates that many of the soils in both the natural and compacted state have textural properties that result in a very low void ratio and a high bulk density. Bulk densities for the Norfolk (fine, loamy, siliceous, thermic Typic Paleudults) soil averaged 1.50, 1.67, and 1.50 g cm^{-3} for the Ap, A2, and B horizons, respectively (Campbell et al., 1974). Bulk densities of some of the natural A2 horizons are as high as 1.9 g cm^{-3}.

Associated with these high bulk densities is a physical impedance to root growth that depends on the soil type and water content. Although the actual impedance encountered by growing roots is difficult to measure, soil probes offer a comparative means of characterizing soil strength in terms of probe resistance. Probe resistance values increase as water content decreases and as initial bulk density increases. Particle size, bulk density, and water content of the A2 horizon directly influence soil strength. Once the field soil settles after tillage and tractor traffic, particle-size distribution and bulk density will change little, so that soil strength then becomes primarily a function of the soil's water content. Campbell et al. (1974), who studied the effects of chiseling some Coastal Plains soils, found that soil strength may have limited root development long before all plant-available water was extracted. Root elongation was limited by the high strength of the A2 horizon and by roots' inability to reach subsoil water needed to meet the plants' transpiration demands. Thus, plants were stressed earlier on shallow-tilled than on the deep-tilled plots.

III. CHISELING AND SUBSOILING

Reicosky et al. (1976) studied the effect of chiseling on the minimum daily leaf water potential. Chiseling caused deeper root penetration that enabled roots to utilize water from the capillary fringe above the water table. Use of this capillary-fringe water resulted in only slightly higher (less negative) minimum water potential of corn leaves, but increased sweet corn ear yields 8 metric tons ha^{-1} that year. Thus, chiseling soil with root-restricting layers can increase corn production, particularly when the water table is near the root zone.

Doty et al. (1975) chiseled a Varina sandy loam to a depth of 38 cm, which sufficiently disrupted the compacted A2 horizon so that water extraction and crop yields were increased. Over a 4-year period the average increase in dry matter yields above the check treatment for millet and sweet corn was 18.5% for an irrigation treatment, 17.1% for a chisel treatment, and 25.6% for both treatments combined. They concluded that chiseling alone to disrupt the A2 horizon resulted in nearly the same net return as irrigation alone.

Besides increasing the rooting depth and thus the amount of water available for plants, chiseling also increased the O_2 content of the soil during extremely wet periods (Campbell & Phene, 1977). During wet periods, chiseled soil permitted water to infiltrate and percolate to greater depths, thus decreasing the degree of saturation in the upper root zone. With sufficient simulated rainfall to maintain the soil matric potential above -400 mbars, and with adequate soil aeration, millet yield was not affected by tillage depth. However, under very high matric potential conditions (from -45 to -87 mbars), chiseling partially alleviated the O_2 stresses associated with wet soil. Apparently, the deeper-rooted plants in the deep-tilled soil were less affected by poor aeration than the shallow-rooted plants in conventionally tilled soil. Millet growth was unaffected when the soil O_2 concentration was $>15\%$, but decreased as O_2 concentrations decreased from 15 to 2%. In the Varina sandy loam soil used in this study, the O_2 concentration decreased as the soil matric potential increased from -80 to -2 mbars.

Research in South Carolina (Peele et al., 1974; Peele & Suman, 1976; Suman & Peele, 1974) indicated that on soils with a hardpan compact enough to seriously interfere with root development, yields of corn, soybeans, and cotton are increased by subsoiling under the plant row in most years. Suman and Peele (1974) found that subsoiling a Dothan loamy sand (fine, siliceous, thermic Plinthic Paleudults) without irrigation increased soybean yields by 249 to 625 kg ha^{-1} in 4 out of 6 years. Subsoiling the same soil increased corn yields an average of 314 kg ha^{-1} with irrigation and 778 kg ha^{-1} without irrigation (Peele & Suman, 1976). Subsoiling increased the rooting depth of corn, enabling it to use more subsoil water.

Similar subsoiling work in the sandy soils of Georgia increased soybean yields. Parker et al. (1976) measured a 60% yield increase from subsoiling in a relatively dry year, but very little increase in a year with adequate soil moisture. They noted deeper penetration of nematodes in the subsoiled plots, which can be a problem unless the proper nematicide is incorporated. In a study of its effect on cotton production on a Marlboro loamy sand (clayey, kaolinitic, thermic Typic Paleudults), subsoiling under the row in front of the planter increased root penetration and moisture and nutrient uptake and helped to overcome the effects of cotton stunt (Boswell et al., 1974).

Cassel et al. (1978) studied the effect of conventional tillage, chisel plowing, and subsoiling to a depth of 20–28 cm for reducing the mechanical impedance of a hardpan on a Norfolk sandy loam in the North Carolina coastal plains. Mechanical impedance was measured with a hydraulically driven penetrometer with readings taken at seven locations perpendicular to the direction of tillage. Chisel plowing reduced impedance to values lower than those measured for moldboard plowing at all seven positions. Of the three treatments, the subsoiled plots had the lowest impedance only in or near the subsoil slot, but had the highest impedance in the traffic area. Cassel et al. (1978) recommended that future research include sufficient measurements of soil mechanical impedance to isolate not only the tillage effects but also the position and depth effects in those profiles where physical properties vary with depth. Further work is needed on the longevity of the effects of subsoiling.

IV. TRAFFIC CONTROL

Techniques have been developed to keep the tractor wheel tracks confined to permanent strips within the field (Dumas et al., 1972). These techniques minimize the wheel traffic compaction that might interfere with plant growth. In one experiment with cotton, yields were increased about 300 kg ha^{-1}, compared with a nonchiseled control, when the soil was chiseled to a depth of 45 cm. However, with subsequent traffic, yield differences decreased rapidly. Seed cotton yields on plots with controlled traffic were about 3,600 kg ha^{-1}, as compared with 3,000 kg ha^{-1} on plots with unrestricted traffic. Deep tillage with controlled traffic resulted in plants 25 cm taller than those from the conventionally tilled and normally compacted plots. The average weekly growth rate of cotton on the deep-tilled and controlled-traffic plots exceeded those on all other plots. Deep tillage resulted in up to 37% greater yields than those on conventionally tilled plots, and the influence of interrow traffic reduced yields significantly.

V. SUBSOIL TOXICITY AND ACIDITY CONTROL

Although chiseling and subsoiling look promising for disrupting hardpans, some genetic horizons in the Coastal Plains are too deep to reach with tillage implements. Even where hardpan can be disrupted physically, many crops do not respond to deep tillage because the soils contain toxic levels of soluble Al, particularly in the acid subsoils. Growth of roots in the acid subsoils and utilization of subsoil water is limited because of Al toxicity (Rios & Pearson, 1964). Soluble Al concentrations are high enough to inhibit root development of sensitive crops even without the compacted layers. Disrupting the compacted layer will not increase yields as much as if the subsoil acidity were not as high.

Extreme acidity in subsoil layers directly affects root growth by increasing the H^+ ion concentration, which increases the solubility of Al. The effect seems to be limited to the roots. The restricted root development reduces water and nutrient uptake. Some characteristic symptoms of Al toxicity include discoloration, loss of turgidity, thickening, distortion of the main roots, and the development of short, stubby laterals on the cotton plant.

The damaging effects of Al on the plants have been widely attributed to interference of P uptake. Using split-root experiments, Rios and Pearson (1964) confirmed the independence of Al toxicity and P uptake.

Effects on plant roots of subsoil pH and liming of acid soils have been easily demonstrated in pot experiments. However, under field conditions, the effects of liming on deep root proliferation are often difficult to define. First, satisfactorily incorporating lime into the subsoil is difficult, and second, criteria for evaluating root behavior in the field are generally inadequate. Rarely have root observations been reported for this type of experiment, and, when results were reported, improvement in plant growth was not generally consistent. The main reason for lack of response to subsoil liming is probably due to inadequate mixing of lime with the subsoil material.

Subsoil infertility has been frequently cited as a probable contributing factor to shallow rooting. However, evidence is meager concerning root requirement for specific nutrients, other than Ca, at the point of root growth. Attempts to correct subsoil acidity and infertility often have not been successful. Adams and Pearson (1969) applied large amounts of neutral N fertilizer ($Ca(NO_3)_2$) to coastal burmudagrass, an acid-tolerant grass. The salt leached into the root zone as the crop removed the NO_3^-, leaving the cation in the root zone. Unfortunately, this procedure requires that an Al-tolerant crop occupy the land for 2–3 years, and is very expensive because of the large amounts of fertilizer required.

The quickest and most practical method of alleviating subsoil acidity is to incorporate the lime from the surface. This incorporation requires a tremendous amount of physical energy to uniformly distribute lime to maximize rooting depth. Some basic cations can be incorporated with the irrigation water, but the depth of incorporation depends on the frequency of water application. Doss and Lund (1975) found that irrigated plots had a uniform pH of 6.0 in the surface layer and a pH range of 4.4–6.4 in the subsoil. Plant heights increased as the subsoil pH increased. However, cotton yield leveled off at the pH of 5.6. They concluded that when water was adequate in a surface layer that was maintained at a pH of 6.0, it was only necessary to lime the surface layer. These results suggested that the main reason for increasing the rooting depths is to provide plants with a larger volume of soil from which to use water during a drought, unless the plants can be irrigated. Most of the plant's nutrient demand can be met by very small soil volumes that the plant's root system utilizes.

VI. WATER-FERTILITY MANAGEMENT

A technique developed by Phene and Beale (1976) for coping with a compacted layer in a soil that restricts rooting is to use high-frequency irrigation techniques for water-nutrient management. This system uses trickle-irrigation techniques designed to prevent plant water and nutrient stresses while maximizing the available water storage capacity to accommodate rainfall. The system was based on the range of soil matric potential that resulted in the optimum O_2 diffusion rate and soil strength. Water desorption characteristics and unsaturated conductivity were used in irrigation-design criteria. The system increased water- and N-use efficiency, and reduced N leaching. The high-frequency application of water and nutrients through the trickle-irrigation system was adjusted to maintain optimum N level in the corn plants. While high-frequency irrigation systems using automatic sensors seem to have some merit, they require careful management and trouble-free equipment operation so that plants are not left without water for any length of time. Any major equipment malfunction could adversely affect yields.

VII. SUMMARY

In the above section, some of those factors that significantly affect water-use efficiency in the southeastern Coastal Plains were discussed. Utilizing soil and water management practices to maximize water-use efficiency to obtain higher yields from a given amount of water presents a major challenge. Even with an annual rainfall of 100–120 cm year^{-1}, the interaction of the soil with environmental factors often limits plant water use and causes plants to be stressed. This results from an imbalance between the soil-supplying capability and the evapotranspiration demand of the plants. While the soil physical properties may limit plant growth, tillage and water-application techniques can be used to increase efficient use of water. Chiseling and subsoiling can increase water use from the subsoil without irrigation in several cases. Controlling wheel traffic and improving irrigation systems will also increase yields. While many of these factors have increased yields, they are small, though important, steps towards maximizing the efficient use of water.

LITERATURE CITED

Adams, F., and R. W. Pearson. 1969. Neutralizing soil acidity under bermudagrass sod. Soil Sci. Soc. Am. J. 33:737–742.

Boswell, F. C., D. A. Ashley, O. L. Brooks, G. W. Bird, T. D. Canerday, F. G. Futral, S. M. McCarter, C. E. Perry, C. E. Rice, C. R. Roland, and R. W. Roncardori. 1974. Cotton stunt in Georgia is investigated by research team. Ga. Agric. Res. 15:4–7.

Buol, S. W. (ed.). 1973. Soils of the southern States and Puerto Rico. Southern Coop. Ser. Bull. no. 174. p. 105. U.S. Government Printing Office, Washington, D.C.

Campbell, R. B., D. C. Reicosky, and C. W. Doty. 1974. Physical properties and tillage of Paleudults in the southeastern Coastal Plain. J. Soil Water Conserv. 29:220–224.

Campbell, R. B., and C. J. Phene. 1977. Tillage, matric potential, oxygen and millet yield relations in a layered soil. Trans. ASAE 20:271–275.

Cassel, D. K., H. D. Bowen, and L. A. Nelson. 1978. An evaluation of mechanical impedance for three tillage treatments on Norfolk sandy loam. Soil Sci. Soc. Am. J. 42:116–120.

Doss, B. D., and Z. F. Lund. 1975. Subsoil pH effects on growth and yield of cotton. Agron. J. 67:193–196.

Doty, C. W., R. B. Campbell, and D. C. Reicosky. 1975. Crop response to chiseling and irrigation in soils with a compact A2 horizon. Trans. ASAE 18:668–672.

Dumas, W. T., A. C. Trouse, L. A. Smith, F. A. Kummer, and W. R. Gill. 1972. Development and evaluation of tillage and other cultural practices in a controlled traffic system for cotton in the southern Coastal Plains. Trans. ASAE 16:872–875, 880.

Parker, M. B., N. A. Minton, O. L. Brooks, and C. E. Perry. 1976. Soybean response to subsoiling and a nematicide. Ga. Agric. Res. 181:1–22.

Peele, T. C., B. J. Gossett, and J. E. Evans. 1974. Minimum tillage studies with soybeans and corn on Piedmont soil. Agron. and Soils Res. Ser. 92. South Carolina Agric. Exp. Stn., Clemson. p. 8.

Peele, T. C., and R. F. Suman. 1976. Limiting agronomic factors in corn production. Bull. no. 591. South Carolina Agric. Exp. Stn., Clemson. p. 27.

Phene, C. J., and O. W. Beale. 1976. High-frequency irrigation for water nutrient management in humid regions. Soil Sci. Soc. Am. J. 40:430–436.

Reicosky, D. C., R. B. Campbell, and C. W. Doty. 1976. Corn plant water stress as influenced by chiseling, irrigation, and water table depth. Agron. J. 68:499–503.

Rios, M. A., and R. W. Pearson. 1964. The effect of some chemical environmental factors on cotton root behavior. Soil Sci. Soc. Am. Proc. 28:232–235.

Suman, R. F., and T. C. Peele. 1974. Limiting agronomic factors in soybean production. Bull. no. 1051. South Carolina Agric. Exp. Stn., Clemson. p. 12.

CHAPTER 11A

Irrigation Options to Avoid Critical Stress: An Overview

EDWARD A. HILER

Texas A&M University
College Station, Texas

TERRY A. HOWELL

Water Management Research Laboratory, USDA-ARS
Fresno, California

I. INTRODUCTION

In many areas of the United States and other parts of the world, available irrigation water supplies are either very low already or being depleted rapidly. As water supplies diminish, population increases, and municipal and industrial water needs increase, optimization of efficiency of irrigation water use in food and fiber production becomes increasingly imperative.

A recent United Nations Food and Agricultural Organization publication (Doorenbos & Kassam, 1979) on yield response to water contains in its preface the following statement in bold-faced lettering:

> The upper limit of crop production is set by the climatic conditions and the genetic potential of the crop. The extent to which this limit can be reached will always depend on how finely the engineering aspects of water supply are in tune with the biological needs for water in crop production. Therefore, efficient use of water in crop production can only be attained when the planning, design and operation of the water supply and distribution system is geared toward meeting in quantity and time, including the periods of water shortages, the crop water needs required for optimum growth and high yields.

This suggests three areas that offer considerable promise for increasing efficiency of water use in irrigated agriculture as related to the subject of this chapter: (i) improved irrigation scheduling techniques, (ii) improved irrigation water application and distribution systems, and (iii) application of system optimization methods to improve the on-farm water allocation to crops. These three areas will be the principal topics considered in this chapter. Off-farm allocation, though important, is beyond the scope of this chapter, but some aspects are discussed in Chapter 12 by van Schilfgaarde and Rawlins.

Limitations to Efficient Water Use in Crop Production.

Any discussion of critical water stresses in crops and irrigation approaches to avoid them should be based upon a sound knowledge of soil-water-atmosphere-plant relationships. See Chapters 6A and 7A for information on this subject. Other excellent presentations providing background directly pertinent to this chapter are given by Kramer (1969), Kozlowski (1968a, 1968b, 1972, 1976), Slatyer (1967), and Levitt (1972). The following minimal discussion of yield responses to water is essential as a background for a meaningful presentation of what follows in this chapter.

It is desirable at this point to distinguish between the terms "water stress" and "water deficit" as they are used in this chapter. "Water stress" refers to the undesirable effect on plant yield caused by excessive plant "water deficit."

Most nonforage crop yields are more sensitive to water deficits at certain growth stages than at other stages (e.g., Salter & Goode, 1967; Shipley & Regier, 1970; Downey, 1972; Hagan & Stewart, 1972; Hiler et al., 1972; Lewis et al., 1974; Doorenbos & Kassam, 1979). Evidence summarized by these authors demonstrates that the extent of yield reduction depends not only on the stage of growth but also on the magnitude of the deficit.

Hiler and Clark (1971) developed the stress day index (SDI) concept as a quantitative means for determining the stress imposed on a crop during its growing season. This concept will be discussed later regarding its utilization in irrigation scheduling. One important component of SDI is the crop susceptibility factor (CS), a term that indicates the plant's sensitivity to water deficit at different growth stages. It provides a numerical answer to the question, "How much will crop yield be reduced by a plant water deficit of known magnitude occurring at a given growth stage?" Crop susceptibility values are determined experimentally as the fractional reduction in yield resulting from a maximum water deficit during a given growth stage. For example:

$$CS_i = (x - x_i)/x, \qquad [1]$$

where x is the marketable yield from a control treatment kept well-watered throughout the season, and x_i is the yield in the treatment that was subjected to deficit only during the ith growth stage. Table 1 shows CS values for various irrigated crops. Doorenbos and Kassam (1979) provide data for a much broader range of crops, which could be used to develop information similar to that in Table 1. All of these data generally support the widely held belief that the most critical growth stage with regard to water deficit is the period shortly before or after anthesis.

Since the reduction in yield depends on the magnitude of the deficit as well as the stage of growth, the CS values in Table 1 represent only one column of a matrix that describes the truer situation. With this in mind, Hiler et al. (1972) conducted an experiment to determine CS values for southern peas (var. Burgundy) in which both growth stage and magnitude of deficit were variables. The 70-day growing season was divided into three stages: (i) vegetative (from germination to the onset of flowering, (ii)

Table 1—Crop susceptibilities (CS) for various irrigated crops (Hiler et al., 1974).

Crop	Growth stage	CS
Corn	Vegetative	0.25
	Silking and tasseling to soft dough	0.50
	After soft dough	0.21
Cotton	Prior to flowering	0.00
	Early flowering	0.21
	Peak flowering	0.32
	Late flowering	0.20
Grain sorghum	Vegetative—six- to eight-leaf stage	0.25
	Boot to heading	0.36
	Heading to soft dough	0.45
	After soft dough	0.25
Peanuts	Vegetative to peak flowering and early pegging (30–50 days after emergence)	0.36
	Peak pegging and nut development (60–80 days)	0.24
	Late nut development and maturation (90–110 days)	0.12
Rice	Vegetative	0.17
	Reproductive and ripening	0.30
Southern peas	Vegetative—prior to flowering	0.13
	Flowering and early pod formation	0.46
	Pod development to maturation	0.43
Soybeans	Vegetative	0.12
	Early-to-peak flowering	0.24
	Late flowering and early pod development	0.35
	Late pod development to maturation	0.13

flowering and early pod formation, and (iii) pod development and maturation. In each stage plants were subjected to soil water deficit that resulted in three levels of leaf water potential: −14, −21, and −28 bars. Crop susceptibility values were calculated for each stage and plant deficit level, as shown in Table 2.

The presentation in Table 2 still does not reflect the complete picture, since effects of conditioning or drought hardening at a given growth stage resulting from deficits at early stages are not taken into account. That added complexity will not be considered in this chapter; conditioning provides a significant opportunity for further research.

Another area, covered in detail in Chapter 11B needing brief treatment here for background, is mathematical models relating water stress to crop yields. Howell and Hiler (1975) reported three models for use in system simulation; a brief description follows.

Jensen (1968) proposed a multiplicative-type model for crop yield given as:

$$Y/Y_{max} = \prod_{i=1}^{n} (ET/ET_o)_i^{\lambda_i}, \qquad [2]$$

where λ = a sensitivity factor, Y = actual yield, Y_{max} = maximum yield, ET = actual evapotranspiration, and ET_o = potential evapotranspiration. For grain sorghum, Jensen proposed sensitivity values of 0.5, 1.5, and 0.5

Table 2—Dry pea and aboveground dry matter yields and crop susceptibilities (Hiler et al., 1972).†

Treatment‡	Dry pea yield	Crop susceptibility	Dry matter yield
	g pot^{-1}		g pot^{-1}
C	12.9	--	27.6
I a	11.3	0.13	28.9
I b	8.9	0.31	20.0
I c	4.3	0.66	14.6
II a	7.0	0.46	15.0
II b	4.6	0.65	12.1
II c	1.3	0.90	7.5
III a	7.3	0.43	13.7
III b	7.0	0.46	12.5
III c	6.2	0.52	12.2

† Dry matter yield includes only the aboveground portion of the plants and does not include the dry peas.

‡ C = control—no deficit during growing season, I = deficit during vegetative growth stage only, II = deficit during flowering and early pod formation stages only, III = deficit during pod development and maturation stages only, and a, b, c are the −14, −21, and −28 bar levels of deficit, respectively.

for the emergence-to-boot, boot-to-milk, and milk-to-harvest stages, respectively.

Based on the additive-type SDI model (Hiler & Clark, 1971), the following equation can be developed for relative yield (Howell & Hiler, 1975):

$$\frac{Y}{Y_{max}} = 1.0 - \frac{A}{Y_{max}} \sum_{i=1}^{n} \left[\alpha_i \left(1.0 - \frac{ET}{ET_o} \right)_i^2 \right], \quad [3]$$

where A and α are sensitivity functions. Minhas et al. (1974) proposed a similar equation, except that it is multiplicative, as follows:

$$\frac{Y}{Y_{max}} = \prod_{i=1}^{n} \left[1.0 - \left(1.0 - \frac{ET}{ET_o} \right)_i^2 \right]^{b_i}, \quad [4]$$

where b is a sensitivity factor. Equations [2], [3], and [4] are similar and, if the same data are used to compute the sensitivity factors, each should produce similar yield prediction results.

Equations [1]–[4] and Tables 1 and 2 will be utilized in the materials that follow in this chapter.

II. EFFICIENT WATER USE THROUGH IMPROVED IRRIGATION SCHEDULING

Irrigation scheduling can be oriented toward maximizing either yield per unit of land area or yield per unit of applied irrigation water. The former objective may be economically justified when water supplies are readily available and irrigation costs are low. Situations such as this are becoming rare in this time of increasing scarcity of water resources for irrigation and rapidly escalating energy costs for pumping irrigation water. Thus the principal emphasis of this section will be on irrigation scheduling techniques for improving yield per unit of applied irrigation water. It should be noted here, however, that Hagan and Stewart (1972) and Taylor (1965) have presented excellent summaries of allowable root-zone depletions between irrigations for near-maximum yields of various classes of crops. This is very useful information when maximum yield is the production objective.

This section is not intended to provide a comprehensive discussion of irrigation scheduling techniques for more efficient water use. This has been done elsewhere by several investigators (e.g., Fleming, 1966; Linacre & Till, 1969; Jensen et al., 1970). Traditional methods of irrigation scheduling are based on a fixed-level deficit throughout the season. Two more recent methods with similar objectives will be discuseed here: the SDI method as proposed by Hiler et al. (1974), and the method of optimal sequencing of ET deficits as proposed by Stewart et al. (1975). These will be followed by a presentation of the computer-based irrigation scheduling based on water balance techniques as originally proposed by Jensen et al. (1970, 1971).

A. Stress Day Index Method

The SDI concept provides a quantitative method for determining the stress imposed on a crop during its growing season. It is based on the fact that crop sensitivities to internal water deficit depend on species and stage of growth of the crop. A detailed general discussion of the SDI concept to characterize effects of water stress on crop yields has been given by Hiler and Clark (1971). As utilized in irrigation timing, the SDI concept is formulated as

$$SDI = CS \times SD, \quad [5]$$

where CS is the crop susceptibility factor as defined in Eq. [1] and SD is the stress day factor. The SD is a measure of plant water deficit. It answers quantitatively the question "How dry is the crop?" at any particular time.

Values of SD can be characterized by several different indicators (Table 3). Crop growth and yield are determined directly by plant water stress rather than by soil water stress (Kramer, 1969). Theoretically, direct measurement of plant water deficit therefore should be a better indicator than such indirect indicators as stem diameter, diffusive resistance, and soil water status. However, the desired indicator in given circumstances will depend on a number of practical considerations as well as theoretical reasoning. The SDI approach to irrigation timing can be used with any of the indicators shown in Table 3.

Table 3—Possible indicators of the stress day factor (Hiler et al., 1974).†

Type of indicator	Indicator
Plant-based	|Leaf water potential|
Plant-based	Leaf-air temperature differential
Plant-based	Leaf diffusion resistance
Plant-based	1 ÷ stem diameter
Soil-based	|Soil water potential|
Soil-based	Percentage of available soil water depleted
Soil- and climate-based	ET_0 × |soil water potential|
Time-based	Irrigation period

† Absolute values are used for convenience to avoid negative numbers.

Irrigation timing with the SDI concept is accomplished by irrigating when the daily SDI (daily SD × CS for the appropriate growth stage) reaches some predetermined critical value, SDI_0. In a production system, the value chosen for SDI_0 depends on the availability and/or cost of irrigation water. For example, if water supplies were severely limited, the value of SDI_0 would be set at a high level and one would irrigate only during the most susceptible periods.

Suppose an irrigator wishes to use the SDI concept for irrigating grain sorghum. Also assume he wants to use a leaf water potential measurement as an indicator of water deficit (SD). First he must determine his desired SDI_0 value to use in the SDI approach. The value of SDI_0 could be determined, as shown in Table 4, as the average of the SDI's using an SD of 12, or $SDI_0 = 4.0$. The last step is to determine the value of SD at irrigation time for each growth stage. This is done by dividing SDI_0 by the appropriate CS value for each growth stage, also as shown in Table 4. Thus the SDI approach would result in values of SD at irrigation time of 16, 11, 9, and 16 bars for the six- to eight-leaf, boot, heading-to-soft-dough, and after-soft-dough growth stages, respectively. With the SDI approach, the crop would be irrigated at low water deficit values (and hence frequently) during periods of high crop susceptibility, and vice versa.

Table 4—Calculation of the critical stress day index value (SDI_o) and determination of the stress day factor (SD) at irrigation time for each growth stage.

Growth stage	CS†	SD‡	SDI	SDI_o§	SD¶ at irrigation time
		bars			
6- to 8-leaf	0.25	12.0	3.00	4.0	16
Boot	0.36	12.0	4.32	4.0	11
Heading to soft dough	0.45	12.0	5.40	4.0	9
After soft dough	0.25	12.0	3.00	4.0	16
			15.72		

† Crop susceptibility (CS) determined from Table 1.
‡ Assuming SD = absolute value of leaf water potential.
§ SDI_o = avg. SDI = 15.72/4 = 3.93 ≈ 4.0 bars.
¶ Determined from SD = SDI_o/CS.

Another irrigator may also wish to use the SDI approach for timing grain sorghum irrigations, but he has equipment for measuring soil water potential and thus desires to use the absolute value of soil water potential in the crop root zone as SD. He has irrigated in the past when soil water potential reached −70 cbars. Following the same procedure as given in Table 4, the value of SDI_0 would be approximately 23 cbars and the values of SD at irrigation time would be 92, 64, 51, and 92 cbars for the respective growth stages.

Still another irrigator may be accustomed to using a bookkeeping (water balance) procedure for timing grain sorghum irrigations and desires to use percentage available soil moisture (ASM) depleted as his criterion for determining when to irrigate, i.e., as his SD indicator. He has previously irrigated when 50% ASM has been depleted from the crop root zone. Again following the pattern given in Table 4, the value of SDI_0 would be approximately 14% ASM depleted and the values of SD at irrigation time would be 56, 40, 31, and 56% ASM depleted for the same four crop growth stages.

As can be seen from the aforementioned examples, with the SDI approach the water deficit indicator is weighted according to the crop sensitivities at different growth stages so that reduced deficit occurs during

Table 5—Practical guide to irrigation timing using the SDI method.

Growth stage	Yield reduction if water-stressed†	To avoid stress, irrigate when soil moisture depleted reaches:‡
	%	%
	Grain sorghum	
Vegetative (6- to 8-leaf stage)	25	65
Boot to heading	36	45
Heading to soft dough	45	35
After soft dough	25	65
	Cotton	
Prior to flowering	00	--
Early flowering	21	60
Peak flowering	32	35
Late flowering	20	60
	Soybeans	
Vegetative	12	80
Early-to-peak flowering	24	45
Late flowering, early pod development	35	30
Late pod development to maturity	13	80
	Corn	
Vegetative	25	65
Silking and tasseling to soft dough	50	30
After soft dough	21	75

† From Table 1.

‡ Values are calculated for irrigation programs in which standard practice is to irrigate when 50% ASM is depleted. To adjust for any other percent ASM, divide values by 50 and multiply by the desired ASM.

critical stages. With this approach, it is hypothesized that increased water-use efficiencies will result. Efficient irrigators are likely using the essence of the SDI approach in a qualitative sense by trying to make sure that crops are well-irrigated during growth periods they know to be critical from past experience; for those irrigators, the material presented here merely allows them to put their irrigation-timing practice on a more quantitative basis.

Table 5 shows an example of a guide to irrigation timing using the SDI method. In this example, percent ASM is used as the deficit criterion, or SD factor, and 50% ASM depletion is used as the "standard" practice. As can be seen from Table 5, this approach results in more frequent irrigation at the critical growth stages and longer intervals between irrigations during the less sensitive periods.

In experimental rain-sheltered lysimeter and field plot studies with grain sorghum (cv. RS-671) and cotton (var. Lankart 611), Bordovsky et al. (1974) and Hiler et al. (1974) have shown that significant improvements in yield per unit of applied water (>50% increases) occur using the SDI method compared with using fixed-level deficit criteria throughout the irrigation season. Irrigation amounts with the SDI method were determined by water balance techniques, similar to other irrigation scheduling methods.

The principal shortcoming of the SDI approach to irrigation timing is that the CS values depend on the SD permitted in their determination. To precisely utilize the SDI approach, experiments are needed for which CS values are determined not only as a function of crop species and growth stage but also as a function of level of water deficit, as indicated by the results shown in Table 2 and the related discussion. Despite this shortcoming, utilization of the first-approximation CS values (Table 1) with the SDI method results in significantly increased yields per unit of applied water, primarily because the method accounts for variable crop sensitivities to water deficit at different growth stages.

B. Optimal Sequencing of Evapotranspiration Deficits

As indicated earlier in this chapter, little quantitative information is available on effects of "preconditioning" on crop sensitivities to water deficits at different growth stages. Stewart et al. (1975) attempted to incorporate this aspect into optimal irrigation scheduling, building further on the notion that crop sensitivities to water deficit vary with growth stage (Jensen, 1968; Hiler & Clark, 1971). They found that an optimal linear yield vs. seasonal ET relationship exists for given crops whenever ET deficits are optimally sequenced, assuming water deficit is the principal limiting yield-reduction factor. Deficits in ET accumulate whenever the actual ET rate falls below the potential rate. These ET deficits cause a reduction in yield below a maximum yield level; however, a minimum yield reduction takes place when ET deficits are optimally sequenced. Hence, the yield level is the maximum attainable for the attained ET accumulation. A suboptimal ET-deficit sequence results in a seasonal ET accumulation that is excessive for

Table 6—Upper limits of ET-deficit intensity in the pollination and grain-filling periods of corn, based on levels of ET deficit experienced by the crop in prior growth periods (from Stewart et al., 1975).

Vegetative period (weeks 5-8, inclusive) → pollination period (weeks 9-12, inclusive)		Pollination period (weeks 9-12, inclusive) → grain-filling period (weeks 13-17, inclusive)		
			Permissible upper limit in this period is	
If expected ET-deficit intensity is†	Permissible‡ upper limit in this period is	If expected ET-deficit intensity is	If veg.-period ET deficit	
			<38%	>38%
		%		
0	0	0	32	65
10	14	14	60	65
20	28	28	87	87
30	42	42	87	87
40	56	56	87	87
50	65	65	87	87

† Evapotranspiration deficits are expressed as percents of seasonal ET_{max}.
‡ "Permissible" means that if these upper limits of ET-deficit intensity are observed, the ET-deficit sequence will be optimal, i.e., the expected yield will be maximized for the available water supply.

the corresponding yield attainment. Stewart et al. (1975) found that optimal and suboptimal ET-deficit sequences are identifiable. Thus, if irrigations can be scheduled to avoid a suboptimal ET-deficit sequence, irrigators can be assured of producing the maximum yield for a given water supply.

To avoid suboptimal ET-deficit sequences, Stewart et al. (1975) have determined the approximate upper limits of ET deficit by growth stage of corn. These data (Table 6) indicate that the optimal yield vs. ET relationship exists if:

1) ET deficits are not >50% of the ET_0 in the vegetative period prior to tassel.
2) ET deficits in the pollination (tassel-blister-kernel) period are not >1.4 times the ET deficit in the prior vegetative period. If no "conditioning" ET deficit occurs in the vegetative period, none is tolerable in the pollination period.
3) ET deficits do not exceed 32% of the potential ET in the grain-filling period when no prior deficit occurs in earlier periods. However, if "conditioned" by a pollination-period deficit exceeding 28%, a deficit up to 87% is tolerable in the grain-filling period.

Table 6 sets forth the upper limits of tolerable ET-deficit intensity in any given growth period based on the levels of ET deficits in previous periods. These data have not been widely tested and may be most applicable to deep soils and warm, arid climates.

Recent works have followed up on the concept of a stress index for scheduling irrigations. Idso et al. (1977) developed a stress degree day concept in which canopy temperatures and albedo measurements have been correlated quantitatively with wheat yields. Sudar et al. (1979) expanded a soil-plant-air-water (SPAW) model to compute crop water stress effects on growth and yield as an integration to climatic, crop, and soil characteristics. Daily water stress index (WSI) values accumulated for the growing season correlated well with the observed corn and soybean yields.

C. Scheduling Irrigations by Water Balance Techniques Using Computers

The advent of commercial irrigation scheduling services occurred in the early 1970's. This approach is expanding as farms become larger, precise management becomes more critical, and computers become commonplace. The general procedure most widely used by commercial irrigation scheduling services is the USDA computer-based method developed by Jensen et al. (1970, 1971). The following discussion of this method follows closely the excellent presentation of it by Stegman et al. (1980), and the reader is referred to that work as well as Kincaid and Heermann (1974) for greater detail.

With the USDA model, soil water depletion is calculated as:

$$D = \sum_{i=1}^{n} (\mathrm{ET} - P_e - I_{dn} + W_d)_i, \quad [6]$$

where D is the root-zone water depletion in mm (after a thorough irrigation, $D = 0$); P_e is the effective daily precipitation (i.e., precipitation minus runoff); I_{dn} is the daily net irrigation; W_d is the daily drainage loss from the root zone or upward movement from a saturated zone (for upward movement the sign is negative); and i is the time increment (usually in days) for water balance summation.

Scheduling services can operate best when amounts and occurrence dates for rainfalls and irrigations are observed locally by each irrigator. Irrigation amounts should be metered to obtain accurate measurement of gross applications. The gross irrigation depths must be adjusted for irrigation efficiency to obtain I_{dn}.

The value of ET in Eq. [6] is estimated, usually either by the Jensen-Haise or a modified Penman equation (Jensen, 1973) for estimating the daily potential or energy-limited evapotranspiration, ET_0. Crop efficiencies are applied to the daily ET_0 for estimating the actual crop ET. Weather data input should come from stations located at representative sites within an irrigated area.

Drainage from unsaturated soil profiles strongly depends on soil water content above field capacity. In applications of Eq. [6], W_d is often estimated to equal the amount by which an irrigation or rainfall exceeds the prevailing root-zone water depletion. The errors associated with this simplification may be quite acceptable in arid and semiarid areas, particularly if irrigations are carefully controlled. In studies by Heermann et al. (1976), center-pivot irrigations were scheduled by this method to maintain relatively high available water levels in the top 30 cm of root zone, and at the same time to permit a programmed depletion of soil water from the lower root zone. This type of irrigation scheduling resulted in little drainage loss and crop yields were not depressed relative to wetter treatments.

With conventional gravity irrigation, however, soil water content in part of the root zone is raised to levels near field capacity at the time of each irrigation. Then, from the time each irrigation is terminated to the time root-zone drainage loss becomes negligible, ET also occurs simultaneously. The ET loss in these periods must be considered as part of the water potentially available to plants. To account for this additional available water, attempts can be made to define an ''effective'' field capacity.

Miller and Aarstad (1971) found that the time to sample a soil profile after irrigation for determining the ''effective'' field capacity is affected by ET rates, soil properties, and the initial penetration depth by the applied water. The most practical alternative is to apply daily estimates of W_d to Eq. [6] after each irrigation until drainage loss is negligible, i.e., ≤ 0.1 mm d^{-1}. Ogata and Richards (1957) showed that the water content for a draining soil profile can be expressed empirically by

$$W = cW_o t^{-m}, \quad [7]$$

where W is the water content (mm), t is the time after irrigation has stopped, W_o is the water content when $t = 1$ day, m is a constant for the given soil profile, and c is a constant with dimensions of t^m. The drainage rate is then given by

$$\mathrm{d}W/\mathrm{d}t = mW(W/cW_o)^{1/m}. \quad [8]$$

For daily time increments, Jensen (1972) determined the cumulative drainage is approximated by:

$$W_c = \sum_{i=1}^{\omega} m[W_{i-1} - \mathrm{ET_i}]\{(W_{i-1} - \mathrm{ET_i})/W_o\}^{1/m}, \quad [9]$$

where W_c = the cumulative drainage (mm), i is the number of days after irrigation, and ET is the evapotranspiration for the day. Miller and Aarstad (1974) reported m values to vary from 0.10 to 0.15 for a range of soil textures.

Use of Eq. [9] requires the field measurement of m and W_o values. As an alternative, Black et al. (1970) demonstrated that when unsaturated hydraulic conductivity data are available W_c can be estimated in sandy soils as:

$$W_c = \sum_{i=1}^{\infty} [K(\theta_a)_i - \mathrm{ET_i}], \quad [10]$$

where $\theta = W/z$ is the average soil water content through depth z on day i.

The value of z includes the deepest rooting depth of the crop. Equation [10] will estimate the drainage loss in finer textured soils if the hydraulic gradient in the root zone remains close to unity during the drainage period.

A negative drainage loss, or upward flow, occurs when the water table is relatively close to the soil surface or close to the bottom of the root zone. Theoretical solutions to flow equations for this process have been given by Philip (1957) and Gardner (1958). For applied scheduling, Jensen (1972) suggested:

$$W_c = \{1 - [(\mathrm{ASW} - a_4)/(100 - a_4)]\}[z_c/(z_w - z_r)]^n\,\mathrm{ET}, \qquad [11]$$

where z_c is the effective height of the capillary fringe above the water table, z_w is the depth to the water table; z_r is the depth of the roots, ASW is the available soil water to z_r in percent; a_4 is a constant (about 25), and n is a constant for the given soil profile (expected to vary between 1 and 3).

With the USDA scheduling program, the days to the next irrigation are computed as

$$I_{\mathrm{days}} = (D_o - D)/(\mathrm{d}D/\mathrm{d}t)_{\mathrm{antic}}, \qquad [12]$$

where I_{days} = the estimated days, D_o = the current optimum depletion of soil water, D = the estimated depletion to date, and $(\mathrm{d}D/\mathrm{d}t)_{\mathrm{antic}}$ = the anticipated rate of soil water depletion in the root zone until the next irrigation is needed.

The anticipated rate of soil water depletion is often estimated from a calculation of the mean ET rate for the previous 3–5 days, or from advance 2- to 5-day weather forecasts. When I_{days} is >5 days into the future, expected $\mathrm{ET_o}$ rates may be estimated from the long-term mean levels of each climatic parameter at given days, t. Assuming no rainfall, the advance estimates of the daily $(\mathrm{d}D/\mathrm{d}t)_{\mathrm{antic}}$ are calculated as

$$\mathrm{ET_i} = K_c\,(\mathrm{antic\ ET_o})_{\mathrm{i}}, \qquad [13]$$

where K_c is the prevailing value of the crop coefficient (Jensen et al., 1971) on each day, and antic ($\mathrm{ET_o}$) is the expected value of potential evapotranspiration.

In climates where rainfall is significant, $(\mathrm{d}D/\mathrm{d}t)_{\mathrm{antic}}$ may include an expected rainfall value (P_e) and is calculated as $(\mathrm{ET} - P_e)_i$. Published functions for estimating expected rainfall are available for many areas in the United States.

The amount of water to apply at irrigation is usually calculated as

$$W_I = D_o/I_e;\ D_o > D \qquad [14]$$

or

$$W_I = D/I_e;\ D > D_o, \qquad [15]$$

where I_e is the attainable irrigation efficiency, expressed as a decimal. These calculations are most applicable to conventional gravity methods and set-type sprinkler systems. With automated or highly mechanized systems such as the center pivot type, irrigations may be applied so that the prevailing D is not entirely eliminated at each water application. When necessary, W_I can also be adjusted for the necessary leaching requirement.

To date, the USDA computer-based scheduling approach does not incorporate procedures to account directly for the fact that crop sensitivities to water deficit vary with growth stage. This could be accomplished by incorporating the SDI method (Hiler et al., 1974) or the optimal sequencing approach (Stewart et al., 1975) into the computer program. Addition of this component into the system should improve yields per unit of applied water and hence be particularly valuable for scheduling irrigations in water-short areas.

III. EFFICIENT WATER USE THROUGH IMPROVED WATER APPLICATION SYSTEMS

The very best irrigation scheduling method is of little value to irrigators if their irrigation systems do not have the capability of meeting the recommended schedule. Scheduling flexibility usually requires a water supply rate considerably larger than the crop ET rate. This section briefly enumerates a few of the recent developments in irrigation application equipment that increase the irrigator's options to avoid critical crop stress.

A. Surface Irrigation

With traditional gravity systems, irrigation management options have been fairly limited in the past. As available irrigation water becomes more limited and expensive, however, more innovative approaches are being adopted. Irrigators have moved toward shorter lengths of run, tailwater recovery, and alternate-furrow irrigation to increase their irrigation management options. Recent developments in laser technology have made possible the extensive use of level-basin irrigation in the western United States, resulting in savings in both water and energy (Dedrick et al., 1978).

Much excellent theoretical and applied research has been done in the area of surface-irrigation hydraulics. However, until recently, the labor requirement to attain the recommended cutback furrow stream sizes was excessive to the point that this was rarely adopted in practice. Recent developments in electronic microprocessor technology should make automation of surface-irrigation systems much easier to achieve and result in considerable improvement of both irrigation management options and efficiency of water application.

B. Sprinkler Irrigation

The major trend in sprinkler irrigation has been toward mechanical-move systems, particularly center-pivot systems. Since most of these systems are designed to supply only the crop ET needs, scheduling flexibility depends on utilizing precipitation effectively. These systems can be managed by adjusting travel speed, system flow rate, sprinkler nozzles, etc., to achieve an efficient, uniform system for irrigating a wide range of soil, topographic, and crop conditions.

Sprinkler irrigation requires a large energy input to pressurize the system. Operating pressures of 3.0–7.8 bars (45–115 psi) are common. Sprinklers with good uniformities at operating pressures near the 2- to 2.4-bars range are being tested and used on many center-pivot systems to reduce operating costs. Low-pressure sprinkler systems tend to have larger instantaneous-application rates; therefore, runoff from low-intake soils have been a disadvantage.

Sprinkler systems have been utilized to apply limited irrigations to crops to avoid critical stresses. Many systems are in use in the southern, eastern, and midwestern United States as insurance investments against short-term droughts. One system can be used to irrigate several fields by moving the system and staggering planting dates or plant genotypes to keep all the fields from experiencing critical stress at once. A single irrigation application or several irrigations have improved yields dramatically in many cases.

C. Trickle Irrigation

Trickle, or drip, irrigation has a major advantage over sprinkler and surface irrigation in that water is applied in a very limited area adjacent to the crop's immediate root zone, and in precise amounts. Thus, because water losses to soil water evaporation, runoff, and percolation can be minimized and low soil water tensions can be maintained in the root zone, the yield per unit of irrigation water can generally be expected to be higher for trickle irrigation than for other application methods. This has been borne out by numerous experimental studies (e.g., DeRemer, 1970; Cole, 1971; Davis & Nelson, 1970; Hiler & Howell, 1973). Emitter clogging has been a problem with trickle irrigation. Trickle irrigation is economical at present for tree crops and high-cash-value horticultural crops. Bubbler systems (Rawlins, 1977) offer a trickle alternative that is not as susceptible to plugging but is limited to relatively flat slopes.

D. Low-Energy, Precision-Application (LEPA) System

A new system that offers much promise for both water and energy savings and is applicable to the major crops grown in the United States is the

LEPA system (Lyle, 1979; Lyle & Bordovsky, 1979). With this system, water is applied from a continuously moving system to microbasins formed by placing dikes across furrows. The small basins retain applied irrigation water as well as provide a means to capture rainfall without allowing runoff. The system will allow precise control of application amounts with uniform distribution and will accomplish this at a much lower energy requirement than that of current nongravity systems.

The overall system consists of two essential components: (i) a soil-surface modification implement, and (ii) the low-energy precision-application irrigation system. Both are necessary to accomplish the stated goal of maximizing irrigation efficiency and rainfall utilization.

The low-energy, precision-application system was developed with the following design criteria:

1) Adaptability to a wide range of flow rates (5–80 liters sec^{-1}) in order to utilize both low- and high-capacity wells.
2) Adaptability to all soil types (requires variable travel speed; when combined with basin tillage, travel speed is less critical).
3) Low-pressure operation (<0.7 bar for energy conservation and use with existing low-pressure pipe).
4) Capability for continuous rectilinear movement.
5) Capability for utilizing low-pressure underground pipe systems already installed on many farms (pipe currently used for gated-pipe irrigation).
6) Adaptability to present production systems with minimum alterations.

All of the design criteria have been met in the construction of a prototype system that is currently being evaluated. Water is delivered to the system from low-pressure underground pipe through portable hose. From the primary overhead supply pipe of the structure the water goes to a manifold system by means of pressure-regulating valves and then through drop tubes and emitters that discharge the water about 80 mm above the furrow and beneath the crop canopy. Preliminary tests are producing operating pressures as low as 0.35 bar while maintaining distribution efficiencies at about 95%.

Basin-tillage furrow-diking is the soil surface modification being used in conjunction with the precision-application system (Lyle & Dixon, 1977). A hydraulically operated implement has been designed and constructed to form the dikes. It was designed to allow flexibility in the control of dike placement and size. Criteria followed in the basin-tillage implement design were as follows:

1) Capability for trouble-free operation.
2) Compatibility with equipment already being used in row-crop production, such as bedders, planters, and cultivators, so it can be attached and not require a separate operation.
3) Capability for high-speed operation.
4) Adjustability of dike spacing and height so that basin size can be regulated.

The basin-tillage implement has proven mechanically successful and is capable of operating at speeds of 8–13 km h^{-1}, with dike-spacing capability of 0.8–7.6 m at these speeds. The practice is also proving to be extremely successful for use with the low-energy, precision-application system, and in retaining rainfall. Rainfall has not run off of test plots in 4 years of tests at Lubbock and Halfway, Tex. (Hudspeth, 1978). The greatest single rainfall event received during this period was an 80-mm rain occurring on test plot slopes ranging from 1.2 to 2.3%. Irrigation water has also been applied to the microbasins at a rate of 105 mm h^{-1} for a total application of 115 mm without causing runoff. This water was applied to a soil with a final intake rate of <2.5 mm h^{-1}; however, the slope was relatively flat (<0.2%).

Evaluation of this combined system for total water utilization is essentially in its initial stages. However, the concept looks extremely promising as a means to obtain efficient total water utilization along with substantial energy conservation in irrigation.

IV. EFFICIENT WATER USE THROUGH OPTIMIZATION OF ON-FARM WATER ALLOCATION

The development of optimization models for improved water management has expanded rapidly since the mid-1960's. Both dynamic and linear programming have been used (e.g., Flinn & Musgrave, 1967; Hall & Butcher, 1968; Dudley et al., 1971; Salcedo & Meier, 1971; Windsor & Chow, 1971). In this section, brief reviews will be presented of three more recent works (Howell et al., 1975; Ahmed et al., 1976; Trava et al., 1977). A broad elaboration of this topic is presented by Jones (Chapter 11C).

Howell et al. (1975) conducted a study to optimize irrigation decisions under high-frequency irrigation so that the maximum yield was obtained with a given quantity of irrigation water. Stochastic dynamic programming was used to maximize yield subject to irrigation-water availability constraints. The proper irrigation decision for each period was predicted if the soil water content and amount of irrigation water available at the start of the period were known. System simulation was utilized to simulate the stochastic state transitions. Results demonstrated that the irrigation amount could be reduced substantially without a large decrease in expected yield if the irrigation water was distributed optimally over the season.

Ahmed et al. (1976) utilized a simulation program to compute the plant water-use rate under a single stochastic regime using Monte Carlo techniques. The dynamic simulation model of the soil-water-atmosphere-plant system was developed as a tool for optimal irrigation decision-making under a stochastic weather regime and when water supplies are limited. Crop yield, foliage development, plant water deficits, and irrigation decisions have been treated as interdependent processes. Yield susceptibility of the crop to water deficit is a function of crop growth stage. The stomatal regulation of transpiration by the crop also has been taken into account. The model is a closed-loop dynamic system in which past irrigation decisions and weather conditions affect the current as well as the future response of the crop system and, consequently, the future irrigation decisions and the overall water-use efficiency. A specific example pertains to grain sorghum grown under weather conditions typical for south-central Texas.

Trava et al. (1977) developed a management computer program for a farm irrigation system covering several crops and soil types that would determine irrigation strategies to minimize labor costs while distributing a limited water supply to minimize crop stress. The USDA scheduling method (discussed earlier in this chapter) (Jensen et al., 1970, 1971; Stegman, 1980; Kincaid & Heerman, 1974) was used along with the Dantzig-Wolfe optimization technique to achieve the desired objective.

V. SUMMARY

In this chapter irrigation options to avoid critical stress are considered through discussion of (i) improved irrigation scheduling techniques, (ii) improved irrigation water application systems, and (iii) application of system-optimization methods to improve on-farm water allocation to crops. Principal emphasis is placed on innovations made in these areas during the past decade. Much excellent work will be required in future decades to meet the demands of society for food production with dwindling irrigation water and energy resources. The application of systems-analysis methods in conjunction with crop simulation models in making real time irrigation management decisions has far-reaching possibilities for further development.

LITERATURE CITED

Ahmed, J., C. H. M. van Bavel, and E. A. Hiler. 1976. Optimization of crop irrigation strategy under a stochastic weather regime: a simulation study. Water Resour. Res. 12:1241–1247.

Black, T. A., W. R. Gardner, and C. B. Tanner. 1970. Water storage and drainage under a row crop on a sandy soil. Agron. J. 62:48–51.

Bordovsky, D. G., W. R. Jordan, E. A. Hiler, and T. A. Howell. 1974. Choice of irrigation timing indicator for narrow row cotton. Agron. J. 66:88–91.

Cole, T. E. 1971. Subsurface and trickle irrigation, a survey of potentials and problems. ORNL-NDIC-9. Oak Ridge Natl. Lab., Nuclear Desalination Info. Ctr. Oak Ridge, Tenn.

Davis, S., and S. D. Nelson. 1970. Subsurface irrigation today and tomorrow in California. p. H1–H8. *In* Proc. Natl. Irrig. Symp., Univ. of Nebraska, Lincoln. 10–13 Nov. 1970. Am. Soc. Agric. Eng., St. Joseph, Mich.

Dedrick, A. R., J. A. Replogle, and L. J. Erie. 1978. On-farm level-basin irrigation—save water and energy. Civ. Eng. 48:60–65.

DeRemer, E. D. 1970. An introduction to drip irrigation for vegetables. Veg. Crop Mgt. June 1970:11–12.

Doorenbos, J., and A. H. Kassam. 1979. Yield response to water. Irrig. and Drainage Paper 33. FAO, United Nations, Rome, Italy.

Downey, L. A. 1972. Water-yield relations for nonforage crops. J. Irr. Drain. Div. Am. Soc. Civ. Eng. 98(IRI):107–115.

Dudley, N. J., D. T. Howell, and W. F. Musgrave. 1971. Optimal intraseasonal irrigation water allocation. Water Resour. Res. 7:770–778.

Fleming, P. M. 1966. Crop water requirements and irrigation. p. 591–618. Proc. World Meteorol. Seminar, Melbourne, Australia. World Meteorol. Org., Melbourne.

Flinn, J. C., and W. F. Musgrave. 1967. Development and analysis of input-output relations for irrigation water. Aust. J. Agric. Econ. 11:1–19.

Gardner, W. R. 1958. Some steady state solutions of the unsaturated moisture flow equation with application to evaporation from a water table. Soil Sci. 85:222–232.

Hagan, R. M., and J. I. Stewart. 1972. Water deficits—irrigation design and programming. J. Irrig. Drain. Div. Am. Soc. Civ. Eng. 98(IR2):215–237.

Hall, W. A., and W. S. Butcher. 1968. Optimal timing of irrigation. J. Irrig. Drain. Div. Am. Soc. Civ. Eng. 94:267–275.

Heermann, D. F., H. R. Haise, and R. H. Mickelson. 1976. Scheduling center pivot sprinkler irrigation for corn production in eastern Colorado. Trans. ASAE 19:284–287.

Hiler, E. A., and R. N. Clark. 1971. Stress day index to characterize effects of water stress on crop yields. Trans. ASAE 14:757–761.

Hiler, E. A., and T. A. Howell. 1973. Grain sorghum response to trickle and subsurface irrigation. Trans. ASAE 16:799–803.

Hiler, E. A., T. A. Howell, R. B. Lewis, and R. P. Boos. 1974. Irrigation timing by the stress day index method. Trans. ASAE 17:393–398.

Hiler, E. A., C. H. M. van Bavel, M. M. Hossain, and W. R. Jordan. 1972. Sensitivity of southern peas to plant water deficit at three growth stages. Agron. J. 64:60–64.

Howell, T. A., and E. A. Hiler. 1975. Optimization of water use efficiency under high frequency irrigation. I. Evapotranspiration and yield relationship. Trans. ASAE 18:873–878.

Howell, T. A., E. A. Hiler, and D. L. Reddell. 1975. Optimization of water use efficiency under high frequency irrigation. II. System simulation and dynamic programming. Trans. ASAE 18:879–887.

Hudspeth, E. B. 1978. Basin tillage for water conservation and maximum dryland cotton production. Cotton Gin Oil Mill Press 11 Feb. 1978.

Idso, S. B., R. D. Jackson, and R. J. Reginato. 1977. Remote sensing of crops. Science 196: 19–25.

Jensen, M. E. 1968. Water consumption by agricultural plants. p. 1–22. *In* T. T. Kozlowski (ed.) Water deficits and plant growth. Vol. II. Academic Press Inc., New York.

Jensen, M. E. 1972. Programming irrigation for greater efficiency. p. 133–161. *In* D. Hillel (ed.) Optimizing the soil physical environment toward greater crop yields. Academic Press Inc., New York.

Jensen, M. E. (ed.). 1973. Consumptive use of water and irrigation water requirements. Tech. Comm. on Irrig. Water Requirements, Irrigation and Drainage Division, Am. Soc. Civ. Eng., New York.

Jensen, M. E., D. C. N. Robb, and C. E. Franzoy. 1970. Scheduling irrigations using climate-crop-soil data. J. Irrig. Drain. Div. Am. Soc. Civ. Eng. 96(IRI):25–38.

Jensen, M. E., J. L. Wright, and B. J. Pratt. 1971. Estimating soil moisture depletion from climate, crop and soil data. Trans. ASAE 14:954–959.

Kincaid, D. C., and D. F. Heermann. 1974. Scheduling irrigations using a programmable calculator. USDA-ARS-NC-12. U.S. Government Printing Office, Washington, D.C.

Kozlowski, T. T. (ed.). 1968a. Water deficits and plant growth. Vol. I. Development, control and measurement. Academic Press Inc., New York.

Kozlowski, T. T. (ed.). 1968b. Water deficits and plant growth. Vol. II. Plant water consumption and response. Academic Press Inc., New York.

Kozlowski, T. T. (ed.). 1972. Water deficits and plant growth. Vol. III. Plant responses and control of water balance. Academic Press, New York.

Kozlowski, T. T. (ed.). 1976. Water deficits and plant growth. Vol. IV. Soil water measurement, plant responses, and breeding for drought resistance. Academic Press, Inc., New York.

Kramer, P. J. 1969. Plant and soil water relations: a modern synthesis. McGraw-Hill Book Co., New York.

Levitt, J. 1972. Responses of plants to environmental stresses. Academic Press, Inc., New York.

Lewis, R. B., E. A. Hiler, and W. R. Jordan. 1974. Susceptibility of grain sorghum to water deficit at three growth stages. Agron. J. 66:589–591.

Linacre, E. T., and M. R. Till. 1969. Irrigation timing and amounts. J. Aust. Inst. Agric. Sci. 35:175–196.

Lyle, W. M. 1979. Low energy precision application irrigation system. p. 25–29. *In* Proc. Western Irrigation Forum. Denver, Colo.

Lyle, W. M., and J. P. Bordovsky. 1979. Traveling low energy precision irrigator. p. 121–131. *In* Irrigation and drainage in the nineteen-eighties. Proc. 1979 ASCE Specialty Conf., Albuquerque, N.M.

Lyle, W. M., and D. R. Dixon. 1977. Basin tillage for rainfall retention. Trans. ASAE 20: 1013–1021.

Miller, D. E., and J. S. Aarstad. 1971. Available water as related to evapotranspiration rates and deep drainage. Soil Sci. Soc. Am. Proc. 35:131–134.

Miller, D. E., and J. S. Aarstad. 1974. Calculations of the drainage component of soil water depletion. Soil Sci. 118:11–15.

Minhas, B. S., K. S. Parikh, and T. N. Srinivasan. 1974. Toward a structure of a production function for wheat yields with dated inputs of irrigation water. Water Resour. Res. 10: 383–393.

Ogata, G., and L. A. Richards. 1957. Water content changes following irrigation of bare-soil that is protected from evaporation. Soil Sci. Soc. Am. Proc. 21:355–356.

Philip, J. R. 1957. The physical principles of soil water movement during the irrigation cycle. Proc. 3rd Int. Congr. of Irrigation and Drainage 8:125–128, 154. CSIRO, Melbourne, Australia.

Rawlins, S. L. 1977. Uniform irrigation with a low-head bubbler system. Agric. Water Manage. 1:167–178.

Salcedo, D., and W. L. Meier, Jr. 1971. Timing of irrigations to maximize economic yields. p. 81–114. *In* Conf., Process of Optimization of Irrigation and Drainage Systems, Lincoln, Nebr. 6–8 Oct. 1971. Am. Soc. Civ. Eng., New York.

Salter, P. J., and J. E. Goode. 1967. Crop responses to water at different stages of growth. Commonwealth Agricultural Bureaux, London.

Shipley, J., and C. Regier. 1970. Water response in the production of irrigated grain sorghum, High Plains of Texas, 1969. Tex. Agric. Exp. Stn. Prog. Rep. 2829. College Station.

Slatyer, R. O. 1967. Plant-water relationships. Academic Press Inc., New York.

Stegman, E. C., J. T. Musick, and J. I. Stewart. 1980. Irrigation water management. Chapter 18. *In* M. E. Jensen (ed.) Design and operation of farm irrigation systems. Am. Soc. Agric. Eng., St. Joseph, Mich.

Stewart, J. I., R. D. Misra, W. O. Pruitt, and R. M. Hagan. 1975. Irrigating corn and grain sorghum with a deficient water supply. Trans. ASAE 18:270–280.

Sudar, R. A., K. E. Saxton, and R. G. Spomer. 1979. A predictive model of water stress in corn and soybeans. Trans. ASAE 24:97–102.

Taylor, S. A. 1965. Managing irrigation water on the farm. Trans. ASAE 8:433–435.

Trava, J., D. F. Heermann, and J. W. Labadie. 1977. Optimal on-farm allocation of irrigation water. Trans. ASAE 20:85–88, 95.

Windsor, J. S., and V. T. Chow. 1971. Model for farm irrigation in humid areas. J. Irr. Drain. Div. Am. Soc. Civ. Eng. 97:369–385.

CHAPTER 11B

Irrigation Options to Avoid Critical Stress: Irrigation Scheduling—Some Applied Concepts

EARL C. STEGMAN

North Dakota State University
Fargo, North Carolina

I. SCHEDULING WITH SIMPLE WATER BALANCE METHODS

Evapotranspiration (ET) estimation via either the Jensen-Haise or Penman equations (see Eq. [6] in Chapter 11A) with computers has made large-scale computation of water balances economically possible for use by commercial irrigation scheduling services. However, opportunities exist for improving efficiencies of water use, even with rather simple methods. Producers should be encouraged to use these methods where commercial services are unavailable or when they choose not to contract with a consultant.

Potentials for efficiency improvement are illustrated in Fig. 1. Three water balance methods with simplified techniques for estimating ET are compared with water balance scheduling where ET estimates were made with the Jensen-Haise equation and associated crop coefficients (Stegman & Ness, 1974). Water balances were simulated over a 37-year period of weather records. Relative performance of each method was compared on the basis of average seasonal ET deficits that accumulated relative to estimated seasonal ET_{max} (ET when water is not limiting) levels.

The simplest method (PRESUP, "precipitation supplementation") illustrates the relative inefficiency of a minimum scheduling procedure that is often used by farmers. With this method relatively fixed irrigation amounts (I) are applied weekly (assumed irrigation system = center pivot) to supplement rainfall during a defined irrigation season. Modifications are sometimes made in the event of high rainfall amounts, however, most rainfall is typically ignored in scheduling of irrigations. Figure 1 shows that when net weekly irrigations (I_{net}) approach 3.1 cm week^{-1} in the climatic setting of this simulation, the average seasonal performance will nearly equal the Jensen-Haise method. But this performance is attainable at the expense of annually applying about 35% more water than with the Jensen-Haise method.

For the FIXET method, the daily ET is assumed fixed at a constant amount (see Woodruff et al., 1972). This scheme keeps the summation of

Limitations to Efficient Water Use in Crop Production.

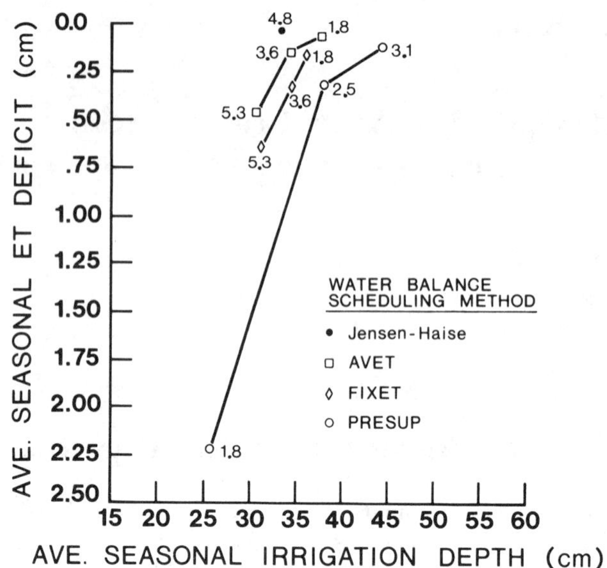

Fig. 1—Potential seasonal performance attainments (as indicated by average seasonal ET deficit) for a range of water-balance scheduling methods. Degree of scheduling sophistication decreases from Jensen-Haise method toward PRESUP method. For the Jensen-Haise, AVET, and FIXET methods the successive numbers for each curve indicate the estimated (as per indicated method) centimeters of soil water depletion at irrigation. For the PRESUP method, the numbers indicate the I_{net} (cm) per week during a defined irrigation season. Simulation assumptions were: irrigation system = center pivot; irrigated area = 53 ha; root-zone available water storage capacity = 10.2 cm; and available system-pumping capacity = 44 L/s.

($ET - P - I_{net}$) within an allowable depletion limit (see limits associated with this curve in Fig. 1). Performance approaches the Jensen-Haise method at about 11.5% more water pumped, thus a considerable improvement in efficiency is likely relative to the simplest PRESUP method. Further efficiency improvement is shown for the AVET method. This scheme utilizes average ET distributions as obtained from average climatic data or crop water-use studies in an area (see Black & Brosz, 1972).

Lundstrom and Stegman (1977) developed a simple tabularized water balance method that responds to day-to-day weather variation and crop development. Evapotransporation is estimated by applying a locally calibrated crop curve to a simplified form of the Jensen-Haise equation, i.e., potential ET (ET_0) is estimated as a function of the daily maximum air temperature and the average climatic solar radiation curve for a given area. This technique then simplifies to tables of ET for each crop, requiring farmers to observe only the daily maximum air temperatures. Lysimeter studies by Dylla et al. (1980) indicate that the efficiency of this method can about equal the Jensen-Haise method.

Other techniques for locally calibrating methods of ET estimation are also available (Pelton & Korven, 1969; Jensen & Middleton, 1970; Sutter & Corey, 1970; Vitosh, 1977). Most important is that these methods are simple to implement and can substantially improve efficiency of water use.

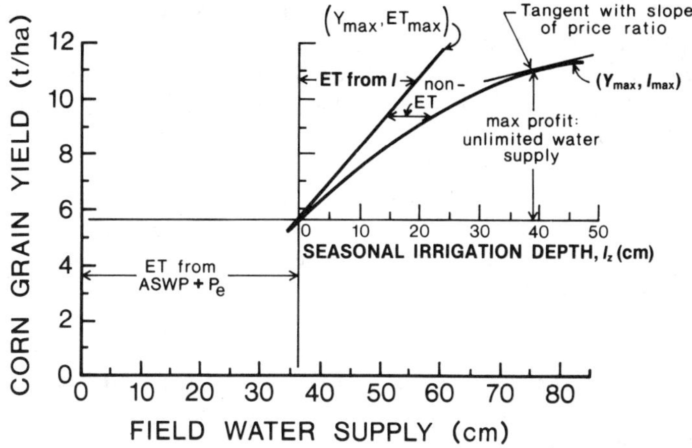

Fig. 2—Relationships of yield (Y) vs. seasonal ET and Y vs. seasonal irrigation depth (I_z), both set within a Y vs. field water supply (FWS) functional context. Also illustrated is the tangency of a price ratio C_I/C_Y to the Y vs. I_z function that locates the point of optimum net return given an unlimited water supply. (From Stewart & Hagan, 1973.)

II. THE CASE FOR ATTAINING YIELDS NEAR THE Y_{max} LEVEL

Stewart and Hagan (1973) demonstrated that crop yields are typically related to seasonal ET and seasonal irrigation depth (I_z) as shown in Fig. 2. In a given season, the available soil water at planting (ASWP) and effective rainfall (P_e) components of the seasonal field water supply allow the crop to reach its dryland yield level. Successive applications of irrigation water increases the (yield) Y vs. ET function above the dryland level. This function rises to a Y_{max}, ET_{max} level when irrigation amounts (I) and frequency are sufficient for plants to meet day-to-day ET demands. Non-ET loss typically increases as I approaches that needed to achieve the Y_{max}, ET_{max} level. The non-ET losses are caused by inefficiencies in water-application methods and the inexactness of irrigation scheduling. This latter effect is illustrated by the comparison of scheduling methods in Fig. 1. As the degree of sophistication in the scheduling procedure decreases, the average annual water pumpage will likely increase to achieve near Y_{max}.

Because Y response to applied I_z is a diminishing-return function, it follows that at some point on this curve further water application cannot be economically justified. A line with slope equal to the price (C) ratio C_I/C_Y will at its point of tangency with the Y vs. I_z function (see example in Fig. 2) locate the optimum economic level of production given an unlimited water supply for a given land area (C_I is the variable cost per unit of water applied and C_Y is the unit crop price). Analyses by Skogerboe et al. (1979) suggest the C_I/C_Y price ratio slope is inherently flat, thus profit will increase with increased water application until Y approaches Y_{max} if non-ET losses do not become excessive. Relative profitability can, however, improve substantially as non-ET loss is reduced by efficient scheduling methods.

One can argue that with increasing energy costs the price ratio will increase, thus lowering the optimal depth of water application. This is true to a degree, but Skogerboe et al. (1979) also argue that the price of water can usually rise only a small amount without a rise in price received for the crop, before irrigation becomes uneconomical. Thus, as price of water application escalates so must crop selling price, or other ways must be found to maintain a rather flat slope of the price ratio. Some means of doing this are by improving scheduling efficiency, reducing pressure or head on pumped water, improving system-application efficiency, maintaining high pump efficiency, reusing tailwater, and producing higher-value crops.

III. SCHEDULING CRITERIA FOR Y_{max} PER UNIT AREA

Water management to achieve near Y_{max} is frequently the most profitable scheme, particularly when water supply for a given land area is unlimited and application efficiency is relatively high. When these conditions occur, *I* should usually be sufficient for plants to meet the day-to-day evaporative demand, and irrigations (particularly in the most stress-sensitive stages of seed crops) should be at a frequency that maintains a high soil water potential, particularly in the upper root zone. This latter requirement relates to studies (Hsiao, 1973; Begg & Turner, 1976) that indicate adequate cell turgor pressures must be maintained to achieve maximum rates of expansive growth. Thus, as high soil water potentials are maintained by frequent irrigation, the daily depression of leaf water potential is minimized and net photosynthesis is optimized within practical limits.

Table 1—Expected mid-afternoon corn leaf water potentials vs. remaining available water in a sandy-loam-textured root zone for the indicated mid-afternoon ambient temperature levels.†

Available water level	Ambient temperatures, °C				
	37.8	32.2	26.7	21.1	15.6
%	predicted leaf water potential, bars				
100	12.5	11.1	9.6	8.2	6.7
90	**12.9**	11.4	10.0	8.5	7.0
80	**13.3**	11.8	10.3	8.9	7.4
70	**13.6**	12.2	10.7	9.2	7.8
60	**14.1**	**12.6**	11.1	9.7	8.2
50	**14.6**	**13.1**	11.6	10.2	8.7
40	**15.1**	**13.6**	12.2	10.7	9.2
30	**15.8**	**14.3**	**12.8**	11.4	9.9
20	**16.6**	**15.1**	**13.7**	12.2	10.7
10	**17.8**	**16.3**	**14.8**	**13.4**	11.9
0	**22.3**	**20.8**	**19.3**	**17.9**	**16.4**

† From Stegman et al. (1976).

‡ Boldface values are yield-reducing depletions. Other values are at safe depletion levels. Studies by Stegman and Aflatouni (1978) indicated yield reductions were likely when leaf water potentials fell below −12.5 bars on several successive days between irrigation periods.

Frequently, Y_{max} is produced in given locations (see summaries by Salter & Goode, 1967) when root-zone water content is near field capacity at planting (to allow rapid early-season development) and subsequent irrigations do not allow depletion of root-zone plant-available water to exceed predetermined limits (Chapter 2A). These limits are subject to modification when, for example, considerations of crop quality and degree of maturity at harvest are important or the practicalities of system operation dictate less-frequent irrigations. Modifications can also be made when the effects of evaporative demand on resulting plant water stress are known. This process is illustrated in Table 1, wherein estimates are given for probable mid-afternoon leaf water potentials of corn (grown in sandy-loam- to loam-textured profiles) as a function of remaining available soil water and prevailing mid-afternoon ambient temperature (Stegman et al., 1976). Data from Stegman and Aflatouni (1978) indicate corn grain yields were not depressed when mid-afternoon leaf water potentials were not allowed to fall below -12.5 bars. Imposition of this limit as a timing criteria to the data of Table 1 delineates the degree of available water depletion that can be tolerated for expected levels of ambient air temperatures.

IV. MAXIMIZING YIELD PER UNIT WATER APPLIED

As illustrated in Fig. 2, the Y vs. I_z function curves away from the Y vs. ET function. Thus, it can be visualized that Y/I_z will increase as non-ET losses are reduced. When water supply is unlimited one should strive to achieve Y in the Y_{max} region (for maximum profitability) and to achieve these with a minimum of non-ET losses. Irrigation systems must be appropriately designed, installed, and operated to minimize deep-percolation and surface-runoff losses. Further improvement in Y/I_z is possible with irrigation scheduling techniques that successfully minimize the seasonal evaporation loss but cause as little reduction in the seasonal potential transpiration as possible.

Evaporation losses can be reduced by minimizing frequency of irrigation in the early ground-cover development period. If the root-zone water content is near field capacity at planting, the early root-zone advance takes place to moisture and a maximum delay of initial irrigation is possible. Once ground cover is complete, many crops enter their reproductive growth stages and irrigations should usually be sufficient to maintain potential transpiration rates. In the latter grain-filling period of seed crops, irrigations can often be reduced to allow some withdrawal of earlier-stored water from the lower root zone. However, root-zone water depletion should not approach levels that significantly limit transpiration rates until most seed crops are near physiologic maturity. The SDI (stress day index) method, as detailed in Chapter 11A, essentially seeks to provide a quantitative basis for this type of scheduling scheme, i.e., to minimize irrigation in the early vegetative period, fully irrigate in the most stress-sensitive growth stages, and then allow a programmed depletion of soil water in the later, less-sensitive growth period.

As an alternative to applying irrigations with variable frequency as implied by the above discussion, high-frequency irrigation concepts (Fonken et al., 1974; Rawlins & Raats, 1975; Miller & Aarstad, 1976) may also be used with irrigation systems capable of uniform water application in controlled amounts. With high-frequency irrigation, the definition of allowable soil water depletion or plant stress limits becomes relatively less important. Given adequate system capacity (pumping rates), the high irrigation frequency can satisfy the day-to-day ET demand, and a high soil water potential is maintained in the upper root zone. Root-zone water deficit can be managed to oscillate at about the deficit existing at the time irrigations are begun, thus maintaining a degree of storage capacity for rainfall. Reduced I, rather than reduced frequency in late season, can achieve a programmed depletion of soil water to maximize storage capacity for off-season precipitation. Near-Y_{max} attainment requires that I plus water extraction from storage matches the seasonal ET requirement not satisfied by rainfall.

If water supply is so limited that Y_{max} potentials are unattainable, Y/I_z will usually be near-maximum when the water management scheme achieves the maximum yield possible for the attainable seasonal ET, i.e., the attained Y plots on or near the optimal Y vs. ET relationship (see Stewart et al., 1975). Optimal sequencing of ET deficits (Chapter 11C) provides a complex scheduling criteria for attaining the optimal Y vs. ET relationship. Thus, Stewart et al. (1977) have also suggested a simplified management criteria (probably most applicable to semiarid and arid climates) by noting that the maximum attainable yield for a given seasonal ET-deficit level tends to occur when deficits are spread as evenly as possible over the growing season. The main factors to consider in implementing this principle are length of the irrigation season (dependent on crop variety and soil characteristics), minimum depth of water that can be uniformly applied with a given irrigation system, and available seasonal I_z. Scheduling is based on the concept of high-frequency irrigation, i.e., applying small depths (or the minimum feasible for gravity methods) per irrigation at essentially evenly timed intervals.

High frequency of irrigation with an inadequate water supply attempts to keep the soil zone of greatest nutrient supply moist and to minimize the magnitude and duration of daily leaf water potential depression. Sufficient turgor recovery is thus more likely (Rawlins & Raats, 1975) in the nighttime and morning hours of most days to permit at least some photosynthetic activity and the subsequent conversion of assimilates to growth. This form of day-to-day plant activity under conditions of limited water supply leads to more efficient water utilization than where drought or water deficits are concentrated in one particular growth stage.

In semiarid and arid areas, continued high-frequency irrigations with a limited water supply may lead to excessive salinity buildup in the root zone. Willardson and Hanks (1976), in leaching tests with a solid-set sprinkler system, concluded that it is probably unrealistic to maintain low leaching fractions of <0.10–0.15 on a field scale. They recommend that it is most efficient to irrigate without necessarily leaching for several seasons, while monitoring the soil salinity status, and then to leach periodically. More complete management guidelines for salinity control are given in a review by Ayers and Westcott (1976) and in Chapter 2C.

LITERATURE CITED

Ayers, R. S., and D. W. Westcott. 1976. Water quality for agriculture, Irrig. and Drain. Paper 29. United Nations FAO, Rome, Italy.

Begg, J. E., and N. C. Turner. 1976. Crop water deficits. Adv. Agron. 28:161–217.

Black, S., and D. Brosz. 1972. Irrigation scheduling through moisture accounting. South Dakota State Univ. Ext. Circ. 686, Brookings.

Dylla, A. S., D. R. Timmons, and H. Shull. 1980. Estimating water used by irrigated corn in west central Minnesota. Soil Sci. Soc. Am. J. 44:823–827.

Fonken, D. W., J. C. Steele, and P. E. Fischbach. 1974. Sprinkler irrigation design criteria for sugarbeets. Trans. ASAE 17:889–891.

Hsiao, T. C. 1973. Plant responses to water stress. Annu. Rev. Plant Physiol. 24:519–570.

Jensen, M. C., and J. E. Middleton. 1970. Scheduling irrigation from pan evaporation. Washington Agric. Exp. Stn. Circ. 527, Pullman.

Lundstrom, D. R., and E. C. Stegman. 1977. Checkbook method of irrigation scheduling. Am. Soc. Agric. Eng. Paper no. NCR77-1001. ASAE, St. Joseph, Mich.

Miller, D. E., and J. S. Aarstad. 1976. Yields and sugar content of sugarbeets as affected by deficit high frequency irrigation. Agron. J. 68:231–234.

Pelton, W. L., and H. C. Korven. 1969. Evapotranspiration estimates from atmometers and pans. Can. J. Plant Sci. 49:615–621.

Rawlins, S. L., and P. A. C. Raats. 1975. Prospects for high frequency irrigation. Science 188: 604–610.

Salter, P. J., and J. E. Goode. 1967. Crop responses to water at different stages of growth. Commonw. Agric. Bur., Farnham Royal, Bucks, England.

Skogerboe, G. V., J. W. Hugh Barrett, B. J. Treat, and D. B. McWhorter. 1979. Potential effects of irrigation practices on crop yields in Grand Valley. U.S. Environ. Proc. Agency Tech. Ser. Rep. no. EPA600/2-79-149. U.S. Government Printing Office, Washington, D.C.

Stegman, E. C., and L. D. Ness. 1974. Evaluation of alternative scheduling schemes for center pivot sprinkler systems. N. Dak. Agric. Exp. Stn. Rep. no. 48, Fargo.

Stegman, E. C., L. H. Schiele, and A. Bauer. 1976. Plant water stress criteria for irrigation scheduling. Trans. ASAE 19:850–855.

Stegman, E. C., and M. Aflatouni. 1978. Corn yield responses to water stress management. Am. Soc. Agric. Eng. Paper no. 78-2558. ASAE, St. Joseph, Mich.

Stewart, J. I., and R. M. Hagan. 1973. Functions to predict effects of crop deficits. J. Irrig. Drain. Div. Am. Soc. Civ. Eng. 99(IR4):421–429.

Stewart, J. I., R. D. Misra, W. O. Pruitt, and R. M. Hagan. 1975. Irrigating corn and grain sorghum with a deficient water supply. Trans. ASAE 18:270–280.

Stewart, J. I., R. J. Hanks, R. E. Danielson, E. B. Jackso, and W. O. Pruitt. 1977. Optimizing crop production through control of water and salinity levels in the soil. Water Res. Lab. Rep. PRWG151-1. Utah State University, Logan.

Sutter, R. J., and G. L. Corey. 1970. Consumptive irrigation requirements for crops in Idaho. Idaho AGric. Exp. Stn. Bull. 516, Moscow.

Vitosh, M. L. 1977. Irrigation scheduling guide for field crops and vegetables. Michigan State Univ. Ext. Bull. E-1110, East Lansing.

Willardson, L. S., and R. J. Hanks. 1976. Irrigation management affecting quality and quantity of return flow. U.S. Environ. Prot. Agency Tech. Ser. Rep. no. EPA600/2-76-226. U.S. Government Printing Office, Washington, D.C.

Woodruff, C. M., M. R. Peterson, D. H. Schnarre, and C. F. Cromwell. 1972. Irrigation scheduling with planned soil moisture depletion. Am. Soc. Agric. Eng. Paper no. 72-722. ASAE, St. Joseph, Mich.

CHAPTER 11C

Irrigation Options to Avoid Critical Stress: Optimization of On-Farm Water Allocation to Crops

JAMES W. JONES

University of Florida
Gainesville, Florida

I. INTRODUCTION

It is recognized that proper scheduling of irrigation can increase the efficiency with which water is used. As discussed by Hiler and Howell in Chapter 11A, the objective of scheduling irrigations to maintain favorable soil moisture conditions and thus to achieve maximum economic yield is justified only if: (i) water is not limited, and (ii) costs of applying water are low. If either or both of these conditions are violated, management policies to effect the efficient use of water are more difficult to define. For example, if irrigation water is limited and deficit irrigation must be practiced, the stage of crop growth and its sensitivity to soil water deficits must be considered. Stewart et al. (1975) experimentally demonstrated that the sequencing of a limited supply of water to corn and grain sorghum is important and that an optimal sequence exists.

In order to study farm irrigation policies when water supply is limited or expensive, optimization techniques have gained increased acceptance during the last 15 years. In this chapter, some relevant optimization techniques and examples are reviewed. Basic characteristics of optimization problem formulation are presented with examples applied to irrigation management. Mathematical models are inherent in the methodology, and because of the recent emphasis on model development by agricultural researchers, the role of models and simulation in irrigation-optimization problems will be discussed. Example applications will be reviewed to demonstrate spatial and temporal considerations in on-farm allocation of water.

II. OPTIMIZATION PROBLEM CHARACTERISTICS

A goal of optimization or operations research techniques is to develop the best possible course of action for management, or the optimal solution

Limitations to Efficient Water Use in Crop Production.

(Hillier & Lieberman, 1967). The optimization process is initiated by observing and formulating the problem. Then a mathematical model that captures the essence of the real problem is developed. It is hypothesized that the model is a sufficiently precise representation of the real system such that conclusions obtained from the model are also valid for the real system. This is a top-down systems approach in which the model formulation, including the detail with which the real system is represented, is determined by the problem formulation and objective.

Optimization studies can be broken into two phases: (i) the formulation of the model in the contexts of the decisionmaker's objective and the entire system, and (ii) the development of a solution that yields the optimal set of decisions or policy. Such efforts require a team approach because of the broad scope encompassed and the availability of many modeling and optimizing techniques. Independent development of these two phases may result in a problem formulation that cannot be solved, or in solution techniques for which adequate models of the real system may not exist.

Most existing studies of optimal on-farm water allocation have applied mathematical programming techniques (dynamic, linear, or integer programming). Here emphasis is placed on the problem-formulation phase for applying such mathematical programming procedures. Solution techniques for these classes of problems are described elsewhere (Hillier & Lieberman, 1967; Lasdon, 1970, Nemhauser, 1966; Garfinkel & Nemhauser, 1972).

Optimization problems can be formulated using four basic steps: (i) define the system, (ii) develop the objective function, (iii) identify and quantify the decision variables, and (iv) develop the problem constraints. The definition of the system is normally accomplished at the same time that the problem objectives are developed. The scope of the system must be carefully stated so that spatial and temporal resolutions can be established. The objective function is a quantitative representation of the decisionmaker's goal. It must be explicit and leave no ambiguity to the objective of the problem. One may wish to maximize yield, net profit, or water-use efficiency, however these objectives are not equivalent and the use of different objectives may result in different solutions.

Several authors have formulated objective functions for maximizing crop yield per unit land area when water is limited (Dudley et al., 1971; Hall & Butcher, 1968; Howell et al., 1975; Windsor & Chow, 1971). Dynamic programming techniques were used to solve these problems. As mathematically stated by Jensen (1968), the Howell et al. (1975) objective function is:

$$\text{Max}\,(Y/Y_o) = \prod_{i=1}^{n} (\text{ET}/\text{ET}_0)_i^{\lambda_i}, \qquad [1]$$

where

Y = grain yield (kg ha^{-1}),
Y_o = grain yield with water not limiting (kg ha^{-1}),
ET = evapotranspiration (mm),
ET_0 = evapotranspiration with water not limiting (mm),

λ_i = sensitivity of the yield to water deficit during plant growth stage i, and

n = number of growth periods considered.

The time resolution for decisions was 25 days in each of the five stages. Equation [1] can be modified so that the objective is to maximize net profit, or

$$\text{Max } P_{\$\text{net}} = [V_\$ \, Y_o \prod_{i=1}^{n} (\text{ET}/\text{ET}_o)^{\lambda_i} - \sum_{i=1}^{n} C_i I_i], \qquad [2]$$

where

$P_{\$\text{net}}$ = net profit (dollars ha^{-1}),

$V_\$$ = value of one unit of crop yield (dollars kg^{-1}),

C_i = cost per unit of water applied during growth stage i (dollars mm^{-1} ha^{-1}), and

I_i = amount of water applied during growth stage i (mm ha^{-1}).

The objective functions in Eq. [1] and [2] are spatially limited to a unit area of land and temporally to one growing season. Windsor and Chow (1971) developed a farm-level objective function for maximizing $P_{\$\text{net}}$ by selecting land area and soil type to be planted in each of two different crops, and type of irrigation system and level of irrigation to use. Their model was static in that a decision was made at only one point in time—before crops were planted. Linear programming was used to solve the model. Their model can also be modified to determine when to plant the crops to take advantage of seasonal rainfall or water availabilities. The in-season scheduling or irrigation on a farm basis (for multiple fields) after crops are planted would require a different formulation. Trava et al. (1977) present one example of such a problem.

Decision variables in the problem are unknowns, with respective values to be quantified by the solution process to identify the optimal set of decisions. The decision variables must be consistent with the objective function, and restrictions on their possible ranges must be quantified. As an example, the decisions may consist of when to apply irrigation water and at what level to a unit area of crop. This could be stated as a discrete set of irrigation decisions to be made each day, or

$$I_t \in [0, 10, 20, \ldots 150], \qquad [3]$$

where I_t = amount of water to apply on day t (mm), and t = any day during the growing season. If more than one crop were being considered, I_t could be replaced by I_{jkt}, where j refers to crop type and k represents the field. Howell et al. (1975) assumed high-frequency (daily) irrigation and defined their decision variables to be a fraction of the ET occurring during a crop growth stage, or

$$I_n \in [0.0, 0.25\text{ET}_n, 0.50\text{ET}_n, 0.75\text{ET}_n, 1.00\text{ET}_n], \qquad [4]$$

where I_n = total amount of water to apply during crop growth stage n (mm), and ET_n = total ET occurring during crop growth stage n (mm).

If the objective is to select crops to grow on a farm where water is limiting, the decision variables may consist of the set of variables X_{jlmn} as defined by Windsor and Chow (1971). The X_{jlmn} represents the number of hectares of crop n to grow in field (or soil type) j, using irrigation practice l, and irrigation system m. The solution would then select X_{jlmn} to maximize $P_{\$net}$ for the farm.

Constraints of optimization problems are mathematical statements of restrictions on the system. These restrictions may be caused by the internal structure of the problem, limitations in resources, and superimposed management policies. Structural constraints maintain internal balances and variable interrelationships. They may relate to physical aspects of the problem, such as the conservation of resources through time. As an example, the Howell et al. (1975) dynamic programming model includes a set of soil water balance constraints, or:

$$SW_n = SW_{n-1} + I_n + R_n + DR_n - ET_n - Q_n - DRP_n, \qquad [5]$$

where

SW_{n-1} = soil water content at beginning of stage n (mm),
I_n = total amount of water to apply during crop growth stage n (mm),
ET_n = total ET occurring during crop growth stage n (mm),
R_n = total rainfall during stage n (mm),
DR_n = water added to root zone by root-zone extension (mm),
Q_n = runoff during stage n (mm), and
DRP_n = profile drainage during stage n (mm).

Also, a balance was required on the remaining irrigation water that can be used during the rest of the season (IR_n), or

$$IR_n = IR_{n-1} - I_{n-1}$$

where

IR_n = irrigation water available at the beginning of stage n, (mm),
IR_{n-1} = irrigation water available at the beginning of stage $n - 1$ (mm), and
I_{n-1} = irrigation water used during growth stage $n - 1$ (mm).

The resource constraints in the Howell et al. (1975) problem consisted of the limitation that the water allocated during period n, I_n, be less than or equal to the remaining water, IR_n, or $I_n \leq IR_n$ for all n. In a farm-level problem, resource constraints express the limitations to water, land area, labor, capital, and other resource availabilities.

Policy constraints may be added to eliminate the possibility that the optimal solution will conflict with operating policies of the manager. For example, a farmer may wish to eliminate Sundays from the set of possible irrigation dates.

The reader should note that the example objective function, Eq. [1], is a mathematical model of crop yield (Jensen, 1968). Efforts to maximize Y or $P_{\$net}$ characteristically require a crop response model within the mathematical statement of the objective function. Furthermore, the soil water status (Eq. [5]) is a set of constraints in the optimization problem. Although not always the case, this illustrates the fact that models that have been used in simulation studies (Hanks, 1974; Jones & Smajstrla, 1980) may be adapted for optimization studies.

Inputs to optimization models depend on the model and its formulation. In the examples of Howell et al. (1975), Dudley et al. (1971), and Windsor and Chow (1971), weather variables were required inputs. Stochastic dynamic programming techniques were used to account for the uncertainties in R_n and ET_o, and the expected or long-term average Y or $P_{\$net}$ was maximized.

Optimization techniques are methods for finding the best or optimal decisions in an organized and efficient manner. Computer programs have been developed for solving optimization problems. In particular, linear programming and mixed-integer programming problems can be solved using the widely available MPSX optimzation package (IBM, 1974). Because of the size of some problems, analysts are developing matrix generators (Huang et al., 1975; Mishoe et al., 1979) to process input data and order it into the format required by available optimization computer programs. Dynamic programming problems usually are unique and generalized computer programs for solving such problems are not readily available.

III. THE ROLE OF SIMULATION

There is a rich tradition in modeling various components in the soil-plant-atmosphere system. Recently, emphasis has been placed on the integration of various process models into dynamic crop growth and yield models. The role of these models is optimizing on-farm water allocation deserves special attention.

Simulation can be used to evaluate specified irrigation policies in an enumerative search for the best policy among those tested. For this approach, models of the soil water status and crop yield responses are required (Ahmed et al., 1976; Barfield et al., 1977; Childs et al., 1977; Jones & Smajstrla, 1980; Tcheschke & Gilley, 1978). By defining several explicit, alternate policies and simulating results for one or more crop seasons, crop yields or net returns can be compared for the different policies and the best policy can then be selected. This procedure will not necessarily produce an optimal solution, but from a practical viewpoint it can provide valuable information to decisionmakers. Variability in climate can be brought into the analysis using the risk-analysis procedures outlined by Haan (1979).

Simulation models may also be used to provide input to optimization models. In the work of Howell et al. (1975) a soil water balance simulation was used to estimate the probabilities of soil water conditions changing from various levels to other levels at a later crop growth stage. Their model required daily rainfall and pan evaporation data, and simulations of each irrigation policy generated the probability tables or stochastic transition matrices and selected the best irrigation policy for a unit crop area based on expected or average yields. If an objective function on a farm level is to select how much of each crop to plant and when, data are needed to estimate the yield of each crop if planted during each week using each specific irrigation policy. Enumeration of unit-area, crop response simulations can be used to provide such responses to couple with farm level objective functions and constraints.

Research efforts that focus on individual processes can be of considerable use if results are integrated into soil-plant-atmosphere models that can subsequently be used to fill gaps of knowledge of plant growth responses to specific soil and aerial environments. For example, the effects of limited soil moisture on snap bean transpiration rate was presented by Millar and Gardner (1972). Soil-plant-transpiration models (e.g., Federer, 1979) can be used to estimate how transpiration rates vary with soil characteristics, root depths and densities, crop leaf areas, and ET_0 rates. An appealing research goal is the development of simplified functions or response surfaces that describe environmental effects on crop growth processes for use in more comprehensive simulation and optimization models.

IV. EXAMPLE APPLICATIONS

Because of the multiplicity of possible objective functions, constraints, and problem-solving methodologies, a standard optimization formulation for water allocation has not evolved. Here, two representative optimization formulations used in water-allocation studies are briefly discussed.

A. Allocation of Water to Crops

Characteristically, dynamic programming problems are decomposed into stages and decisions are required at each stage. The decision at any stage transforms the system state or states and increments the value of the objective function at a particular stage. Changes in the system state may be described by a probability distribution. At a given stage, the optimal policy for all remaining stages depends only on the current system state, regardless of decisions made in previous stages.

In the Howell et al. (1975) dynamic programming formulation, the decision process consisted of whether to irrigate 0.0, 0.25, 0.50, 0.75, or 1.0 times the ET_0 during each of five crop growth stages for grain sorghum. The states consisted of the remaining water to be allocated at each stage and the soil water status, a stochastic state variable.

Equations [1], [4], [5], and [6] collectively form the overall Howell et al. (1975) problem statement. The reader is referred to their paper for the more rigorous recursive formulation used in the dynamic programming solution. The solution to their problem provided the optimal irrigation policy, based on maximizing expected yield, in tabular form. The table can be used at the beginning of the season and at the end of each growth stage. The soil water state and amount of water available must be known, then the table provides the stage-by-stage optimal policy. As the season progresses, R_m and ET cause the soil water and the remaining water supply to vary from year to year. Therefore, at each stage, the irrigator can update the optimal policy, using the same table, to allocate the water during the remaining part of the growing season.

The Howell et al. (1975) solution produces an optimal sequencing of water application based on expected weather patterns and on differential crop sensitivities to water deficits during each growth stage. For example, with an initial soil water storage of 75 mm, and 150 mm of water available during the season, the optimal solution specified no irrigation during stage 1 and yield was 5,740 kg ha^{-1}. However, if 25 or 50 mm of the available water were applied during stage 1, yields were reduced to 5,530 or 5,200 kg ha^{-1}, respectively. Thus, with 150 mm of water available, grain yields per millimeter of applied water were 38.3, 36.9, and 34.7 kg ha^{-1}, respectively, for the optimal and two suboptimal cases described above. Their optimization procedure has the potential for enhancing the efficient use of water by increasing the yield per unit of water applied.

B. Allocation of Water on a Farm

Windsor and Chow (1971) describe a linear programming model for selecting the area of land to allocate to each crop, the irrigation intensity, and the type of irrigation system. Their objective function was:

$$\text{Max}\, Z = \sum_{j=1}^{J} \sum_{l=1}^{L} \sum_{m=1}^{M} \sum_{n=1}^{N} C_{jlmn} X_{jlmn}, \qquad [7]$$

where

Max Z = maximum profit on the farm,
J = number of field (or soil) types on the farm,
L = number of irrigation strategies to consider,
M = number of irrigation systems to consider,
N = number of crops to consider,
C_{jlmn} = profit per unit of crop area for crop n grown in field j using irrigation strategy l and irrigation system m, and
X_{jlmn} = land area of crop n to plant field type j using irrigation strategy l and irrigation system type m.

Constraints consisted of limitations in labor, seasonal water supply, and land area. Corn and soybeans were the crops, and irrigation systems consisted of self propelled and two-line systems. For the parameter values in their problem, the solution indicated that growing corn exclusively and using the more expensive self propelled irrigation system was optimal.

The Windsor & Chow (1971) model was designed for decision analysis prior to planting. A required input is the C_{jlmn}, which includes a crop yield response to various conditions. Dynamic programming was used to estimate crop yield response for optimal unit-area water allocation. It is this level of optimization analysis that can be tailored to the specific conditions and constraints of individual farm enterprises.

V. LIMITATIONS AND RESEARCH NEEDS

Optimization studies require models. In many of the reported studies, the lack of suitable crop response models was listed as a major limitation. Progress has been made during the last 10 years to fill that need. Yield response models of the type used by Hanks (1974) and Howell et al. (1975) are considered state of the art for use in irrigation management studies. These empirical models require calibration and testing for local conditions and estimate relative rather than absolute yields. However, according to Hillier and Lieberman (1967), "It is not even necessary that the absolute magnitude of the measure of effectiveness be approximately correct for the various alternatives, provided their relative values are sufficiently precise." Whereas in research efforts the mechanistic details of a model may be very important to the investigator, the criterion for optimization or management models is whether or not they are sufficiently accurate on a relative basis to permit a sound decision.

Dynamic crop growth models are being developed to predict growth and yield of crops using more theoretical considerations and physiological detail. One goal of such efforts is to provide models for particular crops that can be used in various geographical locations without expensive recalibration. These models are attractive because crop growth stresses, such as nutritient or pest, can be included in addition to those caused by water deficits to provide a more comprehensive tool for crop-production management.

Because of the dependence of optimization results on the models used, it is fitting that major research resources be used in developing and improving the models. In the short run, calibration of the relative-yield-response models should be performed for locations where on-farm water allocation analyses are needed. In the longer run, the dynamic, physiologically based crop growth models should be developed and tested. In addition, models predicting crop water use and the soil water status need to be further developed and tested for local weather, crop, and soil characteristics for application in the optimization studies.

Research is also needed to improve the optimization formulations for on-farm water allocation. The dynamic programming approach is limited from a practical point of view to small problems (one- or two-state variables) because of the solution requirements. Mixed integer programming and stochastic linear programming techniques may provide excellent alternatives.

Finally, it is the challenge of the individual researcher to recognize this top-down, systems-optimization approach to on-farm water allocation and its role in increasing the efficiency with which water is used. Research results and detailed models should be interpreted or translated into simplified relationships and models that can be integrated into management studies. This is the framework in which the state of the art will be advanced.

LITERATURE CITED

Ahmed, J., C. H. M. van Bavel, and E. A. Hiler. 1976. Optimization of crop irrigation strategy under a stochastic weather regime: a simulation study. Water Resour. Res. 12: 1241–1247.

Barfield, B. J., W. G. Duncan, and C. T. Haan. 1977. Simulating the response of corn to irrigation in humid areas. Am. Soc. Agric. Eng. Tech. Paper no. 77-2005. ASAE, St. Joseph, Mich.

Childs, S. W., J. R. Gilley, and W. E. Splinter. 1977. A simplified model of corn growth under stress. Trans. ASAE 20:858–865.

Dudley, N. J., D. T. Howell, and W. F. Musgrave. 1971. Optimal intraseasonal irrigation water allocation. Water Resour. Res. 7:770–788.

Federer, C. A. 1979. A soil-plant-atmosphere model for transpiration and availability of soil water. Water Resour. Res. 15:555–562.

Garfinkel, R. S., and G. L. Nemhauser. 1972. Integer programming. John Wiley & Sons Inc., New York.

Haan, C. T. 1979. Risk analysis in environmental modification. *In* B. J. Barfield and J. F. Gerber (ed.) Modification of the aerial environment of plants. Am. Soc. Agric. Eng. Monogr. no. 2:30–51. ASAE, St. Joseph, Mich.

Hall, W. A., and W. S. Butcher. 1968. Optimal timing of irrigation. J. Irr. Drain. Div. Am. Soc. Civ. Eng. 94(IE2):267–275.

Hanks, R. J. 1974. Model for predicting plant yield as influenced by water use. Agron. J. 60:660–665.

Hillier, F. S., and G. J. Lieberman. 1967. Introduction to operations research. Holden-Day Inc., San Francisco.

Howell, T. A., E. A. Hiler, and D. L. Reddell. 1975. Optimization of water use efficiency under high frequency irrigation: II. System simulation and dynamic programming. Trans. ASAE 18:879–887.

Huang, W.-Y., T. Liang, and I-P. Wu. 1975. Optimizing water utilization through multiple crops scheduling. Trans. ASAE 18:293–298.

International Business Machines Corp. 1974. IBM mathematical programming system extended/370 (MPSX/370). Program Ref. Manual SH 19-1095-0. IBM, White Plains, N.Y.

Jensen, M. E. 1968. Water consumption by agricultural plants. Chapt. 2. *In* T. T. Kozlowski (ed.) Water deficits and plant growth. Vol. 2. Academic Press Inc., New York.

Jones, J. W., and A. G. Smajstrla. 1980. Application of modeling to irrigation management of soybeans. p. 571–599. *In* F. T. Corbin (ed.) World Soybean Research Conf. II. Proc. Westview Press, Boulder, Colo.

Lasdon, L. S. 1970. Optimization theory for large systems. The MacMillan Co., New York.

Millar, A. A., and W. R. Gardner. 1972. Effect of the soil and plant water potentials on the dry matter production of snap beans. Agron. J. 64:559–562.

Mishoe, J. W., J. W. Jones, and G. J. Gascho. 1979. Harvesting scheduling of sugarcane for optimum biomass production. Trans. ASAE 22:1299–1304.

Nemhauser, G. L. 1966. Introduction to dynamic programming. John Wiley & Sons Inc., New York.

Stewart, J. I., R. D. Misra, W. O. Pruitt, and R. M. Hagan. 1975. Irrigating corn and grain sorghum with a deficient water supply. Trans. ASAE 18:270–280.

Trava, J., D. F. Heermann, and J. W. Labadie. 1977. Optimal on-farm allocation of irrigation water. Trans. ASAE 20:85–88, 95.

Tscheschke, P., and J. R. Gilley. 1978. Modification and verification of Nebraska's corn growth model—CORNGRO. ASAE Tech. Paper no. 7804033. Am. Soc. Agric. Eng., St. Joseph, Mich.

Windsor, J. S., and V. T. Chow. 1971. Model for farm irrigation in humid areas. J. Irri. Drain. Div. Am. Soc. Civ. Eng. 97:369–385.

CHAPTER 12

Water Resources Management in a Growing Society

J. VAN SCHILFGAARDE AND S. L. RAWLINS

U.S. Salinity Laboratory, USDA-ARS
Riverside, California

I. INTRODUCTION

The foregoing chapters in this book have presented a competent and up-to-date evaluation of the state of the art in a variety of aspects dealing with the science of water use by crop plants, and its supply and management. This substantial body of knowledge forms the scientific basis for improvements in agricultural water management; it provides the building blocks for developing a strategy for optimal natural resource use in this country and abroad.

Fourteen years have elapsed since a similar symposium considered the technical aspects of efficient water use for crop production (Pierre et al., 1966). During that time our concern has shifted from food surpluses to the question of how to feed an ever-increasing world population. The precipitous rate at which this turn of events has occurred is illustrated in Table 1. Not only has the world's stock of stored grains decreased, but the cushion provided by about 20 million ha of idle land in the United States has disappeared. Whereas all areas of the world but Western Europe exported grain prior to World War II, only North America and Australia now do (Brown, 1975). North America has emerged as the global food supplier.

These fortunes have turned primarily as a consequence of population growth overwhelming the food-producing capacity of one nation after another. Also contributing to the food deficits has been the upgrading of diets in several countries that can afford to import grain for animal feed.

World land resources that could potentially be used for crop production amount to 3.4×10^9 ha, as compared with 1.4×10^9 ha currently used in this manner (NAS, 1977). The theoretical level of production attainable from these resources has been estimated to be about 40 times the present level (Buringh et al., 1975). As part of this potential, the 170×10^6 ha of irrigated land could be expanded to 450×10^6 ha (Buringh et al., 1975, as corrected in NAS, 1977, p. 225). The high unit-production potential implied assumes a very high level of technological and capital- and energy-intensive inputs. Considering a far lower, "labor oriented" level of technology,

Limitations to Efficient Water Use in Crop Production.

Table 1—Index of world food security, 1961–1976 (from Brown, 1975).

Year	Reserves			Reserves as days of world grain consumption
	Grain†	Grain equivalent of idled U.S. cropland	Total	
	metric tons × 10^6			days
1961	163	68	231	105
1962	176	81	257	105
1963	149	70	219	95
1964	153	70	223	87
1965	147	71	218	91
1966	151	78	229	84
1967	115	51	166	59
1968	144	61	205	71
1969	159	73	232	85
1970	188	71	259	89
1971	168	41	209	71
1972	130	78	208	69
1973	148	24	172	55
1974	108	0	108	33
1975	111	0	111	35
1976‡	100	0	100	31

† Based on carry-over stocks of grain at beginning of crop year in individual countries for year shown. The USDA has recently expanded the coverage of reserve stocks to include importing as well as exporting countries, thus the reserve levels are slightly higher than those heretofore published.

‡ Preliminary estimates by USDA.

Buringh and van Heemst (1977) estimated that the potential production on the maximum land base would only feed 20% more than the current world population.

Such global estimates do not take into account distributional problems. Broadly speaking, the new land is not where the people are, although the land that would benefit from irrigation does tend to be in regions of high population density. Neither do these statistics account for weather. Weather variations frequently reduce crop yields drastically below the average expected and can raise havoc, not only with local food supplies, but also with international markets.

Thus, the best estimates available suggest that the potential exists for increasing food production well above present levels, while it is also indicated that the level of technological inputs required to meet demands must be well above those defined by Buringh and van Heemst (1977) as "labor oriented."

We cannot let the matter rest there, because there is far more at stake than just the potential availability of land and water. The food production and delivery system developed in the West is capital-intensive, energy-intensive, and labor-poor. For better or for worse, it developed as an integral part of the evolution of Western civilization. Its export, at least in the short run, would fail completely for social and cultural reasons (Moravosik & Ziman, 1975). As important, it would be technically indefensible and environmentally precarious, if not downright disastrous.

It would be technically indefensible because neither the capital nor the fossil energy is available for such a system. Even if for each joule of energy expended on the farm for crop production 1.5–2.0 J are recovered in food, the total energy input into the food system is more nearly 8 J per Joule recovered (Steinhart & Steinhart, 1974). Energy is a key issue. If the world's estimated population of 7×10^9 at the turn of the century were to grow its food and eat like the present Dutch population, the annual energy demand by agriculture would be 40 EJ (40×10^{18} J) (de Wit, 1975). For comparison, this would exceed by one-quarter the estimated total annual energy production at that time in the United States of some 32 EJ (Hayes, 1979). Obviously, some compromise must be made. By consuming more plant products directly rather than via animals, using energy primarily to increase yields rather than decrease labor, and by using unconventional energy resources such as solar, de Wit (1975) speculated that the energy demand could be reduced substantially.

It could be environmentally disastrous because much of the land potentially available for new or expanded production—such as the Ultisols and Oxisols of the tropics—now maintain a precarious hydrological and nutritional balance that, if disturbed, is likely to lead to extensive erosion, denudation, and long-term loss of fertility (Pereira, 1973; Buringh & van Heemst, 1977). More generally, man's technological innovations have often been directed at short-term gains that, in the longer run, irreparably insult the ecosystem on which he depends (Cole, 1966; Brown, 1975).

This question of the short- vs. the long-run brings us to another crucial issue. Garrett (1978) defined engineering as "the art of organizing men, and directing the forces and materials of nature for the benefit of the human race." This seemingly innocuous and altruistic definition takes on a new aspect when we define the "human race" as including all future generations. Georgescu-Roegen (1975) addressed this question cogently and explicitly, basing his arguments on the second law of thermodynamics. If all future generations are to be included, asked Georgescu-Roegen (1975), who will represent them in bidding for a future world store of fossil energy and mineral resources? In this context, only the solar energy flux can be seen as a renewable resource.

Thomas (1975) also questioned our present focus on the need for increased production "at any cost" with too little attention paid to resource conservation. His conclusion that "perhaps increased awareness will lead to a better understanding of our planet earth, the role of mankind and the limitations of our environment" is a challenge we find difficult to ignore.

The conclusion is not that we must "return to nature" (Wadleigh, 1965), that we must deny the use of technology or shun the application of science to build a better world. We must guard, also, against the elitist's plea to protect a piece of nature for the leisure class (Tucker, 1977). It does mean that we must take seriously the call for alternatives (Clarke, 1973; Schumacher, 1973). In fact, the need is for more science, more ingenuity than ever, but applied in a way consistent with building a conservative society—conservative in the sense that it is sustainable over the long haul.

The foregoing dealt with world more than U.S. agriculture. It suggests that greater (and less-risky) strides are likely to be made by emphasizing increasing unit production on currently cultivated land, rather than increasing the land base; it must be recognized, however, that land development will take place (FAO, 1974; Jornal do Brasil, 1975; Nicholson & Rohter, 1979).

With reference to the United States, the problems are very different in detail but not so different in principle. A key factor in past increases in production has been the tremendous increase in use of fossil fuels. Mechanization has released large land areas needed to grow feed for draft animals and N fertilizer similarly has released land used to grow legumes. Edminster (1976)[1] estimated that today 73×10^6 ha would be required for the former and 57×10^6 ha for the latter, as compared with a total crop land base of 400×10^6 ha. The potential for substantial yield increases is less than in the developing countries, considering the high base (Jensen, 1978); expanding the land base would certainly create hazards of increased resource deterioration, both onsite and off-site, even if less dramatically than in the Amazons; and the opportunities are there for good science to develop agricultural practices more in harmony with the resources exploited.

Water management is an integral and primary part of any agricultural production system. Good soil and water management is prerequisite to effective use of other resource inputs. Agricultural water use, however, must be considered in relation to water demands in other phases of society. Increasing needs for municipal and industrial uses often seem in conflict with agricultural needs; the extent to which various uses can be made compatible, rather than competitive, needs to be evaluated.

II. IRRIGATED AGRICULTURE

Irrigation is an age-old art that predates recorded history. Only about 12% of the world's farmland is irrigated, but a substantial portion of the population depends on it. The percentage of production is substantially higher than that based on area. Even with only adequate management, the unit production can be high and double- or even triple-cropping is often possible. On the other hand, capital costs are generally high, and the energy inputs can be substantial. In the absence of stored food reserves, irrigated agriculture helps insulate against the vagaries of weather.

On-farm water-use efficiency in irrigation tends to be surprisingly low. A value of 30% is often quoted for the United States. Because water lost from one farm can return to the ground-water aquifer or to a stream, and serve as a supply to another, overall irrigation efficiency is often much greater. Irrigation in arid and semiarid regions is always associated with salinity. The salt contained in the water tends to be left behind as pure water transpires or evaporates.

---

[1] Edminster, T. W. 1976. What is the secret of affluence? Adm. Lett., USDA-ARS. August 1976.

To obtain the full benefits of irrigation without adverse effects, it is necessary to have carefully controlled water management. Careful control is required to provide the desired amount of water uniformly at specified times. As pointed out by Rawlins and Raats (1975), to obtain maximum dry matter production, either per unit of water or per unit of land, it is generally desirable to avoid water stress by maintaining a uniformly high water potential in the soil root zone; for some crops, it may be essential to program one or more periods of stress to maximize the harvestable product. On the other hand, an irrigation schedule that always maintains a fully charged root zone does not provide adequate opportunity for taking advantage of rainfall and/or stored soil water (Fischbach & Somerhalder, 1974). Furthermore, if yields are to be maximized, irrigation must also be managed to minimize the salinity of the soil solution.

Technical details have been covered in earlier chapters. Here, our main purpose is to emphasize the need to design and operate irrigation systems so that control of the soil water content is possible. With conventional gravity irrigation systems, labor and operating costs are minimized by decreasing the frequency of irrigation as much as possible; water-use efficiency and crop yield tend to be maximized when frequency is increased. It is not enough to obtain "best" use of land and water; it must also be economically feasible to do so.

Automated solid-set and center-pivot sprinkler systems are obvious candidates for systems with close control; in principle, trickle irrigation is equally suited. Compared with gravity systems, such systems offer greater control, together with decreases in labor requirements, at the expense of increased capital cost and fossil fuel consumption. Gravity systems, however, also can be modified to permit closer control. Laser-controlled precision land leveling allows better areal water distribution over the field and smaller water applications; combined with automation, it has led to dramatic improvements in irrigation efficiency for dead-level, flooded systems (Dedrick et al., 1978). Using closed conduit rather than open waterways for laterals offers the opportunity for off-on control, as well as for capturing gravitational energy for pressurizing delivery systems or controls. In furrow-irrigated areas, furrow length can be reduced—and thus intake distribution improved and tail water eliminated—using a system such as Worstell's (1976) multi-set system. For tree crops, a low-head bubbler system has been developed that provides excellent control, requires minimal pressure—as low as 1.0 m of head—and is competetive in terms of costs (Rawlins, 1977).

Thus, numerous opportunities exist for modifying existing irrigation systems to increase the effectiveness of water use. Few of these have found widespread application to date. At least one reason is the lack of incentive, associated with the typical pricing mechanism for irrigation water in much of the United States and the world; another is that many of these ideas need further work to make them acceptable for routine use.

Another restraint to improved on-farm water use often is the off-farm distribution system. Typically, in large irrigation projects, water is distributed through large canals feeding laterals that deliver water at the farm gate. The design and operation of the canal system is often such that at best water must be ordered some days in advance, or at worse so that water is delivered on a fixed rotation scheme. Reidinger (1974), for example, analyzed the water-rotation system in Northern India: whereas each farmer is to receive his equitable part of the total supply, he has no control over either the timing or the quantity he receives at any given time. Recognizing that efficient on-farm water use requires that, to the extent feasible, water be delivered on demand, the U.S. Bureau of Reclamation initiated its Irrigation Management Service. It is the hope that this program, designed to help schedule irrigations on individual fields, can provide sufficient feedback to improve the operation of the delivery system.

Irrigation also must avoid excess water applications, and all irrigation must be accompanied by drainage. Drainage must remove precipitation and irrigation water applied in excess of crop needs to prevent waterlogging; salts that accumulate in the root zone to avoid salination; and "foreign" water, or water that seeps into the area from leaky canals, excess irrigation on adjacent fields, or similar sources separate from the field under consideration. To meet drainage demands, one depends on the natural drainage rate, supplemented as needed by drainage installations (van Schilfgaarde, 1979).

To avoid yield reductions from salinity, one must maintain a downward flux through the root zone sufficient to avoid concentrating the soil solution in the root zone above harmful levels. This flux is generally expressed as the leaching requirement, which is the fraction of the total surface-applied water that must percolate through the root zone. For a typical U.S. irrigation water containing 1,000 mg L^{-1} of dissolved solids or less, the leaching requirement for most crops is less than 0.05 (van Schilfgaarde et al., 1974; Maas & Hoffman, 1977). Thus, even with the most efficient irrigation management, if uniform areal distribution is obtained, the leaching requirements are likely to be met implicitly. Without uniform distribution some parts of a field always receive too much and some may receive too little water. Drainage requirements, then, are closely related to on-farm irrigation management and to the seepage and spills from the distribution system. The adverse effects of excessive water percolation, however, are not restricted to the irrigated farm. The drainage water tends to be of lower quality than the irrigation water, and ultimately one must dispose of it.

Besides improved on-farm management and better-tailored distribution systems, another major component of irrigation water management concerns conjunctive use of ground water. The supply of ground water is immense, although most of it is beyond economic recovery. Both Ambroggi (1977) and Revelle and Lakshminarayana (1975) stressed the advantages of artificial recharge, from leaking irrigation canals or excessive irrigation, combined with pumping to recover the water thus stored underground. By deliberately drawing the water tables down, space is provided to store excess water available during flood seasons. Pumping of ground water also provides close control over time and place of water availability, thus increasing the potential level of efficiency of use. A serious problem with both their proposals is the question of salt management. Salt accumulation is one of the major problems facing, for example, the Indus Basin of Pakistan. The consumptive use of water of necessity increases the salt content of the remaining water; the better the irrigation management, the higher the salt concentration in the drainage water. Mixing this drainage water with uncontrolled seepage from canals tends to result in water too saline to produce good crops, but too dilute to warrant disposal. Thus, induced recharge into a closed or slowly draining basin—such as the Indus—is bound to lead to salination and, ultimately, disaster. Properly designed and operated recharge systems can, in the right circumstances, aid tremendously in enhanced water management. But to be most efficient, the drainage water not destined for reuse should be kept separate from the water supply if salination is to be avoided.

Both Ambroggi (1977) and Revelle and Lakshminarayana (1975) overlooked the problem of water degradation and the consequent necessity, in a ground-water recharge scheme, of "bleeding off" the salt accumulations. In a mixed disposal and supply system, this requires that drainage out of the basin be at a quality still acceptable for irrigation and, thus, in a relatively large volume of water. Details would vary drastically from situation to situation, but in all cases, this requirement places a drastic set of restraints on conjunctive use schemes with forced recharge.

To illustrate some of the consequences and opportunities in enhancing water use in irrigation, we now turn to some specific examples.

The Colorado River drains an area 1/12th the size of the continental United States. Its basin covers some 64×10^6 ha in seven western states of the United States and in Mexico. Rich in mineral resources, blessed with natural beauty and a wide range of scenery, topography, and climate, the land resources of the basin are extensively used for grazing, wildlife, timber production, recreation, and agriculture. About 1.2×10^6 ha are irrigated, whereas 0.25×10^6 ha are used for nonirrigated cropping. Books have been written about the various potentials and problems in the basin (e.g., Peterson & Crawford, 1978). The Colorado River probably is the most controlled (and litigated) major river in the world. Its water resources are overcommitted by court decree; its water quality has been deteriorating. Studies suggest that the salinity at Lee's Ferry before man's influence was about half the 600 mg L^{-1} now encountered. At Imperial Dam, salinity is now about 850 mg L^{-1}, and is projected to increase with further development in the basin unless ameliorative action is taken; estimates vary, but 1,250 mg L^{-1} is a representative prediction for the year 2000. Roughly a third of the salt load in the river has been attributed to irrigation. United States Public Law 93-320 charges the Department of Interior, in cooperation with others, to develop and implement a plan that will permit further development of the river's water resources without increasing its salinity.

For the present purpose, we shall take a brief look at some highly variable but significant examples of irrigation projects and their relation to water quality in the river.

Some 24,000 ha are irrigated in the Grand Valley of Colorado by gravity, with water supplied through a set of canals that have grown historically into a crazy-quilt pattern. Partly because of the soils, but primarily because of the irrigation systems, diversions are very large relative to consumptive use. Estimates vary, but the valley probably adds about 650,000 Mg of salt to the Colorado River each year. The valley is underlain by a massive, highly saline, Mancos shale formation. Water spilled from canals or laterals and returned to the river through surface channels adds no significant salt, but water that seeps from water courses or percolates through the root zone of agricultural land tends to come to chemical equilibrium with the underlying materials and is ultimately discharged with a salt concentration of about 6,500 mg L^{-1}, the drainage water displacing highly saline ground water. Thus, a reduction in water throughput through the soil system will result in an approximately proportional reduction in salt loading. Increasing on-farm irrigation efficiency and reducing canal and lateral seepage would reduce the salinity of the Colorado River downstream. At the same time, the salination of parts of the lower valley, intensified by high water tables, would be stemmed and reclamation made easier, eliminating, or at least reducing, the need for additional drainage, and enhancing the productivity of the farms in the valley.

Conceptually, a strong case can be made for a more drastic approach: replacing the existing distribution system with a single water conduit along the northern perimeter of the valley, which in turn would distribute water through closed conduit laterals. The advantage would be that the same benefits are realized as from lining the existing system with, in addition, the delivery of water under pressure at all user points. In principle, such a system would lend itself to demand delivery. It would permit adoption of a wider range of irrigation systems on the farm, depending on the site and crop, available technology, or newly developed techniques. It would provide, in many circumstances, sufficient pressure to operate automation devices, bubbler systems for orchards, or even low-pressure sprinkler systems.

Several disadvantages must also be recognized. First and foremost, the cost would no doubt be higher. More immediate is probably the question of acceptability. The concept under consideration would require consolidation of several irrigation and drainage districts, with separate patronage, management, and water rights, into a single unit. It also would likely result in delivery of adequate amounts of water to meet crop needs, but substantially less than established water rights. Suffice it to say that it would be very difficult to obtain broad public acceptance for such a solution.

Farther down the river, the Palo Verde Irrigation District in California diverts water to irrigate 36,000 ha. Again, neither the irrigation water nor the drainage water is pumped. The Palo Verde, however, is an alluvial plain where there are no significant salt deposits that are displaced by the drainage water. The on-farm irrigation efficiency has been estimated at 53%. Clearly, this efficiency could be raised substantially with current technology. Such a change, however, would have essentially no effect on water-use efficiency; it would have minimal effect on energy consumption if N fertilizer leaching were not excessive, and but a minor effect on the salt concentration of the river downstream from the valley, at least after steady state had been reached. What little effect could be anticipated would result from carbonate precipitation in the soil profile. Considering the costs, both real and imagined, associated with improved water management in the district, the obtainable benefits may be extremely small. In an area where, as a first approximation, no "foreign" salts are involved, efforts to improve irrigation efficiency can only be justified on the basis of increased farm income. Providing adequate drainage even for relatively sloppy irrigation is probably cheaper, and certainly is more readily accepted by the farming community; neither is a significant social benefit sacrified, assuming again that irrigation inefficiency does not cause excessive leaching of N fertilizer.

Still farther downstream, the Wellton-Mohawk Irrigation and Drainage District in southwestern Arizona is one of the last units within the United States to draw water from the Colorado River. The District serves about 26,000 ha of irrigated land and, historically, diverts about 640 $\times$ 10^6 m^3 $year^{-1}$ through open canals, and lifts this water some 55 m. A geologically closed basin, salt accumulated in the ground water before drainage provisions were made. Currently, more than 100 wells pump drainage water into a bypass canal that discharges near the ocean in Mexico; in 1973, the drainage water concentration averaged 3,600 mg L^{-1}. Were this drainage water to be mixed with the Colorado River water that the United States delivers to Mexico by treaty, its concentration would be increased by 440 mg L^{-1}. Such an increase would have a significant adverse impact on agriculture in the Mexicali Valley, while spilling the drainage water wastes a valuable resource out of a water-short basin. To resolve the dilemma, the U.S. government plans to construct a desalting plant to treat the drainage water.

A separate chapter could be written easily on this particular situation. Suffice it here to observe that a number of alternate solutions can readily be imagined that are more cost-effective, less energy-intensive, and more in tune with resource conservation. All of those alternatives, however, involve significant social, legal, and institutional complications.

Leaving the Colorado River Basin, we devote a few remarks to the Central Valley in California. The importation of water from northern California has increased the potential for irrigation substantially, while also aggravating a drainage disposal problem. Although the need for drainage has long been recognized and is projected to increase, there are serious objections to the construction of a central drain to transport drainage from the closed valley to the Sacramento Delta, the most accessible outlet. Extensive studies have evaluated alternatives, of which we shall consider only one.

As indicated by Rhoades (1977), the salt concentrations of many drainage waters are such that they are still suitable for production of relatively salt-tolerant crops at essentially full yield if proper management practices are used. Rhoades' projections indicate, for example, that irrigation with water containing 6,300 mg L^{-1} of salt at a leaching fraction of 0.25 should give full yield of cotton. Thus, selective use of such drainage waters to irrigate appropriate crops offers the potential of reducing by three-fourths the volume of drainage water of which one must dispose. Less costly, or less objectionable, means can be found to dispose of this reduced volume, while productive use is made of a larger fraction of the irrigation water.

In the example quoted, based on an actual water supply near Bakersfield, Calif., the drainage water from the cotton would have a salinity of 17,500 mg L^{-1}. It is possible to carry the concept one step further and to use this water, if it can be collected, once again for biomass production. Although currently no crop plants exist that can be grown economically with 17,500 mg L^{-1} water (about 0.4 seawater concentration), there are halophytes that can be grown as wildlife habitats, or as sources for methane production. Thus plants, whether they be conventional crops or species not now used in agriculture, provide the option to make beneficial use of a resource now considered a waste product.

Such a system, however, would require a separate distribution system for the drainage water to be reused, and a series of questions would need to be resolved in terms of water rights and pricing. Again, the institutional as well as technical problems must be coped with.

These examples illustrate the potential, with present technology, for enhancing resource use in irrigated agriculture. With imaginative scientific and engineering research, the technological potential can be expanded considerably. However, such potential innovations will only be put in practice if the institutional framework is appropriate. We only have hinted at the economic, legal, and institutional restraints that are frequently encountered. Science, powerful as it may be, cannot be expected to have its full and potentially constructive impact in the decision process unless its practitioners involve themselves, at least to a degree, in the societal framework where the decisions must be made.

III. RAINFED AGRICULTURE

Rainfed agriculture differs from irrigated agriculture in several key aspects. The obvious difference is that it depends, primarily, on natural rainfall and the related climatic and weather variability. Closely associated is the more acute and prevailing hazard of erosion. Another aspect is that with rainfed agriculture the nature of the competition for limited water resources among alternative uses is more subtle and less direct, even if equally real.

To obtain most effective use of an often drastically variable water supply, a variety of techniques are used that, basically, seek to modulate this variability. They include increasing storage in the soil profile by (no-) tillage practices and various conservation measures, as well as engineering structures such as reservoirs, terraces, drainage, and supplemental irrigation systems. Another approach is to adapt the cropping system to the water available. Changing planting dates, selecting varieties with most favorable maturation periods, adapting fertilizer applications to expected water supply, and breeding or selecting for specific environmental limitations all play a role. So does the introduction of multiple cropping or intercropping.

It is in the latter areas that substantial opportunities well may exist that have not been explored as systematically as they might be. Recent developments in computer technology should make it possible to develop data bases that, together with plant growth models, enable evaluation of a wide range of options in terms of management strategy and selection of germ plasm to significantly increase the effectiveness of field trials, and similarly to improve the transferability of research data from site to site (NAS, 1977).

Some may dismiss these observations as mundane and lacking in scientific elegance. To the contrary, they should provide invaluable feedback for promising leads in the application of science. Plant breeders have selected varieties that performed well in given but ill-defined environments; for example, the selection of small grains for Al tolerance did not come about by design but was recognized after the fact; still later, the genetic mechanism was recognized (Foy, 1974; Moore, 1974). A clearer definition of the environmental parameters should help in breeding to select cultivars that are specifically adapted to the environment specified. Similarly, the potential for multiple cropping, whether relay cropping or intercropping, cannot be realized without far greater understanding of the intricate physical, chemical, and biological interactions in the rhizosphere than exists at present.

Particularly when considering the need to increase crop production in developing countries, we are reminded of Georgescu-Roegen's (1975) exhortations against excessive dependence on external inputs. As Greenland (1975) argued convincingly, "...direct translation of the technical package of fertilizers, pesticides, mechanization, and erosion control structures...to the shifting cultivator of the low-land humid tropics cannot be expected to be successful." Yet he visualized the potential of increasing production from 2- to 16-fold by an integrated package of practices that are suitable under the low-capital, small-holding system of farming that prevails. This package includes zero tillage and use of plant-residue mulches, mixed cropping of high-yielding varieties that are disease- and pest-resistant, fertilizer to replace P losses, legumes with highly active rhizobia to supply N, and control of acidity by means of ash or mulches from deep-rooted species. The key to Greenland's prognosis is that such a system will control erosion and maintain a nutrient balance, so that water-use efficiency is increased. His recommendations reflect the highly sophisticated application of science and technology to modify a primitive farming system without recourse to a high level of mechanization depending on excessive external inputs.

In the United States, over 40 years of an aggressive and extensive soil conservation program has resulted in significant strides towards reduction of erosion and maintenance or restoration of soil productivity; yet, it has been estimated that only 25% of the land is adequately protected today. In parts of Africa and Asia, once-stable, if technologically simple, land-use systems are unable to cope with increasing population pressures. As long as the population density was not excessive, these systems were often surprisingly productive (Hekstra, 1975); in more recent years, however, erosion and consequent loss of productivity have led to increasing problems as land holdings were decreased, less suitable areas put under cultivation, or fallow periods reduced in length. Elsewhere, as in Brazil, the drive for economic development is threatening to lead to disastrous consequences, as large areas of once-undisturbed forest or plain are cleared and broken up, even though the population density is still low. There is little doubt that much of this land can be put to productive use; it is beyond question that it will be used. The challenge is to devise the schemes that will enable such development with a minimum of external inputs and so that productivity can be maintained. As illustrated by Greenland (1975), it can be done.

Any water that is used by plants through evapotranspiration is no longer available for use, although a part will augment precipitation and thus provide a new water source elsewhere. Unfortunately, good conservation practices that reduce flood hazards, control erosion, and thus minimize sediment transport and deposition, also tend to reduce water yield (Pereira, 1973). In regions with marginal precipitation, the goals of modulating streamflow and improving water quality conflict with the objective of maximizing water yield, since generally the best way to protect the land is to maintain a continuous cover of deep-rooted plants. In more humid areas, such conflict is not likely to arise, although there, too, water use by plants would increase. This situation illustrates some of the complexity of the hydrologic cycle, intensified by the demands of man.

IV. SUMMARY

Great strides have been made in recent decades in the understanding of the energetics and physiology of water use by plants. Technology for water management in agriculture has also advanced substantially. It is less clear that we have learned to make use of these advances in knowledge in a manner that is compatible with our needs and with the institutional framework in which we live.

Land and water by themselves are not sufficient to provide mankind with food, not to mention his other wants; they are, however, indispensable—a "conditio sine qua non." Regional shortages, in quantity or quality, are commonplace now; absolute shortages may not be far away.

In Brown's (1975) words, "the scarcity of basic resources required to expand food output, the negative ecological trends that are gaining momentum year by year in the poor countries, and the diminishing return on the use of energy and fertilizer in agriculture in the industrial countries lead me to conclude that a world of cheap, abundant food with surplus stocks and a large reserve of idled cropland may now be history. In the future, scarcity may be more or less persistent. . .."

The factual framework exists for significant enhancement of the effectiveness with which the soil and water resources are utilized to man's purposes. Whether the wisdom is forthcoming to do so is another question. It appears that at least 40 million ha of new irrigation development is planned in the near future; millions of hectares also are slated for agricultural development in Latin America. Will these developments repeat the foibles of the Mangla Dam (Eckholm, 1976), the Colorado River? Or the tragedy of destruction of the land, as in Mississippi (Hilgard, 1860)?

There is an important challenge for research in furthering our knowledge so that new options are opened up for better resource use. The most exciting challenge may well be in learning to apply such understanding so that new systems and ideas are indeed implemented in a responsible manner.

LITERATURE CITED

Ambroggi, R. 1977. Underground reservoirs to control the water cycle. Scient. Am. 236(5): 21–27.

Brown, L. R. 1975. The world food prospect. Science 190:1053–1059.

Buringh, P., and H. D. J. van Heemst. 1977. An estimation of world food production based on labour-oriented agriculture. Centre for World Food Market Research, Wageningen, The Netherlands.

Buringh, P., H. D. J. van Heemst, and G. J. Staring. 1975. Computation of the absolute maximum food production of the world. Agricultural University, Wageningen, The Netherlands.

Clarke, R. 1973. The pressing need for alternative technology. Impact Sci. Soc. 23(4):257–271.

Cole, L. C. 1966. Man's ecosystem. BioScience 16(4):243–248.

Dedrick, A. R., J. A. Replogle, and L. J. Erie. 1978. On-farm level-basin irrigation—save water and energy. Civ. Eng. 48(1):60–65.

de Wit, C. T. 1975. Agriculture's uncertain claim on world energy resources. Span 18(1):2–4.

Eckholm, E. P. 1976. Losing ground. W. W. Norton and Co. Inc., New York. p. 122.

Fischbach, P. E., and B. R. Somerhalder. 1974. Irrigation design requirements for corn. Trans. ASAE 17:162–165.

Food and Agricultural Organization, United Nations. 1974. FAO proposals for action. World Food Conference. FAO, Rome.

Foy, C. D. 1974. Effects of aluminum on plant growth. I. Aluminum toxicity in soils. p. 601–642. *In* E. W. Carson (ed.) The plant root and its environment. University of Virginia Press, Charlottesville.

Garrett, R. 1978. Good engineering produces appropriate technology. p. 136–141. *In* R. C. Dorf and Y. L. Hunter (ed.) Appropriate visions: technology, the environment and the individual. Boyd and Fraser Publishing Co., San Francisco, Calif.

Georgescu-Roegen, N. 1975. Energy and economic myths. Southern Econ. J. 41:347–381.

Greenland, D. J. 1975. Bringing the green revolution to the shifting cultivator. Science 190: 841–844.

Hayes, E. T. 1979. Energy resources available to the United States, 1985 to 2000. Science 203: 233–239.

Hekstra, G. 1975. De landbouw en het kleine boerenbedrijf op Midden-Java. Landbouwkd. Tijdschr. 87:140–146.

Hilgard, E. W. 1860. A report on the geology and agriculture of the state of Mississippi. E. Barksdale, Jackson, Miss.

Jensen, N. F. 1978. Limits to growth in world food production. Science 201:317–320.

Jornal do Brasil. 1975. O decreto do Polocentro. J. Brasil 30:26.

Maas, E. V., and G. J. Hoffman. 1977. Crop salt tolerance—current assessment. J. Irrig. Drain. Div. Am. Soc. Civ. Eng. 103(IR2):115–134.

Moore, D. P. 1974. Physiological effects on pH on roots. I. Effect of pH on root growth. p. 135–152. *In* E. W. Carson (ed.) The plant root and its environment. University of Virginia, Charlottesville.

Moravosik, M. J., and J. M. Ziman. 1975. Paradisia and dominatia: science and the developing world. Foreign Affairs 53:699–724.

National Academy of Sciences. 1977. Supporting papers: World Food and Nutrition Study. Vol. II. NAS, Washington, D.C.

Nicholson, T., and L. Rohter. 1979. The Amazon bonanza. Newsweek XCV(2):67–68, 70–71. 15 Jan. 1979.

Pereira, H. C. 1973. Land use and water resources in temperate and tropical climates. Cambridge University Press, London.

Peterson, D. F., and A. B. Crawford (ed.). 1978. Values and choices in the development of the Colorado River Basin. The University of Arizona Press, Tucson.

Pierre, W. H., D. Kirkham, J. Pesek, and R. Shaw. 1966. Plant environment and efficient water use. Am. Soc. of Agron. and Soil Sci. Soc. of Am., Madison, Wis.

Rawlins, S. L. 1977. Uniform irrigation with a low-head bubbler system. Agric. Water Manage. 1(2):167–178.

Rawlins, S. L., and P. A. C. Raats. 1975. Prospects for high-frequency irrigation. Science 188:604–610.

Reidinger, R. B. 1974. Institutional rationing of canal water in northern India: conflict between traditional patterns and modern needs. Econ. Dev. Cultural Change 23(1):79–104.

Revelle, R., and V. Lakshminarayana. 1975. The Ganges water machine. Science 188:611–616.

Rhoades, J. D. 1977. Potential for using saline agricultural drainage waters for irrigation. Proc. Spec. Conf. on Water Management for Irrigation and Drainage Div., Reno, Nev. 85–116. Irrig. Drain. Div., Am. Soc. Civ. Eng., New York.

Schumacher, E. F. 1973. Small is beautiful. Harper & Row Publishers Inc., New York.

Steinhart, J. S., and C. E. Steinhart. 1974. Energy use in the U.S. food system. Science 184: 33–42.

Thomas, G. W. 1975. World hunger: The new global challenge. J. Soil Water Conserv. 30: 8–11.

Tucker, W. 1977. Environmentalism and the leisure class. Harper's, 225:49–56, 78–80. Dec. 1977.

van Schilfgaarde, J. 1979. Environmental and institutional aspects of irrigation agriculture. Trans. ASAE 22:344–346, 350.

van Schilfgaarde, J., L. Bernstein, J. D. Rhoades, and S. L. Rawlins. 1974. Irrigation management for salt control. J. Irrig. Drain. Div. Am. Soc. Civ. Eng. 100(IR3):321–338.

Wadleigh, C. H. 1965. Coming to terms with nature. J. Soil Water Conserv. 20:43–45.

Worstell, R. V. 1976. An experimental buried multi-set irrigation system. Trans. ASAE 19: 1122–1128.

Glossary of Common and Scientific Names of Plants

Common name	Scientific name
Alfalfa	*Medicago sativa* L.
Alfalfa, yellow	*Medicago falcata* L.
Apple	*Malus* sp.
Asparagus	*Asparagus officinalis* L.
Barley	*Hordeum vulgare* L.
Beans	*Phaseolus vulgaris* L.
	Vicia faba L.
Bermudagrass	*Cynodon dactylon* (L.) Pers.
Birch	*Betula pubescens* (Ehrh.) L.
Birdsfoot trefoil	*Lotus corniculatus* L.
Blue panicgrass	*Panicum antidotale* (Retz.) L.
Bluegrass	*Poa pratensis* L.
Bromegrass	*Bromus* sp.
Bromegrass, smooth	*Bromus inermis* L.
Cabbage	*Brassica oleracea* L.
Celery	*Apium graveolens* L.
Citrus	*Citrus* sp.
Clover	
red	*Trifolium pratense* L.
white	*Trifolium repens* L.
Corn	*Zea mays* L.
Cotton	*Gossypium hirsutum* L.
Cowpeas	*Vigna unguiculata* (L.) Walp.
Creosote bush, spreading	*Larrea divaricata* L.
Crownvetch	*Coronilla varia* L.
Cucumber	*Cucumis sativus* L.
Douglas-fir	*Pseudotsuga menziessi* Franco
Eucalyptus	*Eucalyptus socialis* L.
Fescue	
Kentucky 31	*Festuca arundinaceae* (Schreb.) L.
tall	*Festuca arundinacea* Schreb.
Fig	*Ficus carica* L.
Flax	*Linum usitatissiumum* L.
Grapes	*Vitis* sp.
Guayule	*Parthenium argentatum* (Gray) L.
Hazel	*Corylus avellena* L.
Jojoba	*Simmondsia chinensis* L.
Kentucky bluegrass	*Poa pratensis* L.
Lettuce	*Lactuca sativa* L.

Common name	Scientific name
Lovegrass	
boer	*Eragrostis curvula* (Nees) L.
lehmann	*Eragrostis lehmanniana* (Nees) L.
Maize	*Zea mays* L.
Manioc	*Cassava* L.
	Yucca L.
Melons	*Cucurbitaceae* L.
Millet	*Panicum* sp.
German (foxtail)	*Setaria italica* L.
pearl	*Pennisetum americanum* L.
	Pennisetum glaucum L.
proso	*Panicum milliaceum* L.
Mitchellgrass	*Astrebla lappacea* L.
Oak	*Quercus* sp.
Oats	*Avena sativa* L.
Olive	*Olea europaea* L.
Orange	*Citrus sinensis* L.
Orchardgrass	*Dactylis glomareta* L.
Pangolagrass	*Digitaria decumbens* Stent
Pea	
field	*Pisum arvense* L.
garden	*Pisum sativum* L.
Peanut	*Arachis hypogaea* L.
Pine	
eastern white	*Pinus strobus* L.
red	*Pinus resinosa* (Ait.) L.
Pineapple	*Ananas comosus* L.
Potato	*Solanum tuberosum* L.
Pumpkin	*Cucurbita pepo* L.
Rice	*Oryza sativa* L.
Rye	*Secale cereale* L.
Ryegrass	
annual	*Lolium multiflorum* (Lam.) L.
perennial	*Lolium perenne* L.
Saltbush	*Atriplex* sp.
Sesame	*Sesamum indicum* L.
Sorghum	*Sorghum bicolor* (L.) Moench
	Sorghum vulgare L.
Soybeans	*Glycine max* L.
Spruce, sitka	*Picea sitchensis* (Bong.) Carr.
Squash	*Cucurbita pepo* L.
Stargrass	*Cynodon plectostachyus* (K. Schum) Pilger
Sudangrass	*Sorghum vulgare* Pers. *sudanense* Hitchc.
Sudax	*Sorghum bicolor* (L.) Moench. × *S. sudanese* (Piper)
Sugar beet	*Beta vulgaris* L.
Sugarcane	*Saccharum officinarum* L.
Sunflower	*Helianthus annus* L.
Sweet corn	*Zea mays* L.
Sweet potato	*Ipomoea batatas* (L.) Lam.

Tobacco	*Nicotiana tabacum* L.
Tomato	*Lycopersicon cheesmanii* sp. minor (Hook.) Mull
	Lycopersicon esculentum Mill
Tomato, wild	*Salanum bennellii* L.
Wheat	*Triticum aestivum* L.
Wheat, durum	*Triticum durum* Desf.
Wheatgrass	
crested	*Agropyron desertorum* (Fisch.) Schult.
intermediate	*Agropyron intermedium* (Host) Beauv.
thickspike	*Agropyron dasystachyum* (Hook.) Scribn.
Wildrye, Russian	*Elymus junceus* (Fisch.) L.

SUBJECT INDEX